Click Chemistry for Biotechnology and Materials Science

Click Chemistry for Biotechnology and Materials Science

Edited by

JOERG LAHANN

University of Michigan, Ann Arbor, USA

WILEY

A John Wiley and Sons, Ltd., Publication

Registered office
John Wiley & Sons Ltd, The Atrium, Southern Gate, Chichester, West Sussex, PO19 8SQ, United Kingdom

For details of our global editorial offices, for customer services and for information about how to apply for permission to reuse the copyright material in this book please see our website at www.wiley.com.

Cover concept kindly provided by Himabindu Nandivada (Co-author, Chapter 12).

Library of Congress Cataloging-in-Publication Data

Lahann, Joerg.
Click chemistry for biotechnology and materials science / Joerg Lahann.
p. cm.
Includes bibliographical references and index.
ISBN 978-0-470-69970-6
1. Biotechnology. 2. Materials science. 3. Combinatorial chemistry.
4. Macromolecules–Synthesis. I. Title.
TP248.2.L34 2009
660.6–dc22 2009016549

A catalogue record for this book is available from the British Library.

ISBN(H/B): 978-0-470-69970-6

Typeset in 10/12pt Times by Aptara Inc., New Delhi, India.

For Aimee

Contents

Preface

Over the last few years, click chemistry has taken a dramatic upturn. Since K. Barry Sharpless defined click chemistry as a "set of powerful, highly reliable, and selective reactions for the rapid synthesis of useful new compounds and combinatorial libraries" in 2001, many researchers have recognized the power of this conceptual framework. It was recognized early on that click chemistry is not limited to a specific type of reaction, but stands for a synthetic concept that is built on common reaction trajectories, rather than common reaction mechanisms. As a specific example of a click reaction, Sharpless suggested the copper-catalyzed Huisgen's 1,3-dipolar cycloaddition of azides and terminal alkynes, which has now been used for a wide range of different applications. The area of click chemistry is a highly creative area of research, which has literally exploded over the last few years.

While the concept of click chemistry might have initially been introduced with a firm eye on drug discovery, its applications to materials synthesis and biotechnology have been a startling success story. Thus, as I look ahead toward the advances coming from click chemistry in the next decade, some of the most promising applications are related to materials science and biotechnology. With this book, it is my intention to share some of the excitement surrounding click chemistry by describing the most recent progress with respect to (i) the development of a conceptual framework of click chemistry, (ii) its application to the precise design and synthesis of macromolecules, and (iii) its numerous applications in materials science and biotechnology.

Chapter 1 offers an introduction of the concept of click chemistry and its potential value as a universal ligation strategy for materials science and biotechnology. In the following three chapters, synthetic capabilities and limitations of click chemistry are discussed. Schilling, Jung and Bräse describe common synthons for click chemistry, while Baskin and Bertozzi review recent progress in the rapidly emerging field of copper-free click chemistry. In addition, Broyer, Kolodziej and Maynard describe a very exciting sub group of reactions involving oxime chemistry.

Chapters 5 to 8 outline the versatility of click chemistry for the synthesis of a range of different materials. Starting from the use of click chemistry for polymer synthesis (Lutz and Sumerlin), the journey takes us to the more complex branched polymer structures highlighted by Sinwell, Inglis, Stenzel and Barner-Kowollik. In chapter 7, Sachsenhofer and Binder review supramolecular materials prepared via click chemistry. This extensive survey of recent progress in this very exciting field is nicely complemented by a chapter on dendrimer-related click reactions authored by McNerny, Mullen, Majoros, Banaszak Holl, and Baker Jr (Chapter 8).

An interesting feature of Diels–Alder cycloadditions is their potential to be reversible. In chapter 9 Peterson and Palmese demonstrate that this is a very attractive property, when designing multifunction polymer networks. Moving from polymers to biohybrid and nanomaterials, chapters 10 and 11 authored by Kitto, Lauko, Rutijes and Rowan and van Berkel, Nijhuis, Löwik and van Hest, respectively, highlight truly emerging applications of click chemistry at the boundary of organic chemistry.

This synthesis-focused section of the book is followed by five chapters that describe potential applications of click chemistry in a wide range of important materials and biotechnology areas. In chapter 12, the use of click chemistry for surface engineering is described with a clear focus on biotechnological applications. This chapter is complemented by chapter 13, authored by Dieterich and Link, which surveys recent progress with the use of click chemistry for protein engineering. In chapter 14, LeDroumaguet and Wang describe the benefits of fluorogenic Huisgen's 1,3-dipolar cycloaddition reactions for bioconjugation. This section of the book is concluded by a comprehensive review of recent work in the area of biofunctionalization provided by Schilling, Jung and Bräse. Finally, in chapter 16, Luo, Kim, and Jen highlight unusual electro-optic properties enabled by Diels–Alder reactions of polymers and dendrimers.

When I initially decided to write a book about click chemistry for materials science and biotechnology, it was my main goal to share with a broader readership some of the excitement that was so apparent in discussions with colleagues at conferences and elsewhere. Thanks to my co-authors who have, in 16 loosely-connected chapters, provided a comprehensive view on this rapidly emerging field, I truly feel that this book has delivered on its mission.

Joerg Lahann
Michigan, April 2009

List of Contributors

Baker Jr., James R., Department of Chemistry, University of Michigan, Ann Arbor, MI, USA

Banaszak Holl, Mark M., Department of Chemistry, University of Michigan, Ann Arbor, MI, USA

Barner-Kowollik, Christopher, Institut für Technische Chemie und Polymerchemie, Universität Karlsruhe, Karlsruhe, Germany

Baskin, Jeremy M., Department of Chemistry, Lawrence Berkeley National Laboratory, Berkeley, CA, USA

Bertozzi, Carolyn R., Department of Chemistry, Lawrence Berkeley National Laboratory, Berkeley, CA, USA

Binder, Wolfgang H., Institute of Chemistry, Martin-Luther University, Halle-Wittenberg, Halle, Germany

Bräse, Stefan. Institut für Organische Chemie, Universität Karlsruhe, Karlsruhe, Germany

Broyer, Rebecca M., Department of Chemistry and Biochemistry, University of California, Los Angeles, CA, USA

Dieterich, Daniela C., Department of Neurochemistry and Molecular Biology, Leibniz Institute for Neurobiology, Magdeburg, Germany

Inglis, Andrew J., Institut für Technische Chemie und Polymerchemie, Universität Karlsruhe, Karlsruhe, Germany

Jen, Alex K.-Y., Department of Material Sciences and Engineering, and Insititute of Advanced Materials and Technology, University of Washington, Seattle, WA, USA

Jung, Nicole, Institut für Organische Chemie, Universität Karlsruhe, Karlsruhe, Germany

Kim, Tae-Dong, Department of Advanced Materials, Hannam University, Daejeon, Korea

Kitto, Heather J., Institute for Molecules and Materials, Radboud University Nijmegen, Nijmegen, The Netherlands

Kolodziej, Christopher M., Department of Chemistry and Biochemistry, University of California, Los Angeles, CA, USA

Lahann, Joerg, Department of Chemical Engineering, University of Michigan, Ann Arbor, MI, USA

Lauko, Jan, Institute for Molecules and Materials, Radboud University Nijmegen, Nijmegen, The Netherlands

Le Droumaguet, Céline, Department of Chemistry and Biochemistry and Nanocenter, University of South Carolina, Columbia, SC, USA

Link, A. James, Departments of Chemical Engineering and Molecular Biology, Princeton University, Princeton, NJ, USA

Löwik, Dennis W. P. M., Department of Organic Chemistry, Radboud University Nijmegen, Nijmegen, The Netherlands

Luo, Jingdong, Department of Material Sciences and Engineering, and Insititute of Advanced Materials and Technology, University of Washington, Seattle, WA, USA

Lutz, Jean-François, Research Group Nanotechnology for Life Science, Fraunhofer Institute for Applied Polymer Research, Potsdam, Germany

Majoros, Istvan J., Nanotechnology Institute for Medicine and Biological Sciences, Macromolecular Science and Engineering Center, Ann Arbor, USA

Maynard, Heather D., Department of Chemistry and Biochemistry, University of California, Los Angeles, CA, USA

McNerny, Daniel Q., Department of Chemistry, University of Michigan, Ann Arbor, MI, USA

Mullen, Douglas G., Nanotechnology Institute for Medicine and Biological Sciences, Macromolecular Science and Engineering Center, Ann Arbor, USA

Nandivada, Himabindu, Department of Chemical Engineering, University of Michigan, Ann Arbor, MI, USA

Nijhuis, Arnold W. G., Department of Organic Chemistry, Radboud University Nijmegen, Nijmegen, The Netherlands

Palmese, Giuseppe R., Department of Chemical and Biological Engineering, Drexel University, Philadelphia, PA, USA

Peterson, Amy M., Department of Chemical and Biological Engineering, Drexel University, Philadelphia, PA, USA

Rowan, Alan E., Institute for Molecules and Materials, Radboud University Nijmegen, Nijmegen, The Netherlands

Rutjes, Floris, P. J. T., Institute for Molecules and Materials, Radboud University Nijmegen, Nijmegen, The Netherlands

Sachsenhofer, Robert, Institute of Chemistry, Martin-Luther University, Halle-Wittenberg, Halle, Germany

Schilling, Christine, Institut für Organische Chemie, Universität Karlsruhe, Karlsruhe, Germany

Sinnwell, Sebastian, Institut für Technische Chemie und Polymerchemie, Universität Karlsruhe, Karlsruhe, Germany

Stenzel, Martina H., Centre for Advanced Macromolecular Design (CAMD), School of Chemical Sciences and Engineering, The University of New South Wales, Sydney NSW, Australia

Sumerlin, Brent S., Department of Chemistry, Southern Methodist University, Dallas, Texas, USA

van Berkel, Sander S., Department of Organic Chemistry, Radboud University Nijmegen, Nijmegen, The Netherlands

van Hest, Jan C.M., Department of Organic Chemistry, Radboud University Nijmegen, Nijmegen, The Netherlands

Wang, Qian, Department of Chemistry and Biochemistry and Nanocenter, University of South Carolina, Columbia, SC, USA

Acknowledgments

A book like this one is always the product of a team effort and it is evident that this outcome would not have been possible without the help of many dedicated individuals. First, and most importantly, I can't thank enough the authors of the individual chapters, who took time out of their busy schedules to share and convey their enthusiasm about the emerging field of click chemistry. I felt very privileged to be able to work with some of the best-recognized scientists in the field of click chemistry.

I would also like to thank the highly professional staff at Wiley who made this book a success. I would really like to express my gratefulness to Hima, who contributed to this book through countless stimulating discussions, proof-reading, and, of course, the contribution of one of the book chapters.

Moreover, I want to thank my mentors, Professor Hartwig Höcker, RWTH Aachen, and Professor Robert Langer, Massachusetts Institute of Technology, who taught me everything I know about chemistry, materials, and biotechnology. Without them I would not have become the scientist I am today.

Finally, I would like to thank Aimee, Lucas and Felix, who always supported me and never complained about the extra toll that projects, such as this book, put on our family and their already busy lives. Without their support, it would have been impossible to take on such major challenges in life.

1

Click Chemistry: A Universal Ligation Strategy for Biotechnology and Materials Science

Joerg Lahann

1.1 Introduction

Current advances in our understanding of molecular biology, microelectronics and sensorics have fueled an increased need for tightly defined structural materials and surfaces.[1] The controlled synthesis of such materials, however, imposes major challenges. Moreover, man-made materials have struggled to achieve the superb structural and functional properties of natural macromolecules, such as proteins, DNA or sugars. Recognizing these limitations, researchers in the materials and polymer sciences as well as in biotechnology have been continuously searching for well-defined ligation strategies that can be effectively used in the presence of a wide range of different functional groups typically encountered in these fields. Key requirements for successful ligation strategies include high selectivity, orthogonality to other functional groups, compatibility with water and other protic solvents and, of course, close-to-quantitative yields. To find improved ligation reactions, materials scientists and biotechnologists have increasingly turned towards advanced synthetic organic concepts. In this respect, the most powerful example is undoubtedly the acceptance of the click chemistry concept by the materials science community[2–8] since the first reports on the use of click chemistry for materials science, which appeared as recently as 2004.[9]

In his landmark review published in 2001, Sharpless and coworkers defined click chemistry as a 'set of powerful, highly reliable, and selective reactions for the rapid synthesis of useful new compounds and combinatorial libraries.'[10] Click reactions are driven by a

Click Chemistry for Biotechnology and Materials Science Edited by Joerg Lahann

high thermodynamic driving force (>20 kcal mol^{-1}), which is typically associated with the formation of carbon–heteroatom bonds. Click chemistry is not limited to a specific type of reaction, but stands for a synthetic philosophy that comprises of a range of reactions, with different reaction mechanisms but common reaction trajectories. The prime example of a click reaction is the copper-catalyzed Huisgen's 1,3-dipolar cycloaddition of azides and terminal alkynes.[10,11] This reaction is regioselective, forming only 1,4-substituted products, is insensitive to the solvent, and can be performed at room temperature. Moreover, it proceeds with high yields and is about 10^7 times faster than the uncatalyzed reaction. Another important aspect of the success of this reaction pertaining to materials science and biotechnology is that the starting materials, azides and terminal alkynes, are exceptionally stable and can be introduced in a wide range of macromolecules.

Ever since these initial publications, the area of click chemistry has turned into a highly productive area of research with exponential growth over the last few years. A literature search[12] indicated more than 600 papers and more than 10 000 citations in 2008 that were associated with the term 'click chemistry'. This compares with only about 100 papers and less than 1000 citations in 2005. Interestingly, Sharpless initially introduced the concept of click chemistry with a clear focus on drug discovery. While researchers in the drug discovery field still appear to be somewhat hesitant towards click chemistry,[13] the applicability of click chemistry towards materials science, polymer chemistry and biotechnology has been an astonishing success story, underlining the readiness of these fields for well-defined chemical reactions with the exact reaction profiles developed under the click chemistry framework. In a recent review, Wang and coworkers reported that only 14% of the papers published on click chemistry are actually related to drug discovery, while two-thirds of the click chemistry papers fall into the broad categories of materials science and biotechnology.[13]

1.2 Selected Examples of Click Reactions in Materials Science and Biotechnology

The value of click chemistry for materials synthesis possibly becomes most apparent in the area of polymer chemistry and several recent reviews have described the use of Cu-catalyzed azide–alkyne cycloaddition (CuAAC) for the synthesis of dendritic, branched, linear and cyclic co-polymers.[5–7, 14] Hawker, Sharpless, Fokin and coworkers first introduced CuAAC reaction to polymer chemists for dendrimer synthesis.[9] Triazole-based dendrons were divergently synthesized via CuAAC reaction. These dendrons were then anchored to a variety of polyacetylene cores to generate dendrimers. Since then, the CuAAC reaction has been widely employed to synthesize or modify various dendrimers.[15,16]

The remarkable functional group tolerance of click reactions enables the facile introduction of reactive groups such as hydroxyl and carboxyl through conventional pre-polymerization modification[17,18] or post-polymerization modification.[19] Thus, click reactions have been employed in combination with living polymerization techniques, such as ring-opening polymerization (ROP), ring-opening metathesis polymerization (ROMP), cationic polymerization, nitric oxide-mediated radical polymerization (NMP), atom transfer radical polymerization (ATRP) and reversible addition fragmentation chain transfer polymerization (RAFT).

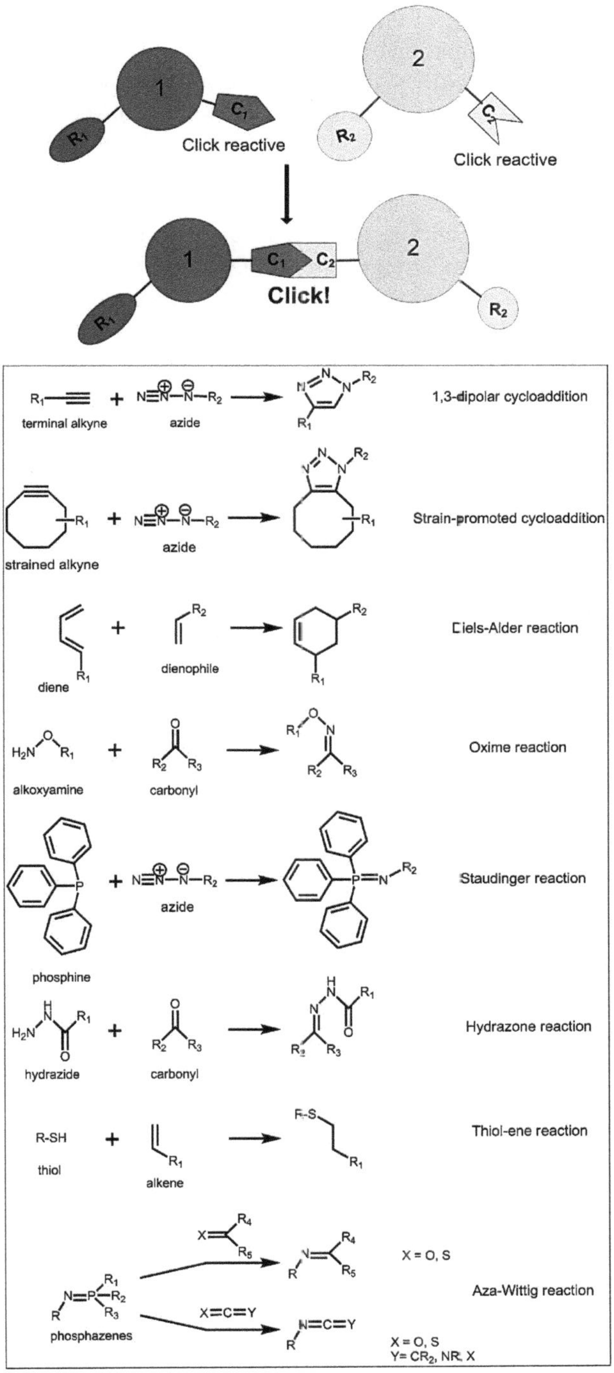

Scheme 1.1 *List of widely exploited chemical reactions that fall within the framework of click chemistry.*

One of the biggest strengths of click chemistry is its utility in conjunction with one or more of these polymerization methods, thus offering a facile access to a broad range of polymeric materials that would be difficult to prepare otherwise. For example, various functional groups (such as carboxyl, olefin and amine groups) were attached to polymers prepared by ATRP and modified by immobilization of biomolecules.[20–22] Beyond synthesis and modifications of functionalized polymers, click reactions have played an important role in the cross-linking of polymeric materials. One such example is the appearance of cross-linked polymeric adhesives synthesized from polyvalent azide and alkyne building blocks, owing their adhesiveness to the strong affinity of triazoles for metal ions and surfaces.[23] In yet another example, poly(vinyl alcohol)-based hydrogels were synthesized by mixing azido-appended PVA and acetylene-appended PVA in the presence of a Cu(I) catalyst.[24]

The controlled decoration of surfaces and design of biointerfaces are other obvious strongholds of click chemistry. Taking advantage of either Huisgen's 1,3-dipolar cycloaddition or Diels–Alder reaction, several advances were made with respect to the preparation of SAMs containing azido groups on well-defined electrode surfaces and subsequent reaction with ethynylferrocene or propynoneferrocene.[25] Moreover, well-defined surface arrays of acetylene-containing oligonucleotides were immobilized via CuAAC reaction onto azide-functionalized SAMs on gold.[26] This chemistry was unaffected by deactivation due to electrophiles or nucleophiles and showed remarkable stability against hydrolysis. Similarly, CuAAC reaction of acetylenyl-terminated SAMs and azide compounds was used as a versatile tool for tailoring surface functionalities under mild conditions.[27] In a universal surface modification approach, alkyne-containing vapor-deposited polymer coatings were shown to possess remarkable reactivity towards azide-functionalized moieties.[28] These reactive coatings, poly(4-ethynyl-*p*-xylylene-*co*-*p*-xylylene), were applied to a wide range of substrates using chemical vapor deposition and modified by subsequent spatially directed CuAAC reaction.

Among the more interesting examples is the application of the CuAAC reaction towards chemical functionalization of nanomaterials, such as single-walled carbon nanotubes (SWNT).[29] In this case, alkyne groups introduced onto the surface of the SWNTs offered a route towards highly specific post-modification. This method granted a greater amount of control on the orientation and the density of the polymer attached to the surface of the nanotube, while reducing the risk of side reactions. Similarly, Diels–Alder reactions were used for selective modification of carbon nanotubes. For instance, *o*-quinodimethane was directly coupled to SWNTs with the help of microwave irradiation.[30] These reactions open up possibilities for enhancing the solubility of carbon nanotubes, as needed in several technological applications.

To take the application of click chemistry even a notch higher, click reactions have been recently proposed for covalent labeling in living systems, such as cells or tissue.[31,32] If biomolecules expressed within cells could be fluorescently labeled, their destiny could be tracked in real time. Moreover, important metabolic studies could be conducted *in vitro* or potentially even *in vivo*. For such concepts to become a reality, the reactions used for covalent labeling must not only fulfill the criteria of an efficient click reaction, such as high yields, selectivity and compatibility with an aqueous environment, but also must be bio-orthogonal. The latter refers to the necessity of exploiting reactants that are 'non-interacting towards the functionalities present in biological systems'.[33] Important bio-orthogonal click reactions, including Staudinger ligation[34] and strain-promoted [3 + 2] heterocycloadditions,[35] have been proven to be effective reactions for protein labeling in living biological systems.

From these few selected examples, it is already apparent why click reactions have been so popular and successful for the synthesis and modification of macromolecules. The need to effectively decorate biomaterials and biointerfaces, the ability to drive covalent reactions inside living organisms and the promise of constructing large macromolecules through complimentary junctions present in their constituents will be important technology drivers for decades to come.

1.3 Potential Limitations of Click Chemistry

In spite of the undisputable success of the concept of click chemistry within just a few years, there are still a few limitations associated with the concept. Because of the stringent criteria that are used to identify click reactions, chemical diversity is intrinsically limited. As a matter of fact, the CuAAC reaction is still by far the most widely used click reaction. However, copper is believed to be cytotoxic and demonstrated side effects associated with excessive copper intake include hepatitis, Alzheimer's disease and neurological disorders.[36] For click reactions to be used in contact with living systems, the copper catalyst must be completely removed or alternatives, such as Staudinger ligation or strain-promoted [3 + 2] heterocycloadditions, must be employed.

Azides, among the prime reactants for Huisgen's 1,3-dipolar cycloaddition reaction, are also often associated with potential toxic side effects, and certain azides may bear a very real explosive potential.[37]

Finally, a more practical limitation is that the supply of clickable starting materials often cannot keep up with the demands of the rapidly emerging application space in materials science and biotechnology. Meanwhile, many of the researchers that work in these fields are not synthetic chemists, who can easily synthesize appropriate starting materials, but must rely on commercial sources for obtaining access to these chemicals. However, as the click chemistry philosophy continues to spread through the area of materials science, polymers and biotechnology, more and more clickable building blocks can be expected to become easily available.

1.4 Conclusions

To address the gap between sophisticated function that is required for future advances in bio- and nanotechnology and the limited chemical control offered by many of the currently available synthetic materials' processes, novel synthetic tools are needed. In spite of the evident differences between small molecules and macromolecules, attempts to extend synthetic concepts from organic chemistry into the nano- and meso-scale dimensions have been increasingly popular. It is mainly for this reason that the fields of materials science and biotechnology enthusiastically embraced the concept of click chemistry as a versatile tool for introducing structural control. Ideally, these efforts will offer molecular-level control during the preparation of nanostructured materials. It is likely that this trend will continue and will ultimately result in an increase in the infusion of concepts from synthetic organic chemistry into materials science and biotechnology.[38]

References

(1) Hawker, C.J. and Wooley, K.L., (2005), The convergence of synthetic organic and polymer chemistries, *Science*, **309**, 1200–1205.
(2) Binder, W.H. and Sachsenhofer, R., (2007), 'Click' chemistry in polymer and materials science, *Macromolecular Rapid Communications*, **28**, 15–54.
(3) Goodall, G.W. and Hayes, W., (2006), Advances in cycloaddition polymerizations, *Chemical Society Reviews*, **35**, 280–312.
(4) Bock, V.D., Hiemstra, H. and van Maarseveen, J.H., (2005), Cu(I)-catalyzed alkyne-azide click cycloadditions from a mechanistic and synthetic perspective, *European Journal of Organic Chemistry*, 51–68.
(5) Nandivada, H., Jiang, X.W. and Lahann, J., (2007), Click chemistry: versatility and control in the hands of materials scientists, *Advanced Materials*, **19**, 2197–2208.
(6) Lutz, J.-F., (2007), 1,3-Dipolar cycloadditions of azides and alkynes: a universal ligation tool in polymer and materials science, *Angewandte Chemie, International Edition*, **46**, 1018–1025.
(7) Fournier, D., Hoogenboom, R. and Schubert, U.S., (2007), Clicking polymers: a straightforward approach to novel macromolecular architectures, *Chemical Society Reviews*, **36**, 1369–1380.
(8) Binder, W.H. and Sachsenhofer, R., (2008), Click chemistry in polymer and material science: an update, *Macromolecular Rapid Communications*, **29**, 952–981.
(9) Wu, P. *et al.*, (2004), Efficiency and fidelity in a click-chemistry route to triazole dendrimers by the copper(I)-catalyzed ligation of azides and alkynes, *Angewandte Chemie – International Edition*, **43**, 3928–3932.
(10) Kolb, H.C., Finn, M.G. and Sharpless, K.B., (2001), Click chemistry: diverse chemical function from a few good reactions, *Angewandte Chemie, International Edition*, **40**, 2004–2021.
(11) Tornoe, C.W., Christensen, C. and Meldal, M., (2002), Peptidotriazoles on solid phase: [1,2,3]-triazoles by regiospecific copper(I)-catalyzed 1,3-dipolar cycloadditions of terminal alkynes to azides, *Journal of Organic Chemistry*, **67**, 3057–3064.
(12) Based on Web of Science.
(13) Hein, C.D., Liu, X.-M. and Wang, D., (2008), Click chemistry, a powerful tool for pharmaceutical sciences, *Pharmaceutical Research*, **25**, 15.
(14) Binder, W.H. and Kluger, C., (2006), Azide/alkyne-'click' reactions: applications in material science and organic synthesis, *Current Organic Chemistry*, **10**, 1791–1815.
(15) Wu, P. *et al.*, (2005), Multivalent, bifunctional dendrimers prepared by click chemistry, *Chemical Communications*, 5775–5777.
(16) Joralemon, M.J. *et al.*, (2005), Dendrimers clicked together divergently, *Macromolecules*, **38**, 5436–5443.
(17) Englert, B.C., Bakbak, S. and Bunz, U.H.F., (2005), Click chemistry as a powerful tool for the construction of functional poly(*p*-phenyleneethynylene)s: comparison of pre- and postfunctionalization schemes, *Macromolecules*, **38**, 5868–5877.
(18) Thibault, R.J. *et al.*, (2006), A versatile new monomer family: functionalized 4-vinyl-1,2,3-triazoles via click chemistry, *Journal of the American Chemical Society*, **128**, 12084–12085.
(19) Thomsen, A.D., Malmstrom, E. and Hvilsted, S., (2006), Novel polymers with a high carboxylic acid loading, *Journal of Polymer Science, Part A: Polymer Chemistry*, **44**, 6360–6377.
(20) O'Reilly, R.K., Joralemon, M.J., Hawker, C.J. and Wooley, K.L., (2006), Facile syntheses of surface-functionalized micelles and shell cross-linked nanoparticles, *Journal of Polymer Science Part a – Polymer Chemistry*, **44**, 5203–5217.
(21) Gao, H. *et al.*, (2005), Gradient polymer elution chromatographic analysis of a,w-dihydroxypolystyrene synthesized via ATRP and click chemistry, *Macromolecules*, **38**, 8979–8982.

(22) Lutz, J.F., Borner, H.G. and Weichenhan, K., (2005), Combining atom transfer radical polymerization and click chemistry: a versatile method for the preparation of end-functional polymers, *Macromolecular Rapid Communications*, **26**, 514–518.
(23) Diaz, D.D. *et al.*, (2004), Click chemistry in materials synthesis. 1. Adhesive polymers from copper-catalyzed azide-alkyne cycloaddition, *Journal of Polymer Science Part a – Polymer Chemistry*, **42**, 4392–4403.
(24) Ossipov, D.A. and Hilborn, J., (2006), Poly(vinyl alcohol)-based hydrogels formed by 'click chemistry', *Macromolecules*, **39**, 1709–1718.
(25) Collman, J.P., Devaraj, N.K. and Chidsey, C.E.D., (2004), 'Clicking' functionality onto electrode surfaces, *Langmuir*, **20**, 1051–1053.
(26) Devaraj, N.K. *et al.*, (2005), Chemoselective covalent coupling of oligonucleotide probes to self-assembled monolayers, *Journal of the American Chemical Society*, **127**, 8600–8601.
(27) Zhang, Y. *et al.*, (2006), Carbohydrate-protein interactions by 'clicked' carbohydrate self-assembled monolayers, *Analytical Chemistry*, **78**, 2001–2008.
(28) Nandivada, H., Chen, H.-Y., Bondarenko, L. and Lahann, J., (2006), Reactive polymer coatings that 'click', *Angewandte Chemie, International Edition*, **45**, 3360–3363.
(29) Li, H., Cheng, F., Duft, A.M. and Adronov, A., (2005), Functionalization of single-walled carbon nanotubes with well-defined polystyrene by 'click' coupling, *Journal of the American Chemical Society; Journal of the American Chemical Society*, **127**, 14518–14524.
(30) Delgado, J.L. *et al.*, (2004), Microwave-assisted sidewall functionalization of single-wall carbon nanotubes by Diels–Alder cycloaddition, *Chemical Communications (Cambridge, UK)*, 1734–1735.
(31) Laughlin, S.T., Baskin, J.M., Amacher, S.L. and Bertozzi, C.R., (2008), *In vivo* imaging of membrane-associated glycans in developing zebrafish, *Science*, **320**, 664–667.
(32) Prescher, J.A. and Bertozzi, C.R., (2005), Chemistry in living systems, *Nature, Chemistry and Biology*, **1**, 13–21.
(33) Baskin, J.M. and Bertozzi, C.R., (2007), Bioorthogonal click chemistry: covalent labeling in living systems, *QSAR and Combinatorial Science*, **26**, 1211–1219.
(34) Lemieux, G.A., de Graffenried, C.L. and Bertozzi, C.R., (2003), A fluorogenic dye activated by the Staudinger ligation, *Journal of the American Chemical Society*, **125**, 4708–4709.
(35) Agard, N.J., Prescher, J.A. and Bertozzi, C.R., (2004), A strain-promoted [3+2] azide–alkyne cycloaddition for covalent modification of blomolecules in living systems, *Journal of the American Chemical Society*, **126**, 15046–15047.
(36) Wang, T. and Guo, Z.J., (2006), Copper in medicine: homeostasis, chelation therapy and antitumor drug design, *Current Medical Chemistry*, **13**, 525–537.
(37) Brase, S., Gil, C., Knepper, K. and Zimmermann, V., (2005), Organic azides: an exploding diversity of a unique class of compounds, *Angewandte Chemie – International Edition*, **44**, 5188–5240.
(38) Dondoni, A., (2008), The emergence of thiol-ene coupling as a click process for materials and bioorganic chemistry, *Angewandte Chemie – International Edition*, **47**, 8995–8997.

2

Common Synthons for Click Chemistry in Biotechnology

Christine Schilling, Nicole Jung and Stefan Bräse

2.1 Introduction – Click Chemistry

The most widely used method for the synthesis of libraries of biologically-active molecular frameworks – particularly for the regioselective synthesis of 1,2,3-triazoles – is the copper-catalyzed 1,3-dipolar cycloaddition (click reaction) of azides[1–4] with alkynes. Meldal *et al.* first published this method in 2002 for triazole synthesis on solid phases,[5,6] closely followed by the report of Sharpless and Fokin *et al.* dealing with the water-based catalyzed reaction with copper sulfate and sodium ascorbate.[7] Many different conditions have been established for cycloaddition reactions.[8] Moreover, this reaction fulfills certain criteria that are particularly advantageous for biological uses: the alkyne group as well as the azide functionality are inert within molecules that exist in living systems, and the click reaction is generally compatible with water-containing systems (aqueous media near physiological pH can be used at ambient temperature and in short reaction times to prevent, for example, the denaturation of proteins). While other reactions include the use of nucleophiles or electrophiles, the components of a click reaction do not react in an undesirable way within biological systems. These advantages facilitate reactions in biological systems and even in living cells. As such, the click reaction is used as an alternative to other established methods described in the literature for the production of biomolecules[9,10] (for example, olefin metathesis of glycosides, intermolecular enyne metathesis of alkynyl and alkenyl glycosides, native chemical ligation, glycosylation of diols and coupling of alkynyl glycosides).

Click Chemistry for Biotechnology and Materials Science Edited by Joerg Lahann

2.2 Peptides and Derivatives

The click reaction has been used to synthesize diverse derivatives of bioactive compounds via substitution of the originally appearing functionality with the resulting triazole unit, in order to improve the activity of the compounds. Gopi *et al.* investigated the structure–activity correlation of peptide conjugates that act as receptor site antagonists of HIV-1 gp120.[11] The group synthesized derivatives of the original bioactive peptide on solid phases through the reaction of an immobilized azide-modified proline residue with alkynes containing different side chains (Scheme 2.1).

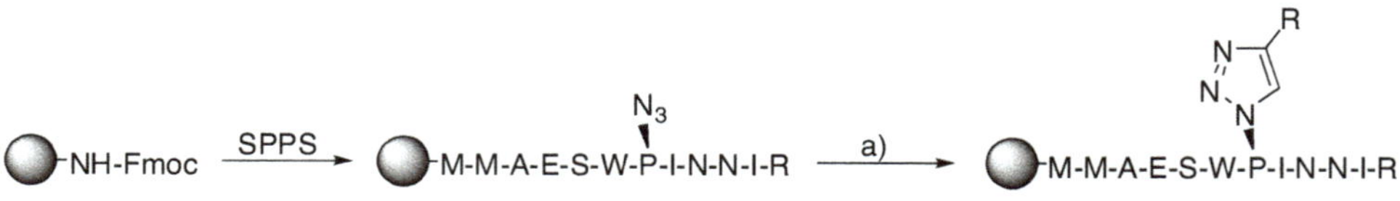

Scheme 2.1 *Synthesis of various triazole-based covalently modified peptide derivatives of an antagonist of HIV-1 gp120 through click chemistry. (a) CH_3CN–H_2O–DIEA–pyridine, alkyne (5 equiv.), CuI (1 equiv.), overnight.*[11]

The introduction of a triazole functionality offers the possibility of producing novel β-turn mimics.[12] The click reaction of an alkyne-containing dipeptide **5** and the azide-functionalized dipeptide **4** generates different tetrapeptides **6** in a convergent synthesis (Scheme 2.2). While the syntheses of Guan *et al.* are based on the reaction of terminally modified azide-containing peptides, there are many approaches using peptidic structures that bear azides in their side chains. Similar to the approach of Gopi *et al.* (Scheme 2.1), these strategies yield peptides with diverse triazole derivatives as side chain substituents.

Following the click strategy, peptides can be turned into glycopeptides as shown by Rutjes *et al.*[13] Different routes to these target glycopeptides have been explored: the coupling of amino acids can be performed traditionally via chemical- or alternatively enzyme-catalyzed coupling. The click reaction itself can be undertaken before, as well as after, peptide formation. Furthermore, azide-modified sugar-moieties and azide-containing peptides **7** have been used to perform the click reaction with alkynylated peptides as well as in the

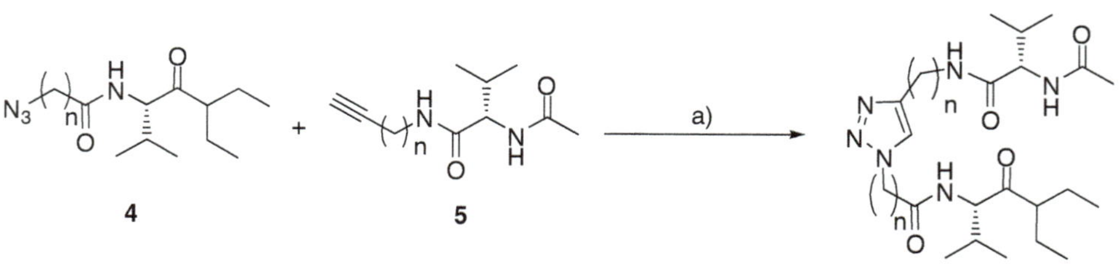

Scheme 2.2 *Synthesis of tetrapeptides* **6** *with n = 1–4 through click reaction. (a) $CuSO_4$ (1 mol %), sodium ascorbate, H_2O/^tBuOH (1:1), r.t.*[12]

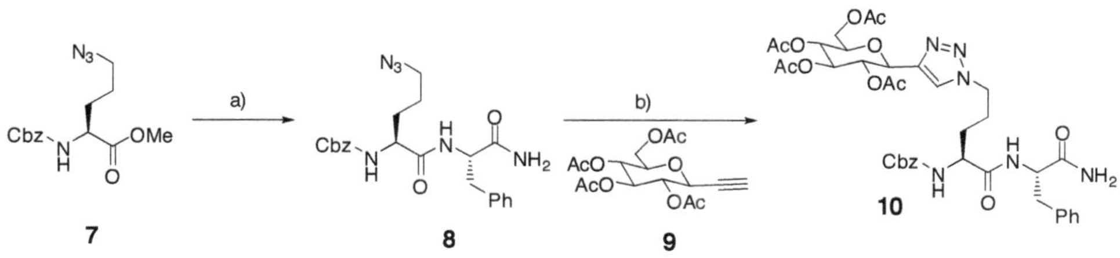

Scheme 2.3 *Enzymatic peptide coupling to give azide-containing dipeptide* **8** *and subsequent click reaction. (a) Phe–NH$_2$ or Gly–NH$_2$,* alcalase, *^{t}BuOH–DMF; (b) CuI, TBTA, NEt$_3$, MeCN, 40 °C.*[13]

inverse manner. It was ascertained that coupling through an enzymatic protocol followed by a click reaction with either the azide- or alkyne-containing sugar yields very good results, in terms of generating the triazole-modified dipeptides **10** (Scheme 2.3).

Aucagne and Leigh[14] demonstrated that the click reaction can be performed chemoselectively (Scheme 2.4). This approach was used to selectively generate two triazole linkages in a one-pot method. The first reaction is traditionally performed in the presence of a copper(I)-species with a dialkynyl-derivative **13**, which contains one free and one TMS-protected triple bond. The second click reaction is carried out in the presence of Cu(I) and Ag(I), which catalyze the deprotection of the second alkynyl functionality. The starting material **12** for those click reactions is easily prepared through the substitution of α-chlorinated peptides **11** with sodium azide.

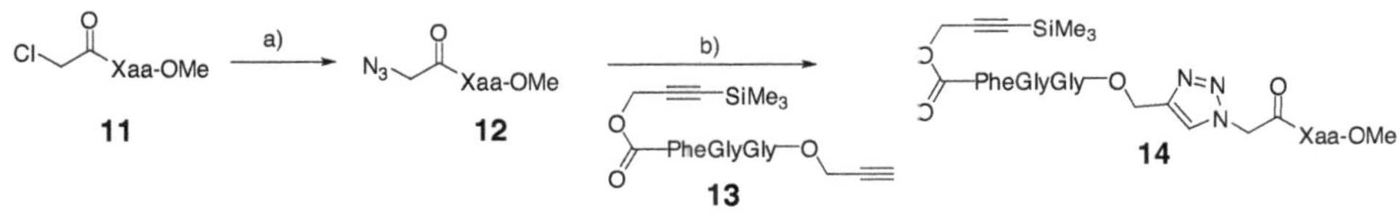

Scheme 2.4 *Synthesis of azide-containing dipeptide analogs* **12** *and chemo-selective click cycloaddition. (a) NaN$_3$, DMF; (b)* **13***, CuSO$_4$, sodium ascorbate, ^{t}BuOH/H$_2$O, 20 °C, 72 h, 98%; Xaa = Phe, Leu, Pro, Lys(Boc).*[14]

Besides the aforementioned methodologies for intermolecular click reactions, there are examples of intramolecular approaches that result in cyclic peptides. One example is found in the research conducted by Maarseveen *et al.*,[15,16] in which intramolecular click reactions were applied to the synthesis of cyclic tetrapeptides (Scheme 2.5). These tetrapeptides are of interest in their function as tyrosinase inhibitors, because the derivatives produced are established bioactive compounds. The synthesis of the triazole-containing cyclic tetrapeptides can theoretically be successful via cyclization through peptide coupling or, on the other hand, through triazole formation in the final step. It has been demonstrated that compound **16** is only accessible through ring closure via click reaction, underlining the importance of click chemistry for the formation of triazole-containing peptidomimetics (Scheme 2.5).

In subsequent investigations of peptide-bond isosteres, the Maarseveen group synthesized cyclotetrapeptide mimics containing two triazole moieties. Although these substrates

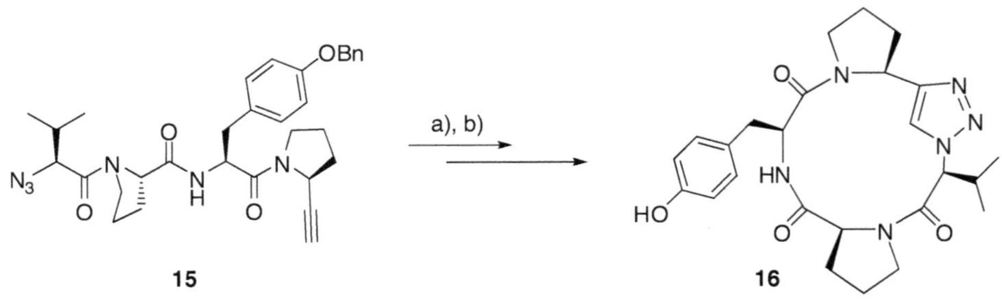

Scheme 2.5 Synthesis of cyclo-[Pro–Val–ψ(triazole)-Pro–Tyr] by click cyclization. (a) CuBr, DBU, toluene, reflux, 70%; (b) H_2, Pd–C, MeOH, 93%.[15,16]

were synthesized in an analogical manner via click cyclization, only moderate yields were attained.

Lokey *et al.*[17] performed similar cyclizations to generate small cyclic peptide derivatives on solid phases. The starting material **17** was initially built up with a propargyl-containing amino acid. After standard peptide coupling on bead, the last amino acid of the linear chain bearing the azide functionality was attached. Cyclization of **17** was performed in the presence of CuBr, sodium ascorbate and a diamine as a ligand, followed by cleavage from the resin achieved with TFA in methylene chloride. The successful synthesis was demonstrated for cyclic peptides **18** containing two to five leucine units. The major drawback of the solid phase synthesis was the appearance of varying amounts of the dimeric peptides as a result of the incomplete conversion of the click reaction.

Another modification on solid phases involves solid-phase peptide synthesis with orthogonal *N*-protected alkylamines as side chains. These can be deprotected, selectively acylated with propiolic acid as shown by Eichler *et al.*, and then 'clicked' via the addition of azide-modified peptides.[18] Another click reaction with resin-bound terminal alkynes has been illustrated by Meldal *et al.*[5,6] This group investigated the formation of triazole derivatives after solid-phase peptide synthesis. In the last step on bead, the terminal alkyne was reacted with different primary, secondary and tertiary alkyl and arylazides and an azido sugar moiety.[19]

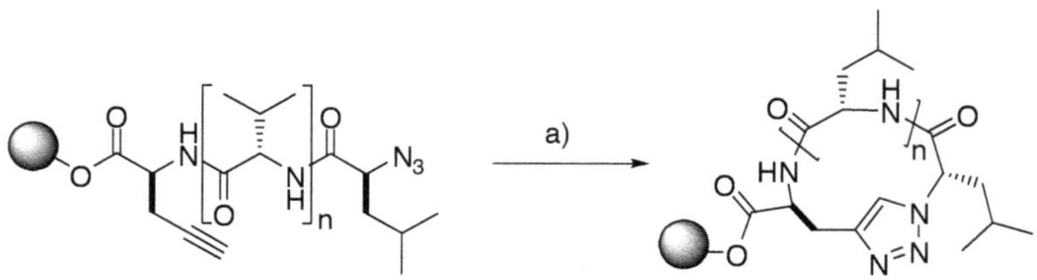

Scheme 2.6 Click macrolactamization of propargyl glycine azido leucine on solid phases. a) CuBr (1 equiv.), sodium ascorbate (3 equiv.), DIEA (10 equiv.), 2,6-lutidine (10 equiv.), DMF, 25 °C, 6 h.[17]

2.3 Peptoids

As triazole-containing mimics of peptides have been extensively investigated, there are other groups who have synthesized peptoids modified through the click reaction of azides with alkynes. Taillefumier *et al.* performed the first synthesis of functionalized β-peptoid macrocycles via linear synthesis of polyalkyne-containing linear β-peptoids and subsequent macrocyclization.[20] The functionalized cyclic peptoids **20** were then modified by the addition of azides to produce polytriazole-containing cycles **21** with different residues on the triazole side chain (Scheme 2.7).

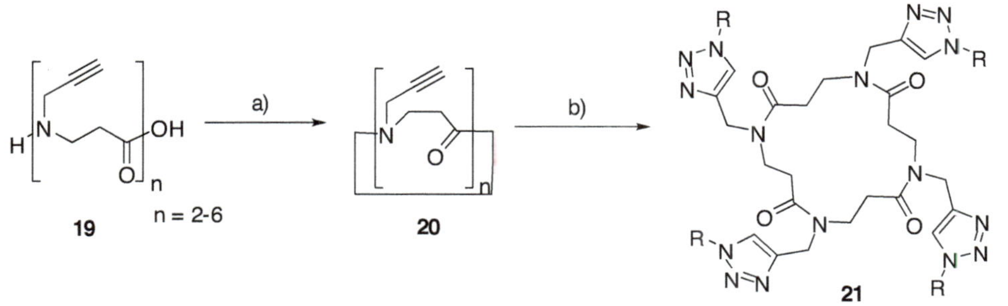

Scheme 2.7 *Macrolactamization of linear peptoids **19** and functionalization via click chemistry. (a) DPPA, $NaHCO_3$, MeCN–DMF; (b) RN_3, $CuSO_4$, sodium ascorbate, TBTA [tris-(benzyltriazolylmethyl)amine].*[20]

Noncyclic peptoid derivatives generated by click reaction can also be found in the literature. Kirshenbaum *et al.* generated *N*-substituted glycine peptoid oligomers **22** on bead that were used for click reactions on solid phases. The group incorporated azides into the oligomer and treated them with acetylenes **23** to generate triazole derivatives **24**. In subsequent steps, the oligomer was elongated and the inverse click reaction was performed, through the immobilization of acetylenes and reaction with azides **26**. Using this conjugation technique, complex peptidomimetic products were synthesized, containing different triazole substituents in specific positions. After peptoid-chain elongation, Kirshenbaum *et al.* were able to isolate 24-meric products and dodecamers containing two different triazole groups and up to four different triazole ring substituents respectively (Scheme 2.8).[21]

2.4 Peptidic Dendrimers

Peptidic dendrimers are biologically relevant structures, for example, as channels for drug delivery.[22] The functionalization of these dendrimers enables the modulation of the surface to generate special properties, which are essential for their specific interactions. Click chemistry has been used by several groups to design novel dendrimeric structures.[23–26] Haridas *et al.* synthesized triazole-based dendrimers **30** from $(Lys)_2$Lys-alkyne **28** and the azide-containing compound **29**. Moreover, the lysine-based dendron **28** has been coupled with cysteine cores and dendron **29** has been attached to a Lys-Asp-scaffold (Scheme 2.9).

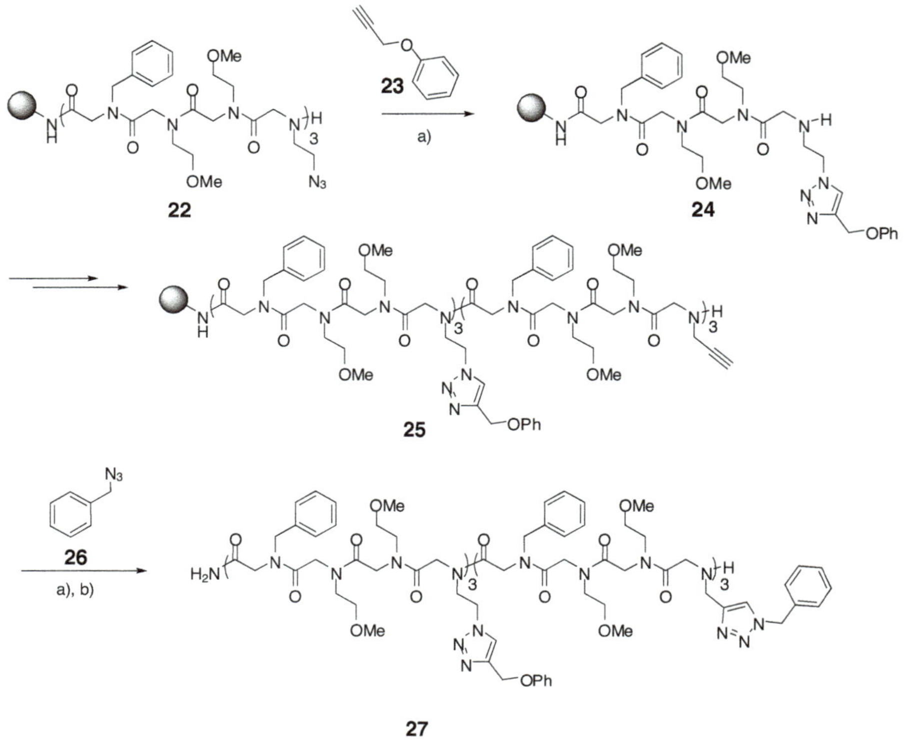

Scheme 2.8 *Modification of peptoid side chains by a sequential series of click reaction and oligomerization. (a) Coupling partner, CuI, sodium ascorbate, DIPEA in DMF–pyridine, r.t., 18 h; (b) 95% TFA in H_2O, r.t., 10 min.*[21]

The groups of Liskamp *et al.*[25] and Pieters *et al.*[23] constructed dendrimers that contain polyalkyne building blocks and coupled those to produce polytriazolic dendrimers. Furthermore, through the application of click chemistry, the latter group successfully synthesized a series of pore-forming compounds in order to build up magainin-based compounds.

2.5 Oligonucleotides

The application capacity of oligonucleotides and their derivatives is seemingly boundless. Modified oligonucleotides have the potential to widen existing diagnostic capabilities, while also fulfilling therapeutic applications. In genomic studies, they act as primers and probes for DNA detection and sequencing,[27,28] as primers in the polymerase chain reaction and as molecular beacons for detecting genetic mutations. As well as being used as anti-sense agents, oligonucleotides serve as probes for measuring gene expression in DNA microarrays and gene chips.[29] The wide variety of structural modifications of oligonucleotides is a result of the substantial number of known bases, sugar residues[30]

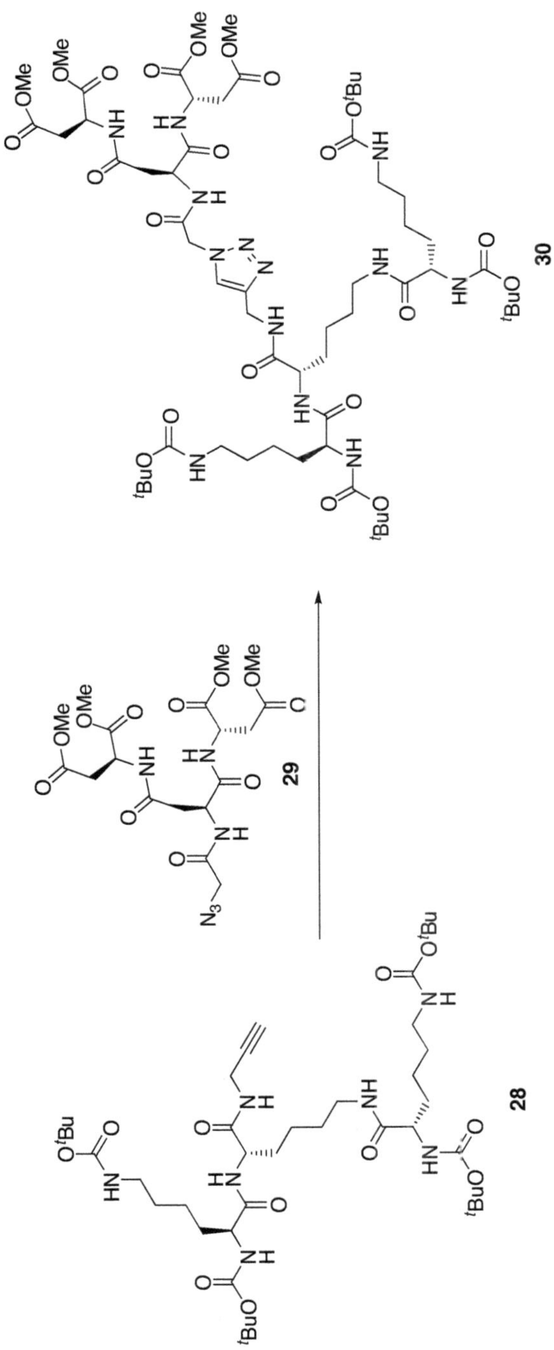

Scheme 2.9 *Synthesis of triazole-based dendrimers by click reaction. (a) 29, Cu(I), DIEA, 68%.*[24]

and altered phosphodiester backbones that can influence hybridization strength, binding affinity, mismatch discrimination, nuclease resistance and cellular uptake.[27,28]

Introduced by Sharpless and coworkers, the click reaction is one of the most powerful methods of labeling oligonucleotides.[31,32] It has been performed not only on phosphoramidate-modified oligonucleotides,[33] but also on triple bonds and azides mostly attached through the 5′- or 3′-terminal conjugation[34] of the oligonucleotides. Because the 5-position of pyrimidines and the 7-position of purines lie in the major groove of the B-DNA duplex and have steric freedom for additional functionalities, most of the reported DNA modifications take place in these positions.[27]

Seela *et al.* reported on some modified nucleosides and oligonucleotides, which were synthesized through the Cu^I-catalyzed azide–alkyne cycloaddition. The group introduced reporter groups as well as duplex stabilizing residues into nucleic acids, one example of which is the cycloaddition of compound **31** and azidothymidine (AZT, **32**) to form the triazole product **33** (Scheme 2.10).[27]

The preceding methods facilitate the synthesis of a variety of lengths and structures of the side chains and nucleobases. Numerous cycloadditions work best using a spacer between the alkyne and the nucleobase.[35,36] Carell *et al.*,[37] for example, reported on the click reaction performed directly on PCR products of alkyne-modified uridine nucleosides **34–37** instead of on natural thymidine triphosphate (Scheme 2.11). The resulting high-density alkyne-modified DNA was subsequently derived to investigate the efficiency of the sugar labeling of DNA under mild conditions, in which galactose azide **41** was 'clicked' onto single- and double-stranded DNA to generate nucleic acids of type **42**.[37]

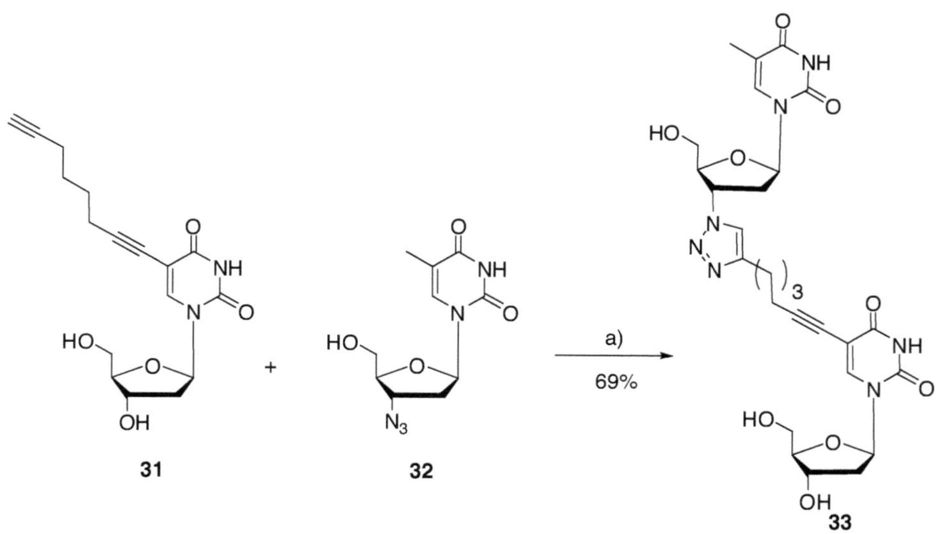

31 32 33

Scheme 2.10 *Click reaction of a thymidine derivative **31** at the 5-position with AZT (**32**): (a) $CuSO_4$, sodium ascorbate, THF/H_2O, r.t., 30 h.*[27]

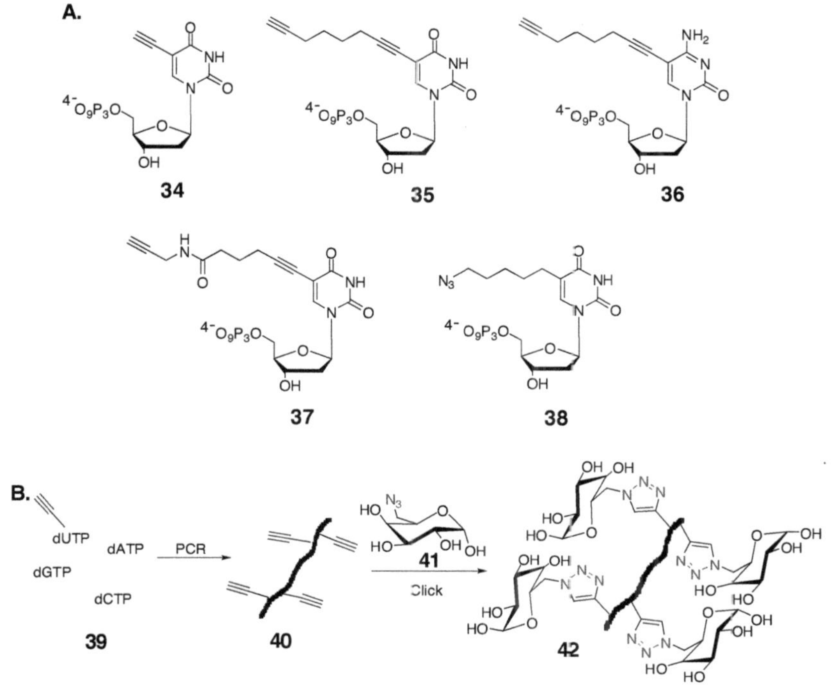

Scheme 2.11 *(A) 5-Modified pyrimidine triphosphates* ***34–38*** *used in this study; (B) schematic description of DNA functionalization using click chemistry.*[37]

The principle of the alkyne-derived formation of nucleic acids shown later on (Scheme 2.12), as well as subsequent click reaction, led to a high potential for nanotechnology, including gold surface and nanoparticle functionalization.[27] Furthermore, in the field of nanotechnology, the magnetic properties and electrical conductivity of the primary sequence and the secondary structure of DNA are currently of particular interest in the preparation of functional self-assembled nanostructures.[38] Beyond the investigations of the Carell group,

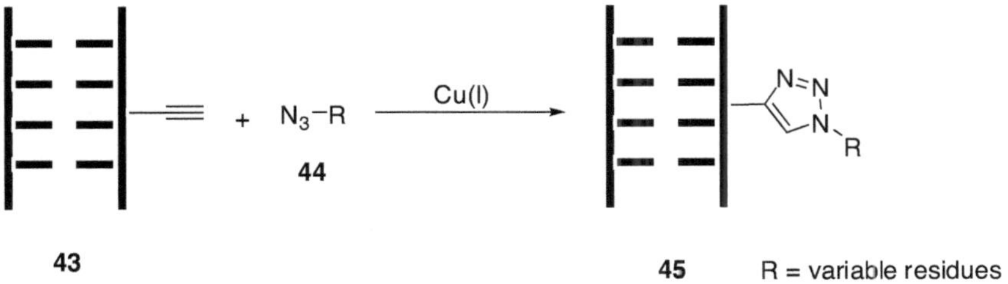

Scheme 2.12 *Schematic view of the cycloaddition reaction performed on duplex DNA.*

Morvan *et al.*[39] developed an approach to sugar-modified oligonucleotides through the use of microwave-assisted click chemistry.

Seela *et al.* converted 5-(octa-1,7-diynyl)-2′-deoxyuridine (**46**)[36] into strongly fluorescent 1*H*-1,2,3-triazole conjugates **49** that incorporated two fluorescent reporters – the pyrrolo[2,3-*d*]pyrimidine derivative (Pyrrolo dC **47**) and the coumarin moiety **48** – via the click reaction, amongst other methods (Scheme 2.13).[40] The photophysical properties of the Cu(I)-catalyzed azide–alkyne cycloaddition product **49** differ from those of pyrdC. Thus, the cycloaddition reaction enabled longer excitation and emission wavelengths (excitation, 340 nm, emission, 466 nm vs excitation, 327 nm, emission, 410 nm for pyrdC). Furthermore, ds-DNA containing the click nucleotide **49** displayed a strong fluorescence increase compared with PyrdC **47** integrated into duplex DNA.[40]

Nielsen and coworkers worked on triazole stacking in the major groove, which produced increased nucleic acid duplex stability.[33] Moreover, Pederson *et al.*[41] reported that incorporated 1,2,3-triazoles stabilize parallel triplexes by twisting intercalating nucleic acids. Furo[2,3-d]pyrimidines conjugated to carbohydrates were synthesized by Ju *et al.* in 2008 via Sonogashira coupling and click chemistry to enhance the efficiency of structural diversity for structure–activity relationship studies.[30] Beyond the approaches concerning the labeling of oligonucleotides, click chemistry has been used to cross-link two ends of a DNA strand, through the reaction of alkynes and azides positioned at the base on the 3′- and 5′-termini.[42] Further examples were illustrated by Ju *et al.* in 2003, such as the synthesis of fluorescent single-stranded DNA (ss-DNA) via 1,3 dipolar cycloaddition between alkynyl 6-carboxyfluorescein (FAM) and 5′-azido-labeled ss-DNA. The resulting fluorescent ss-DNA is applicable as a primer in the Sanger dideoxy chain termination reaction, in order to produce DNA sequencing fragments (for other examples of biomolecule-labeling see Chapter 15).[29]

In the field of nanoarchitectures, nature offers an efficient tool to self-assemble artificial and biological systems via hydrogen bonding. This property was used by Bäuerle *et al.*, who combined semi-conducting oligothiophenes **50** with complementary nucleosides **51** and **52** to create novel recognition-driven self-aggregated superstructures **55** based on the click reaction (Scheme 2.14).[43]

2.6 Carbohydrates

In biology, carbohydrates are a vitally important class of compounds. Oligosaccharides, glycolipids and glycopeptides – displayed on the cell surface – play a crucial role in cellular recognition events, including signal transduction, inflammation, immune response, apoptosis, tumor metastasis and viral and bacterial infections.[44,45] The study of carbohydrate–protein interactions is complicated, hindering efforts to develop a mechanistic understanding of the structure and function of carbohydrates. The weak binding affinities of carbohydrate–protein interactions and the availability of structurally complex carbohydrates remain two major challenges.[46] Thus, it is highly desirable to undertake the construction of mimics of complex saccharides, which can imitate the interaction of a given saccharide with its receptor,[47] as well as rapid screening of target saccharides through the development of modular fluorescent sensors.[48] Interest persists in different research

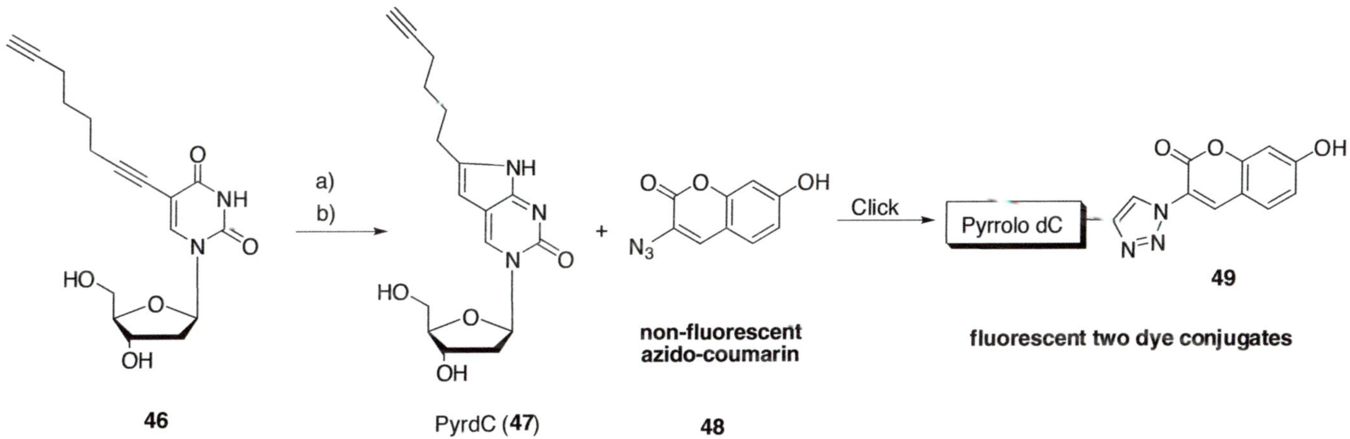

Scheme 2.13 *Synthesis of fluorescent two-dye conjugates via Cu(I)-catalyzed azide–alkyne cycloaddition of 3-azido-7-hydroxycoumarin (**48**) and PyrdC (**47**). (a) Et_3N/MeOH, CuI, reflux 70 °C; (b) 25 % aq. NH_3, overnight, r.t.*[40]

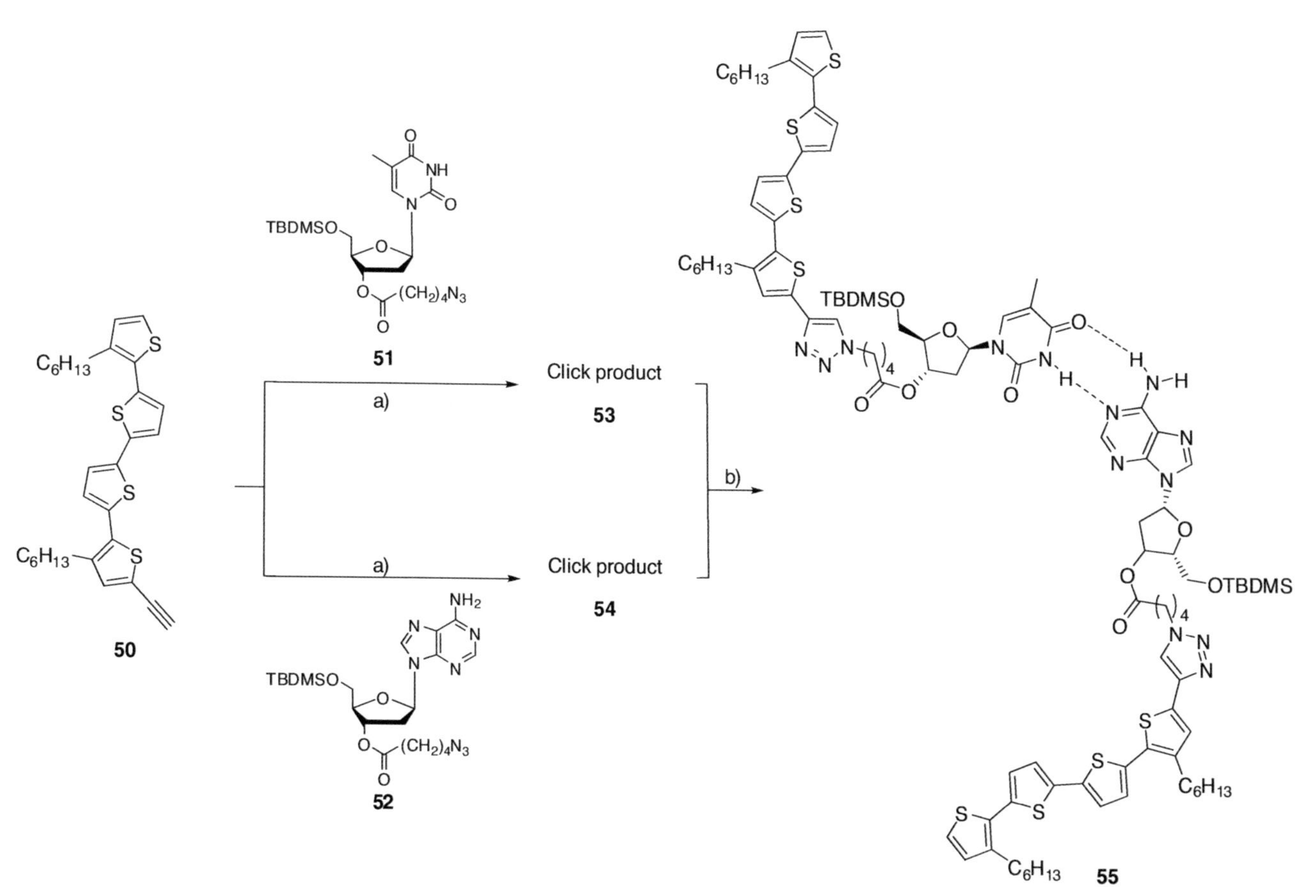

Scheme 2.14 *(a) Click reaction with $Cu(CH_3CN)_4PF_6$–Cu^0, THF; (b) self-assembly of the click products **53** and **54**, thymidine and 2′-desoxyadenosine-functionalized oligothiophenes.*[43]

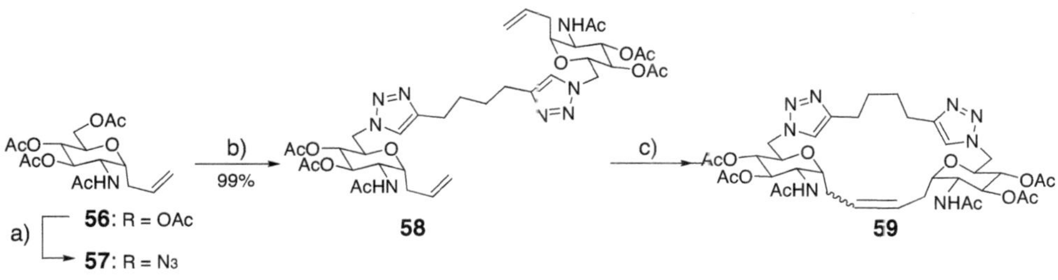

Scheme 2.15 Click-dimerization-ring-closing metathesis for the synthesis of carbohydrate-containing macrocycles. (a) NaOMe, MeOH, r.t.; TsCl, pyridine, 0 °C, r.t. then Ac_2O; NaN_3, DMF, 90 °C; (b) 1,7-octadiyne, ^{t}BuOH, ascorbate, $Cu(OAc)_2$, H_2O, r.t., 12 h; (c) Grubbs I catalyst (5 mol %), reflux, 12–24 h.[49]

groups for dimeric[6] or amino acid glycoconjugates as well as pseudo-oligosaccharides, glycosylated β-peptides,[47] glucosamine derivatives[46] and many more. Beyond the general advantages of click chemistry outlined in the introduction, the formation of triazole mimics of existing carbohydrates is of interest because the heterocyclic ring can participate in hydrogen bonding – an excellent property in the context of biomolecular targets and solubility. 1,2,3-Triazoles containing dimeric saccharides and amino acid glycoconjugates may exhibit a broad spectrum of biological activities, including anti-bacterial, herbicidal, fungicidal, anti-allergic and anti-HIV properties (Scheme 2.15).

Thus, new sugar-containing macrocyclic compounds were developed via click-dimerization-ring-closing metathesis.[49] Following one of the syntheses, the olefinic functions were introduced at the anomeric center either as an *O*-allyl side chain via Königs Knorr-reaction or via Keck-allylation, leading to the *C*-glycosidic compound. The azide functionality of compound **57** was incorporated after deacetylation at the C-6 position of the sugar core through tosylation of the primary hydroxyl group, which was followed by substitution with sodium azide.

The carbohydrate dimers were then generated after the cycloaddition of the corresponding carbohydrate azide **57** and 1,7-octadiyne in *tert*-butanol and after the addition of an aqueous copper(II)acetate–sodium ascorbate mixture. The resulting diolefinic molecules were reacted for 12–24 h in order to accomplish ring-closing metathesis. In addition, 5 mol % of Grubbs' ruthenium catalyst were used at 40 °C to prevent homodimerization. Unfortunately, nonseparable mixtures of both *E/Z*-isomers were observed, thus the double bonds were reduced hydrogenolytically.

Glycodendrimers are mostly used for the study of biological processes that rely on carbohydrate–receptor interactions such as fertilization, pathogen invasion, toxin and hormone mediation, and cell-to-cell interactions, as well as on a variety of inhibitors.[50] In these cases, the syntheses of azido-terminated dendrimers were preferred over those incorporating terminal alkynes, because of the potential affinity of the latter ones to Cu(II)-catalyzed intradendritic oxidative coupling.

Kovensky *et al.* have developed another procedure for the connection of saccharides by click chemistry techniques. With or without carbon chains as spacers, different sugar-bearing azides on C-2, C-5 or C-6 can be used as building blocks.[6,51] This is also true of

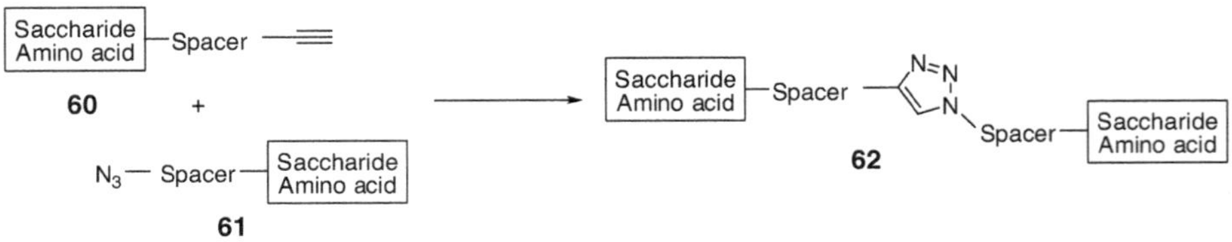

Scheme 2.16 *Synthesis of pseudo-oligosaccharides and amino acid glycoconjugates via 1,3-dipolar cycloaddition (click reaction).*[9]

C-1 alkynylated saccharides or oligosaccharides, in the presence of a range of protecting groups, commonly used in oligosaccharide synthesis.[52]

The efficient coupling of oligosaccharides to oligosaccharide–peptide conjugates **60** and **61** leads to homo- and hetero- dimeric glycoconjugates **62** (Scheme 2.16), which can act as potent reversible cross-linking reagents useful in measuring the distances between carbohydrate binding sites in polyvalent recognition sites. Several pseudo-oligosaccharides **65** and amino-acid glycoconjugates **68** were synthesized using easily accessible carbohydrate **63** and compound **67** in combination with azides **64** and **66** as building blocks for 1,3-dipolar cycloaddition (Scheme 2.17).[9]

Another field of complex oligosaccharides deals with the synthesis of highly glycosylated β-peptides **73** and **76**. Ziegler *et al.* started their synthesis with Fmoc-Asp(*O*-^{t}Bu)OH (**69**) to prepare (*S*)-tert-butyl 3-{[(9H-fluoren-9-yl)methoxy]carbonylamino}-4-azidobutanoate (**71**) in four steps [Scheme 2.18(A)][53] or alternatively with an alkynylated amino acid analog, the *tert*-butyl-protected Fmoc-asparaginic acid propargylamide **74** [Scheme 2.18(B)].[47]

Following the copper-catalyzed coupling reaction, these asparaginic derivatives generated triazole-linked glycosylated amino acids **73** and **76**. Another example was illustrated by Du *et al.*, who described a synthesis for C_3-symmetric glucosamine oligosaccharides using click chemistry for studies of interactions between carbohydrates and proteins.[46]

A new approach, the domino-click synthesis, was reported by Kumar *et al.* (Scheme 2.19), based on a series of consecutive intramolecular organic reactions with resonance-stabilized allenylmagnesium bromide **78**, in order to generate 1,2,3-bistriazoles **81**.[10] By coding mRNAs for proteins in drug discovery, based on size-specific mRNA hairpin loop binding agents, bistriazoles promise to be a good target. The addition of sugar azide **77** to the Grignard reagent **78** resulted in the formation of the novel 5-butynylated triazole **79** in good yield. A subsequent Cu(I)-catalyzed 1,3-dipolar cycloaddition with another equivalent of sugar azide **80a** or **80b** generated novel unsymmetrical bis-1,2,3-triazoles **81a,b**.

The proposed mechanism of this reaction is based on a similar reaction previously published by Sharpless and coworkers. The group investigated the conversion of azides and metalized alkynes into aryl azides through the addition of bromomagnesium acetylides, leading to the desired 1,5-disubstituted triazole regioisomers.[10] Based on these results, the mechanism in the case of the allenylmagnesium bromide addition can be explained via nucleophilic attack of the allenylmagnesium bromide species on the terminal nitrogen of the azide, followed by a simultaneous ring closure through N–C heterocyclization driven

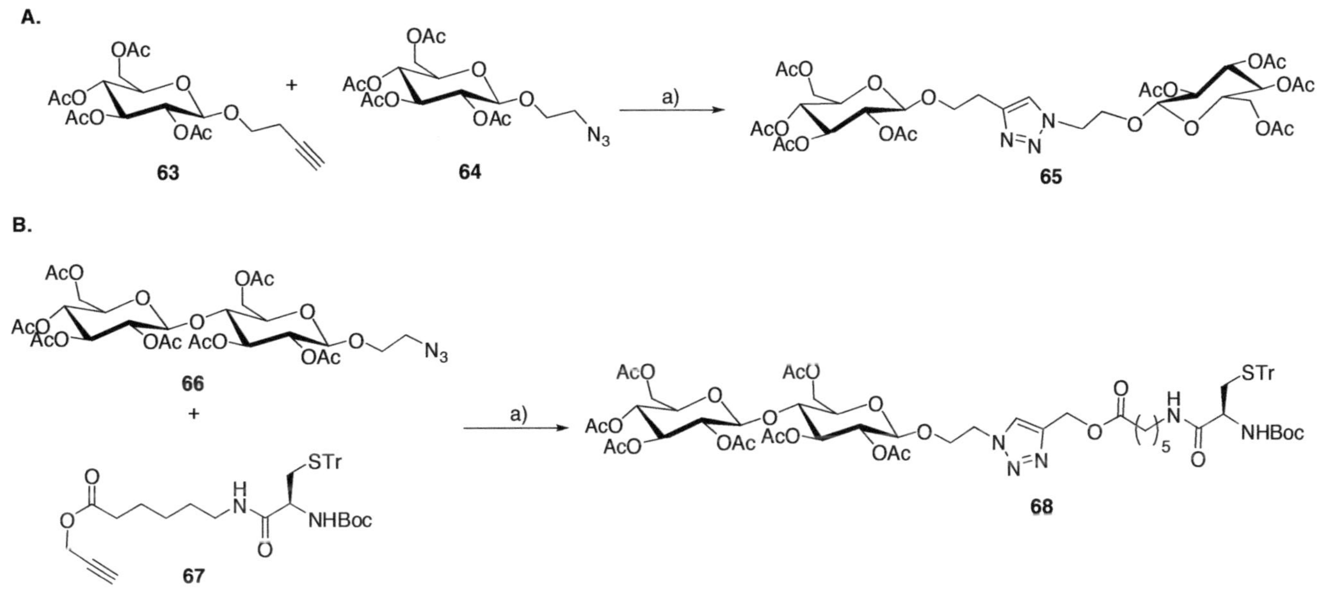

Scheme 2.17 *(a) CuI, EtN^{i}Pr$_2$, MeCN, 2 h. (A) Synthesis of pseudo-oligosaccharide* ***65*** *(96% yield). (B) Conjugation of disaccharide* ***66*** *to amino acid derived alkyne* ***67*** *(91% yield).*[9]

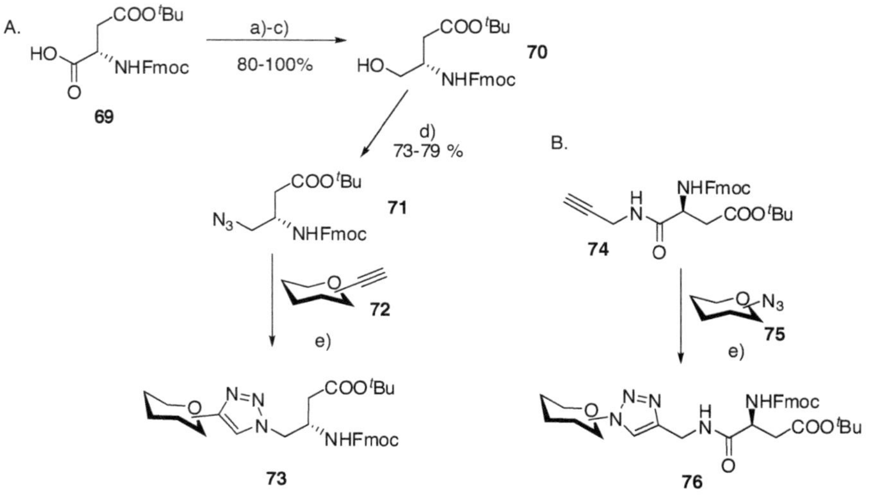

*Scheme 2.18 Synthesis of triazole-glycosylated β-peptides via A. (S)-tert-butyl 3-{[(9H-fluoren-9-yl)methoxy]carbonylamino}-4-azidobutanoate (**71**) and (B) alkynylated compound **74**. (a) NMM; (b) $NaBH_4$; (c) H_2O; (d) PPh_3, DEAD, HN_3; (e) cat. $(EtO)_3P$–CuI, toluene, 80 °C, microwave, 30 min.*[47,53]

by the excess reagent. The final product is generated through the subsequent attack by a second equivalent of allenylmagnesium.

Another way to produce 1,2,3-triazoles is the intramolecular Huisgen-reaction on carbohydrate-derived azido-alkynes. The azido substrates were synthesized by an S_N2 displacement of the corresponding tosylates using NaN_3. Hotha *et al.*[54] and Mandal *et al.*[55] simultaneously illustrated a procedure for intramolecular cycloaddition without using any copper catalysts, ligands or other reagents (Scheme 2.20).

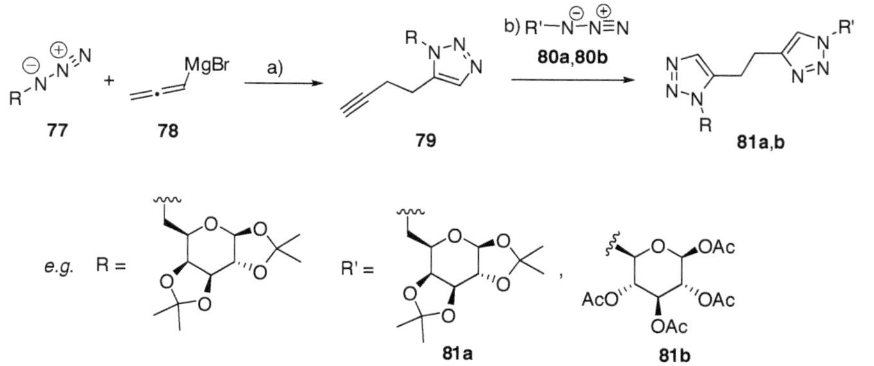

*Scheme 2.19 Domino-click approach to novel bistriazoles **81**. (a) dry THF, r.t.; (b) $CuSO_4{\cdot}5H_2O$, sodium ascorbate, ${}^tBuOH/H_2O$ (1:1), r.t.*[10]

Hotha *et al.*:

Mendal *et al.*:

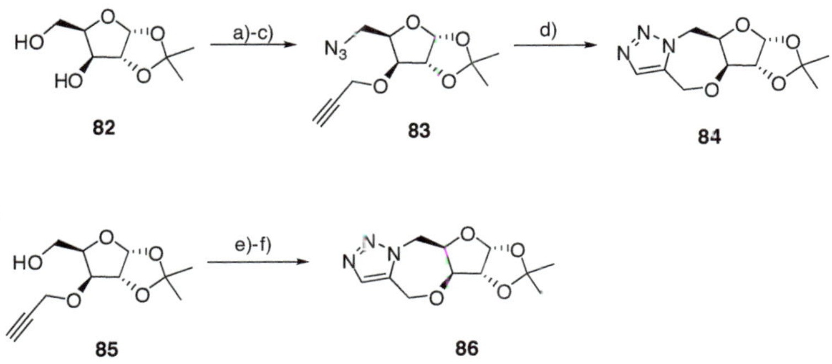

Scheme 2.20 *Hotha et al.: (a) p-TsCl, pyridine, 0 °C, r.t., 10 h, 91%; (b) NaN_3, DMF, 90 °C, 8 h, 95%; (c) NaH, propargyl bromide, DMF, 0 °C, r.t., 2 h, 93%; (d) toluene, 100 °C, 2 h, 95%; Mendal et al.: (e) MsCl, Et_3N, CH_2Cl_2, r.t., 2.5 h; (f) NaN_3, DMF, 90 °C, 6 h, N_2.*[54,55]

Very similar structures were achieved by Bräse *et al.* through the addition of Me_3SiN_3 to a resin-bound thioproparyl ether.[56] Cleavage from the resin and the simultaneous intramolecular click reaction of the liberated azide produced a triazolobenzothiazine in 14% yield over four steps.

The synthesis of a secondary azide for enantiopure carbohydrate mimetics has also been established. The synthesis was performed using Nf-N_3 (nonafluorobutanesulfonyl azide) and copper sulfate for the substitution of an amino function and the *in situ* acetylation of hydroxyl groups (Scheme 2.21).[57]

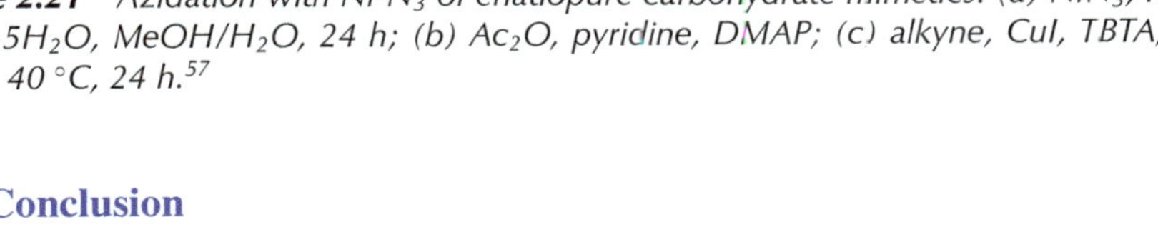

Scheme 2.21 *Azidation with Nf-N_3 of enatiopure carbohydrate mimetics. (a) NfN_3, K_2CO_3, $CuSO_4{\cdot}5H_2O$, MeOH/H_2O, 24 h; (b) Ac_2O, pyridine, DMAP; (c) alkyne, CuI, TBTA, Et_3N, MeCN, 40 °C, 24 h.*[57]

2.7 Conclusion

Click chemistry is a well-established method of generating various derivatives of biomolecules. This article has presented modifications of peptides, peptoids, dendrimers, oligonucleotides and carbohydrates that were synthesized according to their naturally occurring examples. Moreover, several synthetic methods for the formation of the azide functionality on the precursor molecules, as well as applications of the target molecules, have been described herein.

References

(1) S. Bräse, C. Gil, K. Knepper, V. Zimmermann, (2005), Organic azides: an exploding diversity of a unique class of compounds, *Angew. Chem.*, **117**, 5320–5374; *Angew. Chem. Int. Edn*, **44**, 5188–5240.

(2) C. Schilling, N. Jung, S. Bräse, (in press), *Azide Chemistry*, S. Bräse, K. Banert (Eds), Wiley 2009.

(3) S. Bräse, D. Keck, (2007), *Aryl Azides, Science of Synthesis*, Chapter 31.42, Thieme, New York.

(4) S. Bräse, A. Friedrich, M. Gartner, T. Grab, T. Schröder, (2008), Cycloaddition reactions of azides, *Top. Heterocycl. Chem.*, **12**, 45–116.

(5) C. W. Tornøe, C. Christensen and M. Meldal, (2002), Peptidotriazoles on solid phase: [1,2,3]-triazoles by regiospecific copper(I)-catalyzed 1,3-dipolar cycloadditions of terminal alkynes to azides, *J. Org. Chem.*, **67**, 3057–3064.

(6) A. K. Oyelere, P. C. Chen, L. P. Yao, N. Boguslavsky, (2006), Heterogeneous diazo-transfer reaction: a facile unmasking of azide groups on amine-functionalized insoluble supports for solid-phase synthesis, *J. Org. Chem.*, **71**, 9791–9796.

(7) V. V. Rostovtsev, L. G. Green, V. V. Fokin, K. B. Sharpless, (2002), A stepwise Huisgen cycloaddition process: copper(I)-catalyzed regioselective 'ligation' of azides and terminal alkynes, *Angew. Chem., Int. Ed.*, **41**, 2596–2599.

(8) (a) R. A. Evans, (2007), The rise of azide-alkyne 1,3-dipolar click cycloaddition and its application to polymer science and surface modification, *Aust. J. Chem.*, **60**, 384–395; (b) W. H. Binder, R. Sachsenhofer, (2007), Click chemistry in polymer and material science, *Macromol. Sci.*, **28**, 15–54.

(9) S. Hotha, S. Kashyap, (2006), 'Click chemistry' inspired synthesis of pseudo-oligosaccharides and amino acid glycoconjugates, *J. Org. Chem.*, **71**, 364–367.

(10) B. S. Arora, S. Shafi, S. Singh, T. Ismail, H. M. Sampath Kumar, (2008), A novel domino-click approach for the synthesis of sugar based unsymmetrical bis-1,2,3-triazoles, *Carbohydr. Res.*, **343**, 139–144.

(11) H. Gopi, M. Umashankara, V. Pirrone, J. LaLonde, N. Madani, F. Tuzer, S. Baxter, I. Zentner, S. Cocklin, N. Jawanda, S. R. Miller, A. Schoen, J. C. Klein, E. Freire, F. C. Krebs, A. B. Smith, J. Sodroski, I. Chaiken, (2008), Structural determinants for affinity enhancement of a dual antagonist peptide entry inhibitor of human immunodeficiency virus type-1, *J. Med. Chem.*, **51**, 2638–2647.

(12) K. Oh, Z. Guan, (2006), A convergent synthesis of new β-turn mimics by click chemistry, *Chem. Commun.*, 3069–3071.

(13) S. Groothuys, B. H. M. Kuijpers, P. J. L. M. Quaedflieg, H. C. P. F. Roelen, R. W. Wiertz, R. H. Blaauw, F. L. van Delft, F. P. J. T. Rutjes, (2006), Chemoenzymatic synthesis of triazole-linked glycopeptides, *Synthesis*, **18**, 3146–3152.

(14) V. Aucagne, D. A. Leigh, (2006), Chemoselective formation of successive triazole linkages in one pot: 'click–click' chemistry, *Org. Lett.*, **8**, 4505–4507.

(15) V. D. Bock, R. Perciaccante, T. P. Jansen, H. Hiemstra, J. H. Van Maarseveen, (2006), Click chemistry as a route to cyclic tetrapeptide analogs: synthesis of cyclo-[Pro-Val-(triazole)-Pro-Tyr], *Org. Lett.*, **8**, 919–922.

(16) V. D. Bock, D. Speijer, H. Hiemstra, J. H. Van Maarseveen, (2007), 1,2,3-Triazoles as peptide bond isosteres: synthesis and biological evaluation of cyclotetrapeptide mimics, *Org. Biomol. Chem.*, **5**, 971–975.

(17) R. A. Turner, A. G. Oliver, R. S. Lokey, (2007), Click chemistry as a macrocyclization tool in the solid-phase synthesis of small cyclic peptides, *Org. Lett.*, **9**, 5011–5014.

(18) R. Franke, C. Doll, J. Eichler, (2005), Peptide ligation through click chemistry for the generation of assembled and scaffolded peptides, *Tetrahedron Lett.*, **46**, 4479–4482.

(19) C. W. Tornoe, C. Christensen, M. Meldal, (2002), Peptidotriazoles on solid phase: [1,2,3]-triazoles by regiospecific copper(I)-catalyzed 1,3-dipolar cycloadditions of terminal alkynes to azides, *J. Org. Chem.*, **67**, 3057–3064.

(20) O. Roy, S. Faure, V. Thery, C. Didierjean, C. Taillefumier, (2008), Cyclic β-peptoids, *Org. Lett.*, **10**, 921–924.

(21) J. M. Holub, H. Jang, K. Kirshenbaum, (2006), Clickity-click: high functionalized peptoid oligomers generated by sequential conjugation reactions on solid-phase support, *Org. Biomol. Chem.*, **4**, 1497–1502.
(22) K. Kano, M. Liu. J. M. Frechet, (1999), Design of dendritic macromolecules containing folate or methotrexate residues, *Bioconjug. Chem.*, **10**, 1115–1121.
(23) C. J. Arnusch, H. Branderhorst, B. de Kruijff, R. M. J. Liskamp, E. Breukink, R. J. Pieters, (2007), Enhanced membrane pore formation by multimeric/oligomeric antimicrobial peptides, *Biochemistry*, **46**, 13437–13442.
(24) V. Haridas, K. Lal, Y. K. Sharma, (2007), Design and synthesis of triazole-based peptide dendrimers, *Tetrahedron Lett.*, **48**, 4719–4722.
(25) D. T. S. Rijkers, G. W. van Esse, R. Merkx, A. J. Brouwer, H. J. F. Jacobs, R. I. Pieters, R. M. I. Liskamp, (2005), Efficient microwave-assisted synthesis of multivalent dendrimeric peptides using cycloaddition reaction (click) chemistry, *Chem. Commun.*, **36**, 4581–4583.
(26) A. Gopin, S. Ebner, B. Attali, D. Shabat, (2006), Enzymatic activation of second-generation dentritic prodrugs: conjugtion of self-immolative dendrimers with poly(ethylene glycol) via click chemistry, *Bioconjug. Chem.*, **17**, 1432–1440.
(27) F. Seela. V. R. Sirivolu, (2006), DNA containing side chains with terminal triple bonds: base-pair stability and functionalization of alkynylated pyrimidines and 7-deazapurines, *Chem. Biodiv.*, **3**, 509–514.
(28) F. Seela, V. R. Sirivolu, (2007), Nucleosides and oligonucleotides with diynyl side chains: base pairing and functionalization of 2'-deoxyuridine derivatives by the copper(I)-catalyzed alkyne-azide 'click' cycloaddition, *Helv. Chim. Acta*, **90**, 535–552.
(29) T. S. Seo, Z. Li, H. Ruparel, J. Ju, (2003), Click chemistry to construct fluorescent oligonucleotides for DNA sequencing, *J. Org. Chem.*, **68**, 609–612.
(30) (a) X. Jin, H. Ding, R. Yang, Q. Xiao, Y. Ju, (2008), Synthesis of carbohydrate-conjugated furo[2,3-d]pyrimidine by 'click chemistry', *Synthesis*, **6**, 865–870; (b) X. Jin, R. Yang, P. Jin, Q. Xiao, Y. Ju, (2007), Synthesis of carbohydrate-conjugated dT analogs using 'click chemistry', *Synthesis*, **19**, 2967–2972.
(31) H. C. Kolb, M. G. Finn, K. B. Sharpless, (2001), Click chemistry: diverse chemical function from a few good reactions, *Angew. Chem.*, **113**, 2056–2075; *Angew. Chem. Int. Ed.*, **40**, 2004–2021.
(32) H. C. Kolb, K. B. Sharpless, (2003), The growing impact of click chemistry on drug discovery, *Drug Discov. Today*, **8**, 1128–1137.
(33) P. Kočalka, N. K. Andersen, F. Jensen, P. Nielsen, (2007), Synthesis of 5-(1,2,3-triazol-4-yl)-2'-deoxyuridines by a click chemistry approach: stacking of triazoles in the major groove gives increased nucleic acid duplex stability, *ChemBioChem*, **8**, 2106–2116.
(34) R. L. Weller, S. R. Rajski, (2005), DNA methyltransferase-moderated click chemistry, *Org. Lett.*, **7**, 2141–2144.
(35) G. A. Burley, J. Gierlich, M. R. Mofid, H. Nir, S. Tal, Y. Eichen, T. Carell, (2006), Directed DNA metallization, *J. Am. Chem. Soc.*, **128**, 1398–1399.
(36) J. Gierlich, G. A. Burley, P. M. E. Gramlich, D. M. Hammond, T. Carell, (2006), Click chemistry as a reliable method for the high-density postsynthetic functionalization of alkyne-modified DNA, *Org. Lett.*, **8**, 3639–3642.
(37) C. T. Wirges, P. M. E. Gramlich, K. Gutsmiedl, J. Gierlich, G. A. Burley, T. Carell, (2007), Pronounced effect of DNA hybridization on click reaction efficiency, *QSAR Comb. Sci.*, **26**, 1159–1164.
(38) J. Gierlich, K. Gutsmiedl, P. M. E. Gramlich, A. Schmidt, G. A. Burley, T. Carell, (2007), Synthesis of highly modified DNA by a combination of PCR with alkyne-bearing triphosphates and click chemistry, *Chem. Eur. J.*, **13**, 9486–9494.
(39) C. Bouillon, A. Meyer, S. Vidal, A. Jochum, Y. Chevolot, J.-P. Cloarec, J.-P. Praly, J.-J. Vasseur, F. Morvan, (2006), Microwave assisted 'click' chemistry for the synthesis of multiple labeled-carbohydrate oligonucleotides on solid support, *J. Org. Chem.*, **71**, 4700–4702.
(40) F. Seela. V. R. Sirivolu, (2008), Pyrrolo-dC oligonucleotides bearing alkynyl side chains with terminal triple bonds: synthesis, base pairing and fluorescent dye conjugates prepared by the azide-alkyne 'click' reaction, *Org. Biomol. Chem.*, **6**, 1674–1687.

(41) I. Géci, V. V. Filichev, E. B. Pedersen, (2007), Stabilization of parallel triplexes by twisted intercalating nucleic acids (TINAs) incorporating 1,2,3-triazole units and prepared by microwave-accelerated click chemistry, *Chem. Eur. J.*, **13**, 6379–6386.
(42) M. Nakane, S. Ichikawa, A. Matsuda, (2008), Triazole-linked dumbbell oligodeoxynucleotides with NK-κB binding ability as potential decoy molecules, *J. Org. Chem.*, **73**, 1842–1851.
(43) A. Jatsch, A. Kopyshev, E. Mena-Osteritz. P. Bäuerle, (2008), Self-organizing oligothiophene-nucleoside conjugates: versatile synthesis via 'click'-chemistry, *Org. Lett.*, **10**, 961–964.
(44) S. G. Gouin, E. Vanquelef, J. M. Garcia Fernández, C. Ortiz Mellet, F.-Y. Dupradeau, J. Kovensky, (2007), Multi-mannosides based on a carbohydrate scaffold: synthesis, force field development, molecular dynamics studies, and binding affinities for lectin Con A, *J. Org. Chem.*, **72**, 9032–9045.
(45) N. Xu, F.-Z. Lu, F.-S. Du, Z.-C. Li, (2007), Synthesis of saccharide-terminated poly(ε-caprolactone) via Michael addition and 'click' chemistry, *Macromol. Chem. Phys.*, **208**, 730–738.
(46) Q. Chen, F. Yang, Y. Du, (2005), Synthesis of a C3-symmetric (1$\rightarrow$6)-N-acetyl-β-D-glucosamine octadecasaccharide using click chemistry, *Carbohydrate Res.*, **340**, 2476–2482.
(47) N. Pietrzik, C. Schips, T. Ziegler, (2008), Efficient synthesis of glycosylated asparaginic acid building blocks via click chemistry, *Synthesis*, **4**, 519–526.
(48) D. K. Scrafton, J. E. Taylor, M. F. Mahon, J. S. Fossey, T. D. James, (2008), 'Click-fluors': modular fluorescent saccharide sensors based on a 1,2,3-triazole ring, *J. Org. Chem.*, **73**, 2871–2874.
(49) S. Doerner, B. Westermann, (2005), A short route for the synthesis of 'sweet' macrocycles via a click-dimerization-ring-closing metathesis approach, *Chem. Commun.*, **22**, 2852–2854.
(50) S. G. Gouin, L. Bultel, C. Falentin, J. Kovensky, (2007), A simple procedure for connecting two carbohydrate moieties by click chemistry techniques, *Eur. J. Org. Chem.*, **7**, 1160–1167.
(51) (a) L. Marmuse, S. A. Nepogodiev, R. A. Field, (2005), 'Click chemistry' en route to pseudo-starch, *Org. Biomol. Chem.*, **3**, 2225–2227; (b) S. A. Nepogodiev, S. Dedola, L. Marmuse, M. T. De Oliveira, R. A. Field, (2007), Synthesis of triazole-linked pseudo-starch fragments, *Carbohydr. Res.*, **342**, 529–540.
(52) J. Zhang, M. Garrossian, D. Gardner, A. Garrossian, Y.-T. Chang, Y. K. Kim, C.-W. T. Chang, (2008), Synthesis and anticancer activity studies of cyclopamine derivatives, *Bioorg. Med. Chem. Lett.*, **18**, 1359–1363.
(53) A. Boeijen, J. van Ameijde, R. M. J. Liskamp, (2001), Solid-phase synthesis of oligourea peptidomimetics employing the Fmoc protection strategy, *J. Org. Chem.*, **66**, 8454–8462.
(54) S. Hotha, R. I. Anegundi, A. A. Natu, (2005), Expedient synthesis of 1,2,3-triazole-fused tetracyclic compounds by intramolecular Huisgen ('click') reactions on carbohydrate-derived azido-alkynes, *Tetrahedron Lett.*, **46**, 4585–4588.
(55) S. Tripathi, K. Singha, B. Achari, S. B. Mandal, (2004), *In situ* 1,3-dipolar azide cycloaddition reaction: synthesis of functionalized D-glucose based chiral piperidine and oxazepine analogues, *Tetrahedron*, **60**, 4959–4965.
(56) M. Kreis, C. F. Nising, M. Schroen, K. Knepper, S. Bräse, (2005), Efficient solid phase synthesis of benzo[1,2,3]thiadiazoles and related structures, *Org. Biomol. Chem.*, **3**, 1835–1837.
(57) S. Yekta, V. Prisyazhnyuk, H.-U. Reissig, (2007), Simple modifications of enantiopure 1,2-oxazines leading to building blocks for carbohydrate and peptide mimetics, *Synlett*, **13**, 2069–2072.

3

Copper-free Click Chemistry

Jeremy M. Baskin and Carolyn R. Bertozzi

3.1 Introduction

Biological systems exhibit astounding chemical complexity. A single mammalian cell contains thousands of proteins, DNA and RNA, glycans, lipids, myriad small-molecule metabolites, and metal ions, all in an aqueous environment. Each of these components displays many chemical functionalities, including nucleophiles, electrophiles, oxidants, and reductants. Within this environment, enzymes choreograph the innumerable chemical transformations that together constitute the life of the cell: catabolic conversion of metabolites into energy, biosynthesis and posttranslational modification of proteins, replication of DNA, and many other biochemical processes.

Chemical biologists have striven to study the molecular intricacies of living systems by labeling individual components or groups of components – within the complexity of the living system – with probes such as fluorophores and affinity tags. This approach permits both the tracking of biomolecules within the living cell by imaging and also the determination of their exact molecular identities and compositions after purification from a cell lysate. A critical aspect of any strategy for labeling a target biomolecule inside a living cell or organism using a chemical reaction is that the reaction must be exquisitely chemoselective.

The term 'bio-orthogonal' – defined as noninteracting with biology – encompasses all of the characteristics of such a chemical reaction. A bio-orthogonal ligation is thus a chemical reaction in which two functional groups selectively react with one another to form a covalent linkage in the presence of all of the functionality in biological systems (Figure 3.1).[1] In order to maximize labeling efficiency, the reaction should display the properties of a 'click' reaction as outlined by Kolb, Finn and Sharpless: rapid kinetics, high

Click Chemistry for Biotechnology and Materials Science Edited by Joerg Lahann

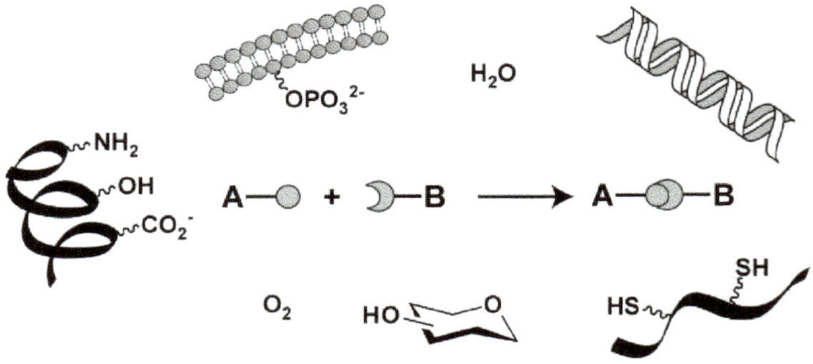

Figure 3.1 Schematic of a bio-orthogonal reaction. Two molecules (A and B) are ligated together in the presence of diverse biological functionality by virtue of two complementary bio-orthogonal functionalities (gray circle and crescent) that selectively react only with each other to form a covalent bond.

yield, lack of unwanted by-products, broad solvent compatibility (including water), and readily accessible starting materials.[2]

Arguably the most widely used click reaction is the Cu-catalyzed azide–alkyne cycloaddition reported simultaneously in 2002 by Sharpless and coworkers and Meldal and coworkers.[3,4] This reaction, which utilizes a Cu(I) catalyst to effect the 1,3-dipolar cycloaddition of azides and terminal alkynes to form 1,4-disubstituted 1,2,3-triazoles, displays the characteristics of an ideal click reaction mentioned above (Figure 3.2). It has thus been widely employed throughout medicinal chemistry, materials science and chemical biology,[5–7] applications that are summarized in other chapters in this book. However, the strict requirement of a Cu(I) catalyst precludes the use of CuAAC for labeling within living systems, due to the cytotoxicity of copper.

Many copper-free, and hence nontoxic, alternatives to CuAAC exist. This chapter will begin by describing these copper-free click chemistries, which include condensations of ketones with hydrazide and aminooxy reagents, Staudinger ligation of phosphines and azides, strain-promoted [3 + 2] cycloaddition of cyclooctynes with azides, and various bio-orthogonal ligations of alkenes. After outlining the chemistries, we will present selected applications of copper-free click chemistries in chemical biology in the context of live cells and whole animals.

3.2 Bio-orthogonal Ligations

Amidst the plethora of organic transformations that form a covalent bond between two reactants, very few fit the criteria of bio-orthogonality.[8] Reactions must take place in aqueous

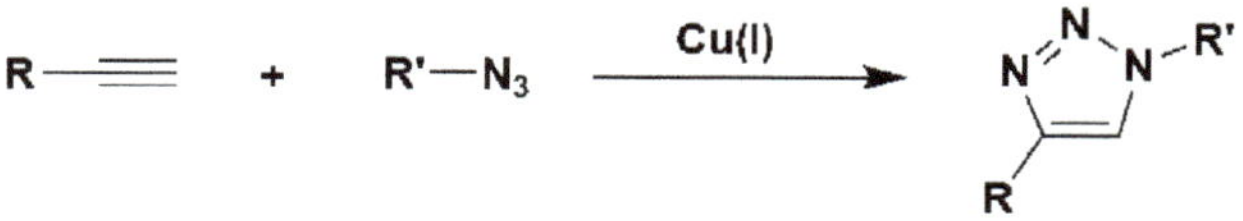

Figure 3.2 Cu-catalyzed azide–alkyne cycloaddition. Azides and terminal alkynes react in the presence of a Cu(I) catalyst to form 1,4-disubstituted 1,2,3-triazole products.

solvent, involve no toxic catalysts or reagents, and – most importantly – involve two functional groups that are not present in and which do not cross-react with any functionality present in biological systems. Classic transformations such as bimolecular substitution reactions of alkyl halides and various nucleophiles, as well as the coupling of amines and carboxylic acids to form amides, are not bioorthogonal, as the reagents would have significant cross-reactivity with many amino acid side chains, other cellular nucleophiles and water. Neither are more modern metal-promoted cross-coupling reactions, as they involve toxic catalysts. Even standard protein bioconjugation reactions (e.g. Michael addition of thiols to maleimides) do not qualify as bio-orthogonal, as thiols are ubiquitous within a cellular environment.

An important principle that has emerged in roughly a decade of bio-orthogonal reaction development is that these reactions do not fit a uniform mold. Early work focused on highly selective nucleophile/electrophile coupling reactions, first the condensation of ketones with heteroatom-bound amine reagents (Section 2.1) and later a modified Staudinger reaction of phosphines and azides (Section 2.2). Much work in recent years has shifted to pericyclic reactions, notably 1,3-dipolar cycloadditions of azides and alkynes (Section 2.3). Finally, an emerging area is the development of bio-orthogonal ligations of alkenes (Section 2.4).

3.2.1 Condensations of Ketones and Aldehydes with Heteroatom-bound Amines

Historically, the first bio-orthogonal ligations involved ketone–aldehyde condensation reactions. While ketones and aldehydes can form reversible imine adducts with many amines found in biological systems, this process is thermodynamically unfavorable in water. The use of hydrazides and aminooxy reagents, often called 'α-effect amines' because the heteroatom-bound amine is much more nucleophilic than simple amines, shifts the equilibrium dramatically to the hydrazone and oxime products, respectively (Figure 3.3). For example, the equilibrium constant for the condensation of acetone and hydroxylamine in water is 1×10^6.[9] This reaction is not an optimal bio-orthogonal ligation for live cell applications, however, because it proceeds most efficiently at pH values of 3–6, well below the physiological level (pH 7.4).[9] The kinetics of these condensation reactions can be improved considerably both at acidic and neutral pH by the addition of aniline-based nucleophilic

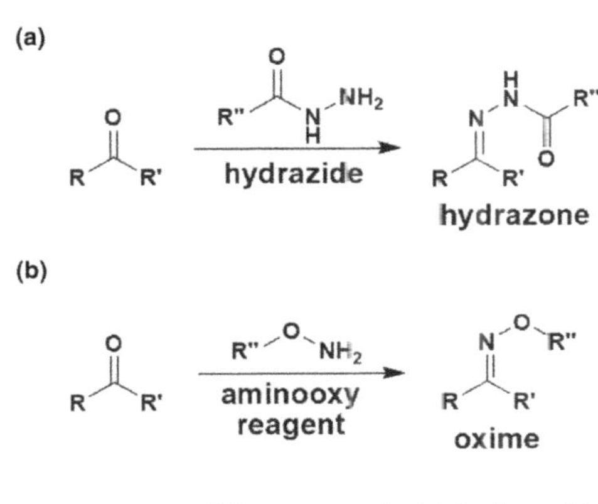

Figure 3.3 Condensation reactions of ketones and aldehydes with (a) hydrazides to form hydrazones or (b) aminooxy reagents to form oximes.

catalysts, although the requisite high concentrations of these reagents (up to 0.1 M) might preclude their use in living systems, wherein toxicity becomes a concern.[10] Further, although ketones and aldehydes are absent from the cell surface and from macromolecules within the cell, these functional groups are present within many intracellular metabolites (e.g. glucose, pyruvate) and can thus be thought of as semi-bioorthogonal. For these reasons, the use of ketone/aldehyde ligations for labeling of biomolecules within living systems has been somewhat limited.

3.2.2 Staudinger Ligation of Phosphines and Azides

A major breakthrough occurred in 2000, with the introduction of the azide as a functional handle for bio-orthogonal chemical reactions. Unlike the ketone, the azide is truly bio-orthogonal – it is essentially unreactive with any biological functional groups under physiological conditions. The slow reaction of alkyl azides with thiols at physiological pH is usually insignificant on the timescale of biological experiments.[11] The azide is also absent from virtually all biological systems. The lone exception reported to date is an azide-containing natural product isolated from the dinoflagellate *Gymnodinium breve*, a species of red algae that is responsible for the production of toxic red tides along the Gulf Coast of Florida.[12,13]

Despite its kinetic stability, the azide is thermodynamically a high-energy species prone to specific reactivity both as a soft electrophile and as a 1,3-dipole.[14] The Staudinger ligation of phosphines and azides takes advantage of the former type of reactivity. This reaction is a modification of the classical Staudinger reduction of azides with phosphines, in which an aza-ylide intermediate collapses in water to amine and phosphine oxide products. However, construction of a triarylphosphine reagent with an ester positioned *ortho* to the phosphorus atom enabled trapping of the aza-ylide intermediate, in aqueous conditions, as an amide, thereby covalently 'ligating' the two molecules (Figure 3.4).[15] The Staudinger ligation has many positive attributes: like azides, phosphines are absent from biological systems, and due to the relative dearth of soft electrophiles in biology, triarylphosphines have no crossreactivity with endogenous biomolecules. Furthermore, the phosphine reagents – derivatized as conjugates of many different epitopes such as biotin, fluorophores, and peptides – demonstrate no toxicity.

The mechanism of the Staudinger ligation allows for a wide variety of modifications and clever applications. For example, inversion of the orientation of the ester has enabled the development of a so-called 'traceless' Staudinger ligation in which an amide is formed and the phosphine oxide byproduct is expelled (Figure 3.5).[16–18] As this variant forms native amides, it is a powerful tool for peptide ligation; however, lower yields and slower kinetics have hindered its use as a bioorthogonal reaction for sensitive detection of azides in living systems.

Further, the Staudinger ligation can be used to create 'smart' phosphine reagents that become fluorescent upon reaction with azides. Such fluorogenic reagents can be tremendously advantageous for biological imaging experiments because they permit dynamic monitoring of labeling reactions and eliminate problems associated with washing away unreacted phosphine probe. Two different approaches have been taken to tackle this challenge, each exploiting a different aspect of the reaction mechanism (Figure 3.6). First, a fluorogenic coumarin–phosphine reagent was synthesized that is nonfluorescent due to quenching by

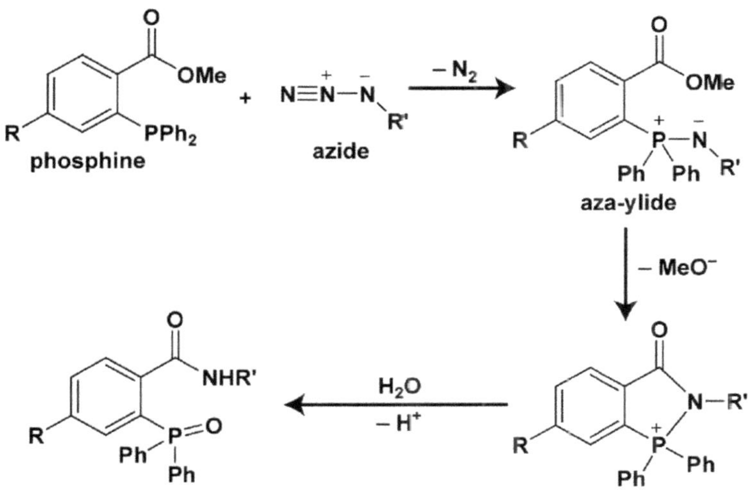

Figure 3.4 The Staudinger ligation of azides and triarylphosphines. The mechanism begins with nucleophilic attack of the phosphine on the terminal nitrogen atom of the azide, liberating N_2. The resultant aza-ylide intermediate is trapped intramolecularly by the methyl ester, releasing methanol and forming an amide and the phosphine oxide.

the phosphine lone pair of electrons. Upon Staudinger ligation with azides, the conversion to a phosphine oxide eliminates this quenching and the fluorophore 'turns on'.[9] Unfortunately, this reagent has not seen widespread use because nonspecific air oxidation of the phosphine also results in fluorophore turn-on, and oxygen cannot be avoided in most biological systems. A second approach utilizes intramolecular fluorescence resonance energy transfer (FRET) quenching. Here, a phosphine–fluorophore conjugate contains a FRET

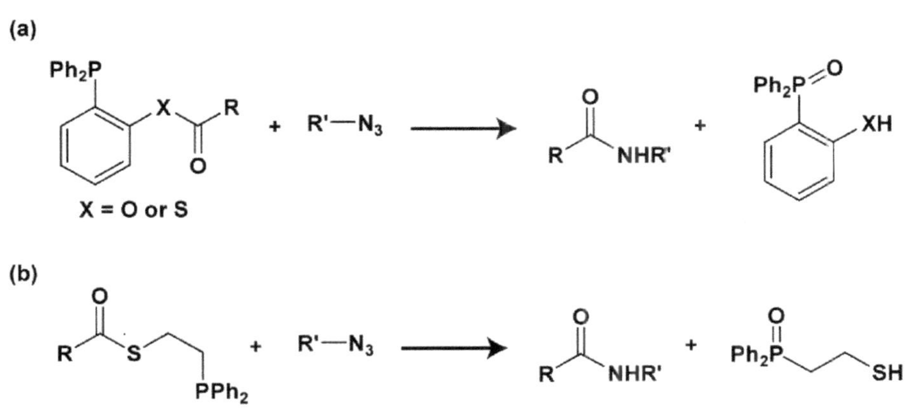

Figure 3.5 'Traceless' Staudinger ligation strategies that enable amide formation concomitant with release of the phosphine oxide by-product.

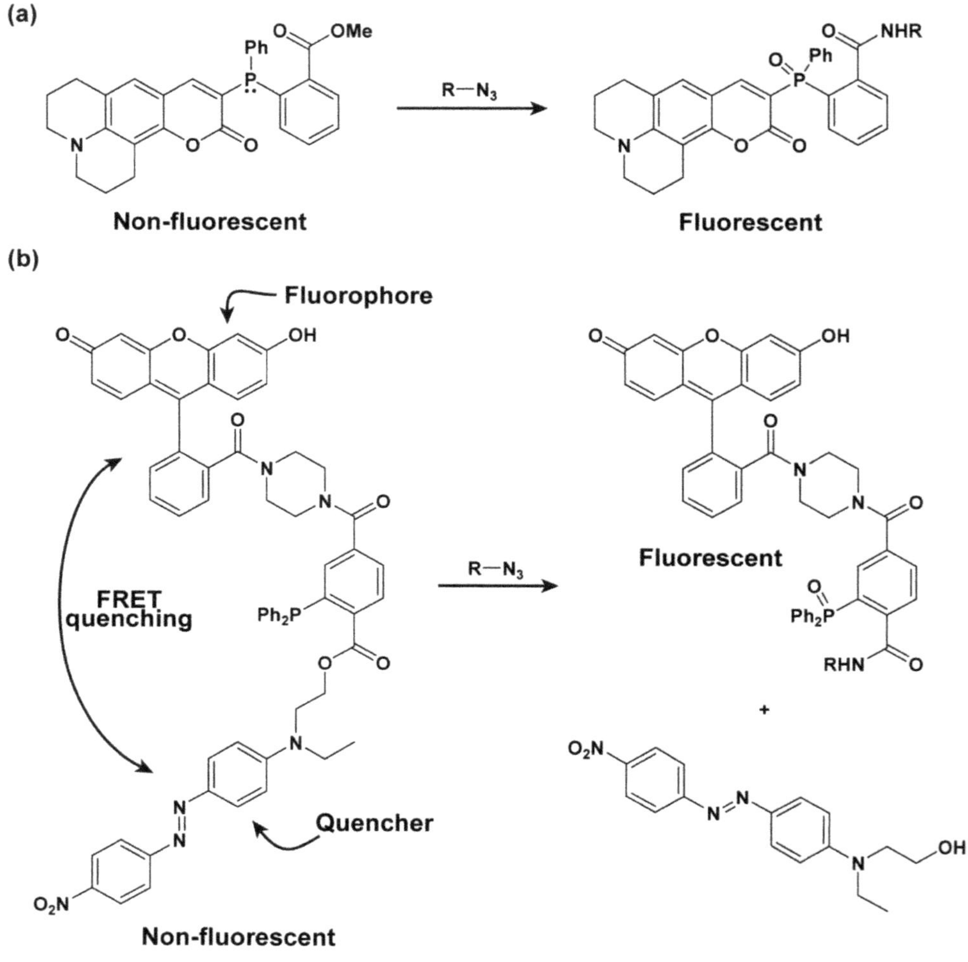

Figure 3.6 *Two approaches to the design of fluorogenic phosphine probes. (a) A coumarin phosphine conjugate is nonfluorescent due to lone-pair quenching of the fluorophore; upon Staudinger ligation with azides, the product is fluorescent because the phosphine is oxidized. (b) A FRET-based fluorescein phosphine conjugate is nonfluorescent due to the action of an intramolecular FRET quencher (Disperse Red 1), attached via an ester. Upon Staudinger ligation, the quencher is liberated and the product is fluorescent.*

quencher attached via the ester and is hence nonfluorescent. Upon reaction with azides, the ester is cleaved, liberating the quencher and turning on fluorescence. As nonspecific phosphine oxidation generates a silent (though nonfunctional) reagent, this probe has been successfully employed for imaging azide-functionalized glycans in living cells.[20]

Despite the myriad applications that the Staudinger ligation enables by virtue of its exquisite selectivity, it suffers from relatively slow reaction kinetics.[21] Accordingly, cell labeling reactions can often require an hour or longer. Thus, for studying dynamic biological processes that occur on faster timescales, improvements to intrinsic reaction kinetics have been sought. A mechanistic study revealed that, for triarylphosphine and alkyl azide reactants, the rate-determining step is the initial nucleophilic attack of the phosphine on the terminal nitrogen atom of the azide.[21] Electron-rich phosphine reagents that could accelerate this step have been designed. Although these reagents performed faster in the Staudinger ligation, they were also highly susceptible to nonspecific air oxidation, which both lowered reaction yields and made the reagents difficult to handle and store.[21] Fundamentally, it is difficult, if not impossible, to decouple improved kinetics in reactions with azides from increased propensity toward nonspecific air oxidation, as both properties ultimately arise from the nucleophilic character of the phosphine. Therefore, for the study for dynamic biological processes, it became necessary to investigate the alternative mode of reactivity of the azide as a 1,3-dipole.

3.2.3 Copper-free Azide–Alkyne Cycloadditions

In addition to being a soft electrophile, the azide is also a 1,3-dipole, and as such, it can undergo [3 + 2] cycloadditions with alkynes as first reported by Michael in 1893[22] and studied extensively by Huisgen.[23] The thermal cycloaddition of azides and unactivated alkynes to form 1,2,3-triazoles is a highly exergonic process ($\Delta G \approx -61$ kcal/mol), although it is kinetically hindered by an activation barrier of 26 kcal/mol.[24] Thus, elevated temperatures or pressures are necessary to accelerate the reaction of azides and simple alkynes. As mentioned earlier, the use of copper catalysis is a highly effective method to lower the activation energy and effect room-temperature triazole formation between azides and terminal alkynes; however, for biological labeling applications, CuAAC is not an ideal bio-orthogonal ligation due to the cytotoxicity of copper.

In an effort to activate the alkyne component for [3 + 2] cycloaddition with azides in a biocompatible manner, we explored the use of ring strain. In 1961, Wittig and Krebs demonstrated that cyclooctyne, the smallest stable cycloalkyne, reacts with azides to form the corresponding 1,2,3-triazole (Figure 3.7).[25] This reaction's fast kinetics – the authors wrote that it 'proceeded like an explosion' – are due to roughly 18 kcal/mol of ring strain

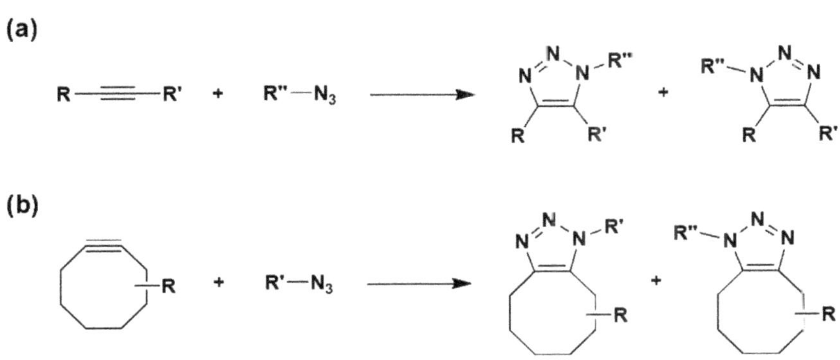

Figure 3.7 *Huisgen 1,3-dipolar cycloadditions of azides and alkynes to form regioisomeric mixtures of 1,2,3-triazoles. (a) Cycloaddition involving azides and linear alkynes. (b) Copper-free, strain-promoted cycloaddition between azides and cyclooctynes.*

in the cyclooctyne starting material, much of which is released in the transition state of the reaction.[26] Building upon this precedent, we synthesized a biotinylated cyclooctyne conjugate and demonstrated that it could selectively label azide-modified proteins *in vitro* and on live cell surfaces.[27] Unlike the case of CuAAC, which exclusively forms the 1,4-disubstituted triazoles, this 'strain-promoted cycloaddition' forms a roughly 1:1 mixture of regioisomeric 1,2,3-triazoles. Importantly, this reagent required no copper catalyst and displayed no toxicity; however, its kinetics were still considerably slower than those of CuAAC and comparable to those of the Staudinger ligation.[27]

We thus set out to improve the kinetics of this strain-promoted cycloaddition by installing LUMO-lowering, electron-withdrawing fluorine atoms adjacent to the cyclooctyne, producing monofluorinated and difluorinated reagents that were roughly 2-fold and 40-fold faster than the original reagent, respectively (Figure 3.8).[28,29] In particular, this last reagent, termed DIFO (for difluorinated cyclooctyne), labels azide-bearing proteins with similar kinetics to CuAAC and has been particularly useful for imaging cultured cells and live animals such as developing zebrafish embryos (see Section 3.3). Other recent additions to the cyclooctyne toolkit include second-generation DIFO reagents with more facile synthetic routes[30] and a highly water-soluble azacyclooctyne reagent designed to eliminate problems associated with nonspecific binding of the more hydrophobic cyclooctyne reagents to

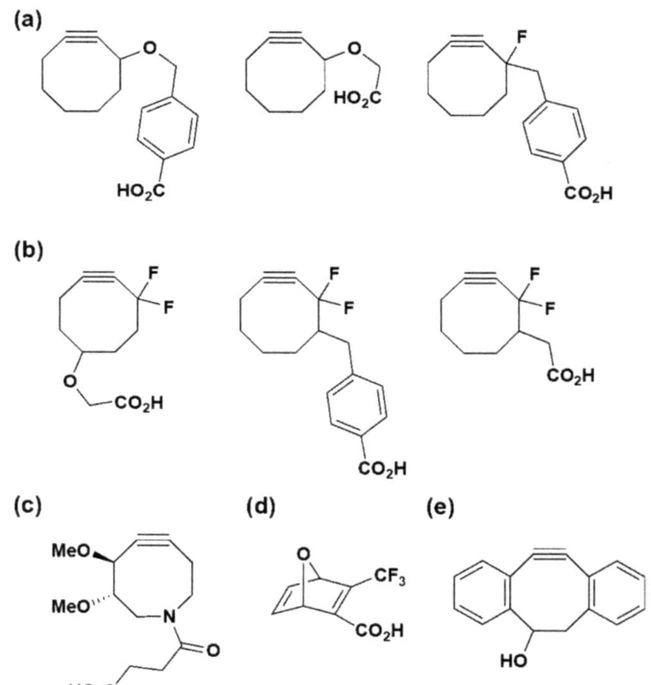

Figure 3.8 Structures of strained alkynes or alkenes for Cu-free [3 + 2] cycloadditions with azides in biological systems. These include (a) simple cyclooctyne, (b) difluorinated cyclooctyne (DIFO), (c) azacyclooctyne, (d) oxanorbornadiene, and (e) dibenzocyclooctyne probes.

biological membranes and serum proteins *in vivo* (Figure 3.8).[31] In addition, strained oxanorbornadiene and dibenzocyclooctyne reagents have emerged for selectively labeling azido biomolecules (Figure 3.8).[32,33] Their kinetic rate constants range from comparable to the original cyclooctyne for the former and comparable to DIFO for the latter. Collectively, these studies suggest a rich future for the development of Cu-free reagents for sensitive detection of azides in biological systems.

3.2.4 Bioorthogonal Ligations of Alkenes

Ideally, a slew of bio-orthogonal ligations would exist that are each orthogonal to one another. This scenario would permit the simultaneous use of multiple reactions to label different components in the same system. Thus, to complement the plethora of azide-specific bio-orthogonal ligations, researchers have been recently looking to alternatives. One emerging example is so-called 'photoclick' chemistry. This reaction involves the light-induced decomposition of 2,5-diaryltetrazoles to form dipolar nitrile imine intermediates, which undergo subsequent [3 + 2] cycloaddition with alkenes to yield stable pyrazoline adducts (Figure 3.9). Like the [3 + 2] azide–alkyne cycloaddition, this reaction was first explored by Huisgen many decades ago and recently revisited in the context of bio-orthogonal labeling. Lin and coworkers have demonstrated that various diaryltetrazole probes can, upon irradiation with 302 nm light, selectively label alkene-containing proteins *in vitro* and in living bacterial cells with rapid kinetics.[34,35] Although the irradiation time was limited to a few minutes, this relatively high-energy ultraviolet light can be damaging to living cells. Thus, Lin and coworkers developed second-generation diaryltetrazole reagents that are activated at a more biocompatible 365 nm;[36] in the future, this approach may be extended to biomolecule labeling in mammalian cells.

Davis and coworkers demonstrated that olefin metathesis can be used to selectively target allyl sulfides on isolated proteins.[37] To accomplish this task, cysteine residues were first converted to the corresponding allyl sulfides, and alkene-functionalized probes (e.g. sugars, polyethyleneglycol) were appended via cross-metathesis using the Hoveyda–Grubbs second-generation catalyst. The efficacy of this reaction in living systems has not yet been established, however, and toxicity is always a concern when dealing with organometallic catalysts.

Other bio-orthogonal ligations involving alkenes include the ligation of strained alkenes such as *E*-cyclooctene or norbornene with tetrazine reagents.[38,39] These inverse electron-demand [4 + 2] Diels–Alder reactions proceed with rapid kinetics and in one case displays

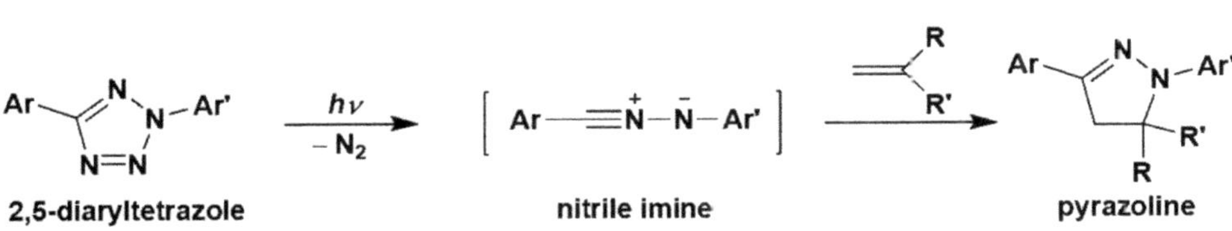

Figure 3.9 'Photoclick' chemistry. Light-induced decomposition of 2,5-diaryltetrazoles creates dipolar nitrile imine intermediates, which react in situ *with alkenes to form pyrazoline products.*

bioorthogonality in the context of cell-surface labeling.[39] However, both components in this reaction (i.e. the strained alkene and the tetrazine reagent) are quite large compared with an azide or a ketone. This characteristic presents a potential limitation for metabolic labeling experiments, where biosynthetic enzymes often tolerate only minor structural changes to their natural substrates (see Sections 3.3.2–3.3.4).

3.3 Applications of Copper-free Click Chemistries

In order to take advantage of the click reactions for biomolecule labeling, one of the two bioorthogonal functional groups that participates in the click reaction (e.g. ketone, azide, etc.), often termed a 'chemical reporter', must first be introduced into biomolecules. Various methods have been developed for this purpose.[1] This section will discuss the use of activity-based inhibitors for labeling classes of enzymes (Section 3.1), genetically encoded peptide tags and unnatural amino acids for labeling proteins (Section 3.2), unnatural monosaccharides for metabolic labeling of glycans (Section 3.3) and unnatural lipids and nucleotide precursors for labeling lipids and nucleic acids (Section 3.4).

3.3.1 Activity-based Profiling of Enzymes

Covalent inhibitors of enzyme activity have seen widespread use as both pharmaceutical agents and tools in basic research. In many cases, inhibitors are not selective for a specific enzyme but instead can label entire classes of enzymes (e.g. serine proteases). Cravatt has pioneered the use of functionalized enzyme inhibitors to probe enzymes based on their activity. This technique, termed activity-based protein profiling (ABPP), involves the covalent labeling of active enzymes within living cells or lysates, and even within live animals, with a mechanism-based inhibitor derivatized either directly with an affinity tag or fluorophore or with a bio-orthogonal chemical reporter (Figure 3.10).[40,41] In the latter case, a second step, the bio-orthogonal click reaction, can be employed to append the desired tag or probe.

This strategy has been highly successful for labeling classes of enzymes with well-characterized covalent, active site-directed inhibitors (e.g. serine hydrolases[42,43] and glycosidases[44]). In early studies using ABPP, active enzymes were targeted from cell or tissue lysates and characterized by mass spectrometry following affinity capture. However,

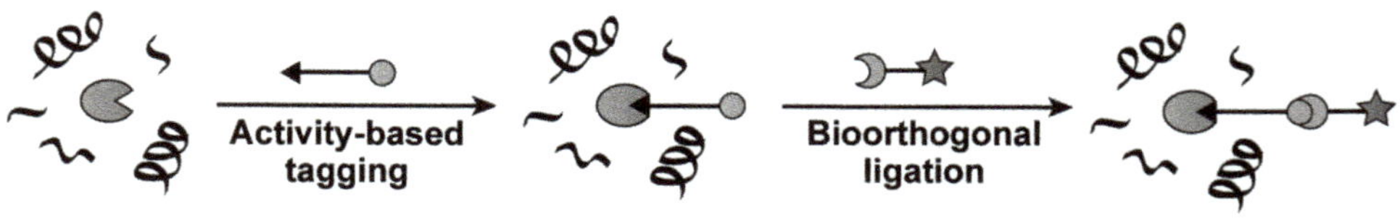

Figure 3.10 Activity-based protein profiling. An enzyme is selectively labeled within a live cell or cell lysate with an activity-based probe that contains a mechanism-based 'warhead' for covalent labeling of the enzyme (triangle) and a bio-orthogonal chemical reporter (circle). The labeled enzymes can be detected in a subsequent bio-orthogonal ligation with a conjugate of a suitable reaction partner (crescent) and a biophysical probe (star).

labeling active enzymes with imaging probes within intact cells or whole organisms could provide critical spatial information that is lost in cell and tissue lysates. Ploegh and coworkers demonstrated the use of an azido probe and the Staudinger ligation to visualize active cathepsins, an important class of cysteine proteases, in cells.[45] In principle, any enzyme class can be studied using ABPP provided that it can be targeted with a selective inhibitor.

3.3.2 Site-specific Labeling of Proteins

For dynamic imaging of enzymes, activity-based approaches – which by design involve the abolition of catalytic activity – may not be desirable, as they irreversibly perturb the physiological system under study. As well, many proteins involved in important biological processes are not enzymes, including many cell-surface receptors and transporters, transcription factors, and structural proteins. Thus, more general and less invasive approaches to visualizing proteins are necessary. Classically, proteins can be labeled at the genetic level by fusion to fluorescent proteins.[46] Although these probes have been widely used for biological imaging studies, their large size can interfere with the functions of many proteins.

As an alternative to fluorescent proteins, many small peptide-based methods have been developed for site-specific labeling of proteins.[47,48] These methods typically exploit short peptide sequences that can direct the chemical or enzymatic attachment of probes. In addition to the small size of the peptide and small molecule probe, other advantages of these approaches include temporally controlled labeling, multicolor labeling and the use of imaging modalities other than fluorescence. Here, we will discuss peptide tags that employ bioorthogonal chemistry as a central component of the labeling strategy.

In pioneering work, Tsien and coworkers demonstrated that bioorthogonality could be achieved by arranging natural amino acids to form a unique chemical environment. They showed in 1998 that proteins containing an engineered tetracysteine motif (CCXXCC) could be selectively tagged in live cells with biarsenical derivatives of the organic fluorophores.[49] The two most common probes used in this technology, known as FlAsH (fluorescein arsenical hairpin binder) and ReAsH (resorufin arsenical hairpin binder), are initially nonfluorescent but become fluorescent upon chelation to tetracysteine-tagged proteins in live cells.[50] Treatment of cells expressing a tetracysteine-containing protein first with FlAsH and then with ReAsH in a pulse-chase manner enables identification of 'old' and 'new' populations of the same protein.[51] Inspired by this work, Schepartz and coworkers recently developed an analogous technology for selectively labeling tetraserine motifs on engineered proteins within living cells using a fluorogenic, bis-boronic acid derivative of rhodamine.[52]

The FlAsH/ReAsH tetracysteine method has been widely used in biological studies, including the imaging of mRNA translation,[53] G-protein-coupled receptor activation,[54] amyloid formation,[55] viral trafficking patterns,[56] bacterial secretion systems[57] and membrane protein conformational changes.[58] In addition to its fluorescent properties, ReAsH can initiate the photoconversion of diaminobenzidine, which ultimately provides contrast for electron microscopy, permitting a high-resolution image of the labeled proteins in fixed cells.[50] In some instances, however, application of the biarsenical dyes can lead to background staining due to a moderate affinity for endogenous monothiol and dithiol motifs found within the cell.[59] Thus, alternative techniques have emerged to address this important

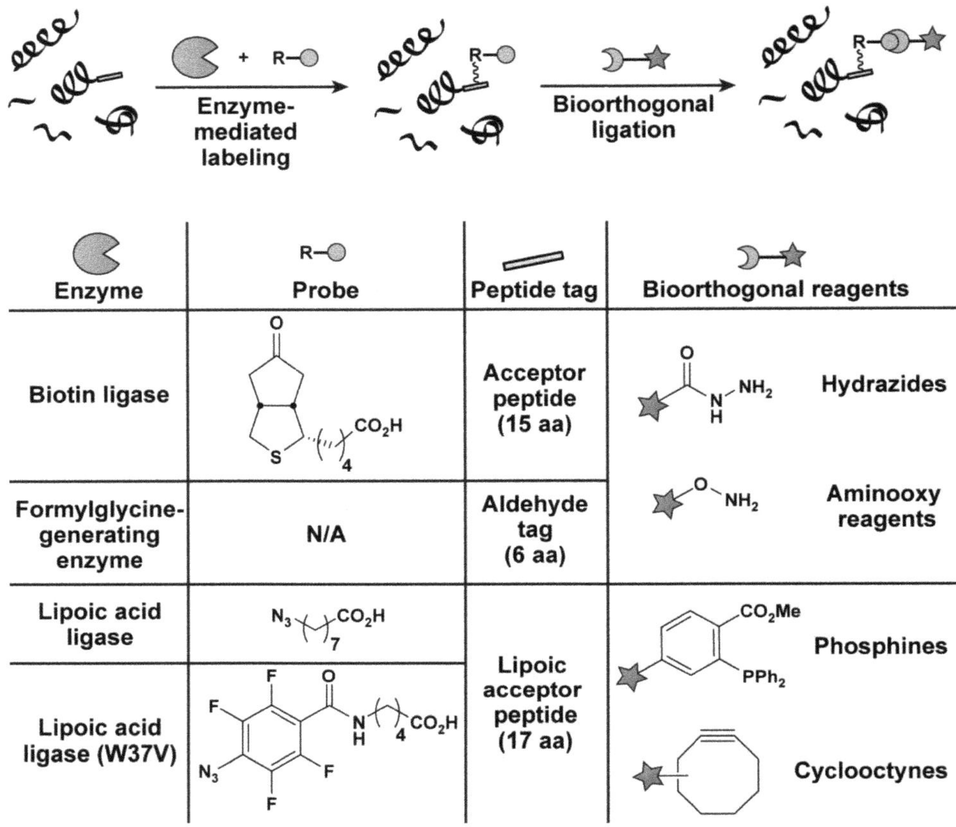

Figure 3.11 Site-specific labeling of proteins via recombinantly fused peptide tags. Recombinant proteins bearing a genetically fused peptide tag are selectively endowed with bioorthogonal functionality within a live cell or cell lysate by the action of a specific ligase or enzyme. Shown are enzyme-based methods for site-specifically labeling proteins with ketones, aldehydes, and azides.

issue of specificity, and critical to the success of these techniques is the use of bioorthogonal ligations discussed in Section 2.

Ting and coworkers have developed two systems that capitalize on the ability of a bacterial cofactor ligase to covalently label a specific lysine residue within in a short peptide acceptor sequence with unnatural versions of the cofactor (Figure 3.11). The first approach utilizes biotin ligase, an enzyme that can attach biotin via its carboxylic acid to lysine side chains within a 15-residue consensus sequence. The target protein of interest is genetically modified to contain this 15-amino acid 'acceptor peptide', to which biotin ligase catalyzes the attachment of a synthetic ketone-containing biotin isostere.[60] The ketone biotin ligase technology has been used to image receptor dynamics in live cells.[60] A more recent addition to the peptide-labeling toolkit is the use of lipoic acid ligase, which accepts a variety of azide and alkyne-containing analogs of lipoic acid, a naturally occurring cofactor.[61] This approach is an improvement over the biotin ligase technology as it allows the use of the superior bioorthogonal ligations of azides; further, rational design of lipoic acid ligase has enabled use of larger reporters such as aryl azides, which can be used as photoaffinity labels for dissecting protein–protein interactions.[62]

Both of these enzyme-mediated methods require the chemical synthesis of an unnatural cofactor, although the short ω-azido fatty acids recognized by lipoic acid ligase are simple to prepare. Another approach involves direct conversion of a specific amino acid side chain within a consensus peptide to a bioorthogonal functionality; this concept is embodied in the 'aldehyde tag' technology. This method takes advantage of the formylglycine-generating enzyme (FGE), an enzyme whose natural function is to co-translationally convert cysteine to an aldehyde-containing formylglycine residue in the active site of sulfatases.[63] FGE recognizes a six-residue consensus sequence (L<u>C</u>TPSR), which we discovered can be modified in the context of heterologous proteins. Thus, expression of recombinant proteins bearing this six-residue 'aldehyde tag' sequence leads to the production of proteins bearing bio-orthogonal aldehyde residues in both bacterial and mammalian systems.[64,65]

An alternative approach to site-specific protein labeling is to re-engineer the protein synthesis machinery to incorporate an unnatural amino acid at precisely one position in the protein (Figure 3.12). To accomplish this, Schultz and others have made use of the 'amber' stop codon and its corresponding tRNA, termed the amber suppressor. Mutants of amino acid tRNA-synthetase (aaRS) enzymes were evolved that charge the amber suppressor tRNA with various unnatural amino acids instead of the cognate amino acid.[66] This approach has been highly successful for the site-specific incorporation of unnatural amino acids that possess fluorophores[67–69] and photochemical crosslinkers,[70,71] as well as bio-orthogonal chemical reporters such as ketones, azides, and terminal alkynes.[70,72–74] This method for site-specific protein labeling, widely applied in bacteria and yeast, has recently been expanded for use in mammalian systems.[72,75]

To achieve site-specific protein labeling, the use of the amber suppressor technology requires considerable genetic manipulation. By contrast, a 'residue-specific' method for globally replacing one amino acid with an unnatural surrogate is much simpler to employ (Figure 3.12)[76]. In this approach, unnatural analogs of methionine[77,78] or phenylalanine[79,80] bearing azides and alkynes (as well as numerous other unnatural functional groups) are recognized by the cell's endogenous translational machinery and utilized for protein synthesis when their natural counterparts are in short supply. Pioneered by Tirrell and coworkers,[78,81] residue-specific protein labeling has been used to create novel protein-based biomaterials.[76] Additionally, it can be used to monitor global *de novo* protein synthesis. Termed bio-orthogonal noncanonical amino acid tagging (BONCAT), this technique can be used to visualize[82] and identify[81] newly synthesized proteins in mammalian cells and has been applied to generate static snapshots in fixed cells using CuAAC. By employing the nontoxic copper-free [3 + 2] cycloaddition with cyclooctyne probes, Tirrell and coworkers used the technique in living bacterial cells to discover novel catalytic activities through a series of *in vivo* evolution experiments.[83]

3.3.3 Metabolic Labeling of Glycans

Residue-specific protein labeling essentially constitutes the metabolic labeling of proteins with unnatural amino acids, and analogous technologies exist for probing other classes of

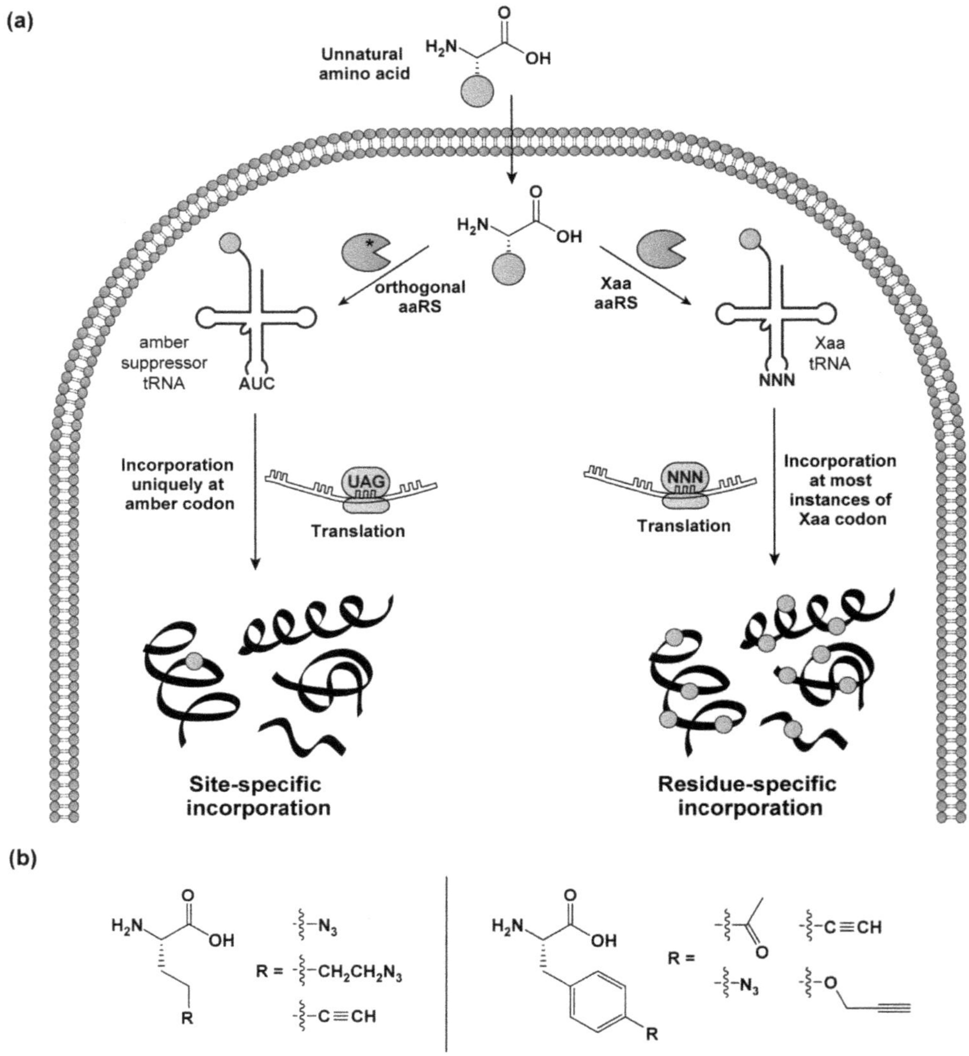

Figure 3.12 Incorporation of unnatural amino acids into newly synthesized proteins. (a) General schematic for site-specific incorporation of unnatural amino acids using amber suppressor technology (left side) or residue-specific incorporation of unnatural amino acids (right side). (b) Structures of unnatural amino acids containing ketones, azides, and alkynes for incorporation into proteins using these methods.

biomolecules as well. Glycans, which are linear or branched chains of sugars, constitute a diverse class of biomolecules that are often found as posttranslational modifications of proteins or covalently bonded to lipids. These biopolymers, found both within the cell and at the cell surface, participate in many physiological processes, including organ development,[84] cancer[85] and host–pathogen interactions.[86]

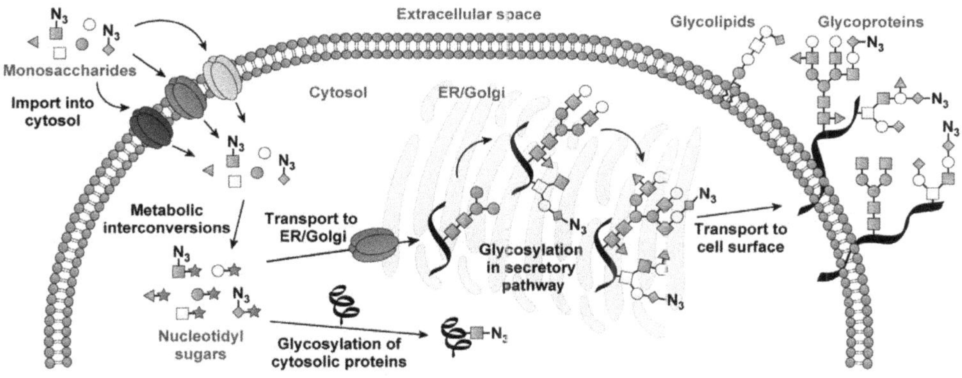

Figure 3.13 *Metabolic oligosaccharide engineering for labeling glycans with unnatural sugars. Monosaccharides are imported into the cytosol (or biosynthesized* de novo*) and metabolically converted to activated nucleotidyl sugars. Unnatural monosaccharides are represented here for clarity as containing the azide group (N_3), but many other unnatural monosaccharides have been employed for metabolic labeling of glycans (see Figure 3.14). These activated sugars are either directly attached to cytosolic proteins or transported into the ER and/or Golgi apparatus and then appended onto protein and lipid scaffolds by the action of glycosyltransferase enzymes. The majority of the resultant glycoproteins and glycolipids are then transported to the cell surface, although some are retained in the ER and Golgi apparatus or trafficked to organelles such as the lysosome, and others still are secreted into the extracellular space (not shown).*

Glycans can be labeled with unnatural monosaccharides bearing chemical reporters in a process termed 'metabolic oligosaccharide engineering'.[87] This approach enables both the visualization of glycans on cells and tissues as well as the profiling of glycosylation at the proteomic level. Analogous to residue-specific protein labeling, synthetic keto, azido or alkynyl sugars can hijack the glycan biosynthetic machinery and label glycans both on cell surfaces and intracellularly (Figure 3.13). Much work in this area has been devoted to the study of cell-surface sialic acids, which can be modified by metabolism of keto,[88] azido[15] and alkynyl[89] analogs of two natural precursors, *N*-acetylmannosamine (ManNAc) and *N*-acetylneuraminic acid (Figure 3.14).[90,91] Other classes of glycoconjugates that have been labeled with unnatural sugars bearing chemical reporters include fucosylated glycans,[89,92,93] mucin-type *O*-glycans[94,95] and cytosolic/nuclear *O*-GlcNAcylated proteins[96] (Figure 3.14).

Imaging of glycans has been performed using a variety of fluorophore conjugates of phosphines[20,97] and cyclooctynes.[29] In particular, the rapid kinetics of the Cu-free [3 + 2] cycloaddition using DIFO enabled us to measure, for the first time in living cells, the dynamics of glycan trafficking (Figure 3.15).[29] We have also extended metabolic oligosaccharide engineering to living animals. Mice and zebrafish treated with appropriate azidosugars can be metabolically labeled at their sialic acid residues and mucin-type *O*-glycans.[98–100] In zebrafish, changes in glycosylation during embryonic development were imaged using DIFO-fluorophore reagents.[100] Further, the Staudinger ligation and strain-promoted [3 + 2] azide–alkyne cycloaddition proceed readily within living mice, suggesting future *in vivo* imaging efforts using this powerful model organism for human disease.[29,98]

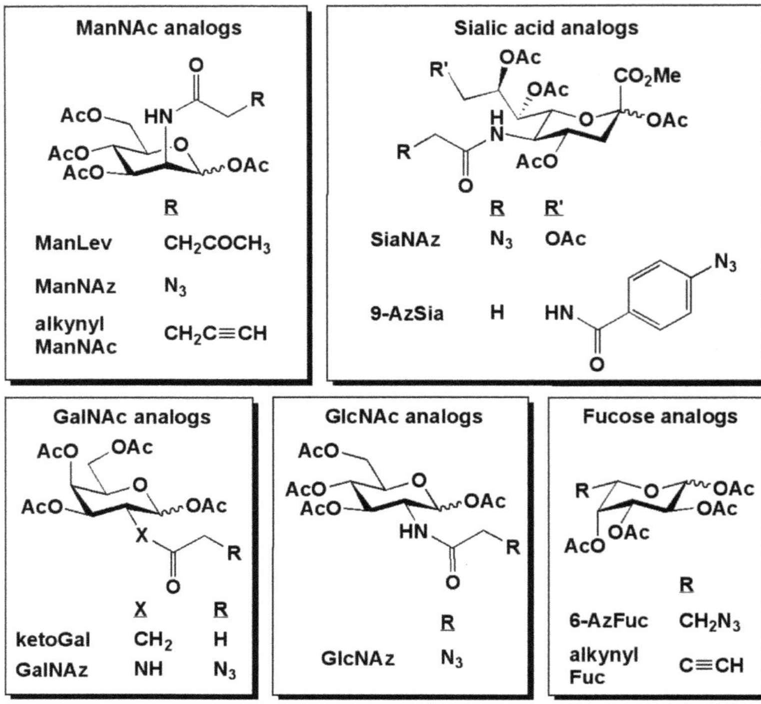

Figure 3.14 *Structures of unnatural monosaccharides containing ketone, azide or alkyne functionality used as substrates for metabolic labeling of glycans.*

3.3.4 Metabolic Targeting of Other Biomolecules with Chemical Reporters

In addition to proteins and glycans, other classes of biomolecules are being targeted using the bio-orthogonal chemical reporter strategy. Most notable among these are lipid-modified proteins, which possess *N*-myristoylation of *N*-terminal glycine residues, or *S*-palmitoylation or farnesylation of cysteine residues. Various groups have demonstrated metabolic labeling of lipidated proteins with azidolipid precursors, which allows for probing the modification in living systems.[101–104] Another protein posttranslational modification that has been probed using chemical reporters is the pantetheinylation. Burkart and coworkers have employed azido and alkynyl pantetheine analogs for metabolic labeling of acyl carrier proteins in living cells.[105] Rajski and coworkers have used azide-derivatized, aziridine-containing *S*-adenosylmethionine analogs for probing substrates of DNA methyltransferases *in vitro*.[106] Finally, Salic and Mitchison reported unnatural nucleotides, ethynyl and azido deoxyuridine, that are incorporated into replicating DNA within living cells and tissues.[107] These probes, after subsequent derivatization with CuAAC reagents, permit the mild detection of dividing cells within fixed cells and tissues and serve as a modern replacement for the bromodeoxyuridine (BrdU) assay that is routinely used to detect dividing cells. Using a similar approach, Jao and Salic were able to probe RNA synthesis in cells and animal tissues using ethynyl uridine. This unnatural nucleotide was incorporated into

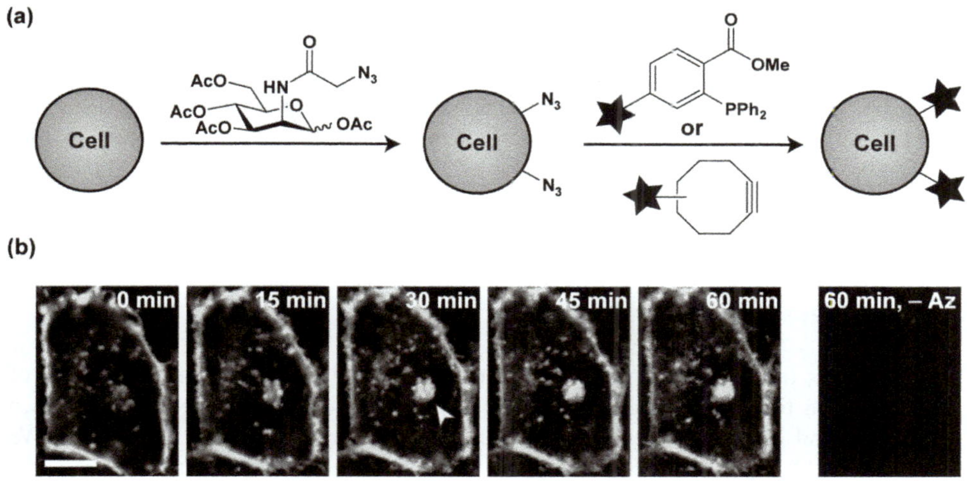

Figure 3.15 Dynamic imaging of glycans in living cells using metabolic labeling and bioorthogonal chemistries. (a) Experimental setup: cells are metabolically labeled with an azidosugar (ManNAz is shown here as an example), rinsed and then reacted with a fluorescent phosphine or cyclooctyne conjugate. (b) Dynamic imaging of sialic acids using ManNAz and DIFO-Alexa Fluor 488. Chinese hamster ovary cells were metabolically labeled with ManNAz for 3 days, rinsed, reacted with DIFO-Alexa Fluor 488 for 1 min, and imaged by epifluorescence microscopy. Images were acquired every 15 min for 1 h. The image on the right (60 min, –Az) indicates a separate sample metabolically labeled with the control sugar ManNAc. The arrowhead indicates a population of glycans that have undergone endocytosis after the Cu-free click labeling reaction. Scale bar: 5 μm.

newly synthesized RNA during transcription, and its presence was detected by CuAAC using azido fluorophores.[108]

3.4 Summary and Outlook

In this chapter, we have discussed the development and implementation of bio-orthogonal ligations for labeling biomolecules in living systems. The strategy relies both on the incorporation of a chemical reporter into biomolecules and the subsequent detection of the reporter using bio-orthogonal chemistries. Initial efforts were centered on condensation reactions of ketones and aldehydes, and more recent work has focused on reactions of azides with triarylphosphines and activated alkynes. In particular, copper-free [3 + 2] cycloaddition with strained alkynes presents a promising area for immediate application and future reagent development using the principles of physical organic chemistry. Reagents with faster kinetics could enable more sensitive detection of azides *in vivo*, and fluorogenic alkynes for copper-free [3 + 2] cycloaddition would permit real-time visualization of biological events.

Despite the numerous advances propelled by these azide-specific chemistries, there is a need for alternatives to azides. Currently, methods to image multiple biomolecules simultaneously are limited by the number of ligation chemistries that are orthogonal to one another. Fertile ground for the search for new biocompatible, copper-free click chemistries includes other pericyclic reactions and photochemistry. The initial reports on 'photoclick' chemistry, a cross-section of these two areas, represent the tip of the iceberg, as a vast literature of classic organic chemistry can be revisited in the modern context of bioorthogonal ligations. This expanded toolkit of copper-free click chemistries should see wide application in chemical biology in applications ranging from biomaterial design to *in vivo* imaging.

Acknowledgments

C.R.B. thanks the National Institutes of Health (GM058867) for financial support of the work from the authors' laboratory, and J.M.B. was supported by National Science Foundation and National Defense Science and Engineering predoctoral fellowships. We thank Pamela Chang and Matthew Hangauer for critical reading of the chapter.

References

(1) J.A. Prescher and C.R. Bertozzi, (2005), Chemistry in living systems, *Nat. Chem. Biol.*, **1**, 13–21.
(2) H.C. Kolb, M.G. Finn and K.B. Sharpless, (2001), Click chemistry: diverse chemical function from a few good reactions, *Angew. Chem. Int. Edn*, **40**, 2004–2021.
(3) V.V. Rostovtsev, L.G. Green, V.V. Fokin and K.B. Sharpless, (2002), A stepwise Huisgen cycloaddition process: copper(I)-catalyzed regioselective 'ligation' of azides and terminal alkynes, *Angew. Chem. Int. Edn*, **41**, 2596–2599.
(4) C.W. Tornoe, C. Christensen and M. Meldal, (2002), Peptidotriazoles on solid phase: [1,2,3]-triazoles by regiospecific copper(I)-catalyzed 1,3-dipolar cycloadditions of terminal alkynes to azides, *J. Org. Chem.*, **67**, 3057–3064.
(5) H.C. Kolb and K.B. Sharpless, (2003), The growing impact of click chemistry on drug discovery, *Drug Discov. Today*, **8**, 1128–1137.
(6) P. Wu and V.V. Fokin, (2007), Catalytic azide–alkyne cycloaddition: reactivity and applications, *Aldrichimica Acta*, **40**, 7–17.
(7) J.F. Lutz, (2007), 1,3-Dipolar cycloadditions of azides and alkynes: a universal ligation tool in polymer and materials science, *Angew. Chem. Int. Edn*, **46**, 1018–1025.
(8) J.M. Baskin and C.R. Bertozzi, (2007), Bioorthogonal click chemistry: covalent labeling in living systems, *QSAR Comb. Sci.*, **26**, 1211–1219.
(9) W.P. Jencks, (1959), Studies on the mechanism of oxime and semicarbazone formation, *J. Am. Chem. Soc.*, **81**, 475–481.
(10) A. Dirksen, T.M. Hackeng and P.E. Dawson, (2006), Nucleophilic catalysis of oxime ligation, *Angew. Chem. Int. Edn*, **45**, 7581–7584.
(11) I.L. Cartwright, D.W. Hutchinson and V.W. Armstrong, (1976), The reaction between thiols and 8-azidoadenosine derivatives, *Nucl. Acids Res.*, **3**, 2331–2339.
(12) M.B. Hossain, D. Vanderhelm, R. Sanduja and M. Alam, (1985), Structure of 6-azidotetrazolo[5,1-a]phthalazine, $C_8H_4N_8$, isolated from the toxic dinoflaggelate Gymnodinium breve, *Acta Cryst.*, **C41**, 1199–1202.
(13) R.J. Griffin, (1994), The medicinal chemistry of the azido group, *Prog. Med. Chem.*, **31**, 121–232.
(14) S. Brase, C. Gil, K. Knepper and V. Zimmermann, (2005), Organic azides: an exploding diversity of a unique class of compounds, *Angew. Chem. Int. Edn*, **44**, 5188–5240.

(15) E. Saxon and C.R. Bertozzi, (2000), Cell surface engineering by a modified Staudinger reaction, *Science*, **287**, 2007–2010.
(16) E. Saxon, J.I. Armstrong and C.R. Bertozzi, (2000), A 'traceless' Staudinger ligation for the chemoselective synthesis of amide bonds, *Org. Lett.*, **2**, 2141–2143.
(17) B.L. Nilsson, L.L. Kiessling and R.T. Raines, (2000), Staudinger ligation: a peptide from a thioester and azide, *Org. Lett.*, **2**, 1939–1941.
(18) B.L. Nilsson, L.L. Kiessling and R.T. Raines, (2001), High-yielding Staudinger ligation of a phosphinothioester and azide to form a peptide, *Org. Lett.*, **3**, 9–12.
(19) G.A. Lemieux, C.L. De Graffenried and C.R. Bertozzi, (2003), A fluorogenic dye activated by the Staudinger ligation, *J. Am. Chem. Soc.*, **125**, 4708–4709.
(20) M.J. Hangauer and C.R. Bertozzi, (2008), A FRET-based fluorogenic phosphine for live-cell imaging with the Staudinger ligation, *Angew. Chem. Int. Edn*, **47**, 2394–2397.
(21) F.L. Lin, H.M. Hoyt, H. Van Halbeek, R.G. Bergman and C.R. Bertozzi, (2005), Mechanistic investigation of the Staudinger ligation, *J. Am. Chem. Soc.*, **127**, 2686–2695.
(22) A. Michael, (1893), Ueber die einwirkung von diazodenzolimid auf acetylendicarbonsauremethylester, *J. Prakt. Chem.*, **48**, 94–95.
(23) R. Huisgen, (1963), 1.3-Dipolare Cycloadditionen: Ruckschau Und Ausblick, *Angew. Chem. Int. Edn*, **75**, 604–637.
(24) F. Himo, T. Lovell, R. Hilgraf, V.V. Rostovtsev, L. Noodleman, K.B. Sharpless and V.V. Fokin, (2005), Copper(I)-catalyzed synthesis of azoles. DFT study predicts unprecedented reactivity and intermediates, *J. Am. Chem. Soc.*, **127**, 210–216.
(25) G. Wittig and A. Krebs, (1961), Zur Existenz Niedergliedriger Cycloalkine 1, *Chem. Ber.*, **94**, 3260–3275.
(26) R.B. Turner, P. Goebel, B.J. Mallon and A.D. Jarrett, (1973), Heats of hydrogenation 9. Cyclic acetylenes and some miscellaneous olefins, *J. Am. Chem. Soc.*, **95**, 790–792.
(27) N.J. Agard, J.A. Prescher and C.R. Bertozzi, (2004), A strain-promoted [3 + 2] azide–alkyne cycloaddition for covalent modification of blomolecules in living systems, *J. Am. Chem. Soc.*, **126**, 15046–15047.
(28) N.J. Agard, J.M. Baskin, J.A. Prescher, A. Lo and C.R. Bertozzi, (2006), A comparative study of bioorthogonal reactions with azides, *ACS Chem. Biol.*, **1**, 644–648.
(29) J.M. Baskin, J.A. Prescher, S.T. Laughlin, N.J. Agard, P.V. Chang, I.A. Miller, A. Lo, J.A. Codelli and C.R. Bertozzi, (2007), Copper-free click chemistry for dynamic *in vivo* imaging, *Proc. Natl Acad. Sci. USA*, **104**, 16793–16797.
(30) J.A. Codelli, J.M. Baskin, N.J. Agard and C.R. Bertozzi, (2008), Second-generation difluorinated cyclooctynes for copper-free click chemistry, *J. Am. Chem. Soc.*, ASAP Article.
(31) E.M. Sletten and C.R. Bertozzi, (2008), A hydrophilic azacyclooctyne for Cu-free click chemistry, *Org. Lett.*, **10**, 3097–3099.
(32) S.S. van Berkel, A.T. Dirks, M.F. Debets, F.L. van Delft, J.J. Cornelissen, R.J. Nolte and F.P. Rutjes, (2007), Metal-free triazole formation as a tool for bioconjugation, *Chembiochem*, **8**, 1504–1508.
(33) X. Ning, J. Guo, M.A. Wolfert and G.J. Boons, (2008), Visualizing metabolically labeled glycoconjugates of living cells by copper-free and fast Huisgen cycloadditions, *Angew. Chem. Int. Edn*, **47**, 2253–2255.
(34) W. Song, Y. Wang, J. Qu, M.M. Madden and Q. Lin, (2008), A photoinducible 1,3-dipolar cycloaddition reaction for rapid, selective modification of tetrazole-containing proteins, *Angew. Chem. Int. Edn*, **47**, 2832–2835.
(35) W. Song, Y. Wang, J. Qu and Q. Lin, (2008), Selective functionalization of a genetically encoded alkene-containing protein via 'photoclick chemistry' in bacterial cells, *J. Am. Chem. Soc.*, **130**, 9654–9655.
(36) Y. Wang, W.J. Hu, W. Song, R.K. Lim and Q. Lin, (2008), Discovery of long-wavelength photoactivatable diaryltetrazoles for bioorthogonal 1,3-dipolar cycloaddition reactions, *Org. Lett.*
(37) Y.A. Lin, J.M. Chalker, N. Floyd, G.J. Bernardes and B.G. Davis, (2008), Allyl sulfides are privileged substrates in aqueous cross-metathesis: Application to site-selective protein modification, *J. Am. Chem. Soc.*, **130**, 9642–9643.

(38) M.L. Blackman, M. Royzen and J.M. Fox, (2008), Tetrazine ligation: fast bioconjugation based on inverse-electron-demand Diels–Alder reactivity, *J. Am. Chem. Soc.*, **130**, 13518–13519.
(39) N.K. Devaraj, R. Weissleder and S.A. Hilderbrand, (2008), Tetrazine-based cycloadditions: application to pretargeted live cell imaging, *Bioconjugate Chem.*, **19**, 2297–2299.
(40) A.E. Speers and B.F. Cravatt, (2004), Chemical strategies for activity-based proteomics, *Chembiochem*, **5**, 41–47.
(41) M.J. Evans and B.F. Cravatt, (2006), Mechanism-based profiling of enzyme families, *Chem. Rev.*, **106**, 3279–3301.
(42) N. Jessani and B.F. Cravatt, (2004), The development and application of methods for activity-based protein profiling, *Curr. Opin. Chem. Biol.*, **8**, 54–59.
(43) A.M. Sadaghiani, S.H. Verhelst and M. Bogyo, (2007), Tagging and detection strategies for activity-based proteomics, *Curr. Opin. Chem. Biol.*, **11**, 20–28.
(44) D.J. Vocadlo and C.R. Bertozzi, (2004), A strategy for functional proteomic analysis of glycosidase activity from cell lysates, *Angew. Chem. Int. Edn*, **43**, 5338–5342.
(45) H.C. Hang, J. Loureiro, E. Spooner, A.W. van der Velden, Y.M. Kim, A.M. Pollington, R. Maehr, M.N. Starnbach and H.L. Ploegh, (2006), Mechanism-based probe for the analysis of cathepsin cysteine proteases in living cells, *ACS Chem. Biol.*, **1**, 713–723.
(46) B.N.G. Giepmans, S.R. Adams, M.H. Ellisman and R.Y. Tsien, (2006), The fluorescent toolbox for assessing protein location and function, *Science*, **312**, 217–224.
(47) I. Chen and A.Y. Ting, (2005), Site-specific labeling of proteins with small molecules in live cells, *Curr. Opin. Biotechnol.*, **16**, 35–40.
(48) K.M. Marks and G.P. Nolan, (2006), Chemical labeling strategies for cell biology, *Nat. Meth.*, **3**, 591–596.
(49) B.A. Griffin, S.R. Adams and R.Y. Tsien, (1998), Specific covalent labeling of recombinant protein molecules inside live cells, *Science*, **281**, 269–272.
(50) S.R. Adams, R.E. Campbell, L.A. Gross, B.R. Martin, G.K. Walkup, Y. Yao, J. Llopis and R.Y. Tsien, (2002), New biarsenical ligands and tetracysteine motifs for protein labeling *in vitro* and *in vivo*: synthesis and biological applications, *J. Am. Chem. Soc.*, **124**, 6063–6076.
(51) G. Gaietta, T.J. Deerinck, S.R. Adams, J. Bouwer, O. Tour, D.W. Laird, G.E. Sosinsky, R.Y. Tsien and M.H. Ellisman, (2002), Multicolor and electron microscopic imaging of connexin trafficking, *Science*, **296**, 503–507.
(52) T.L. Halo, J. Appelbaum, E.M. Hobert, D.M. Balkin and A. Schepartz, (2009), Selective recognition of protein tetraserine motifs with a cell-permeable, pro-fluorescent bis-boronic acid, *J. Am. Chem. Soc.*, **131**, 438–439.
(53) A.J. Rodriguez, S.M. Shenoy, R.H. Singer and J. Condeelis, (2006), Visualization of mRNA translation in living cells, *J. Cell. Biol.*, **175**, 67–76.
(54) C. Hoffmann, G. Gaietta, M. Bunemann, S.R. Adams, S. Oberdorff-Maass, B. Behr, J.P. Vilardaga, R.Y. Tsien, M.H. Ellisman and M.J. Lohse, (2005), A FlAsH-based FRET approach to determine G protein-coupled receptor activation in living cells, *Nat. Meth.*, **2**, 171–176.
(55) M.J. Roberti, C.W. Bertoncini, R. Klement, E.A. Jares-Erijman and T.M. Jovin, (2007), Fluorescence imaging of amyloid formation in living cells by a functional, tetracysteine-tagged alpha-synuclein, *Nat. Meth.*, **4**, 345–351.
(56) N. Arhel, A. Genovesio, K.A. Kim, S. Miko, E. Perret, J.C. Olivo-Marin, S. Shorte and P. Charneau, (2006), Quantitative four-dimensional tracking of cytoplasmic and nuclear HIV-1 complexes, *Nat. Meth.*, **3**, 817–824.
(57) F. Senf, J. Tommassen and M. Koster, (2008), Polar secretion of proteins via the Xcp type II secretion system in *Pseudomonas aeruginosa*, *Microbiology*, **154**, 3025–3032.
(58) S. Chaumont and B.S. Khakh, (2008), Patch-clamp coordinated spectroscopy shows P2X2 receptor permeability dynamics require cytosolic domain rearrangements but not Panx-1 channels, *Proc. Natl Acad. Sci. USA*, **105**, 12063–12068.
(59) K. Stroffekova, C. Proenza and K.G. Beam, (2001), The protein-labeling reagent FLASH-EDT2 binds not only to CCXXCC motifs but also non-specifically to endogenous cysteine-rich proteins, *Pflugers Arch.*, **442**, 859–866.
(60) I. Chen, M. Howarth, W. Lin and A.Y. Ting, (2005), Site-specific labeling of cell surface proteins with biophysical probes using biotin ligase, *Nat. Meth.*, **2**, 99–104.

(61) M. Fernandez-Suarez, H. Baruah, L. Martinez-Hernandez, K.T. Xie, J.M. Baskin, C.R. Bertozzi and A.Y. Ting, (2007), Redirecting lipoic acid ligase for cell surface protein labeling with small-molecule probes, *Nat. Biotechnol.*, **25**, 1483–1487.
(62) H. Baruah, S. Puthenveetil, Y.A. Choi, S. Shah and A.Y. Ting, (2008), An engineered aryl azide ligase for site-specific mapping of protein–protein interactions through photo-cross-linking, *Angew. Chem. Int. Edn.*
(63) T. Dierks, B. Schmidt, L.V. Borissenko, J. Peng, A. Preusser, M. Mariappan and K. von Figura, (2003), Multiple sulfatase deficiency is caused by mutations in the gene encoding the human C(alpha)-formylglycine generating enzyme, *Cell*, **113**, 435–444.
(64) I.S. Carrico, B.L. Carlson and C.R. Bertozzi, (2007), Introducing genetically encoded aldehydes into proteins, *Nat. Chem. Biol.*, **3**, 321–322.
(65) P. Wu, W. Shui, B.L. Carlson, N. Hu, D. Rabuka, J. Lee and C.R. Bertozzi, (2009), Site-specific chemical modification of recombinant proteins produced in mammalian cells using the genetically encoded aldehyde tag, *Proc. Natl Acad. Sci. USA*, **106**, 3000–3005.
(66) J.W. Chin, T.A. Cropp, J.C. Anderson, M. Mukherji, Z.W. Zhang and P.G. Schultz, (2003), An expanded eukaryotic genetic code, *Science*, **301**, 964–967.
(67) J. Wang, J. Xie and P.G. Schultz, (2006), A genetically encoded fluorescent amino acid, *J. Am. Chem. Soc.*, **128**, 8738–8739.
(68) D. Summerer, S. Chen, N. Wu, A. Deiters, J.W. Chin and P.G. Schultz, (2006), A genetically encoded fluorescent amino acid, *Proc. Natl Acad. Sci. USA*, **103**, 9785–9789.
(69) Z.W. Zhang, L. Alfonta, F. Tian, B. Bursulaya, S. Uryu, D.S. King and P.G. Schultz, (2004), Selective incorporation of 5-hydroxytryptophan into proteins in mammalian cells, *Proc. Natl Acad. Sci. USA*, **101**, 8882–8887.
(70) J.W. Chin, S.W. Santoro, A.B. Martin, D.S. King, L. Wang and P.G. Schultz, (2002), Addition of *p*-azido-L-phenylalanine to the genetic code of Escherichia coli, *J. Am. Chem. Soc.*, **124**, 9026–9027.
(71) J.W. Chin, A.B. Martin, D.S. King, L. Wang and P.G. Schultz, (2002), Addition of a photocrosslinking amino acid to the genetic code of *Escherichia coli*, *Proc. Natl Acad. Sci. USA*, **99**, 11020–11024.
(72) W. Liu, A. Brock, S. Chen, S. Chen and P.G. Schultz, (2007), Genetic incorporation of unnatural amino acids into proteins in mammalian cells, *Nat. Meth.*, **4**, 239–244.
(73) A. Deiters, T.A. Cropp, M. Mukherji, J.W. Chin, J.C. Anderson and P.G. Schultz, (2003), Adding amino acids with novel reactivity to the genetic code of *Saccharomyces cerevisiae*, *J. Am. Chem. Soc.*, **125**, 11782–11783.
(74) A. Deiters and P.G. Schultz, (2005), *In vivo* incorporation of an alkyne into proteins in *Escherichia coli*, *Bioorg. Med. Chem. Lett.*, **15**, 1521–1524.
(75) K. Sakamoto, A. Hayashi, A. Sakamoto, D. Kiga, H. Nakayama, A. Soma, T. Kobayashi, M. Kitabatake, K. Takio, K. Saito, M. Shirouzu, I. Hirao and S. Yokoyama, (2002), Site-specific incorporation of an unnatural amino acid into proteins in mammalian cells, *Nucl. Acids Res.*, **30**, 4692–4699.
(76) A.J. Link, M.L. Mock and D.A. Tirrell, (2003), Non-canonical amino acids in protein engineering, *Curr. Opin. Biotechnol.*, **14**, 603–609.
(77) K.L. Kiick, J.C. van Hest and D.A. Tirrell, (2000), Expanding the scope of protein biosynthesis by altering the methionyl-tRNA synthetase activity of a bacterial expression host, *Angew. Chem. Int. Edn*, **39**, 2148–2152.
(78) K.L. Kiick, E. Saxon, D.A. Tirrell and C.R. Bertozzi, (2002), Incorporation of azides into recombinant proteins for chemoselective modification by the Staudinger ligation, *Proc. Natl Acad. Sci. USA*, **99**, 19–24.
(79) D. Datta, P. Wang, I.S. Carrico, S.L. Mayo and D.A. Tirrell, (2002), A designed phenylalanyl-tRNA synthetase variant allows efficient *in vivo* incorporation of aryl ketone functionality into proteins, *J. Am. Chem. Soc.*, **124**, 5652–5653.
(80) K. Kirshenbaum, I.S. Carrico and D.A. Tirrell, (2002), Biosynthesis of proteins incorporating a versatile set of phenylalanine analogues, *Chembiochem*, **3**, 235–237.
(81) D.C. Dieterich, A.J. Link, J. Graumann, D.A. Tirrell and E.M. Schuman, (2006), Selective identification of newly synthesized proteins in mammalian cells using bioorthogonal

noncanonical amino acid tagging (BONCAT), *Proc. Natl Acad. Sci. USA*, **103**, 9482–9487.
(82) K.E. Beatty, J.C. Liu, F. Xie, D.C. Dieterich, E.M. Schuman, Q. Wang and D.A. Tirrell, (2006), Fluorescence visualization of newly synthesized proteins in mammalian cells, *Angew. Chem. Int. Edn*, **45**, 7364–7367.
(83) A.J. Link, M.K.S. Vink, N.J. Agard, J.A. Prescher, C.R. Bertozzi and D.A. Tirrell, (2006), Discovery of aminoacyl-tRNA synthetase activity through cell-surface display of noncanonical amino acids, *Proc. Natl Acad. Sci. USA*, **103**, 10180–10185.
(84) R.S. Haltiwanger and J.B. Lowe, (2004), Role of glycosylation in development, *Annu. Rev. Biochem.*, **73**, 491–537.
(85) D.H. Dube and C.R. Bertozzi, (2005), Glycans in cancer and inflammation – potential for therapeutics and diagnostics, *Nat. Rev. Drug. Discov.*, **4**, 477–488.
(86) P.M. Rudd, T. Elliott, P. Cresswell, I.A. Wilson and R.A. Dwek, (2001), Glycosylation and the immune system, *Science*, **291**, 2370–2376.
(87) D.H. Dube and C.R. Bertozzi, (2003), Metabolic oligosaccharide engineering as a tool for glycobiology, *Curr. Opin. Chem. Biol.*, **7**, 616–625.
(88) L.K. Mahal, K.J. Yarema and C.R. Bertozzi, (1997), Engineering chemical reactivity on cell surfaces through oligosaccharide biosynthesis, *Science*, **276**, 1125–1128.
(89) T.L. Hsu, S.R. Hanson, K. Kishikawa, S.K. Wang, M. Sawa and C.H. Wong, (2007), Alkynyl sugar analogs for the labeling and visualization of glycoconjugates in cells, *Proc. Natl Acad. Sci. USA*, **104**, 2614–2619.
(90) S.J. Luchansky, S. Goon and C.R. Bertozzi, (2004), Expanding the diversity of unnatural cell-surface sialic acids, *Chembiochem*, **5**, 371–374.
(91) S. Han, B.E. Collins, P. Bengtson and J.C. Paulson, (2005), Homomultimeric complexes of CD22 in B cells revealed by protein–glycan cross-linking, *Nat. Chem. Biol.*, **1**, 93–97.
(92) D. Rabuka, S.C. Hubbard, S.T. Laughlin, S.P. Argade and C.R. Bertozzi, (2006), A chemical reporter strategy to probe glycoprotein fucosylation, *J. Am. Chem. Soc.*, **128**, 12078–12079.
(93) M. Sawa, T.L. Hsu, T. Itoh, M. Sugiyama, S.R. Hanson, P.K. Vogt and C.H. Wong, (2006), Glycoproteomic probes for fluorescent imaging of fucosylated glycans *in vivo*, *Proc. Natl Acad. Sci. USA*, **103**, 12371–12376.
(94) H.C. Hang and C.R. Bertozzi, (2001), Ketone isosteres of 2-*N*-acetamidosugars as substrates for metabolic cell surface engineering, *J. Am. Chem. Soc.*, **123**, 1242–1243.
(95) H.C. Hang, C. Yu, D.L. Kato and C.R. Bertozzi, (2003), A metabolic labeling approach toward proteomic analysis of mucin-type *O*-linked glycosylation, *Proc. Natl Acad. Sci. USA*, **100**, 14846–14851.
(96) D.J. Vocadlo, H.C. Hang, E.J. Kim, J.A. Hanover and C.R. Bertozzi, (2003), A chemical approach for identifying *O*-GlcNAc-modified proteins in cells, *Proc. Natl Acad. Sci. USA*, **100**, 9116–9121.
(97) P.V. Chang, J.A. Prescher, M.J. Hangauer and C.R. Bertozzi, (2007), Imaging cell surface glycans with bioorthogonal chemical reporters, *J. Am. Chem. Soc.*, **129**, 8400–8401.
(98) J.A. Prescher, D.H. Dube and C.R. Bertozzi, (2004), Chemical remodelling of cell surfaces in living animals, *Nature*, **430**, 873–877.
(99) D.H. Dube, J.A. Prescher, C.N. Quang and C.R. Bertozzi, (2006), Probing mucin-type *O*-linked glycosylation in living animals, *Proc. Natl Acad. Sci. USA*, **103**, 4819–4824.
(100) S.T. Laughlin, J.M. Baskin, S.L. Amacher and C.R. Bertozzi, (2008), *In vivo* imaging of membrane-associated glycans in developing zebrafish, *Science*, **320**, 664–667.
(101) Y. Kho, S.C. Kim, C. Jiang, D. Barma, S.W. Kwon, J. Cheng, J. Jaunbergs, C. Weinbaum, F. Tamanoi, J. Falck and Y. Zhao, (2004), A tagging-via-substrate technology for detection and proteomics of farnesylated proteins, *Proc. Natl Acad. Sci. USA*, **101**, 12479–12484.
(102) H.C. Hang, E.J. Geutjes, G. Grotenbreg, A.M. Pollington, M.J. Bijlmakers and H.L. Ploegh, (2007), Chemical probes for the rapid detection of fatty-acylated proteins in mammalian cells, *J. Am. Chem. Soc.*, **129**, 2744–2745.
(103) D.D. Martin, G.L. Vilas, J.A. Prescher, G. Rajaiah, J.R. Falck, C.R. Bertozzi and L.G. Berthiaume, (2008), Rapid detection, discovery, and identification of post-translationally

myristoylated proteins during apoptosis using a bio-orthogonal azidomyristate analog, *FASEB J.*, **22**, 797–806.

(104) M.A. Kostiuk, M.M. Corvi, B.O. Keller, G. Plummer, J.A. Prescher, M.J. Hangauer, C.R. Bertozzi, G. Rajaiah, J.R. Falck and L.G. Berthiaume, (2008), Identification of palmitoylated mitochondrial proteins using a bio-orthogonal azido-palmitate analogue, *FASEB J.*, **22**, 721–732.

(105) J.L. Meier, A.C. Mercer, H. Rivera, Jr and M.D. Burkart, (2006), Synthesis and evaluation of bioorthogonal pantetheine analogues for *in vivo* protein modification, *J. Am. Chem. Soc.*, **128**, 12174–12184.

(106) L.R. Comstock and S.R. Rajski, (2005), Conversion of DNA methyltransferases into azidonucleosidyl transferases via synthetic cofactors, *Nucl. Acids Res.*, **33**, 1644–1652.

(107) A. Salic and T.J. Mitchison, (2008), A chemical method for fast and sensitive detection of DNA synthesis *in vivo*, *Proc. Natl Acad. Sci. USA*, **105**, 2415–2420.

(108) C.Y. Jao and A. Salic, (2008), Exploring RNA transcription and turnover *in vivo* by using click chemistry, *Proc. Natl Acad. Sci. USA*, **105**, 15779–15784.

4

Protein and Peptide Conjugation to Polymers and Surfaces Using Oxime Chemistry

Heather D. Maynard, Rebecca M. Broyer and Christopher M. Kolodziej

4.1 Introduction

Aldehydes and ketones react with alkoxyamines to form a stable oxime species (Figure 4.1). The process occurs in aqueous solution in the presence of a wide variety of functional groups without the addition of other reagents. Oxime chemistry is compatible with biomolecules. As a result, there has been increasing interest in employing this bond formation in conjugation reactions. This chapter focuses on studies that use oxime chemistry to conjugate proteins and peptides to polymers and surfaces.

Oxime bond formation provides a convenient method to produce site-specific conjugates.[1,2] Proteins and peptides often contain multiple amine groups, which also form Schiff bases with aldehydes and ketones. Typically these are unstable in aqueous solution because the equilibrium favors the oxo version. For *O*-hydroxylamine compounds, the equilibrium favors the oxime. This is significant because site-specific conjugation is important for retention of bioactivity, a required feature for most applications.

Aldehyde, ketone and aminooxy groups are not among the side-chains of naturally occurring amino acids. Because these functionalities may be easily incorporated into peptides and proteins, conjugation can be restricted to only the desired site on the protein or peptide.[3–5] Peptides are readily synthesized with either aminooxy groups or other oxo moieties using solid-phase methods.[6] A facile method to install α-oxoamides at the *N*-termini of proteins was reported by Dixon.[7–11] A transamination reaction in the presence of sodium

Click Chemistry for Biotechnology and Materials Science Edited by Joerg Lahann

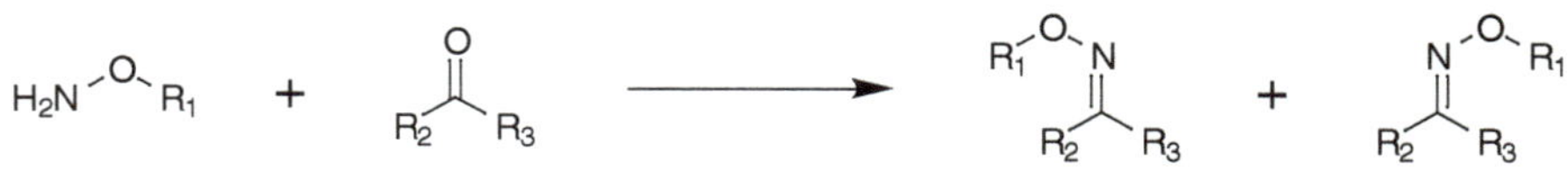

Figure 4.1 *Oxime bond formation.*

glyoxylate and copper sulfate forms the desired group (Figure 4.2), which, in turn, can be used to form conjugates with amiooxy-functionalized polymers. More recently Francis and coworkers reported that most *N*-terminal amino acids can be subjected to a transamination reaction using pyridoxal-5-phosphate (PLP).[12] They did note that the reaction was not successful for *N*-terminal serine, cysteine, proline, tryptophan or threonine residues. An artificial amino acid containing the appropriate moiety can also be introduced through recombinant methods or by enzymatic *N*-terminal addition.[13] Introduction of N^{ε}-levulinyl lysine residues site-specifically in proteins and peptides can also be achieved by solid-phase peptide synthesis or native chemical ligation.[3,4] Clearly, all of these strategies can be used to form biomolecules for reaction with functionalized polymers or surfaces.

4.2 Protein/Peptide–Polymer Conjugates

Covalent attachment of polymers to proteins and peptides has also been shown to improve properties such as stability, biocompatibility and performance compared with the naturally derived materials.[14,15] Moreover, polymers can confer new properties such as externally switchable phase behavior or self-assembly properties. Indeed, peptide–polymer and protein–polymer bioconjugates are hybrid materials that make up a large class of biopharmaceuticals.[16–18] A second class of conjugates is made from ‘smart’ polymers. ‘Smart’ polymers respond to external stimuli, such as changes in pH, temperature and light, and can impart this response to a conjugated peptide or protein.[19]

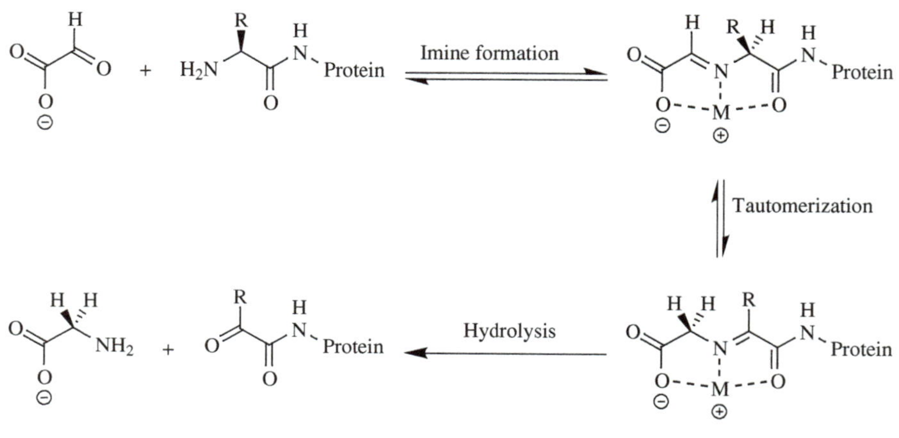

Figure 4.2 *N-terminal oxidation via transamination with sodium glyoxylate.*

There are a number of different synthetic strategies available to prepare these hybrids.[20–22] The more traditional approach involves synthesis of protein-reactive polymers, and covalent attachment of the polymer to the side chains of the proteins post-polymerization. Another strategy that we recently developed involves the synthesis of polymers directly from proteins that have been modified with polymerization initiators. The final method involves conjugation of proteins or peptides to side chain-reactive polymers.

The first reported use of oxime bond formation to achieve protein–drug conjugates was by Webb and Kaneko in 1990.[23] Their approach was to conjugate a drug molecule to monoclonal antibodies (MAb) via a bifunctional linker containing a hydroxylamine and pyridyl disulfide. The pyridyl disulfide allowed for conjugation to the free cysteine of the MAb, while the aminooxy allowed for formation of an oxime bond with a model drug. The linkage was found to be highly stable, requiring exposure to pH ~2 to hydrolyze (Figure 4.3).

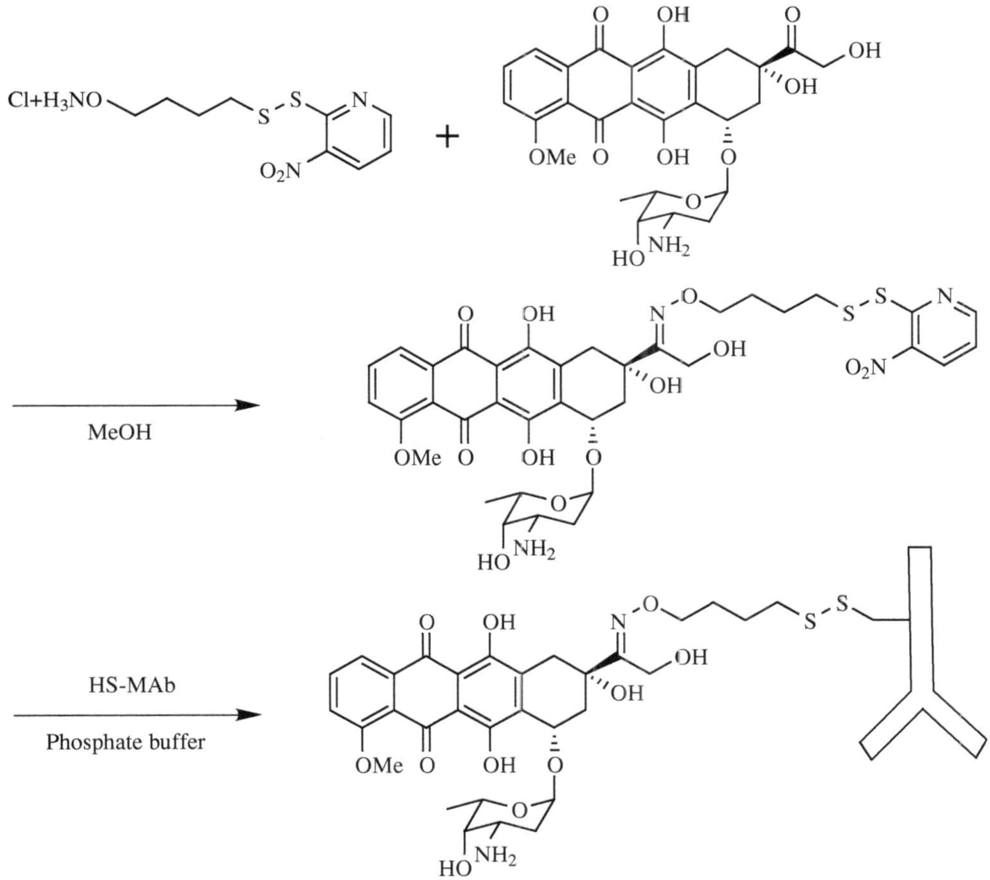

Figure 4.3 *Conjugation of adriamycin to monoclonal antibody via oxime bond formation with a bifunctional linker. Reprinted with permission from Webb, R. R., Kaneko, T., (1990),* Bioconjugate Chemistry, ***1****, 96–99. Copyright 1990 American Chemical Society.*

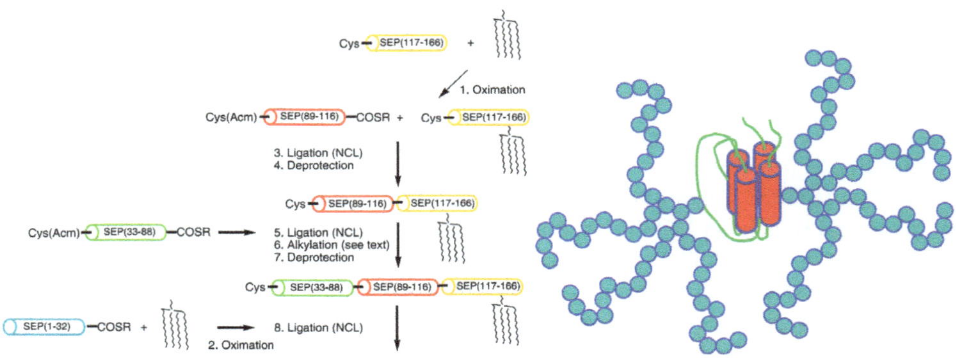

Figure 4.4 *Selective conjugation of branched PEG to synthetic erythropoietin. Reprinted with permission from Kochendoerfer, G. G.* et al.*, (2003),* Science*,* **299***, 884–887. Copyright 2003 AAAS.*

The utility of oxime chemistry in preparing protein–polymer conjugate drugs was first demonstrated in 2003 in the synthesis of PEGylated erythropoietin (Epo).[3] Kochendoerfer and coworkers prepared a synthetic Epo that incorporated two N^{ε}-levulinyl-modified lysine residues (Figure 4.4). These ketone-functionalized residues provided specific sites for polymer attachment. Aminooxy-functionalized four-arm PEG was conjugated to the ketone-bearing subunits, which were then assembled to form the final drug via native chemical ligation. The results demonstrated the orthogonality of oxime chemistry to other functionalities in proteins and peptides. The pharmacokinetics of the conjugate was also studied, and found to be superior to those of the native protein. Remarkably, at the same time the conjugate displayed equivalent bioactivity. This suggested that the oxime bond and polymer chains did not interfere with protein function.

The same group further demonstrated with the use of oxime chemistry by preparing PEGylated CCL-5 (RANTES).[4] Lys-45 of the CCL-5 subunit (34–67) was modified with isopropylidine-protected aminooxyacetic acid. Following native chemical ligation with the second subunit, the aminooxy functionality was revealed through oxime exchange with methoxyamine (Figure 4.5). Linear PEG–aldehyde was then conjugated to the protein. This result showed that either aminooxy or aldehyde functionality can be incorporated into proteins for conjugation. The conjugate was found to be more active against HIV, a property attributed to the polymer blocking aggregation through the GAG binding site.

CCL-5 has also been modified with branched PEG via oxime bond formation.[24] Tumelty and coworkers altered Lys-67 to contain a 1,3-dithiolane-protected N^{ε}-levulinyl lysine. Removal of the dithiolane protecting group was accomplished in the presence of free cysteines by treatment with silver triflate. Following deprotection, the ketone-modified protein was conjugated to an aminooxy-functionalized four-arm star PEG (Figure 4.6).

Francis and coworkers exploited their reported PLP method to specifically modify proteins at the *N*-terminus to introduce aminooxy-PEG onto proteins.[12] The method was also used with expressed protein ligation (EPL) to create a protein with reactive groups at both termini.[25] eGFP was expressed in *Escherichia coli* as an intein–chitin fusion and a cysteine piperidone amide was attached by EPL.[26] The protein was then subjected to PLP to create a

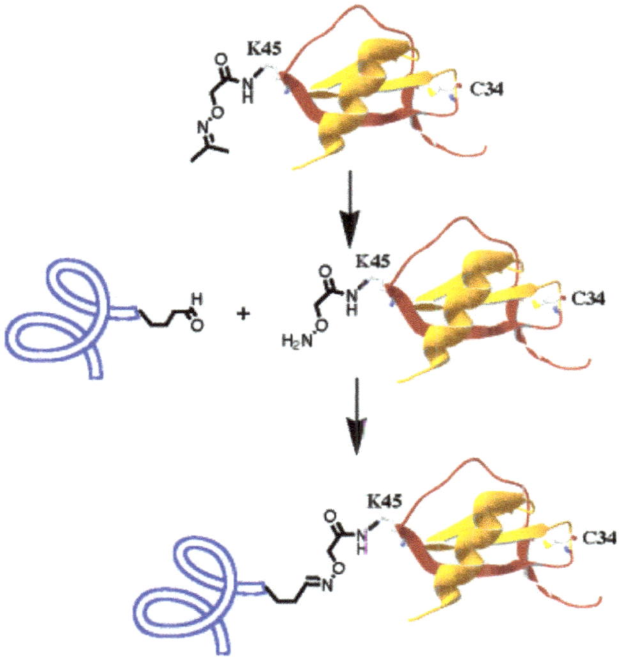

Figure 4.5 *Conjugation of PEG to aminooxy-modified CCL-5. Reprinted with permission from Shao, H., et al., (2005),* Journal of the American Chemical Society, ***127****, 1350–1351. Copyright 2005 American Chemical Society.*

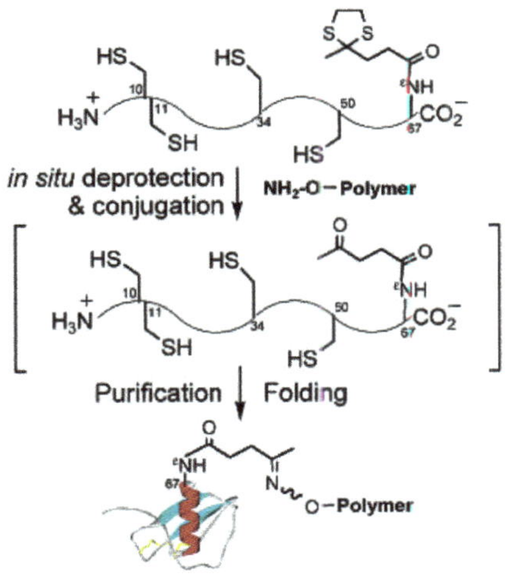

Figure 4.6 *Conjugation of PEG to ketone-modified CCL-5. Reprinted with permission from Tumelty, D. et al., (2003),* Journal of the American Chemical Society, ***125****, 14238–14239. Copyright 2003 American Chemical Society.*

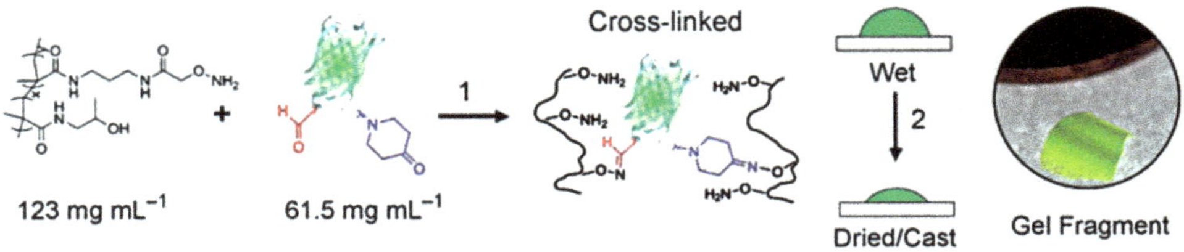

Figure 4.7 Preparation of protein–polymer hydrogels. N*-Oxo-C-keto GFP forms oxime crosslinks with a polymer bearing aminooxy side-chains. Reprinted with permission from A. P. Esser-Kahn, M. B. Francis, (2008),* Angewandte Chemie – International Edition in English, ***47****, 3751–3754. Copyright 2008 Wiley-VCH.*

protein modified for conjugation to aminooxy functionalities at both termini. In this work, the authors synthesized an alkoxyamine-*co*-hydroxypropyl methacrylate for conjugation to the protein. Hydrogels were formed by mixing the activated protein with the polymer (Figure 4.7). Furthermore, the hydrogel was found to undergo a structural denaturation from 60 to 80 °C, which was consistent with the properties of eGFP. This work demonstrated the ability to create a bulk material with properties of both the protein and polymer and should be useful to create a number of active biomaterials.

Peschke and coworkers developed a complementary method for modifying proteins for oxime bond formation utilizing a transpeptidation reaction to incorporate ketone functionality at the *C*-terminus.[27] An hGH subunit (178–191) was modified at the *C*-terminus with an N^{ε}-ketone-modified lysine derivative. The ketone-modified peptide was then conjugated to an aminooxy-functionalized two-arm PEG (Figure 4.8). The one disadvantage of this method was that the oxime formation required 10 days. However, this appeared to be unique to this system.

We have focused on developing straightforward synthetic methods to produce aminooxy-end functionalized polymers in a single step. For this purpose, controlled radical polymerization was used to prepare aminooxy end-functionalized polymers suitable for conjugation to ketone- or aldehyde-functionalized proteins.[28] Two initiators for atom transfer radical polymerization (ATRP) were prepared via reaction of Boc-aminooxyacetic acid. One, a bromoisobutyrate, was efficient for ATRP of methacrylate monomers, and the other, a chloropropionate, was suitable for ATRP of acrylamide or styrenyl monomers. Poly (*N*-isopropylacrylamide), prepared by ATRP from the chloropropionate initiator, was conjugated to bovine serum albumin (BSA) modified with N^{ε}-levulinyl lysine residues (Figure 4.9). This method has advantages over previous methods to produce the polymers, which required installing the aminooxy moiety post-polymerization.

Oxime chemistry can also be used to prepare side-chain-functionalized polymers. We polymerized 3,3′-diethyoxypropyl methacrylate (DEPMA) by reversible addition–fragmentation chain transfer (RAFT) polymerization.[29] Following acid deprotection, the resulting polymer with aldehyde-functionalized side chains was conjugated to a mixture of aminooxyacetic acid and *N*-terminal aminooxy-RGD (Figure 4.10). The latter is a ligand for cell surface integrins. The polymer composition was found to be identical to that of the feed solution, indicating that the oxime formation was efficient and that copolymers can be readily prepared utilizing this technique.

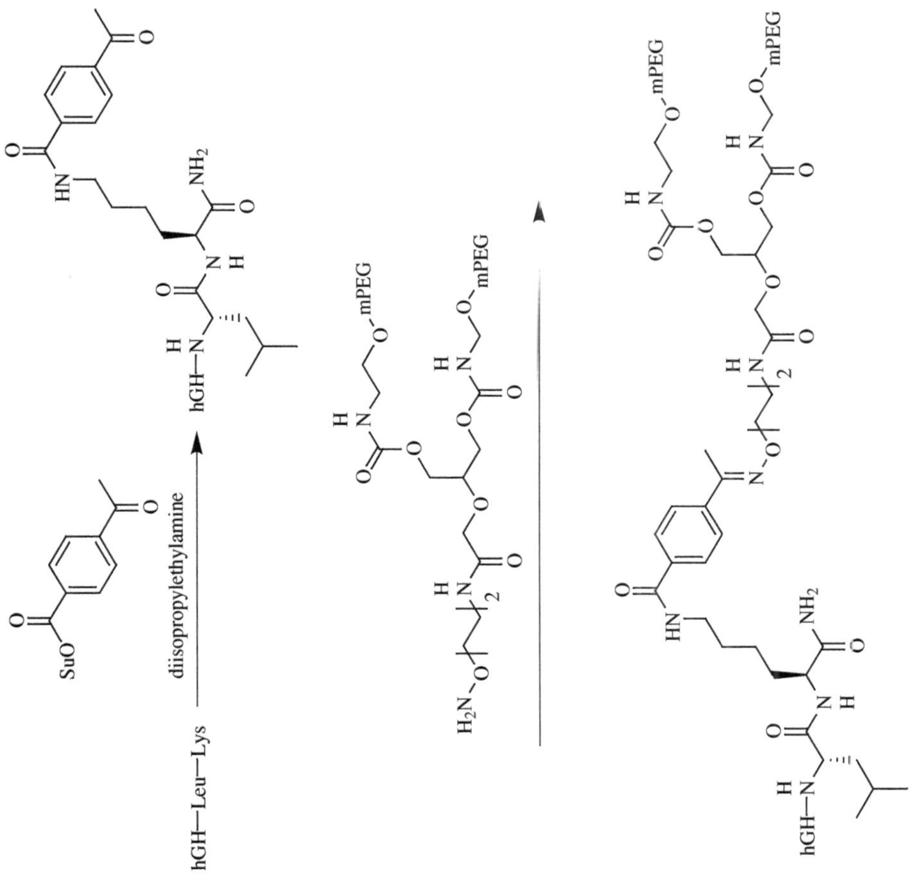

Figure 4.8 *C-terminal ketone modification of hGH, and subsequent conjugation to a branched aminooxy-PEG. Reprinted with permission from B. Peschke et al., (2007),* Bioorganic and Medicinal Chemistry, ***15****, 4382–4395. Copyright 2007 Elsevier.*

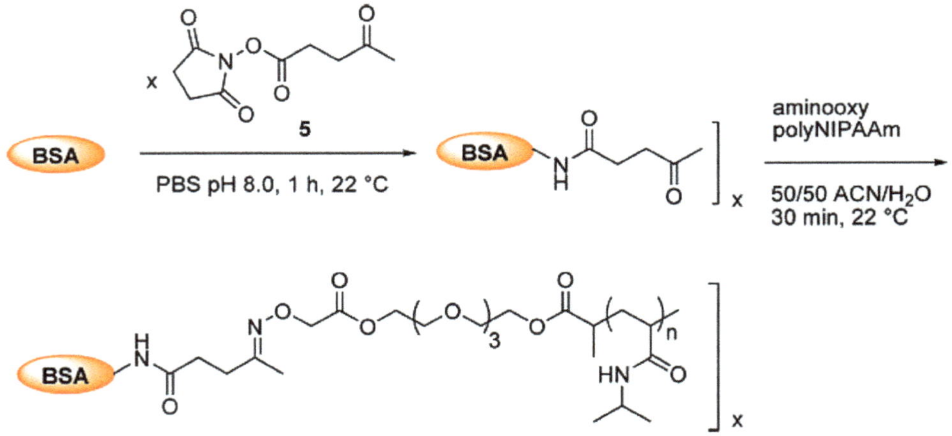

Figure 4.9 *Ketone modification of BSA and subsequent conjugation to aminooxy-pNIPAAm. Reprinted with permission from Heredia, K. L.* et al., *(2007),* Macromolecules, ***40****, 4772–4779. Copyright 2007 American Chemical Society.*

4.3 Immobilization of Proteins and Peptides on Surfaces

Peptide and protein arrays are widely used in the fields of biomaterials, medicine and biotechnology for applications that include diagnostics and microarray technology, as well as cell and tissue engineering.[19,22] Thus, there is a growing need for versatile synthetic strategies to immobilize the biomolecules on surfaces.[30–33] A number of different chemistries have been demonstrated.[34] This section focuses on the use of oxime chemistry to chemically attach proteins and peptides to surfaces. In the beginning, oligonucleotide and carbohydrate conjugations are also described for introductory purposes.

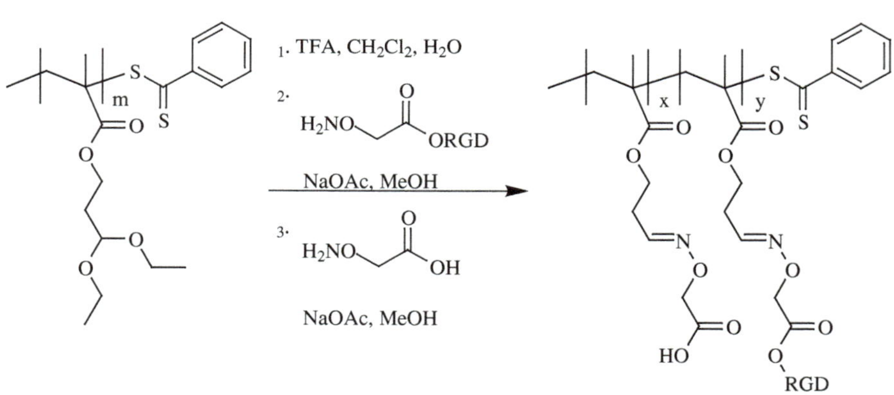

Figure 4.10 *Conjugation of aminooxy-RGD to a polymer bearing aldehyde side-chains. Reprinted with permission from Hwang, J. Y.* et al., *(2007),* Journal of Controlled Release, ***122****, 279–286. Copyright 2007 Elsevier.*

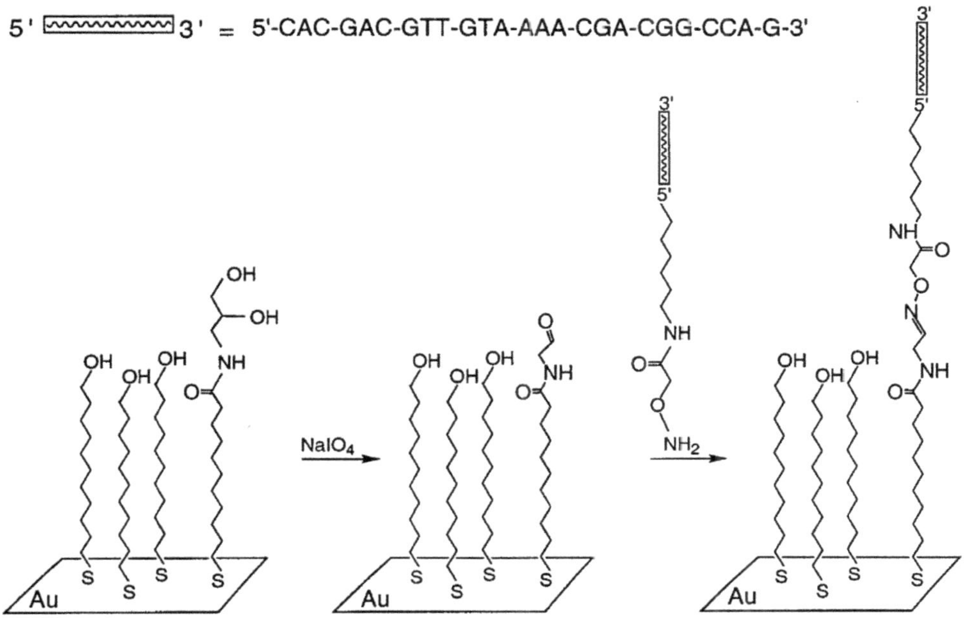

Figure 4.11 *Immobilization of oligonucleotides via oxime bond formation. Reprinted with permission from Boncheva, M.,* et al., *(1999),* Langmuir, ***15****, 4317–4320. Copyright 1999 American Chemical Society.*

The first example of oxime chemistry for surface immobilization of biomolecules was demonstrated by Boncheva *et al.* in 1999.[35] In this work, oligonucleotides were coupled to mixed alkanethiol-on-gold self-assembled monolayers (SAMs) to achieve arrays (Figure 4.11). The surface density of the DNA arrays was optimized to allow for high hybridization efficiency. The materials were characterized by surface plasmon resonance (SPR) and attenuated total reflectance-FTIR spectroscopy. This first example nicely demonstrated the utility of oxime chemistry for immobilization of biomolecules on surfaces.

This chemoselective immobilization technique was later used by a number of other groups to create oligonucleotide arrays[36–38] as well as immobilized microarrays of carbohydrates.[39,40] Oxime chemistry has also been used to immobilize oligonucleotides on micropatterned glass substrates[36,37] and inside glass capillaries.[38] Carbohydrates immobilized on polymer films using oxime bonds were demonstrated by Onodera and coworkers.[40] Specifically, a Boc-protected aminooxy monomer was copolymerized with methyl methacrylate and the polymer was cast onto a 96-well plate. The hydroxylamines were deprotected using 20% TFA and the carbohydrates were immobilized on the film in aqueous HCl (pH 2). The sugar trapping efficiencies in this work were reported to be about 80%. This was attributed to the fact that the pH required for glycoblotting was lower than the optimal pH for oxime bond formation. Tully *et al.* demonstrated synthetic chrondroitin sulfate (CS) gylcosaminoglycan microarrays utilizing oxime chemistry.[39] A series of synthetic CS molecules were synthesized with allyl end groups. Upon ozonolysis and subsequent

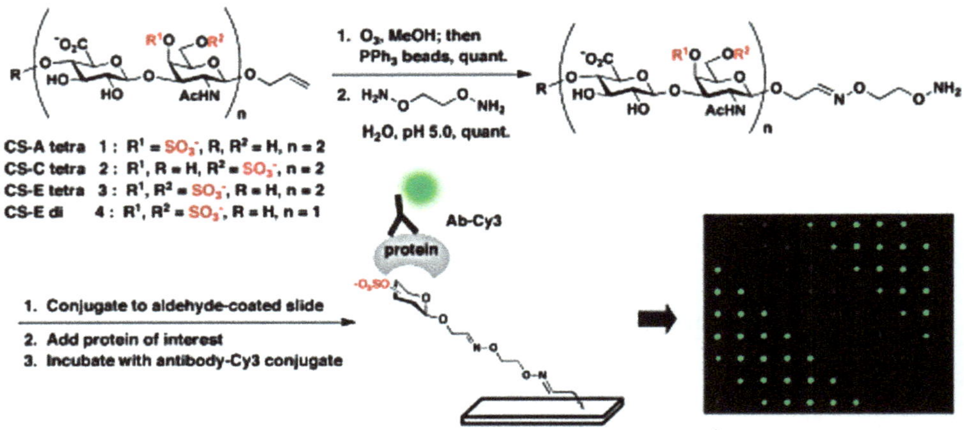

Figure 4.12 Formation of carbohydrate arrays via oxime bond formation. Reprinted with permission from Tully, S. E. et al., *(2006),* Journal of the American Chemical Society, ***128**, 7740–7741. Copyright 2006 American Chemical Society.*

treatment with 1,2-(bisaminooxy)ethane, aminooxy groups were created for conjugation to aldehyde functionalized glass slides (Figure 4.12).

The use of peptides in synthetic chemistry has increased significantly since the introduction of solid-phase peptide synthesis in the 1960s. Oxime bond formation was first used to immobilize peptides by Falsey and coworkers in 2001.[41] In this work, peptide microarrays were fabricated (Figure 4.13) for cell adhesion and functional assays. Commercially

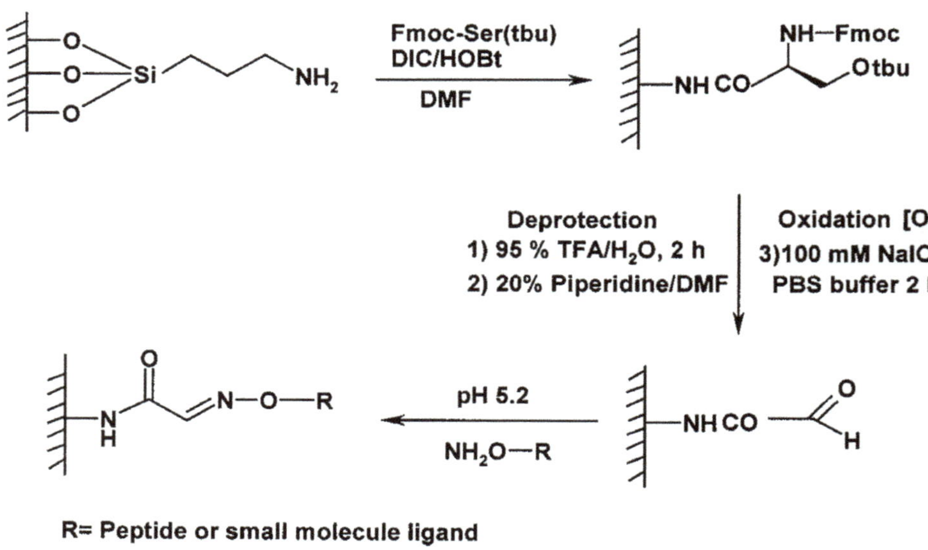

Figure 4.13 Aldehyde functionalization of glass substrates, and subsequent immobilization of aminooxy-peptides. Reprinted with permission from Falsey, J. R. et al., *(2001),* Bioconjugate Chemistry, ***12**, 346–353. Copyright 2001 American Chemical Society.*

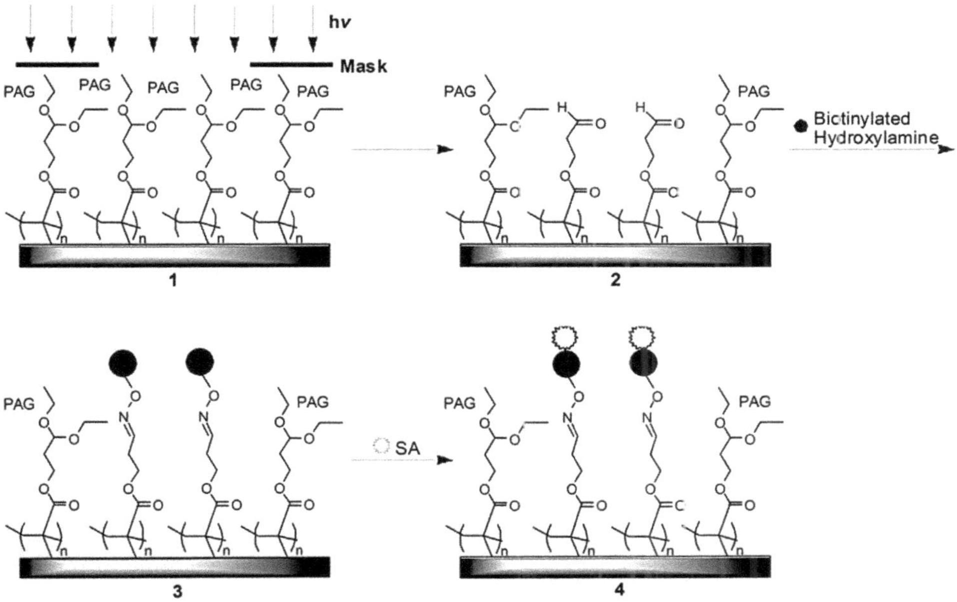

Figure 4.14 *Fabrication of streptavidin micropatterns via binding to a biotinylated surface prepared by oxime bond formation with an aldehyde-patterned polymer. Reprinted with permission from Christman, K. L., Maynard, H. D., (2005),* Langmuir, ***21****, 8389–8393. Copyright 2005 American Chemical Society.*

available microscope slides were first incubated with (3-aminopropyl)triethoxysilane to create amine-functionalized glass slides. The slides were subsequently converted to glyoxyl derivates by two different routes: coupling with Fmoc-protected serine, followed by subsequent deprotection and oxidation to the aldehydes or coupling with protected glyoxylic acid and subsequent deprotection using HCl. The surfaces were treated to remove nonspecific binding by co-spotting with steric acid in a ratio of 1:4 glyoxylic:steric acid.

We reported a method of creating oxime micropatterns using a pH-responsive polymer, poly(3,3′-diethoxypropyl methacrylate) (PDEPMA).[42] The polymer was spin-coated onto Si–SiO_2 substrates. Upon exposure to acid, the acetal groups of PDEPMA were converted aldehydes for site-specific conjugation of proteins. In this work, chemical deprotection was achieved in two ways: incubation with aqueous HCl or photochemical deprotection using the photoacid generator (PAG) triphenylsulfonium triflate and deep UV light. Micropatterns of selectively deprotected acetals were achieved by exposing PDEMPMA and PAG to deep UV light through a 1000-mesh Ni TEM grid (Figure 4.14). In order to demonstrate that proteins and other biomolecules could also be immobilized on these micropatterns, the surface was incubated with biotinylated hydroxylamine, thus forming streptavidin (SA)-reactive micropatterns. Because streptavidin has four binding sites for biotin, the array may be used as a platform to immobilize other biotinylated proteins or antibodies.

In an extension of this work, we showed that submicron patterns of streptavidin could be produced for protein assembly.[43] In this study, the photoacid generator diphenyliodonium-9, 10-dimethoxyanthracene-2-sulfonate (DIAS), which is excited at 365 nm, was spin-coated

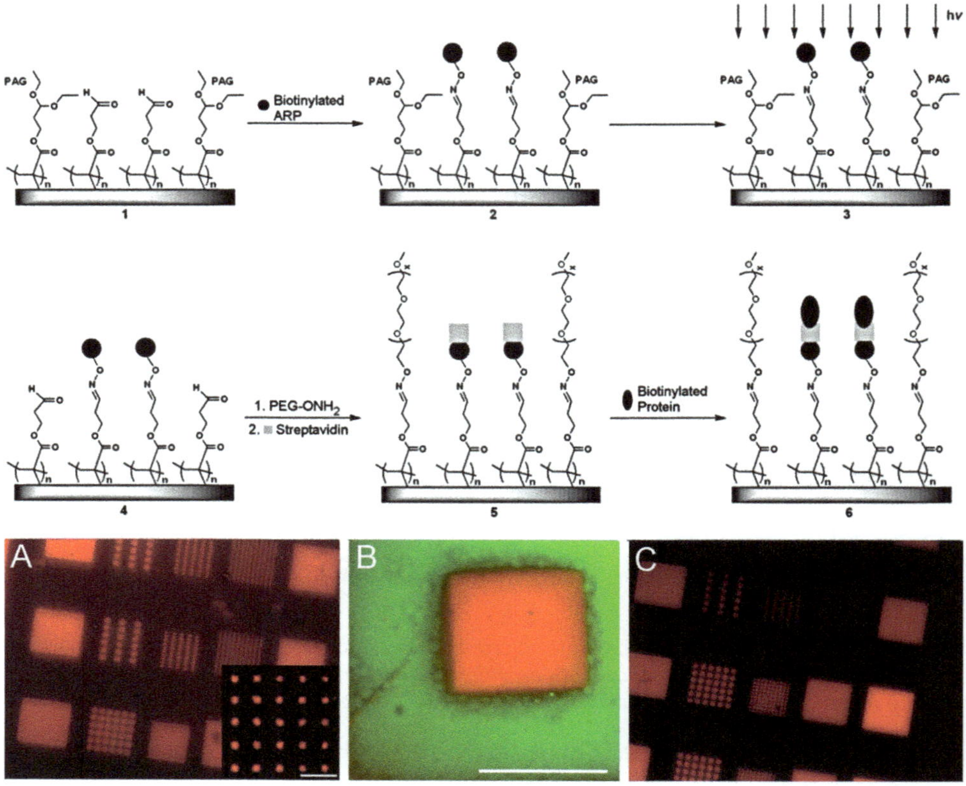

Figure 4.15 *Fabrication of micropatterned protein assemblies using streptavidin as a general linker. Reprinted with permission from Christman, K. L.* et al., *(2006),* Langmuir, ***22****, 7444–7450. Copyright 2006 American Chemical Society.*

with the PDEPMA polymer. An i-line wafer stepper and a chrome-on-quartz mask were used to produce aldehyde features ranging from 500 nm to 40 μm. These aldehydes were then conjugated to SA via a biotinylated hydroxylamine, in a similar fashion to the previous report.[42] In an effort to reduce nonspecific binding of proteins, after biotinylation the background was hydrolyzed to aldehydes and conjugated to an aminooxy-terminated PEG (Figure 4.15). The PEG background was found to reduce the nonspecific absorption of SA by approximately 98%. To demonstrate the stepwise assembly of proteins, biotinylated anthrax toxin receptor-1 (ANTXR-1) and the green fluorescent protective antigen component of anthrax toxin were assembled on the patterns. This was the first example of submicron protein patterning by photolithography.

Site-specific immobilization through oxime bond formation was also demonstrated utilizing a polymer with protected aminooxy groups.[44] A copolymer consisting of HEMA and Boc-protected aminooxy tetra(ethylene glycol) methacrylate was synthesized by free-radical polymerization. The polymer was designed with a PEG group in the monomer to provide protein resistance. The polymer was covalently attached to the native oxide of

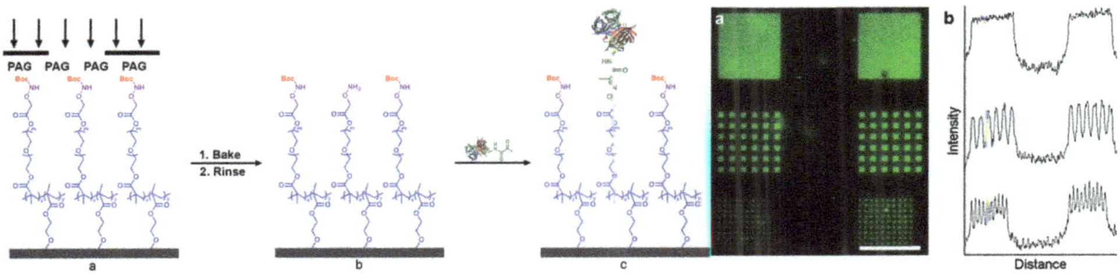

Figure 4.16 *Immobilization of N-terminal aldehyde-modified streptavidin on aminooxy micropatterns via oxime bond formation. A fluorescence image of the patterns (a), and intensity profile (b) are shown. Reprinted with permission from Christman, K. L. et al., (2007),* Journal of Materials Chemistry, ***17**, 2021–2027. Copyright 2007 The Royal Society of Chemistry.*

silicon by spin-coating a film of the copolymer on the surface. After baking at 110 °C the hydroxyl groups were covalently linked to the surface. An i-line sensitive PAG was then used to create patterns of aminooxy groups (Figure 4.16). Finally, *N*-terminal α-ketoamide-modified SA was immobilized to the surface. This strategy could be used to attach numerous proteins via the *N*-termini to patterned surfaces.

Park and Yousaf demonstrated use of an interfacial oxime reaction to immobilize ligands and cells in patterns and gradients to photoactive surfaces.[45] In this work UV-active nitroveratryloxycarbonyl (NVOC)-protected aminooxy alkanethiol SAMs were prepared. Alkoxyamine micropatterns were formed via exposure to UV irradiation through a micropatterned mask (Figure 4.17). The micropatterns were incubated with ketone-functionalized RGD to form cell-adhesive micropatterned substrates. Cell adhesion was observed only in those areas which had been deprotected.

A complementary method has been reported by the same group, whereby an aldehyde-functionalized surface was employed to conjugate aminooxy-functionalized peptides.[46]

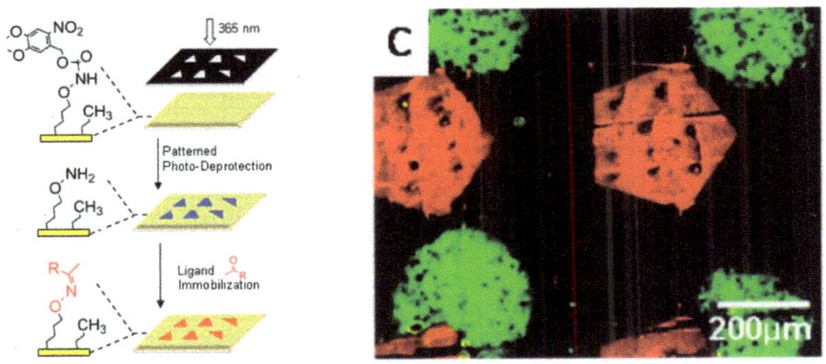

Figure 4.17 *Preparation of micropatterned aminooxy surfaces by photodeprotection, and subsequent ligand immobilization via oxime bond formation. Reprinted with permission from S. Park, M. N. Yousaf, (2008),* Langmuir, ***24**, 6201–6207. Copyright 2008 American Chemical Society.*

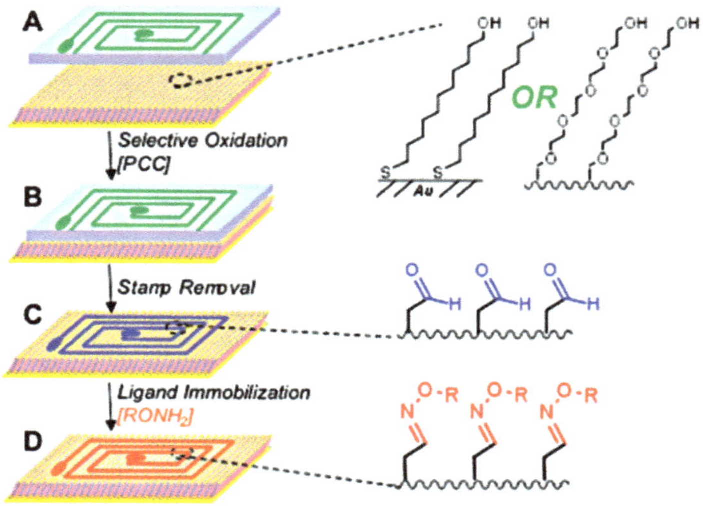

Figure 4.18 *Preparation of aldehyde micropatterns via selective oxidation within microfluidic channels and subsequent immobilization of aminooxy-functionalized ligands. Reprinted with permission from Westcott, N. P.* et al., *(2008),* Langmuir, ***24****, 9237–9240. Copyright 2008 American Chemical Society.*

PEGylated SAMs were prepared, and covered with a patterned poly(dimethylsiloxane) (PDMS) stamp to form microchannels. The SAM exposed to the microchannels was chemically oxidized to form micropatterned aldehydes (Figure 4.18). The PDMS was then removed, and the patterns were incubated with aminooxy-functionalized RGD to form cell-adhesive micropatterned substrates.

4.4 Conclusions

Protein– and peptide–polymer conjugates and surface hybrids are important in biomaterials and medicine. One crucial aspect to consider while preparing these materials is that the reaction site or sites be controlled. Oxime linkages provide a means to site-specific conjugations with retention of bioactivity. Moreover, installation of the reactive groups into the biomolecules is facile and can be achieved in aqueous solutions.

Acknowledgments

The authors thank the National Science Foundation (CAREER CHE-0645793) for funding. C.M.K. appreciates the NSF IGERT: MCTP (grant no. DGE-0114443) and the CNSI for a pre-doctoral fellowship. H.D.M. thanks the Alfred P. Sloan Foundation and Amgen (New Faculty Award) for additional funding.

References

(1) Lemieux, G. A., Bertozzi, C. R., (1998), *Trends in Biotechnology*, **16**, 506–513.
(2) Kochendoerfer, G. G., (2005), *Current Opinions in Chemistry and Biology*, **9**, 555–560.
(3) Kochendoerfer, G. G., Chen, S. Y., Mao, F., Cressman, S., Traviglia, S., Shao, H., Hunter, C. L., Low, D. W., Cagle, E. N., Carnevali, M., Gueriguian, V., Keogh, P. J., Porter, H., Stratton, S. M., Wiedeke, M. C., Wilken, J., Tang, J., Levy, J. J., Miranda, L. P., Crnogorac, M. M., Kalbag, S., Botti, P., Schindler-Horvat, J., Savatski, L., Adamson, J. W., Kung, A., Kent, S. B., Bradburne, J. A., (2003), *Science*, **299**, 884–887.
(4) Shao, H., Crnogorac, M. M., Kong, T., Chen, S. Y., Williams, J. M., Tack, J. M., Gueriguian, V., Cagle, E. N., Carnevali, M., Tumelty, D., Paliard, X., Miranda, L. P., Bradburne, J. A., Kochendoerfer, G. G., (2005), *Journal of the American Chemical Society*, **127**, 1350–1351.
(5) Schlick, T. L., Ding, Z. B., Kovacs, E. W., Francis, M. B., (2005), *Journal of the American Chemical Society*, **127**, 3718–3723.
(6) Dirksen, A., Dirksen, S., Hackeng, T. M., Dawson, P. E., (2006), *Journal of the American Chemical Society*, **128**, 15602–15603.
(7) Wu, P. G., Brand, L., (1997), *Fluorescence Spectroscopy*, **278**, 321–330.
(8) Dixon, H. B. F., (1964), *Biochemical Journal*, **92**, 661.
(9) Duan, X. J., Zhao, Z., Ye, J. P., Ma, H. M., Xia, A. D., Yang, G. Q., Wang, C. C., (2004), *Angewandte Chemie – International Edition*, **43**, 4216–4219.
(10) Dixon, H. B., (1964), *Biochemistry Journal*, **90**, 2C–3C.
(11) Dixon, H. B. F., Fields, R., (1972), *Methods in Enzymology*, **25**, 409–419.
(12) Gilmore, J. M., Scheck, R. A., Esser-Kahn, A. P., Joshi, N. S., Francis, M. B., (2006), *Angewandte Chemie – International Edition in English*, **45**, 5307–5311.
(13) Connor, R. E., Piatkov, K., Varshavsky, A., Tirrell, D. A., (2008), *Chembiochem*, **9**, 366–369.
(14) Caliceti, P., Veronese, F. M., (2003), *Advanced Drug Delivery Reviews*, **55**, 1261–1277.
(15) Vandermeulen, G. W. M., Klok, H. A., (2004), *Macromolecular Bioscience*, **4**, 383–398.
(16) Duncan, R., (2006), *Nature Reviews Cancer*, **6**, 688–701.
(17) Pasut, G., Veronese, F. M., (2007), *Progress in Polymer Science*, **32**, 933–961.
(18) Maeda, H., Greish, K., Fang, J., (2006), *Polymer Therapeutics II: Polymers as Drugs, Conjugates and Gene Delivery Systems*, **193**, 103–121.
(19) Hoffman, A. S., Stayton, P. S., (2004), *Macromolecular Symposia*, **207**, 139–151.
(20) Heredia, K. L., Maynard, H. D., (2007), *Organic and Biomolecular Chemistry*, **5**, 45–53.
(21) Gauthier, M. A., Klok, H. A., (2008), *Chemical Communications (Cambridge)*, 2591–2611.
(22) Duncan, R., (2003), *Nature Reviews Drug Discovery*, **2**, 347–360.
(23) Webb, R. R., 2nd, Kaneko, T., (1990), *Bioconjugate Chemistry*, **1**, 96–99.
(24) Tumelty, D., Carnevali, M., Miranda, L. P., (2003), *Journal of the American Chemical Society*, **125**, 14238–14239.
(25) Esser-Kahn, A. P., Francis, M. B., (2008), *Angewandte Chemie – International Edition in English*, **47**, 3751–3754.
(26) Muir, T. W., Sondhi, D., Cole, P. A., (1998), *Proceedings of the National Academy of Sciences of the USA*, **95**, 6705–6710.
(27) Peschke, B., Zundel, M., Bak, S., Clausen, T. R., Blume, N., Pedersen, A., Zaragoza, F., Madsen, K., (2007), *Bioorganic and Medicinal Chemistry*, **15**, 4382–4395.
(28) Heredia, K. L., Tolstyka, Z. P., Maynard, H. D., (2007), *Macromolecules*, **40**, 4772–4779.
(29) Hwang, J. Y., Li, R. C., Maynard, H. D., (2007), *Journal of Controlled Release*, **122**, 279–286.
(30) Cairo, C. W., Gestwicki, J. E., Kanai, M., Kiessling, L. L., (2002), *Journal of the American Chemical Society*, **124**, 1615–1619.
(31) Mann, B. K., West, J. L., (2002), *Journal of Biomedical Materials Research*, **60**, 86–93.
(32) Irvine, D. J., Mayes, A. M., Griffith, L. G., (2001), *Biomacromolecules*, **2**, 85–94.
(33) Drumheller, P. D., Hubbell, J. A., (1994), *Analytical Biochemistry*, **222**, 380–388.
(34) Christman, K. L., Enriquez-Rios, V. D., Maynard, H. D., (2006), *Soft Matter*, **2**, 928–939.
(35) Boncheva, M., Scheibler, L., Lincoln, P., Vogel, H., Akerman, B., (1999), *Langmuir*, **15**, 4317–4320.

(36) Defrancq, E., Hoang, A., Vinet, F., Dumy, P., (2003), *Bioorganic and Medicinal Chemistry Letters*, **13**, 2683–2686.
(37) Dendane, N., Hoang, A., Defrancq, E., Vinet, F., Dumy, P., (2008), *Bioorganic and Medicinal Chemistry Letters*, **18**, 2540–2543.
(38) Dendane, N., Hoang, A., Guillard, L., Defrancq, E., Vinet, F., Dumy, P., (2007), *Bioconjugate Chemistry*, **18**, 671–676.
(39) Tully, S. E., Rawat, M., Hsieh-Wilson, L. C., (2006), *Journal of the American Chemical Society*, **128**, 7740–7741.
(40) Onodera, T., Niikura, K., Iwasaki, N., Nagahori, N., Shimaoka, H., Kamitani, R., Majima, T., Minami, A., Nishimura, S. I., (2006), *Biomacromolecules*, **7**, 2949–2955.
(41) Falsey, J. R., Renil, M., Park, S., Li, S. J., Lam, K. S., (2001), *Bioconjugate Chemistry*, **12**, 346–353.
(42) Christman, K. L., Maynard, H. D., (2005), *Langmuir*, **21**, 8389–8393.
(43) Christman, K. L., Requa, M. V., Enriquez-Rios, V. D., Ward, S. C., Bradley, K. A., Turner, K. L., Maynard, H. D., (2006), *Langmuir*, **22**, 7444–7450.
(44) Christman, K. L., Broyer, R. M., Tolstyka, Z. P., Maynard, H. D., (2007), *Journal of Materials Chemistry*, **17**, 2021–2027.
(45) Park, S., Yousaf, M. N., (2008), *Langmuir*, **24**, 6201–6207.
(46) Westcott, N. P., Pulsipher, A., Lamb, B. M., Yousaf, M. N., (2008), *Langmuir*, **24**, 9237–9240.

5

The Role of Click Chemistry in Polymer Synthesis

Jean-François Lutz and Brent S. Sumerlin

5.1 Introduction

The expected increased demand for novel nanomaterials is directly accompanied by a need for efficient routes to prepare such materials. The realm of macromolecular science has always been intimately entwined with studies in the nanosciences, and a great deal of attention has been recently devoted to increase the diversity of available polymeric materials. Despite the enormous progress made over the last few decades, synthetic macromolecules generally remain rather undefined in comparison to many biopolymers, such as proteins or nucleic acids. Such limitations of polymer chemistry may seem surprising to many nonspecialists since modern organic synthesis provides a plethora of solutions for the preparation of chemo-, regio- and stereo-controlled low-molecular-weight compounds.[1] However, these chemical tools are not always readily transferable to the macromolecular scale. While the last 50 years has resulted in the development of a multitude of new synthetic polymerization and polymer modification strategies, many of these methods were plagued by their complexity and narrow focus. Ironically, it is often the case that the more complex a macromolecular structure is, the more important it is to have a simple pathway for its preparation. Whether the goal is to ensure complete consumption of a plurality of reactive functionalities or to specifically functionalize a single site on a polymer chain, efficient and/or orthogonal synthetic strategies are crucial.

While Nature has optimized its chemistry through evolution to select robust chemical tools perfectly adapted to Earth's environmental conditions, the discovery and selection of simpler and more universal synthetic methods is essential for advancement in polymer

Click Chemistry for Biotechnology and Materials Science Edited by Joerg Lahann

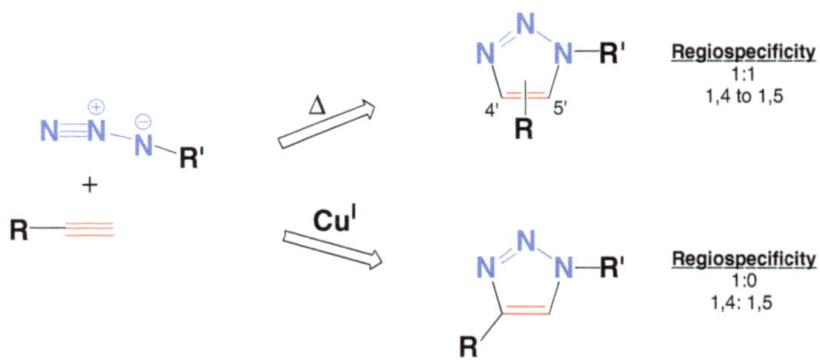

Scheme 5.1 *Thermal and copper-catalyzed cycloaddition of azides and alkynes.*

science. Selection of the most versatile chemical tools is the essence of click chemistry, an appealing concept proposed by Sharpless and coworkers.[2] Click chemistry is not a scientific discipline, but rather a synthetic philosophy focused on *function* and inspired by the simplicity and efficiency of the chemistry utilized by Nature. Highly complex biological systems rely on a modest library of monomers joined by only a few efficient reactions. Similarly, the objective of click chemistry is to establish an ideal set of straightforward and highly selective reactions in synthetic chemistry. For instance, the archetypal example of click chemistry is undoubtedly the copper-catalyzed Huisgen 1,3-dipolar cycloaddition of azides and terminal alkynes.[3–6] In the absence of an appropriate catalyst, this reaction is generally slow and lacks regiospecificity, but in the presence of copper(I), which binds to terminal alkynes to form intermediate copper acetylides, this cycloaddition reaction is dramatically accelerated, regioselective and highly efficient (Scheme 5.1). Moreover, the copper-catalyzed azide–alkyne cycloaddition (CuAAC) can be performed in a variety of solvents (including water) and in the presence of numerous other functional groups.[5,6]

Although click chemistry was initially postulated as a general concept for organic synthesis, this strategy also has enormous potential in materials science.[1,7] The first example to illustrate this point was reported by the groups of Hawker, Fokin and Sharpless.[8] Afterwards, the visibility of click chemistry within the materials community grew.[7–26] The goal of this chapter is not to be a comprehensive review of the application of click chemistry in materials science, as many of these topics will inevitably be covered in other chapters of this volume. Rather, by highlighting some specific examples, many of which represent the first instances of click chemistry being employed in macromolecular synthesis and functionalization, we hope to give context to the click concept and its particular utility in polymer science.

5.2 Polymerization via CuAAC

High conversion is fundamental requirement for obtaining high molecular weight polymer by step-growth polymerization mechanisms. Therefore, it is logical that many of the first applications of the click concept in polymer synthesis involved the polymerization of azide- (or nitrile-) and alkyne-containing monomers to form linear polymers, dendrimers

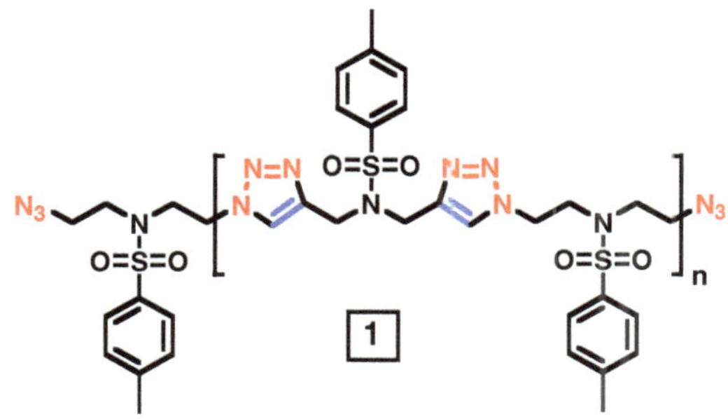

1

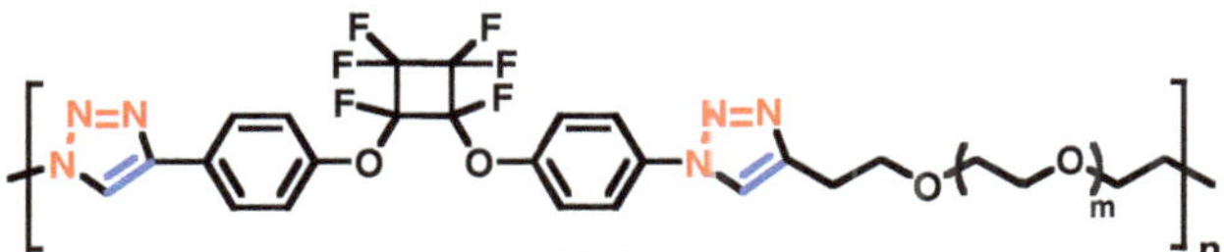

2

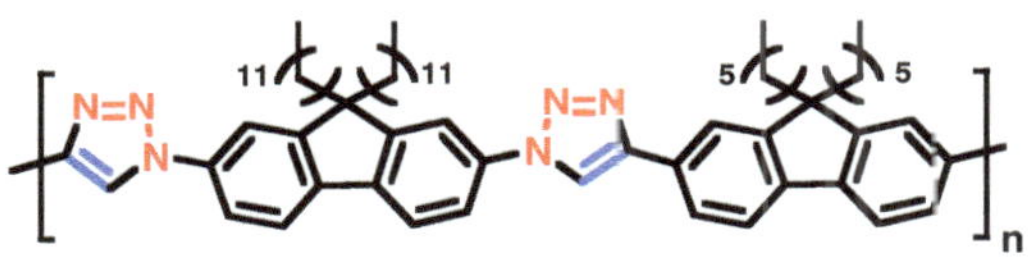

3

Figure 5.1 *Examples of CuAAC being employed to prepare linear polymer structures by polymerization of azide and alkyne functional monomers.*[9,27,29] *Reprinted with permission from J.-F. Lutz, (2007), 1,3 Dipolar Cycloadditions of Azides and Alkynes: A Universal Ligation Tool in Polymer and Materials Science,* Angew. Chem. Int. Ed.*,* ***46*** *(7), 1018–1025. Copyright 2007 Wiley-VCH.*

and rotaxanes.[8,9,12,20,27–29] In these cases, the resulting structures contain multiple triazole or tetrazole repeating units and constitute a novel class of macromolecules with potentially interesting properties. Hawker, Fokin and Sharpless first reported CuAAC for the convergent synthesis of dendrimers.[8] The application of click strategies for the preparation of dendrimers proved highly beneficial, as there is perhaps no other area of polymer synthesis that relies so heavily on near-quantitative reaction conversions. This method proved to be a straightforward strategy for large-scale synthesis of near-perfect triazole-containing dendrimers. Shortly thereafter, Finn *et al.* extended CuAAC to prepare linear polymer chains (Figure 5.1, **1**) and macromolecular networks.[9] Both cases resulted in a prominent presence of triazole functionalities within the polymer structure, and the latter proved to be remarkable copper–copper adhesives as a result of the ability of these triazole rings to coordinate to miniscule amounts of copper(I) inevitably present on the surface of copper metal.

Fleury and coworkers recently conducted in-depth investigations regarding the polymerization of difunctional azide and alkyne monomers.[30] 1,6-Diazidohexane and a,w-bis(*O*-propargyl)diethylene glycol were polymerized with a $Cu(PPh_3)_3Br$ catalyst to yield triazole-containing copolymers. Kinetic studies were conducted, and polymerizations at various temperatures allowed the activation energy for the polyaddition process to be calculated as $E_a = 45 \pm 5$ kJ/mol. Poly(alkyl aryl) ethers containing 1,2,3-triazolyl and perfluorocyclobutyl units (Figure 5.1, **2**) were prepared by Qing *et al.* via click poly(cycloaddition).[29] These novel macromolecules exhibited a rather interesting thermal stability and melting fluidity. On the other hand, polymers containing tetrazole units were found to be significantly less thermally stable. Polymers containing multiple tetrazole side-groups can be prepared by reacting well-defined polyacrylonitrile precursors with sodium azide, leading to tetrazole-functionalized polymers that decompose at temperature as low as 120 °C.[31] In addition to polymerization of low molecular weight azides and alkynes, Matyjaszewski and coworkers demonstrated that linear polymers with alkynyl and/or azido end groups could be further polymerized, with the chain extension leading to high molecular weight homopolymers[32] or multiblock copolymers.[33]

Because of the aromaticity of the resulting triazoles, CuAAC has also proven efficient for the synthesis of conjugated polymers. Both Reek *et al.* (Figure 5.1, **3**) and Bunz *et al.* reported the preparation of poly(fluorenylene-triazolene) via the reaction of diazido-fluorene monomers with various diynes.[27,34] Without relying on a copper catalyst, Bunz and coworkers also reported that these polymers could be prepared by local heating with an AFM tip.[34]

Overall, Huisgen cycloadditions seem to be straightforward and efficient reactions for building novel polymer structures. However, so far the prime focus of the aforementioned works has been on synthesis. Further studies are needed to assess comprehensively the physical behavior of these novel triazole- or tetrazole-containing polymers. Nonetheless, some of these early works already suggested some interesting properties (e.g. solubility, swelling, metal adhesion).

5.3 Post-polymerization Modification via Click Chemistry

Macromolecular engineering involves the synthesis of complex macromolecular structures with defined composition, microstructure, functionality and architecture (e.g. telechelic polymers, block copolymers, macromolecular brushes, stars, networks), using covalent chemistry approaches.[35] In addition to versatile polymerization chemistry, the synthesis of such complex macromolecules often requires the use of efficient and specific post-polymerization modification techniques to incorporate functionality potentially incompatible with the polymerization, characterization, or processing conditions.[36] Click reactions are especially suited for such advanced macromolecular design. Indeed, modification of multiple points on a polymer chain requires a highly efficient reaction mechanism, since the unreacted byproducts cannot be simply separated, as is the case is most small molecule reactions. Click strategies have served as a complementary tools for most of the major synthetic polymerization techniques, such as cationic or anionic ring opening polymerization (ROP),[37–40] ring opening metathesis polymerization (ROMP),[41] polycondensation,[42] conventional free-radical polymerization,[10,43] nitroxide-mediated polymerization (NMP),[11,14] reversible addition-fragmentation transfer polymerization (RAFT)[44–49] and atom transfer radical polymerization (ATRP).[19,31,32,50–65] The latter method has been coupled with CuAAC more than any of the others, and the reader is directed to a review specifically highlighting the combination of ATRP and click chemistry.[66] ATRP is a facile and versatile polymerization technique and one of the most employed tools in modern polymer chemistry.[35,67] Despite its inherent versatility, the range of macromolecular structures available by ATRP can be further broadened by click strategies (Figure 5.2).

A first important application of CuAAC in polymer chemistry is undeniably the synthesis of functional polymers (either end-functional or pendant-functional). Telechelic polymers (i.e. polymers with functional end groups) can be efficiently prepared via a straightforward ATRP/CuAAC combination. The halogen end groups of polymers prepared using ATRP can be easily transformed into azide moieties[68–70] and subsequently involved in CuAAC with functional alkynes (Figure 5.2).[32,51,53,63,64] Alternatively, azide- or alkyne-functional initiators can be also utilized, though in the latter case protection of the alkyne moiety is often required to prevent its consumption during polymerization.[32,54,55] Functional telechelic polymers can also be prepared by RAFT polymerization, as first demonstrated by the groups of Barner-Kowollik and Stenzel[46] and Sumerlin *et al.*[45] In this case, the azide or alkyne moiety is typically incorporated in the chain transfer agent (CTA) structure prior to polymerization.

Polymers with multiple functional side chains have been prepared by CuAAC using precursors built with alkyne-functional monomers.[10,14,37,40,42] For example, Fréchet and coworkers constructed dendronized polymers (i.e. linear polymer chains with bulky side dendrons) via side-chains cycloadditions (Scheme 5.2).[10] Hawker and coworkers exploited this concept even further via cascade side-chain functionalization of macromolecules (Scheme 5.3).[14] Mantovani and Haddleton recently demonstrated that glycopolymers could be prepared in a one pot-process by simultaneous ATRP of an alkynyl monomer and CuAAC with azido-functionalized sugars.[71] The relative rates of polymerization and click functionalization were conveniently tunable by varying the catalyst concentration, solvent and temperature.

Azide-containing monomers and related polymer precursors have been also studied for preparing macromolecules with functional side-groups.[38,41,43,56] For example, after preparing homopolymers and block copolymers of 3-azidopropyl methacrylate (AzPMA), Sumerlin *et al.* functionalized the resulting azido-functionalized polymer (PAzPMA) by coupling with a variety of low molecular weight alkynyl species (Scheme 5.4).[56] Interestingly, the rate of azide–alkyne coupling of this polymer was observed to be significantly higher than that for the corresponding monomer, an effect attributed to autoacceleration by anchimeric assistance. Previous reports had demonstrated polytriazoles to be excellent ligands for copper(I).[72] Thus, during functionalization of PAzPMA in the absence of additional ligand, triazoles formed along the polyAzPMA backbone were believed to complex the catalyst, leading to a higher local copper(I) concentration in the immediate vicinity of neighboring unreacted azido groups. Similar autocatalytic results were reported by Fokin and Finn.[73]

Although efficient, such strategies are potentially experimentally risky (like any approach involving molecules with a high density of azide groups) due to the potentially explosive character of organic azides and should be investigated with extreme care.[74,75] However, an alternative route was reported by Matyjaszewski *et al.* in which poly(glycidyl methacrylate) (PGMA) copolymers were prepared by ATRP, and subsequent ring opening of the pendant epoxide with NaN_3 led to efficient incorporation of azide groups on each PGMA repeat unit (Scheme 5.5).[76] The resulting 1-hydroxy-2-azido copolymers were reacted with alkyne-terminated poly(ethylene oxide) (PEO) to yield graft copolymers.

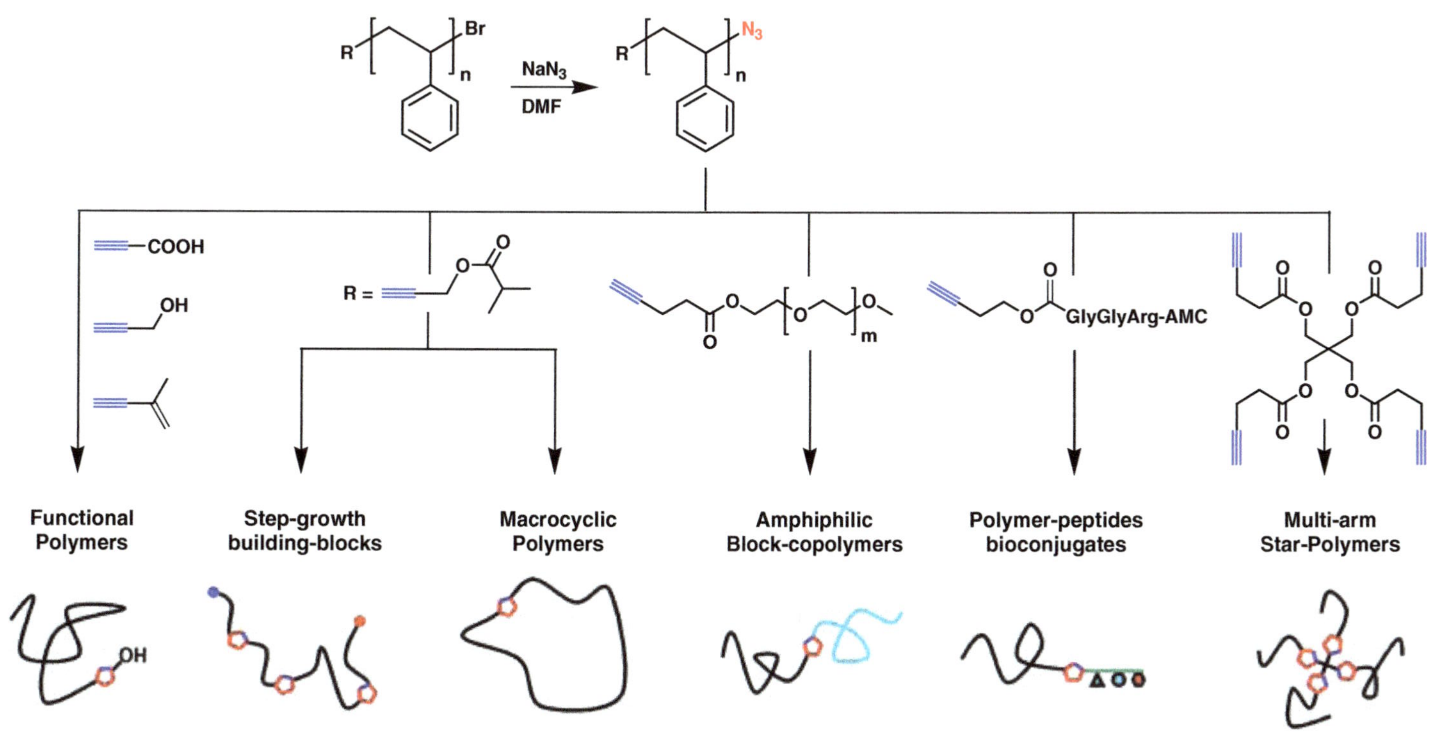

Figure 5.2 *Example macromolecular architectures accessible via the click modification of well-defined polystyrene prepared by ATRP.*[32,50,53,55,58,61] *Reprinted with permission from J.-F. Lutz, (2007), 1,3 Dipolar Cycloadditions of Azides and Alkynes: A Universal Ligation Tool in Polymer and Materials Science,* Angew. Chem. Int. Ed., ***46*** *(7), 1018–1025. Copyright 2007 Wiley-VCH.*

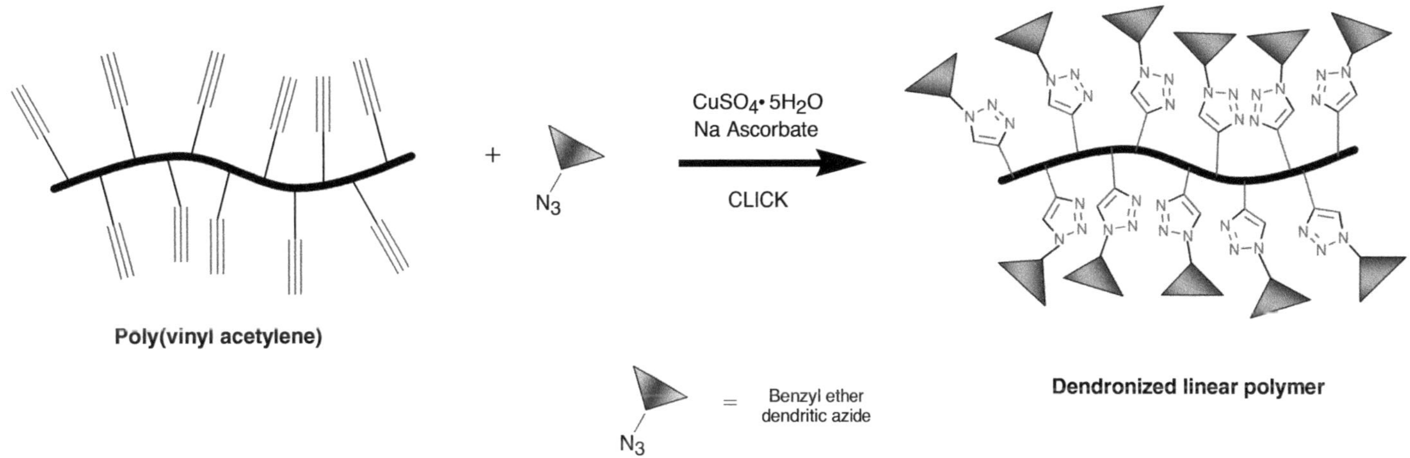

Scheme 5.2 *Synthesis of dendronized linear polymers by grafting-to via CuAAC.*[10]

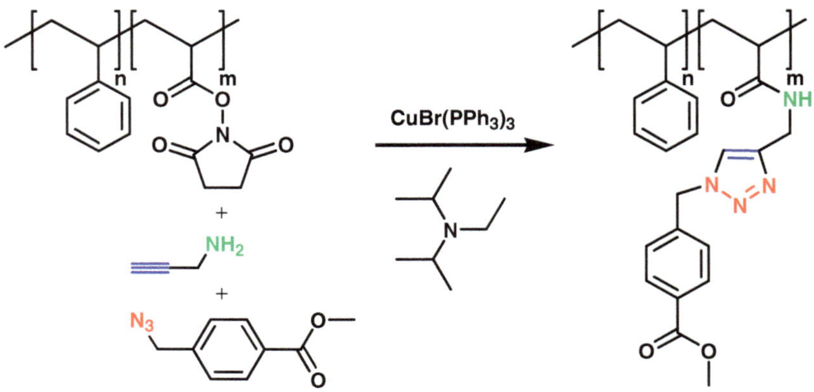

Scheme 5.3 *Example of cascade functionalization (i.e. amide formation and cycloaddition) of synthetic macromolecules.*[14] *Reprinted with permission from J.-F. Lutz, (2007), 1,3 Dipolar Cycloadditions of Azides and Alkynes: A Universal Ligation Tool in Polymer and Materials Science,* Angew. Chem. Int. Ed., ***46*** *(7), 1018–1025. Copyright 2007 Wiley-VCH.*

In addition to proving useful for the synthesis of pendant functional and telechelic macromolecules, click strategies have also been employed to build polymers with complex architectures. One of the first examples of architectural control using was reported by the team of Van Hest, which synthesized amphiphilic block copolymers by coupling azide- and alkyne- functional homopolymers (Figure 5.1).[55] Such ligation strategy has been shown to be quite efficient for linking segments of various natures. Barner-Kowollik and Stenzel later used a similar approach to couple azido- and alkyne-terminated polymers prepared by RAFT.[46] Self-ligation of a-alkyne-w-azido-telechelic polymers has proven to be an elegant strategy for preparing macrocycles, an often difficult synthetic challenge (Figure 5.1).[32,61,77] Additionally, other nonlinear architectures such as stars,[39,58] miktoarm stars,[57,65,78] graft copolymers,[64,79] networks[15,19] and hyperbranches[80] were constructed using click chemistry. All of these reports relied on CuAAC, with the exception of the approach reported by Yagci and coworkers for preparing graft copolymers by Diels–Alder chemistry.[79]

As mentioned previously, CuAAC has been used to prepare triazole-containing dendrimers. Alternatively, CuAAC can be used to functionalize the outer-shells of dendrimers or for linking preformed dendrons (e.g. polyamidoamine or polyester-based dendrons).[13,17,81,82] The latter is an interesting route for preparing unsymmetrical dendrimers with distinct hemispheres.[17]

5.4 Polymer–Biomacromolecule Conjugation

The combination of macromolecules of both natural and synthetic origin is an appealing strategy to prepare hybrid materials that combine the advantages of standard synthetic polymers with advanced biological functions (e.g. molecular recognition, programmed self-organization, biological targeting, enzymatic activity).[83] Fortunately, the high degree of selectivity of click reactions makes them particularly attractive for modifying biomolecules

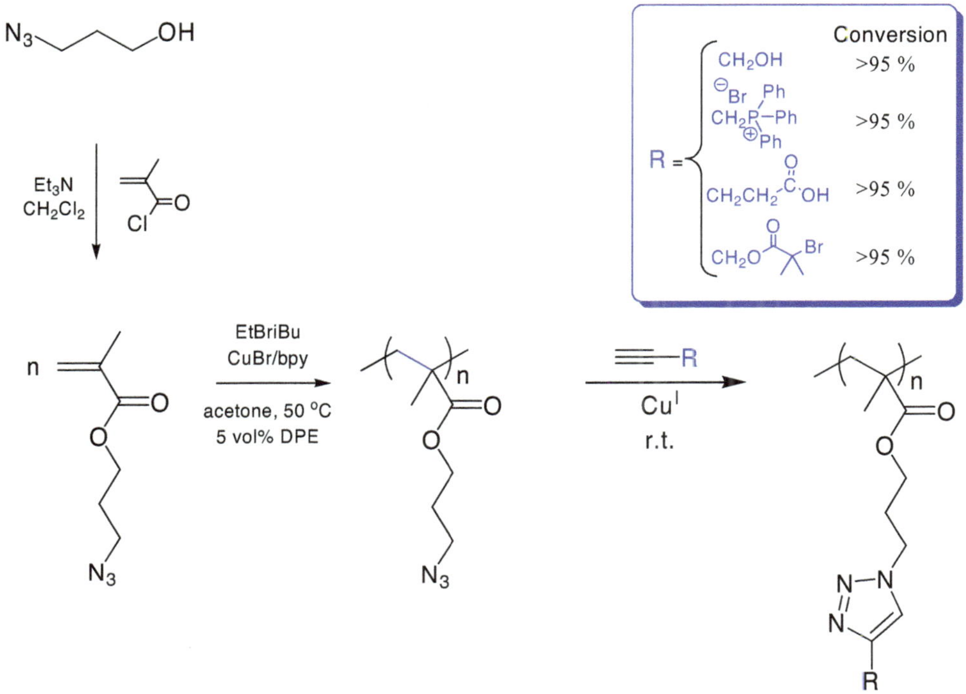

Scheme 5.4 *Click functionalization of azido-pendant polymers with various functional alkynes.*[56]

that often contain a wide range of reactive functionalities. CuAAC was reported to be an excellent method to modify biological polymers such as nucleic acids or polysaccharides (the strategies for functionalizing biopolymers are conceptually similar to those described above for synthetic polymers).[84–86] More specifically, CuAAC is a highly versatile tool for covalently functionalizing biological molecules with synthetic polymers.[87] For instance, several reports indicated that sequence-defined oligopeptides can be linked to synthetic macromolecules using click ligation.[37,50,63,88,89] In particular, Nolte *et al.* described the

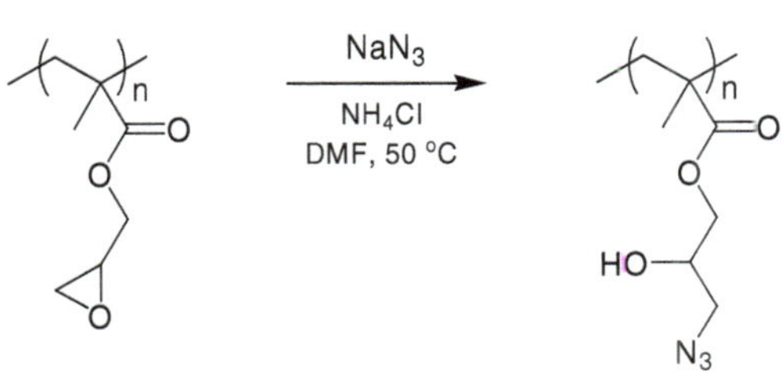

Scheme 5.5 *Alternative strategy for preparing azido-pendant polymers by ring opening of epoxy functionalities in PGMA.*[76]

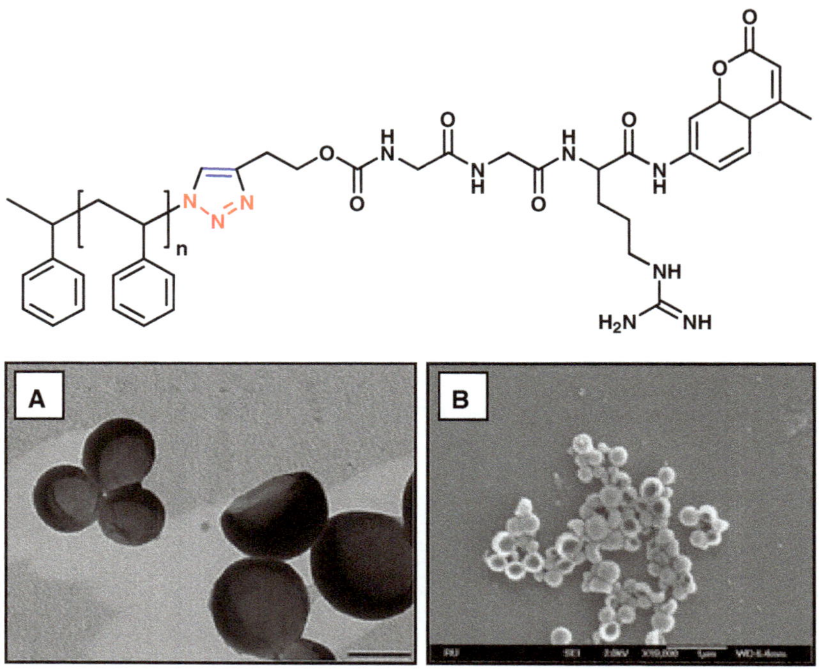

Figure 5.3 *Molecular structure of an amphiphilic polymer bioconjugate polystyrene*-b-*oligopeptide synthesized partly by click chemistry (top) and visualization by electron microscopy of its aggregates in water [bottom: TEM (A) and SEM (B)]. Reprinted with permission from A. J. Dirks* et al.*, (2005),* Chem. Commun.*,* ***33****, 4172–4174. Copyright 2005 The Royal Society of Chemistry.*

synthesis and self-assembly in aqueous medium of bio-hybrid amphiphiles composed of a hydrophobic polystyrene segment coupled to a hydrophilic oligopeptide (Figure 5.3). Lutz and coworkers studied the click cycloaddition of short peptides with well-defined synthetic polymers synthesized by ATRP (Figure 5.4).[63,89] Typically, the ω-bromine chain-ends of ATRP polymers were transformed into azides by nucleophilic substitution and subsequently reacted with alkyne-functionalized peptides. This allowed azido-terminated polystyrene to be coupled with an alkyne-labeled protein transduction domain, specifically the short arginine-rich sequence GRKKRRQRRR that is known for enhancing the intracellular delivery of low molecular weight drugs, proteins, oligonucleotides, and DNA plasmids (Figure 5.4).[89] A similar approach resulted in conjugation of poly[oligo(ethylene glycol) acrylate] to oligopeptides containing the particularly useful arginine–glycine–aspartic acid (RGD) sequence.[63] In these studies, the protecting side-groups of the amino acids were not cleaved after solid-phase synthesis to allow sufficient solubility of the peptide segment to facilitate size exclusion chromatography in organic eluents (Figure 5.4).

This precaution was not necessary for efficient synthesis since CuAAC is a highly chemoselective reaction. Indeed, unprotected peptides can be directly employed in click reactions.[37,50,88] Nevertheless, oligopeptides may also be deprotected after performing a

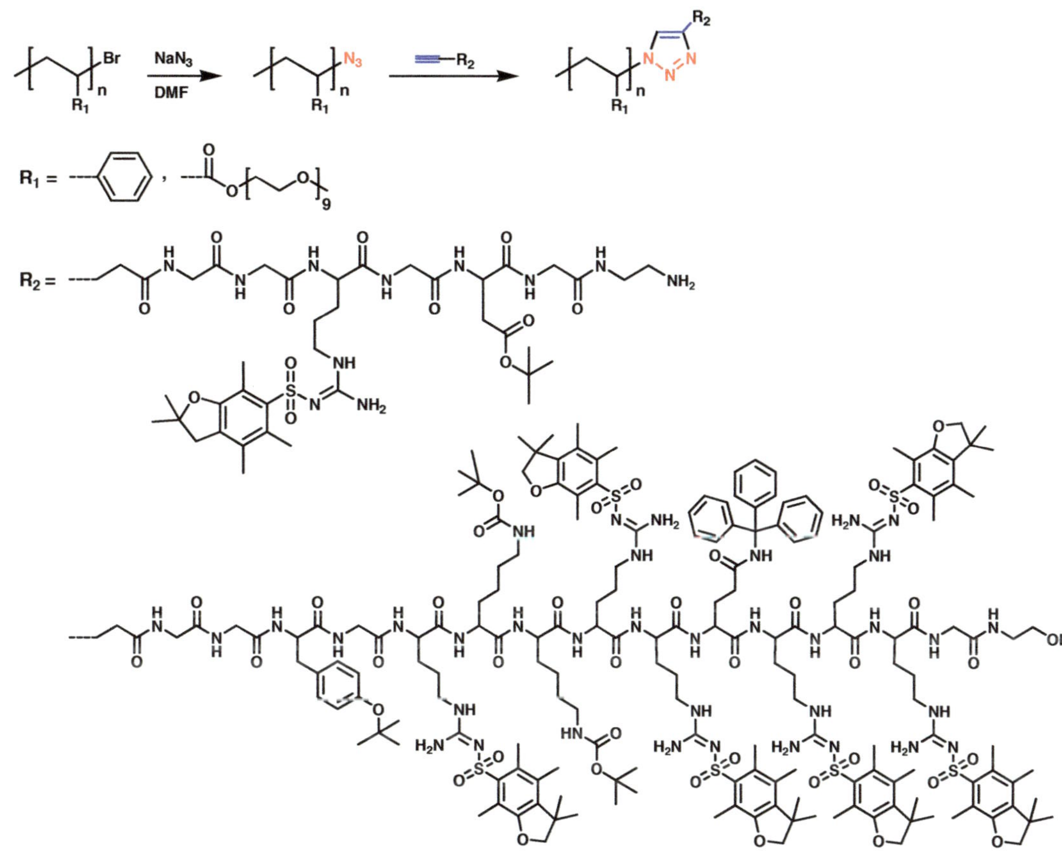

Figure 5.4 Examples of click bioconjugation of synthetic polymers prepared by ATRP with sequence-defined oligopeptides.[63,89] Reprinted with permission from J.-F. Lutz, H. G. Börner, (2008), Modern trends in polymer bioconjugates design, Progr. Polym. Sci., **33** (1), 1–39. Copyright 2008 Elsevier.

CuAAC as triazole rings are generally not damaged in concentrated trifluoroacetic acid.[16,20] Besides peptide bioconjugation, CuAAC has also been employed to functionalize linear or dendritic synthetic macromolecules with carbohydrates (e.g. mannose, galactose, fucose or lactose moieties)[17,60,90] or cancer-targeting folate ligands.[91,92]

More complex biological entities such as proteins, enzymes, viruses, bacteria or cells may also be functionalized by azide–alkyne chemistry.[93–98] For example, Finn and coworkers efficiently modified the surface of the cowpea mosaic virus using CuAAC.[96] Such reactions can be performed with experimental conditions compatible with biological environments (e.g. aqueous medium and room temperature). However, such chemical modifications of biological assemblies should be cautiously characterized since many reactants or catalysts may induce denaturation or disassembly.[96] Additionally, the team of Nolte reported interesting examples of protein conjugation [either transport proteins such as bovine serum albumin (BSA) or enzymes such as lipases] using CuAAC to couple polymers prepared by ATRP.[50,99] Sumerlin and coworkers later employed a similar concept by labeling BSA with an alkyne by reaction of its lone free cysteine with propargyl maleimide.[100] The resulting activated protein was coupled with azido-terminated PNIPAM, prepared by RAFT, to yield well-defined polymer–protein bioconjugates capable of temperature-responsive self-assembly (Scheme 5.6). Alternatively, Cornelissen and Nolte combined click chemistry and a cofactor reconstitution approach for building polymer-protein bioconjugates. Well-defined diblock copolymers PEG-*b*-PS were linked to the heme cofactor via azide–alkyne cycloaddition and subsequently reconstituted with either myoglobin or horse radish peroxidase (Figure 5.5).[101] This site-specific cofactor strategy was more versatile than the direct click coupling approach since various proteins can be functionalized with the same cofactor and there was no requirement for the transition metal catalysts in the presence of the proteins. Depending on the nature of the protein and the length of the polymer segment, a wide diversity of solution aggregate morphologies were accessible, including rods, vesicles, toroids, figure eight structures, stars and lamellar spheres.[101]

Many of the first reports detailing protein functionalization by CuAAC were by the teams of Schultz and Tirrell.[93,94,98,102] Their modification approaches relied on the use of nonnatural amino acids containing azido moieties (e.g. *para*-azidophenylalanine, azidohomoalanine, azidonorvaline or azidonorleucine), which were incorporated into

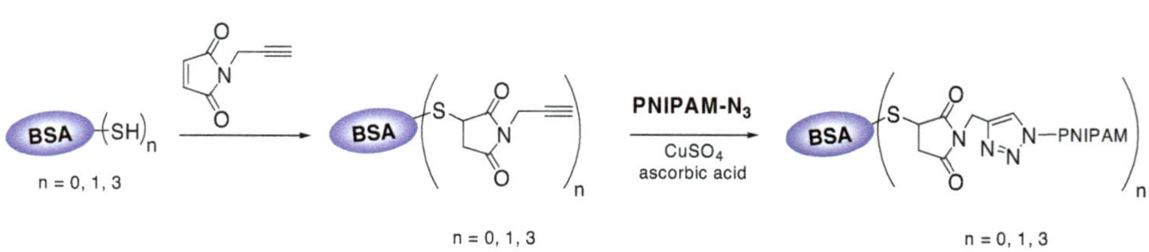

Scheme 5.6 *Polymer–protein bioconjugates prepared by CuAAC of alkyne-functionalized bovine serum albumin and azido-terminated PNIPAM.*[100] *Reprinted with permission from M. Li* et al.*, (2008), Responsive Polymer–Protein Bioconjugates Prepared by RAFT Polymerization and Copper-Catalyzed Azide–Alkyne Click Chemistry,* Macromol. Rapid Commun.*,* ***29****, 12–13, 1172–1176. Copyright 2008 Wiley-VCH.*

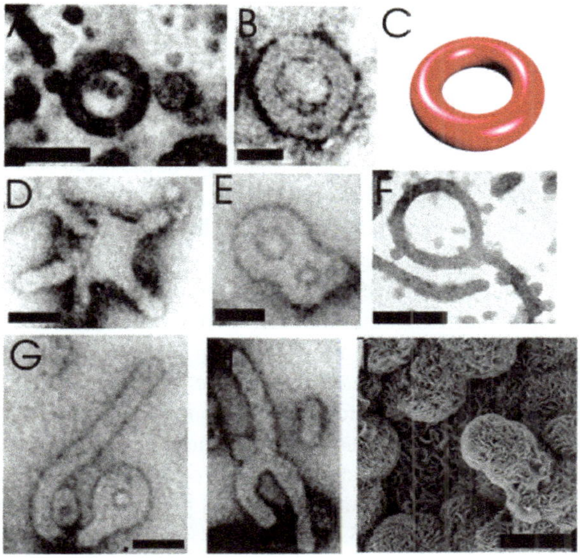

Figure 5.5 *Transmission electron microscopy images of solution aggregates composed of myoglobin conjugated with a* PS_{144}-b-PEG_{113} *block copolymer. (A, B) toroids, (C) schematic figure of a toroid, (D) octopi, (E) figure eights, and (F, G and H) micellar aggregates. (I) Scanning electron microscopy images of spherical aggregates consisting of lamellae. Bars represent 100 nm for A–H and 500 nm for I. Reprinted with permission from I. C. Reynhout,* et al.*, (2007), Self-assembled architectures from Biohybrid Triblock Copolymers,* J. Am. Chem. Soc.*,* ***129*** *(8), 2327–2332. Copyright 2007 American Chemical Society.*

mutant proteins by either genetic engineering or the metabolic replacement of a natural amino acid by a noncanonical substitute.[93,103] The resulting azido-functional proteins were subsequently reacted with various functional alkynes. Such ligations were also directly performed on cell surfaces after mutation of membrane proteins.[94,98] The click strategy of Tirrell *et al.* was also applied to distinguish recent proteins from old proteins in mammalian cells.[104] In this approach, only the newly synthesized proteins contained azidohomoalanine and were therefore selectively labeled by an alkyne affinity tag.

5.5 Functional Nanomaterials

In addition to proving highly relevant in the fields of polymer synthesis and polymer-modified biological materials, highly efficient and selective click reactions have made an enormous impact in nanoscience. As compared with macromolecular products prepared by covalent means, modern nanomaterials often rely on fragile supramolecular construction, and are therefore not easily purified or isolated. In this context, straightforward *in situ* reactions are invaluable in materials science. Thus, the versatility of CuAAC that allows it to be performed at room temperature, in multiple solvents, with stoichiometric amounts of reactants and in the presence of other functional materials has led to it being an attractive tool for nanomaterial synthesis and modification.

CuAAC has proven to be a versatile tool for functionalizing or crosslinking colloidal objects such as polymer, lipid or inorganic nanoparticles.[11,105–109] Wooley *et al.* obtained shell-crosslinked polymeric micelles by reacting alkyne moieties present in the hydrophilic outer-shell of the micelles with azide-functional first generation dendrimers.[11] Schuber and Kros *et al.* reported elegant pathways for efficiently functionalizing lipid vesicles, employing alkyne-functional surfactants that were incorporated in the lipid bilayers and further reacted with azide-containing molecules.[105,107] The mild conditions used in the reaction allowed the lipid membranes to not be damaged during the functionalization process, and only the outer surfaces of the vesicles were modified. Additionally, a few routes were reported for the click functionalization of inorganic nanoparticles with polymers.[108,109] In particular, Turro and coworkers coated the surface of maghemite particles with either azide- or alkyne functional ligands, which were evidenced to be highly versatile platforms for further functionalization.[109] Besides spherical particles, anisotropic objects such as nanotubes were also modified by click chemistry. Adronov *et al.* reported an efficient route to functionalize single-walled carbon nanotubes with polymers to promote their colloidal dispersion in organic solvents (Figure 5.6).[52]

The azide–alkyne ligation has also been employed to prepare various types of bulk materials.[15,16,18,21,110] The teams of Hawker and Hilborn synthesized poly(ethylene glycol) or poly(vinyl alcohol) hydrogels that were crosslinked by triazoles.[15,110] Similar chemistry was used for attaching ligands onto gel bead surfaces to be utilized in affinity chromatography or electrophoresis.[16,21]

In addition to nanoparticles and nanotubes, CuAAC has proven to be an excellent tool for functionalizing flat surfaces. Collman and Chidsey were pioneers in this area and reported several important examples of self-assembled monolayers (SAMs) functionalized by triazole linkages.[111–114] Their work primarily focused on gold surfaces but was extended by other groups to different types of substrates such as silicon wafers or glass slides.[115,116]

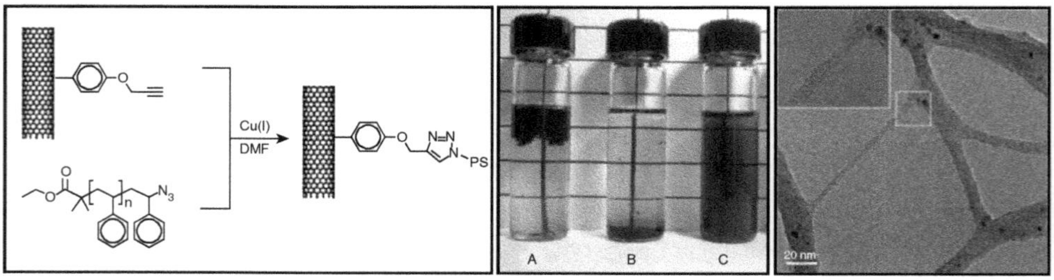

Figure 5.6 *Preparation of polystyrene-modified single walled-carbon nanotubes using click chemistry (left). The middle image shows THF solutions of either pristine (A), alkyne-functional (B) or polymer modified (C) carbon nanotubes. The right image shows transmission electron micrographs of the polymer-nanotubes organic/inorganic hybrid structures. Reprinted with permission from the H. Li,* et al., *(2005), Functionalization of Single-Walled Carbon Nanotubes with Well-Defined Polystyrene by Click Coupling,* J. Am. Chem. Soc., ***127*** *(41), 14518–14524. Copyright 2005 American Chemical Society.*

Overall, a very wide variety of functional molecules (synthetic or biological) have been already attached to SAMs using an azide–alkyne strategy, thus opening a wide range of opportunities for applications such as molecular electronic, catalysis or bio-sensors.[111,112,115–118] Besides the functionalization of SAMs, the Huisgen cycloaddition of alkynes and azides was also utilized for constructing polymer modified surfaces. For example, Caruso and coworkers described a click layer-by-layer approach using alternating layers of alkyne- and azide-functional polymers to construct defined polymer films on gold, silicon, and quartz surfaces.[44]

In addition to the traditional CuAAC conditions being employed for surface functionalization, Lahann *et al.* described the spatial control of click cycloadditions on flat surfaces using microcontact printing.[119,120] Defined biotin functional surface patterns were prepared using a poly(dimethylsiloxane) (PDMS) stamp that was inked with a solution of copper sulfate. The stamp was used to locally catalyze the cycloaddition of an adsorbed alkyne polymer and biotin azide.[119] Alternatively, Reinhoudt and coworkers used an alkyne-inked PDMS stamp for creating a variety of functional patterns on azido SAMs (Figure 5.7).[120] Interestingly, this approach did not necessitate a metal catalyst. Because of the high local concentration of reactants in the confined regions between the stamp and the substrate, the azide–alkyne cycloaddition occurred spontaneously within a short period of time.

5.6 Summary and Outlook

Since the renaissance of Huisgen's azide–alkyne cycloaddition was initiated by Meldal and Sharpless, CuAAC has become one of favorite ligation tools of polymer and materials scientists, proving particularly useful in areas as diverse as polymer synthesis, molecular biology and nanoelectronics. While not all of the examples above employ conditions that truly meet the criteria of click chemistry (especially the requirement for equal reactant stoichiometry), it is clear that CuAAC in particular has provided a versatile polymerization/functionalization procedure that is widely applicable, exceedingly efficient and highly orthogonal. Perhaps in no other area of the chemical sciences are these aspects as important as in polymer science. It may be true that the lasting legacy of the click concept in this area is that it has refocused the community to encourage the use of only the most efficient routes available to prepare complex macromolecular structures.

Nevertheless, there are still limitations of the CuAAC reaction that must be considered before considering its use. While perhaps mostly a problem of perception, the use of a copper-based catalyst could pose problems in some sensitive and highly regulated applications. Thus, the further development and enhanced applicability of optimized catalytic methods[59] or metal-free reactions are important issues in this field.[121–124] Similarly, while the majority of reports of click chemistry being used in polymer science have certainly involved the CuAAC reaction, many other reactions offer the efficiency and selectivity necessary to enhance synthetic capabilities. Thus, the search for and further development of alternative versatile reactions are necessary to provide a more complete toolbox of click reactions for the future.

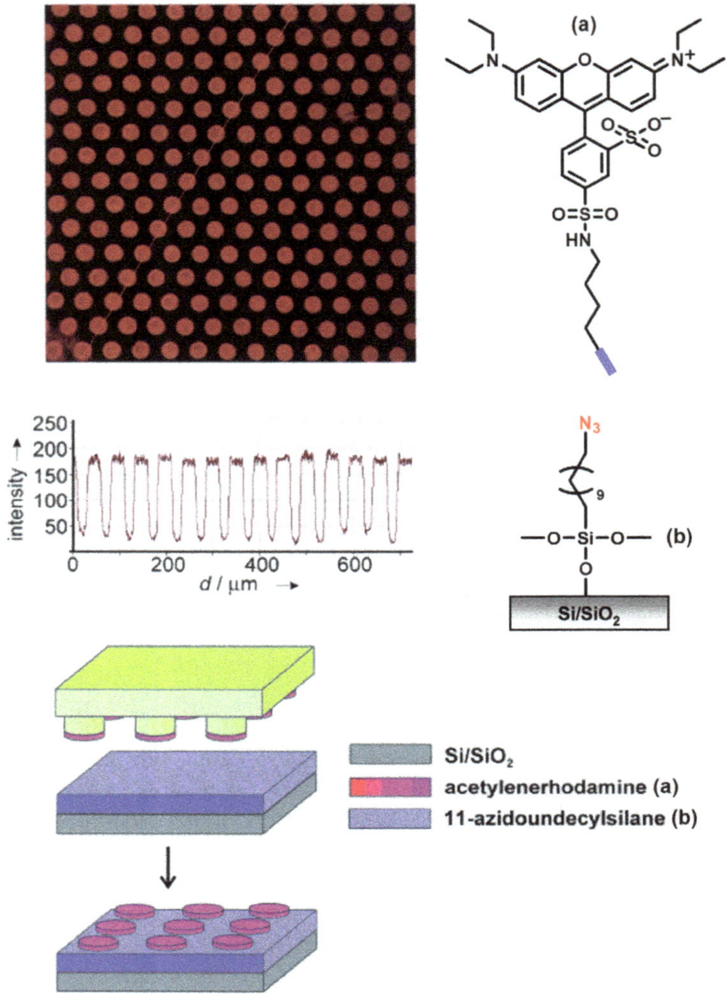

Figure 5.7 *Microcontact click printing in the absence of a copper catalyst. The image on top left shows the visualization of the surface pattern by fluorescence microscopy (image width is 700 mm). Reprinted with permission from D. I. Rozkiewicz,* et al., *(2006), Click Chemistry by Microcontact Printing,* Angew. Chem., ***45*** *(32), 5292–5296. Copyright 2006 Wiley-VCH.*

References

(1) Hawker, C. J., Wooley, K. L., (2005), *Science*, **309**, 1200.
(2) Kolb, H. C., Finn, M. G., Sharpless, K. B., (2001), *Angew. Chem. Int. Ed.*, **40**, 2004.
(3) Huisgen, R., (1963), *Angew. Chem. Int. Ed. Engl.*, **2**, 565.
(4) Huisgen, R., (1963), *Angew. Chem. Int. Ed. Engl.*, **2**, 633.

(5) Rostovtsev, V. V., Green, L. G., Fokin, V. V., Sharpless, K. B., (2002), *Angew. Chem. Int. Ed.*, **41**, 2596.
(6) Tornoe, C. W., Christensen, C., Meldal, M., (2002), *J. Org. Chem.*, **67**, 3057.
(7) Binder, W. H., Kluger, C., (2006), *Curr. Org. Chem.*, **10**, 1791.
(8) Wu, P., Feldman, A. K., Nugent, A. K., Hawker, C. J., Scheel, A., Voit, B., Pyun, J., Frechet, J. M. J., Sharpless, K. B., Fokin, V. V., (2004), *Angew. Chem. Int. Ed.*, **43**, 3928.
(9) Díaz, D. D., Punna, S., Holzer, P., McPherson, A. K., Sharpless, K. B., Fokin, V. V., Finn, M. G., (2004), *J. Polym. Sci. Part A: Polym. Chem.*, **42**, 4392.
(10) Helms, B., Mynar, J. L., Hawker, C. J., Frechet, J. M. J., (2004), *J. Am. Chem. Soc.*, **126**, 15020.
(11) Joralemon, M. J., O'Reilly, R. K., Hawker, C. J., Wooley, K. L., (2005), *J. Am. Chem. Soc.*, **127**, 16892.
(12) Joralemon, M. J., O'Reilly, R. K., Matson, J. B., Nugent, A. K., Hawker, C. J., Wooley, K. L., (2005), *Macromolecules*, **38**, 5436.
(13) Malkoch, M., Schleicher, K., Drockenmuller, E., Hawker, C. J., Russell, T. P., Wu, P., Fokin, V. V., (2005), *Macromolecules*, **38**, 3663.
(14) Malkoch, M., Thibault, R. J., Drockenmuller, E., Messerschmidt, M., Voit, B., Russell, T. P., Hawker, C. J., (2005), *J. Am. Chem. Soc.*, **127**, 14942.
(15) Malkoch, M., Vestberg, R., Gupta, N., Mespouille, L., Dubois, P., Mason, A. F., Hedrick, J. L., Liao, Q., Frank, C. W., Kingsbury, K., Hawker, C. J., (2005), *Chem. Commun.*, 2774.
(16) Punna, S., Kaltgrad, E., Finn, M. G., (2005), *Bioconjugate Chem.*, **16**, 1536.
(17) Wu, P., Malkoch, M., Hunt, J. N., Vestberg, R., Kaltgrad, E., Finn, M. G., Fokin, V. V., Sharpless, K. B., Hawker, C. J., (2005), *Chem. Commun.*, 5775.
(18) Diaz, D. D., Rajagopal, K., Strable, E., Schneider, J., Finn, M. G., (2006), *J. Am. Chem. Soc.*, **128**, 6056.
(19) Johnson, J. A., Lewis, D. R., Diaz, D. D., Finn, M. G., Koberstein, J. T., Turro, N. J., (2006), *J. Am. Chem. Soc.*, **128**, 6564.
(20) Li, C., Finn, M. G., (2006), *J. Polym. Sci. Part A: Polym. Chem.*, **44**, 5513.
(21) Slater, M., Snauko, M., Svec, F., Frechet, J. M. J., (2006), *Anal. Chem.*, **78**, 4969.
(22) Thibault, R. J., Takizawa, K., Lowenheilm, P., Helms, B., Mynar, J. L., Frechet, J. M. J., Hawker, C. J., (2006), *J. Am. Chem. Soc.*, **128**, 12084.
(23) Binder, W. H., Sachsenhofer, R., (2007), *Macromol. Rapid Commun.*, **28**, 15.
(24) Evans, R. A., (2007), *Aust. J. Chem.*, **60**, 384.
(25) Lutz, J.-F., (2007), *Angew. Chem. Int. Ed.*, **46**, 1018.
(26) Lutz, J.-F., Schlaad, H., (2008), *Polymer*, **49**, 817.
(27) van Steenis, D. J. V. C., David, O. R. P., van Strijdonck, G. P. F., van Maarseveen, J. H., Reek, J. N. H., (2005), *Chem. Commun.*, 4333.
(28) Dichtel, W. R., Miljanic, O. S., Spruell, J. M., Heath, J. R., Stoddart, J. F., (2006), *J. Am. Chem. Soc.*, **128**, 10388.
(29) Zhu, Y., Huang, Y., Meng, W.-D., Li, H., Qing, F.-L., (2006), *Polymer*, **47**, 6272.
(30) Binauld, S., Boisson, F., Hamaide, T., Jean-Pierre Pascault, Drockenmuller, E., Fleury, E., (2008), *J. Polym. Sci. Part A: Polym. Chem.*, **46**, 5506.
(31) Tsarevsky, N. V., Bernaerts, K. V., Dufour, B., Du Prez, F. E., Matyjaszewski, K., (2004), *Macromolecules*, **37**, 9308.
(32) Tsarevsky, N. V., Sumerlin, B. S., Matyjaszewski, K., (2005), *Macromolecules*, **38**, 3558.
(33) Golas, P. L., Tsarevsky, N. V., Sumerlin, B. S., Matyjaszewski, K., (2007), *Aust. J. Chem.*, 60.
(34) Bakbak, S., Leech, P. J., Carson, B. E., Saxena, S., King, W. P., Bunz, U. H. F., (2006), *Macromolecules*, **39**, 6793.
(35) Matyjaszewski, K., (2005), *Prog. Polym. Sci.*, **30**, 858.
(36) Sumerlin, B. S., Tsarevsky, N. V., Gao, H., Golas, P., Louche, G., Lee, R. Y., Matyjaszewski, K., (2006), Click functionalization of well-defined copolymers prepared by atom transfer radical polymerization. In *Controlled/Living Radical Polymerization: From Synthesis to Materials*, ACS: Washington, DC, Vol. 944.
(37) Parrish, B., Breitenkamp, R., Emrick, T., (2005), *J. Am. Chem. Soc.*, **127**, 7404.

(38) Riva, R., Schmeits, S., Stoffelbach, F., Jerome, C., Jerome, R., Lecomte, P., (2005), *Chem. Commun.*, 5334.
(39) Hoogenboom, R., Moore, B. C., Schubert, U. S., (2006), *Chem. Commun.*, 4010.
(40) Luxenhofer, R., Jordan, R., (2006), *Macromolecules*, **39**, 3509.
(41) Binder, W. H., Kluger, C., (2004), *Macromolecules*, **37**, 9321.
(42) Englert, B. C., Bakbak, S., Bunz, U. H. F., (2005), *Macromolecules*, **38**, 5868.
(43) Wang, X.-Y., Kimyonok, A., Weck, M., (2006), *Chem. Commun.*, 3933.
(44) Such, G. K., Quinn, J. F., Quinn, A., Tjipto, E., Caruso, F., (2006), *J. Am. Chem. Soc.*, **128**, 9318.
(45) Gondi, S. R., Vogt, A. P., Sumerlin, B. S., (2007), *Macromolecules*, **40**, 474.
(46) Quémener, D., Davis, T. P., Barner-Kowollik, C., Stenzel, M. H., (2006), *Chem. Commun.*, 5051.
(47) O'Reilly, R. K., Joralemon, M. J., Hawker, C. J., Wooley, K. L., (2006), *Chem. Eur. J.*, **12**, 6776.
(48) O'Reilly, R. K., Joralemon, M. J., Hawker, C. J., Wooley, K. L., (2006), *J. Polym. Sci. Part A: Polym. Chem.*, **44**, 5203.
(49) Li, M., De, P., Gondi, S. R., Sumerlin, B. S., (2008), *J. Polym. Sci. Part A: Polym. Chem.*, **46**, 5093.
(50) Dirks, A. J., vanBerkel, S. S., Hatzakis, N. S., Opsten, J. A., vanDelft, F. L., Cornellissen, J. J. L. M., Rowan, A. E., vanHest, J. C. M., Rutjes, F. P. J. T., Nolte, R. J. M., (2005), *Chem. Commun.*, **333**, 4172.
(51) Gao, H., Louche, G., Sumerlin, B. S., Jahed, N., Golas, P., Matyjaszewski, K., (2005), *Macromolecules*, **38**, 8979.
(52) Li, H., Cheng, F., Duft, A. M., Adronov, A., (2005), *J. Am. Chem. Soc.*, **127**, 14518.
(53) Lutz, J.-F., Börner, H. G., Weichenhan, K., (2005), *Macromol. Rapid Commun.*, **26**, 514.
(54) Mantovani, G., Ladmiral, V., Tao, L., Haddleton, D. M., (2005), *Chem. Commun.*, 2089.
(55) Opsteen, J. A., van Hest, J. C. M., (2005), *Chem. Commun.*, 57.
(56) Sumerlin, B. S., Tsarevsky, N. V., Louche, G., Lee, R. Y., Matyjaszewski, K., (2005), *Macromolecules*, **38**, 7540.
(57) Altintas, O., Hizal, G., Tunca, U., (2006), *J. Polym. Sci. Part A: Polym. Chem.*, **44**, 5699.
(58) Gao, H., Matyjaszewski, K., (2006), *Macromolecules*, **39**, 4960.
(59) Golas, P., Tsarevsky, N. V., Sumerlin, B. S., Matyjaszewski, K., (2006), *Macromolecules*, **39**, 6451.
(60) Ladmiral, V., Mantovani, G., Clarkson, G. J., Cauet, S., Irwin, J. L., Haddleton, D. M., (2006), *J. Am. Chem. Soc.*, **128**, 4823.
(61) Laurent, B. A., Grayson, S. M., (2006), *J. Am. Chem. Soc.*, **128**, 4238.
(62) Liu, Q., Chen, Y., (2006), *J. Polym. Sci. Part A: Polym. Chem.*, **44**, 6103.
(63) Lutz, J.-F., Börner, H. G., Weichenhan, K., (2006), *Macromolecules*, **39**, 6376.
(64) Vogt, A. P., Sumerlin, B. S., (2006), *Macromolecules*, **39**, 5286.
(65) Whittaker, M. R., Urbani, C. N., Monteiro, M. J., (2006), *J. Am. Chem. Soc.*, **128**, 11360.
(66) Golas, P. L., Matyjaszewski, K., (2007), *QSAR Comb. Sci.*, **26**, 1116.
(67) Matyjaszewski, K., Xia, J., (2001), *Chem. Rev.*, **101**, 2921.
(68) Coessens, V., Matyjaszewski, K., (1999), *J. Macromol. Sci., Pure Appl. Chem.*, **A36**, 667.
(69) Coessens, V., Nakagawa, Y., Matyjaszewski, K., (1998), *Polym. Bull.*, **40**, 135.
(70) Coessens, V., Pintauer, T., Matyjaszewski, K., (2001), *Prog. Polym. Sci.*, **26**, 337.
(71) Geng, J., Lindqvist, J., Mantovani, G., Haddleton, D. M., (2008), *Angew. Chem. Int. Ed.*, **47**, 4180.
(72) Chan, T. R., Hilgraf, R., Sharpless, K. B., Fokin, V. V., (2004), *Org. Lett.*, **6**, 2853.
(73) Rodionov, V. O., Fokin, V. V., Finn, M. G., (2005), *Angew. Chem. Int. Ed.*, **44**, 2210.
(74) Hassner, A., Stern, M., *Angew. Chem. Int. Ed. Engl.* (1986), **25**, 478.
(75) Bräse, S., Gil, C., Knepper, K., Zimmermann, V., (2005), *Angew. Chem. Int. Ed.*, **44**, 5188.
(76) Tsarevsky, N. V., Bencherif, S. A., Matyjaszewski, K., (2007), *Macromolecules*, **40**, 4439.
(77) Goldmann, A. S., Quemener, D., Millard, P.-E., Davis, T. P., Stenzel, M. H., Barner-Kowollik, C., Mueller, A. H. E., (2008), *Polymer*, **49**, 2274.
(78) Rao, J., Zhang, Y., Zhang, J., Liu, S., (2008), *Biomacromolecules*, **9**, 10.1021/bm800462q.

(79) Gacal, B., Durmaz, H., Tasdelen, M. A., Hizal, G., Tunca, U., Yagci, Y., Demirel, A. L., (2006), *Macromolecules*, **39**, 5330.
(80) Vogt, A. P., Gondi, S. R., Sumerlin, B. S., (2007), *Aust. J. Chem* , **60**, 396.
(81) Lee, J. W., Kim, J. H., Kim, B.-K., (2005), *Tetrahedron Lett.*, **47**, 2683.
(82) Lee, J. W., Kim, J. H., Kim, B.-K., Kim, J. H., Shin, W. S., Jin, S -H., (2006), *Tetrahedron*, **62**, 9193.
(83) Lutz, J.-F., Börner, H. G., (2008), *Prog. Polym. Sci.*, **33**, 1.
(84) Seo, T. S., Li, Z., Ruparel, H., Ju, J., (2003), *J. Org. Chem.*, **68**, 609.
(85) Burley, G. A., Gierlich, J., Mofid, M. R., Nir, H., Tal, S., Eichen, Y., Carell, T., (2006), *J. Am. Chem. Soc.*, **128**, 1398.
(86) Liebert, T., Hänsch, C., Heinze, T., (2006), *Macromol. Rapid Commun.*, **27**, 208.
(87) Droumaguet, B. L., Velonia, K., (2008), *Macromol. Rapid Commun.*, **29**, 1073–1089.
(88) Rijkers, D. T. S., van Esse, G. W., Merkx, R., Brouwer, A. J., Jacobs, H. J. F., Pieters, R. J., Liskamp, R. M. J., (2005), *Chem. Commun.*, 581.
(89) Lutz, J.-F., Börner, H. G., Weichenhan, K., (2007), *Aust. J. Chem.*, **60**, 410.
(90) Fernandez-Megia, E., Correa, J., Rodriguez-Meizoso, I., Riguera, R., (2006), *Macromolecules*, **39**, 2113.
(91) McCormick, C. L., Sumerlin, B. S., Lokitz, B. S., Stempka, J. E., (2008), *Soft Matter*, DOI: 10.1039/b719577j.
(92) De, P., Gondi, S. R., Sumerlin, B. S., (2008), *Biomacromolecules*, **9**, 1064.
(93) Deiters, A., Cropp, T. A., Mukherji, M., Chin, J. W., Anderson, J. C., Schultz, P. G., (2003), *J. Am. Chem. Soc.*, **125**, 11782.
(94) Link, A. J., Tirrell, D. A., (2003), *J. Am. Chem. Soc.*, **125**, 11164.
(95) Speers, A. E., Adam, G. C., Cravatt, B. F., (2003), *J. Am. Chem. Soc.*, **125**, 4686.
(96) Wang, Q., Chan, T. R., Hilgraf, R., Fokin, V. V., Sharpless, K. B., Finn, M. G., (2003), *J. Am. Chem. Soc.*, **125**, 3192.
(97) Agard, N. J., Prescher, J. A., Bertozzi, C. R., (2004), *J. Am. Chem. Soc.*, **126**, 15046.
(98) Link, A. J., Vink, M. K. S., Tirrell, D. A., (2004), *J. Am. Chem. Soc.*, **126**, 10598.
(99) Brennan, J. L., Hatzakis, N. S., Tshikhudo, T. R., Dirvianskyite, N., Razumas, V., Patkar, S., Vind, J., Svendsen, A., Nolte, R. J. M., Rowan, A. E., Brust, M., (2006), *Bioconjugate Chem.*, **17**, 1373.
(100) Li, M., De, P., Gondi, S. R., Sumerlin, B. S., (2008), *Macromol. Rapid Commun.*, **29**, 1172.
(101) Reynhout, I. C., Cornelissen, J. J. L. M., Nolte, R. J. M., (2007), *J. Am. Chem. Soc.*, **129**, 2327.
(102) Deiters, A., Cropp, T. A., Summerer, D., Mukherji, M., Schultz, P. G., (2004), *Bioorg. Med. Chem. Lett.*, **14**, 5743.
(103) Kiick, K. L., Saxon, E., Tirrell, D. A., Bertozzi, C. R., (2002), *Proc. Natl Aacad. Sci. USA*, **99**, 19.
(104) Dieterich, D. C., Link, A. J., Graumann, J., Tirrell, D. A., Schuman, E. M., (2006), *Proc. Natl Aacad. Sci. USA*, **103**, 9482.
(105) Cavalli, S., Tipton, A. R., Overhand, M., Kros, A., (2006), *Chem. Commun.*, 3193.
(106) O'Reilly, R. K., Joralemon, M. J., Hawker, C. J., Wooley, K. L., (2006), *Chem. Eur. J.*, **12**, 6776.
(107) Said Hassane, F., Frisch, B., Schuber, F., (2006), *Bioconjugate Chem.*, **17**, 849.
(108) Sun, E. Y., Josephson, L., Weissleder, R., (2006), *Molecular Im.*, **5**, 122.
(109) White, M. A., Johnson, J. A., Koberstein, J. T., Turro, N. J., (2006), *J. Am. Chem. Soc.*, **128**, 11356.
(110) Ossipov, D. A., Hilborn, J., (2006), *Macromolecules*, **39**, 1709.
(111) Collman, J. P., Devaraj, N. K., Chidsey, C. E. D., (2004), *Langmuir*, **20**, 1051.
(112) Devaraj, N. K., Miller, G. P., Ebina, W., Kakaradov, B., Collman, J. P., Kool, E. T., Chidsey, C. E. D., (2005), *J. Am. Chem. Soc.*, **127**, 8600.
(113) Collman, J. P., Devaraj, N. K., Eberspacher, T. P. A., Chidsey, C. E. D., (2006), *Langmuir*, **22**, 2457.
(114) Devaraj, N. K., Decreau, R. A., Ebina, W., Collman, J. P., Chidsey, C. E. D., (2006), *J. Phys. Chem. B.*, **110**, 15955.
(115) Lummerstorfer, T., Hoffmann, H., (2004), *J. Phys. Chem. B.*, **108**, 3963.

(116) Sun, X.-L., Stabler, C. L., Cazalis, C. S., Chaikof, E. L., (2006), *Bioconjugate Chem.*, **17**, 52.
(117) Lee, J. K., Chi, Y. S., Choi, I. S., (2004), *Langmuir*, **20**, 3844.
(118) Zhang, Y., Luo, S., Tang, Y., Yu, L., Hou, K.-Y., Cheng, J.-P., Zeng, X., Wang, P. G., (2006), *Anal. Chem.*, **78**, 2001.
(119) Nandivada, H., Chen, H.-Y., Bondarenko, L., Lahann, J., (2006), *Angew. Chem. Int. Ed.*, **45**, 3360.
(120) Rozkiewicz, D. I., Janacuteczewski, D., Verboom, W., Ravoo, B. J., Reinhoudt, D. N., (2006), *Angew. Chem. Int. Ed.*, **45**, 5292.
(121) Baskin, J. M., Prescher, J. A., Laughlin, S. T., Agard, N. J., Chang, P. V., Miller, I. A., Lo, A., Codelli, J. A., Bertozzi, C. R., (2007), *Proc. Natl Acad. Sci. USA*, **104**, 16793.
(122) Baskin, J. M., Bertozzi, C. R., (2007), *QSAR Comb. Sci.*, **26**, 1211.
(123) Sletten, E. M., Bertozzi, C. R., (2008), *Org. Lett.*, **10**, 3097.
(124) Johnson, J. A., Baskin, J. M., Bertozzi, C. R., Koberstein, J. T., Turro, N. J., (2008), *Chem. Commun.*, 3064.

6

Blocks, Stars and Combs: Complex Macromolecular Architecture Polymers via Click Chemistry

Sebastian Sinnwell, Andrew J. Inglis, Martina H. Stenzel and Christopher Barner-Kowollik

6.1 Introduction

The formation of complex macromolecular architectures, such as blocks, stars and combs,* has been a constant theme in synthetic polymer chemistry over the last few decades. In the past, anionic polymerization was the method of choice for the synthesis of well-defined complex (co)polymer architectures, yet under relatively demanding reaction conditions.[1] More recently, the development of living/controlled radical polymerization protocols have made complex macromolecular construction a far less demanding feat. Amongst the most prominent living/controlled radical polymerization methods are the reversible addition-fragmentation transfer (RAFT) process,[2–5] nitroxide-mediated polymerization (NMP)[6] as well as atom transfer radical polymerization (ATRP).[7] RAFT and ATRP are the most employed techniques, with RAFT arguably being the most versatile in terms of monomer range and functionality tolerance.

All of the above three control methodologies on their own can not only be employed to construct linear block copolymers, but also more complicated macromolecular architectures such stars and comb-shaped entities.[6,8,9] Typically, multifunctional molecular scaffolds are employed for this purpose, carrying bromine endgroups (for ATRP), nitroxides (for NMP)

* The term *comb* is taken to mean a polymer which consists of a main chain and a plurality of long side chains.

Click Chemistry for Biotechnology and Materials Science Edited by Joerg Lahann

or thiocarbonly thio entities (for RAFT). Theoretically, it should be possible to generate any combination of block copolymers via these techniques. Unfortunately, the process of generating block structures and star polymers via such approaches has its limits and not all desired structures can be obtained. The reasons for the limitations in block copolymer formation are multipronged: (i) the reactivity and type of the monomers from which the polymeric building blocks are synthesized may be very different. In using one of the forms of living/controlled radical polymerization, for example, the reactivities of two monomers must be comparable in order to achieve block structures through chain extension. The consequences of vastly different monomer reactivities can be illustrated by the example of the preparation of a (hypothetical) block copolymer made from vinyl acetate and styrene. The combination of the highly reactive vinyl acetate derived radical and the relatively unreactive styrene derived radical precludes the use of a single controlling agent or initiating system to achieve a well-defined block structure. To date, the most convenient method to polymerize vinyl acetate in a controlled fashion is by the RAFT process, using a xanthate controlling agent.[10] However, this methodology is not appropriate for polymerizing most other monomers. Additionally, in order to synthesize polymer conjugates from different classes of monomers (e.g. vinylic monomers and lactones), polymerization initiators/controlling agents must be specifically designed to perform different techniques of polymerization, e.g. an NMP initiator equipped with a hydroxyl moiety to perform ring-opening polymerization (ROP).[11] (ii) It is often desirable to generate block copolymers that are amphiphilic, i.e. featuring a hydrophobic and a hydrophilc component. Such amphiphilic block copolymers can be assembled in solution into micellar structures or vesicles, which in turn can serve as containers for the target delivery of pharmaceuticals.[12–19] Alternatively, these hydrophobic/hydrophilic structures may be self-assembled in the solid state. However, due to the bipolar nature of amphiphilic block copolymers, their preparation is often limited by the identification of a common solvent in which the block copolymer synthesis can be carried out. (iii) The formation of nonlinear block copolymers (i.e. block-stars or combs) is not only beset with the above two listed problems but also with additional complications, depending upon the method employed for their construction. While in ATRP and NMP processes, radical propagation (i.e. chain growths) always occurs from the multifunctional core (and thus can lead to core-core coupling and other undesirable termination reactions), the RAFT process allows (via its so-called Z-group approach) arm growths without the interference of coupling processes. Both approaches (radical core vs RAFT Z-group approach) have disadvantages. When the core itself carries the radical functionalities, conversions have to be limited as otherwise termination products contribute significantly to the product distribution. While the RAFT-Z group approach yields multiarm polymers free from terminated impurities, the molecular weights that may be reached are limited in many cases due to the growing inaccessibility of the thiocarbonyl thio functions located at the core.

The separate preparation of the individual building blocks and their subsequent coupling to achieve the desired complex polymer structure provides an elegant solution to problems (i) and (iii). However, such an approach of post-polymerization coupling requires chemical transformations of high efficiency as well as the tolerance against a variety of functional groups and reaction conditions. These requirements are perfectly facilitated by the characteristics of click chemistry. In fact, although the concept of click chemistry was initially conceived for use in low molecular weight organic synthesis in 2001,[20] it has seen a great increase of application in polymer science since its first appearance there in 2004.[21–23] In particular, the copper(I)-catalyzed azide–alkyne cycloaddition (CuAAC) has been proven

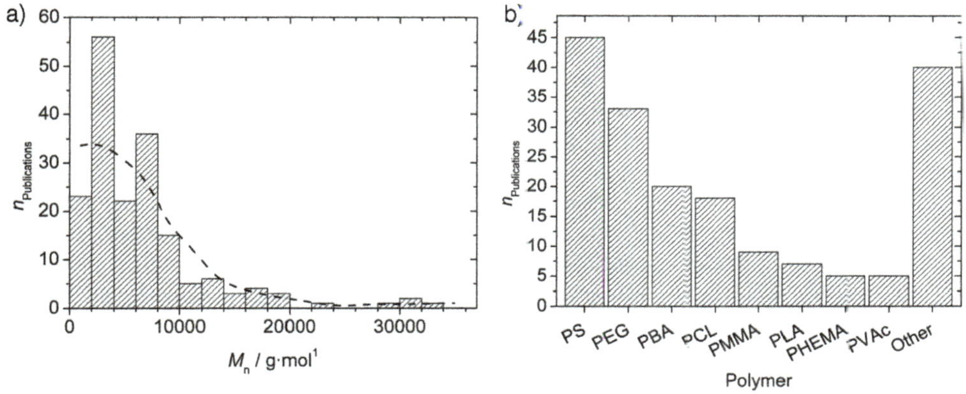

Figure 6.1 *Frequency with which particular molecular weights (a) and polymers (b) have been utilized in the literature to prepare complex architectures via click methodologies.*

as an ideal candidate to fulfill the click criteria[20] and was therefore the reaction of choice in the vast majority of publications. However, Diels–Alder cycloadditions, nucleophilic substitution chemistry of stained rings and additions to carbon–carbon multiple bonds (thiol–ene reaction) have also been shown as efficient orthogonal synthetic strategies and have been used alternatively to or in combination with the CuAAC.

Considering the above-mentioned potential of click chemistry in polymer science, one would expect a great versatility to have emerged in the synthesis of complex macromolecular architectures. It is therefore an interesting exercise to inspect the literature and to statistically explore two aspects of click chemistry for the construction of complex polymer structures. Firstly, it is instructive to establish what molecular weight ranges have thus far been utilized as building blocks in the generation of larger and more complicated structures. Figure 6.1(a) shows the number of studies that have used click chemistry to achieve block, star and comb structures and the molecular weights of the building blocks used. The figure includes the use of all types of pericyclic click reactions; however, the main preparation method has been the CuAAC. Secondly, it is interesting to take note of the types of polymers that have been employed in click reactions. As such, Figure 6.1(b) depicts the frequency in which particular polymers have been utilized in the literature to prepare complex architectures via click methodologies. By a close inspection of Figure 6.1(a and b), it is observed that the vast majority of studies concern themselves with the use of polystyrene (PS) and poly(ethylene glycol) (PEG) polymers with molecular weights below 10 000 g mol^{-1}. This may suggest that the vast majority of publications are concerned with proving the concept rather than expanding the number of available materials. In the following, we will highlight and summarize the most modern approaches that are employed to generate complex macromolecular architectures via orthogonal, selective and rapid cycloadditions.

6.2 Block Copolymers

It seems apt to start our foray into the construction of complex macromolecular architectures with block copolymers, as it was in this area where the first steps using click chemistry for the construction of complex polymers were made. Early in the introduction of click chemistry to polymer science, the synthesis of the most simple of complex architectures, blocks, was a convenient avenue towards proving the concept that click chemistry may have the potential to be useful in the construction of more complicated architectures. Furthermore, block copolymer structures and, in particular, amphiphilic block copolymers are of high interest in polymer science for their self assembly properties, (both in solution and in the solid state) as well as for their potential application in the construction of nano containers that may be of benefit in fields such as organic synthesis and targeted drug delivery.

As mentioned in the introduction, block copolymers may be synthesized through direct polymerization via chain extension in controlled radical polymerization or by the use of multifunctional initiators, in which two separate polymerization reactions occur from a single initiator. The difficulties in this strategy center around the lack of freedom in the choice of monomers that may be used. For example, monomers of significantly different reactivity cannot be used in a chain extension approach. An alternative strategy is to use highly efficient conjugation reactions (click chemistry) to link two or more polymer chains together in a post-polymerization approach.

The greatest advantage of constructing complex macromolecular architectures via click conjugations is that it allows for the complete segregation of synthetic techniques by which individual building blocks are generated. In the field of creating block structures in polymeric systems, such efficient conjugation techniques have been proven to be an invaluable addition to the collection of tools, which has been at the disposal of the organic chemist. Since one of the earliest reports on the synthesis of block structures by click reactions by van Hest *et al.*,[24] the use of click chemistry has opened an entirely new playing field in which a much wider variety of such structures may be synthesized with unprecedented ease. For example, it enables the creation of polymer blocks from monomers of very disparate reactivities such as vinyl acetate and styrene.[25] In the following, we explore how the click concept has been applied to the construction of a wide variety of block copolymer structures that have been otherwise very difficult, if not impossible, to achieve.

6.2.1 Polymer Preparation for Click Conjugations

The first step in generating block structures is to equip the individual building blocks with the complementary functional groups, which react together to establish a linkage. There are essentially two ways in which this may be performed: (i) functionalization of the initiator by which the polymer chain is generated; or (ii) post-polymerization functionalization of a polymer end-group.

Pre-polymerization Functionalization

The use of clickable initiators warrants consideration of the compatibility of the click moiety with the polymerization process. Looking at the most widely used form of click chemistry, the CuAAC, one's attention is immediately drawn to the alkyne moiety. The triple bond is susceptible to addition reactions, under conditions of polymerization reactions. For example, in anionic polymerizations, the acidic proton of terminal alkynes interferes with the highly nucleophilic initiators used for the polymerization of styrene derivatives.[26] The predominant method of generating alkyne terminated linear polymers, via pre-polymerization functionalization, has been the use of ATRP.[24,27–30] The catalyst system by which ATRP

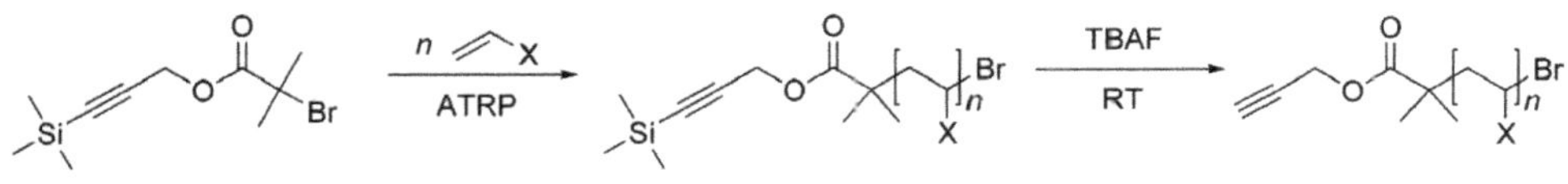

Scheme 6.1 *Pre-polymerization strategy for equipping polymer chains with alkyne moieties.*

is performed is very similar and in many instances identical to that, which is used in the CuAAC. The mechanism by which the latter proceeds involves complexation of the alkyne moiety with the copper(I) catalyst, thus the same may occur during ATRP with an alkyne equipped initiator. Therefore, it would appear necessary to introduce some form of protective chemistry to the alkyne moiety. Such protection of the alkyne moiety has mostly been achieved through the use of a trimethylsilyl (TMS) group, which can readily be removed post-polymerization, e.g. by treatment with tetrabutylammonium fluoride (TBAF) (1–10 equivalents) at room temperature (Scheme 6.1).[24,25,29] Where higher-order linear blocks, such as triblocks, are synthesized by α-azido-ω-alkyne-functionalized polymers, completely protected alkynes is essential. However, van Hest *et al.* has reported that, during an ATRP using a protected alkyne initiator, 70% of the TMS groups were removed when using the conventional copper(I) bromide–*N*,*N*,*N′*,*N″*,*N″*-pentamethyldiethylenetriamine (PMDETA) catalyst system.[29] The nucleophilic attack of one of the nitrogen atoms of the PMDETA on the silyl-group was postulated as a possible side-reaction. In an attempt to circumvent this problem, the less nucleophilic copper(I) bromide–2,2′-bipyridyl (BPy) ATRP catalyst system was trialed, which showed an improvement in that only 10% of the protective groups were removed. In another attempt to circumvent this problem, the TMS group was replaced by a triisopropyl silane group. This variant proved not to be affected by the ATRP process and, consequently, continued to protect the alkyne group.

Chain transfer agents used in RAFT polymerizations have also been equipped with alkyne groups in much the same way as ATRP initiators in the generation of linear block structures.[25,31] Here, the TMS group was also used as the protective group and was removed after the polymerization in the same manner as mentioned previously.

Whilst most authors claim the necessity of using protective chemistry in their alkyne-functionalized initiators, there have been some reports on the use of nonprotected initiators/chain transfer agents. The use of a nonprotected alkyne ATRP initiator was reported by Matyjaszewski *et al.* in the synthesis of α,ω-bifunctional polystyrene.[32] Here, a ^{1}H-NMR spectrum is provided that clearly shows the presence of the alkyne α-protons in the polymer chain; however, a quantitative analysis is not provided. Furthermore, Nasrullah *et al.*[33] and Ranjan and Brittain[34] report the use of RAFT agents equipped with nonprotected alkyne moieties. The resulting polymers were shown to bear the alkyne moiety and were successfully used in subsequent CuAAC reactions. It is also possible to use alkyne based initiators for ROP. Here, propargyl alcohol is a convenient choice and no protective chemistry is required.[35,36]

One of the hallmark features of the CuAAC is the benign characteristic of the azide group in that it may be incorporated into a polymerization initiator or chain transfer agent without it reacting during any process except for the click reaction itself. Thus, there have been many occurrences in the literature of a pre-polymerization functionalization strategy being used to generate polymers bearing an azide moiety.[25,27,31,35] A selected example of such

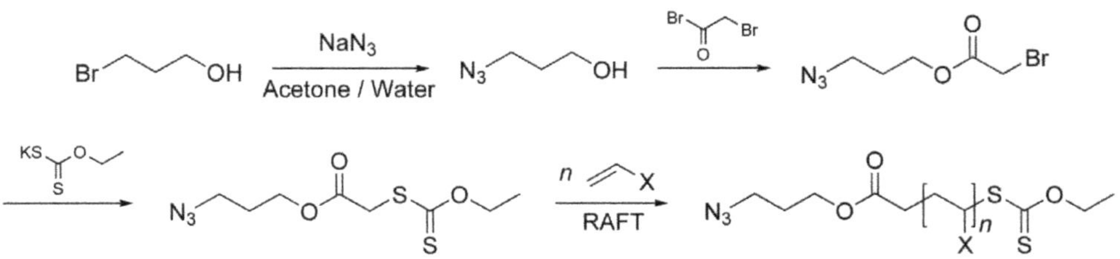

Scheme 6.2 *Pre-polymerization strategy for equipping polymer chains with azide moieties via the use of an azide functionalized RAFT (xanthate) agent.*

strategy for the preparation of an azide functionalized RAFT (xanthate) agent is presented in Scheme 6.2.[25]

However, there have been instances in the literature in which a decrease in the azide content has been observed during the polymerization process.[37–39] Most recently, Perrier *et al.* showed in a series of on-line nuclear magnetic resonance (NMR) experiments that the azide moiety can undergo a 1,3-cycloaddition with the double bond of various monomers.[40] The electron-withdrawing character of the monomers used was identified as key in determining the extent to which the loss of azide occurred. Under the conditions applied (60 °C, 20 h), the reaction of an azide with methyl acrylate yielded a 95% conversion of the azide whereas reaction with the more stable styrene achieved only 5% conversion. Therefore, when considering using an azide equipped radical polymerization initiator, the monomer that is to be used must also be considered and, as recommended by Perrier *et al.*, the limiting of these side reactions of the azide may be achieved by using short reaction times and low temperatures.[40] However, by utilizing a post-polymerization strategy, as has been the case in the vast majority of reports,[24,28,29,33,41,42] this problem has not been more widely reported.

Although the CuAAC has clearly been dominating the realm of click chemistry, other methodologies have been employed that have proven to be equally efficient and useful. The Diels–Alder cycloaddition between anthracene- and maleimide-functionalized polymers has successfully been used to generate numerous block structures.[41,42] In the pre-polymerization approach to equip linear polymer chains with these functional groups, 9-anthracene methanol is a useful anthracence derivative. Although behaving as a reactive diene in the Diels–Alder cycloaddtion, anthracene derivatives are stable against radical attack, thus no protective chemistry is required under such conditions. Maleimides, on the other hand, require protection if they are to be incorporated pre-polymerization. Maleimide functional groups have been protected as the Diels–Alder adduct with furan and, unlike the protective chemistry used with alkynes, the deprotection of the maleimide occurs through an *in situ* retro-Diels–Alder reaction during the click conjugation step, which will be elaborated upon in a subsequent section.

Most recently, the RAFT–hetero Diels–Alder (HDA) concept has efficiently demonstrated the pre-polymerization functionalization strategy in that the RAFT agents used in the synthesis of the precursor, linear polymer chains require no additional functional groups to perform the conjugation reaction.[43,44] The dithioester end-group that is inherent of the RAFT process is sequentially used for the controlled polymerization process and as the reactive heterodienophile in a HDA reaction. The RAFT agents that have been successfully utilized in this context are shown in Scheme 6.3. The principal criterion for a RAFT

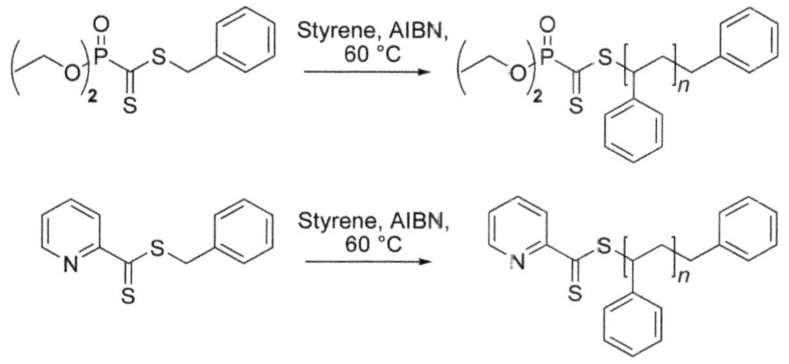

Scheme 6.3 *Appropriate RAFT agents for use in the RAFT–HDA concept.*

agent to be useful in the RAFT–HDA concept is an electron-withdrawing Z group. The complementary moiety with which the RAFT end-group reacts to form a linkage is a diene. For the purposes of the RAFT–HDA concept, *trans*,*trans*-2,4-hexadien-1-ol has been used successfully as an initiator for the ROP of CL to produce diene terminated PCL.[44]

Post-polymerization Functionalization

The second method by which linear polymer chains may be equipped with the functionalities required to perform a click conjugation is by modifying the polymer chain, typically through some form of substitution chemistry. By inspection of the literature, the most widely used form of post polymerization functionalization in the context of click chemistry has been the nucleophilic substitution of the terminal bromide of a polymer that has been prepared via ATRP, typically with sodium azide in *N*,*N*-dimethylforamide (DMF) or azido-trimethylsilane in tetrahydrofuran (THF) in an overnight, room-temperature reaction (Scheme 6.4).[24,28–30,33,36,42,45,46] This methodology circumvents any potential side reaction of the azide during the polymerization process and its only downfall is the fact that, for higher degrees of polymerization, the number of polymer chains bearing the bromide end-group decreases, thus limiting the number of polymer chains that can potentially be transformed into a 'clickable' species.

The major advantage of the post-polymerization alkyne functionalization is that it circumvents the issues with protective chemistry of the alkyne moiety. The incorporation of

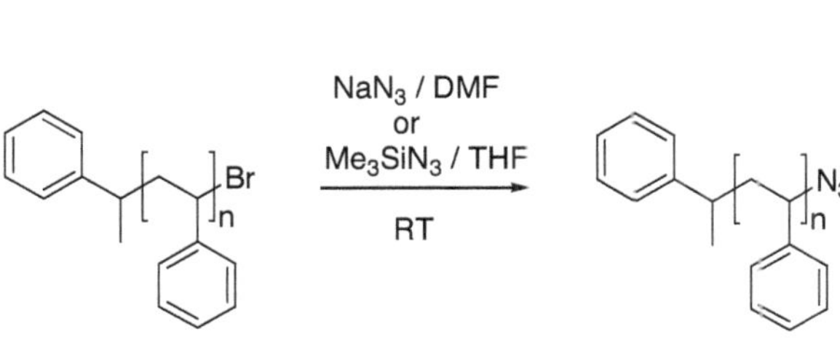

Scheme 6.4 *Post-polymerization bromide substitution to prepare azide terminated polymers.*

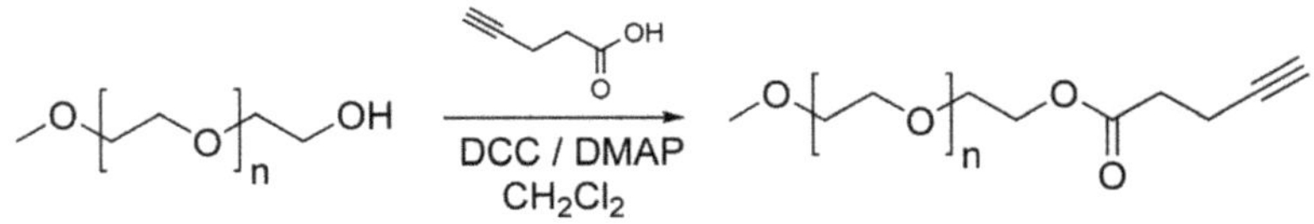

Scheme 6.5 *Functionalization of a polymer chain with an alkyne moiety.*

an alkyne moiety onto the end of a polymer chain has mostly been achieved through a simple *N,N′*-dicyclohexylcarbodiimide (DCC) coupling of low molecular weight derivatives such as 4-pentyonic acid to the target polymer (Scheme 6.5).[42,46–48] Other methodologies include the use of Grignard chemistry[49] or substitution chemistry.[50]

The anthrancene–maleimide route also lends itself easily to the post-polymerization functionalization strategy in that the maleimide functionality can be incorporated into a commercially available PEG chain through simple DCC couplings.[41,42]

6.2.2 The Click Reaction: Methodologies and Isolation

There are a variety of ways in which the CuAAC may be performed to create a linkage between two polymer blocks. The fundamental components are the azide-functionalized block, the alkyne-functionalized block, a source of copper(I) and an appropriate ligand to solubilize the copper source if necessary. The most widely used catalyst system has been the combination of copper(I) bromide and PMDETA in either THF or DMF solvent.[27–30,33,35,36,42,45,51] Other catalyst systems that have been reported include the combination of copper(I) iodide and 1,8-diazabicyclo[5.4.0]undec-7-ene (DBU),[24,25] as well as copper(I) bromide with BPy.[46,47] For polymeric systems that are sensitive to strong bases, such as those containing imide groups, the catalyzing copper complex tris(triphenylphosphine)copper(I) bromide in DMF may also be used.[48,49] Surprisingly, catalyst systems containing copper(II) sulfate–sodium ascorbate, which allows for the *in situ* formation of a copper(I) source, have not been utilized in the synthesis of block copolymer structures.

Scheme 6.6 illustrates the use of the CuAAC for the construction of polymeric blocks. The CuAAC is typically performed in an overnight reaction at a temperature ranging from ambient to 50 °C. Exceptions to this are observed when performing a one-pot conjugation

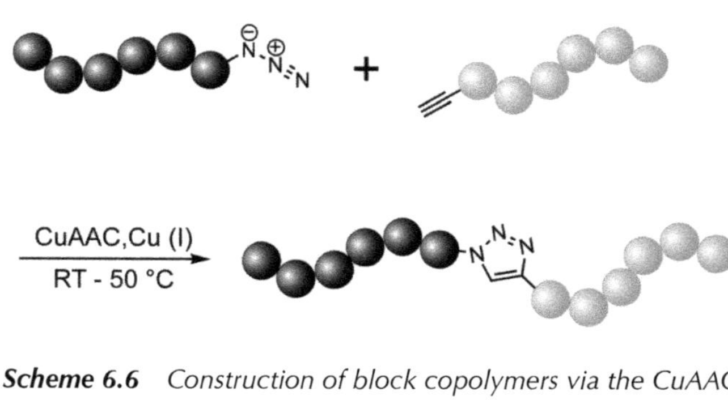

Scheme 6.6 *Construction of block copolymers via the CuAAC.*

that utilizes two different forms of click coupling. For example, Tunca *et al.* synthesized various triblock copolymers of PEG, PS, poly(methyl methacrylate) (PMMA) and poly (ϵ-caprolactone) (PCL) by a one-pot combination of the CuAAC and the anthrance–maleimide Diels–Alder cycloaddition.[42] Here, as will be seen shortly, the Diels–Alder reaction requires high temperatures (110–120 °C); hence the one-pot reaction was also performed at this temperature. As another example Huang *et al.* achieved a series of triblocks from PEG, poly(*tert*-utyl acrylate) (PtBA), PS and PCL from a 90 °C, one-pot reaction involving the CuAAC and an atom transfer nitroxide radical coupling.[28]

While there have been some occurrences of authors using a 1:1 ratio of azide-functionalized polymer to alkyne-functionalized polymer in the CuAAC,[25,33] most reports of block formation have involved using an excess of either the azide or the alkyne.[24,27–30,35,36,42,47–50] The vast majority of polymers investigated in the construction of block structures have been prepared by some variety of controlled free radical polymerization technique. As such, despite obtaining narrowly disperse molecular weight distributions, there still remains the inevitable presence of polymer chains that do not bear a reactive end-group that would lend itself to being used in a conjugation reaction. It is for this reason that several authors have utilized an excess of one reagent to drive the click reactions to completion. The disadvantage here is that it is often necessary to remove the excess reagent in some additional purification strategy.

The nature of the purification strategy is dependent upon the nature of the polymers in the system. For example, the coupling of alkyne-functionalized PEG with azide-functionalized PS or PMMA may be driven to completion by using an excess of the PEG segment, which can then be removed by a washing stage with methanol.[24] Alternatively, if the CuAAC product may be precipitated into methanol, an excess of the PEG chain would remain in solution.[41] Another example of a simple purification technique was performed by Du Prez *et al.*[30] Here, an excess alkyne-functionalized poly(1-ethoxyethyl acrylate) (PEEA) was removed from the click product with poly(isobornyl acrylate) by selective precipitation in cold methanol.

However, there have been some reports on the use of more elaborate purification strategies to remove excess reagent. In the synthesis of PS-*b*-PMMA, van Hest *et al.* used an excess of the alkyne-functionalized PMMA, which was removed by passing the crude product mixture through a column of azidomethyl polystyrene resin.[24] Yet another more complicated strategy involves removal of an excess of azide by first converting the excess to amines by reaction with triphenylphosphine followed by column chromatography.[29] Of course, in situations where a simple precipitation cannot be performed, these above-mentioned more complicated strategies still produce good results.

Whilst on the topic of purification, the use of the CuAAC bears the requirement of the removal of the copper catalyst. This may be achieved by passing the click reaction solution over through a column of basic or neutral alumina.[28,42,47] Despite this simplicity, it is still a technique that is really only feasible in the laboratory and not on an industrial scale. In the context of block formation, there have been two reported examples, however, of the removal of the copper catalyst without the use of a column. The catalyst, under the right circumstances, may be removed by a series of precipitations. For water-insoluble polymers, it is possible to precipitate the polymer and isolate by filtration, thus separating the polymer from the copper salt.[35] Furthermore, purification of the polymer product by dialysis against water may also achieve an effective removal of the copper catalyst.[31]

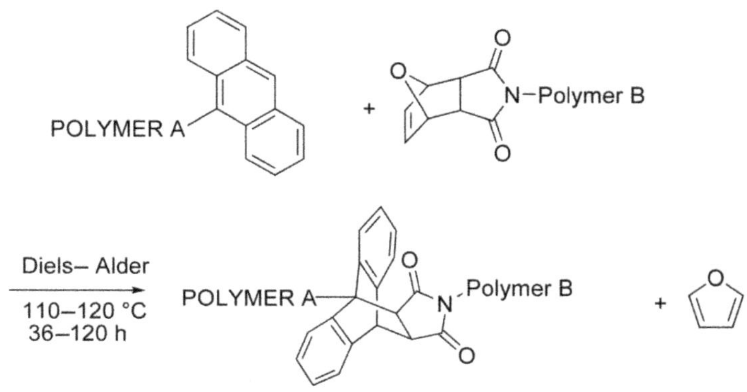

Scheme 6.7 *Construction of block copolymers via the Diels–Alder cycloaddition of anthracence and maleimide.*

The Diels–Alder cycloaddition of anthracene and maleimide has the distinct advantage that no catalyst is required. Where the efficiency of the CuAAC, whilst dependent upon the system in which it is used, can range from 80% to close to quantitative, the said Diels–Alder approach has proven to be just as efficient (97%).[41] The anthracene–maleimide conjugation can be performed by simply heating a solution of the two functionalized blocks in a high-boiling solvent, such as toluene. The concept is illustrated in Scheme 6.7.

In a similar fashion to the CuAAC, the maleimide moiety requires protective chemistry for the stages leading to the final click reaction. However, whereas the protection used for the alkyne moiety in the CuAAC requires a separate deprotection step, the furan-protected maleimide undergoes an *in situ* retro Diels–Alder reaction during the coupling step.[41,42] Furthermore, this Diels–Alder click chemistry requires no catalyst, thus isolation of the formed block copolymers is a simple matter of precipitation. Although, in these two areas, the anthracene–maleimide Diels–Alder cycloaddition appears to be more convenient than the CuAAC, high temperatures (110–120 °C) and long reaction times (36–120 h) are the drawback.

Being one of the most recent forms of efficient conjugation chemistries, the RAFT–HDA concept is an atom-economical approach to block copolymer synthesis (Scheme 6.8). The entire concept is based upon the electron-withdrawing dithioester end-group that is inherent in polymers prepared with specially chosen controlling agents in RAFT polymerization. Unlike the CuAAC and the anthracene–maleimide Diels–Alder approach, the RAFT agent used does not require additional functionalities to be used in the post-polymerization conjugation step. The reaction is quite simply performed by keeping a solution of the dithioester-terminated polymer, diene-functionalized polymer and catalyst at 50 °C for between 10 and 24 h, depending upon the nature of the dithioester end-group.[44]

The reaction is catalyzed by the addition of zinc chloride in the case of the diethoxyphosphoryl dithioester end-group or a simple Brønsted acid such as trifluoroacetic acid in the case of the pyridinyl dithioester end-group. The role of these catalysts is to coordinate with specific atoms on the dithioester end-group in order to enhance the electron withdrawing effect on the thiocarbonyl bond. This serves to lower the energy of the lowest unoccupied molecular orbital of the thiocarbonyl so as to enhance its reactivity towards a

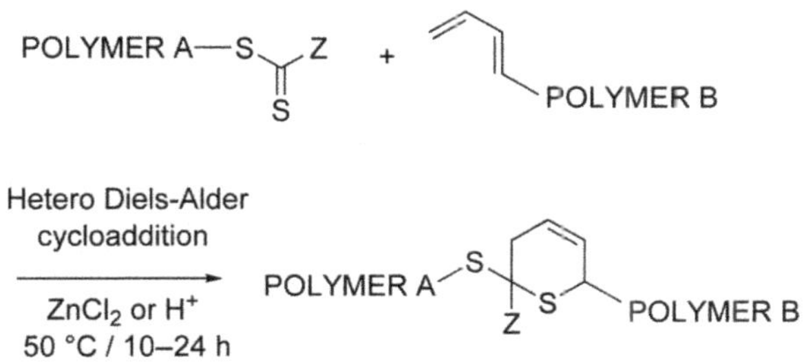

Scheme 6.8 The construction of block copolymers via the RAFT–HDA approach.

hetero Diels–Alder cycloaddition with a diene. This technique has been proven to be just as efficient as the CuAAC and anthracene–maleimide Diels–Alder cycloaddtion.[43,44] It circumvents the requirement of using a copper catalyst and also circumvents the requirement of high temperatures.

6.2.3 Polymer Characterization

As polymeric architectures become more and more complex, the use of a wide variety of characterization techniques is necessary. As such, amongst the most important and useful techniques that have been used to characterize complex macromolecules include gel permeation chromatography (GPC), NMR, Fourier transform infrared (FT-IR), UV–vis as well as mass spectrometry (MS) methods.

GPC is convenient in that the molecular weight distributions of the individual blocks may be compared with that of the coupled product. In many respects, useful qualitative information concerning the success of the click reaction may be elucidated by a quick inspection of the product chromatogram. For example, the presence of shoulders or a bimodal distribution can indicate the presence of unreacted starting material. Figure 6.2 displays three GPC analyses of block formation via the CuAAC, the anthracene–maleimide Diels–Alder (DA) reaction and the RAFT–HDA concept. It may be observed that each effectively showed a complete shift of the molecular weight distribution (either to lower retention times or higher molecular weight) from the individual building blocks to the final coupled product. It is important to note, however, that the molecular weight of the block structure, as determined by GPC measurements, may not necessarily be in agreement with the arithmetic sum of the building blocks. Although not being widely used in this context in the surveyed literature, more advanced techniques such as GPC with triple detection may allow better insights into the molecular weight of the obtained species.

When analyzing more complicated structures via GPC, such as stars or combs, one must also take into consideration two important factors. Firstly, the relatively compact structure of such architectures leads to an under-estimation of M_n as determined by GPC measurements. As such, increases in the observed molecular weight become less pronounced with increasing arm number. Secondly, the determination of M_n by this technique takes into consideration any remaining linear precursor material as well as any material that has not completely been converted, which tends to skew the observed M_n to lower values and leads to broader polydispersity index (PDI) values. This is particularly noticeable in

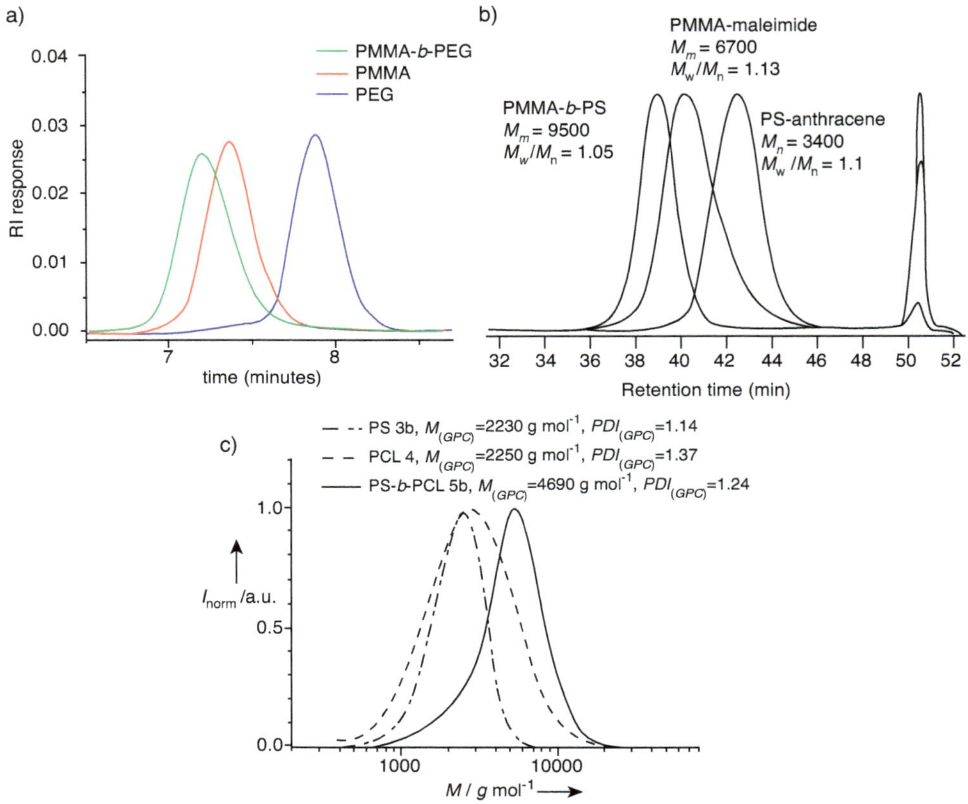

Figure 6.2 *GPC spectra for block copolymer formation via (a) CuAAC (Reprinted with permission from ref.[24]. Copyright 2005 Royal Society of Chemistry), (b) anthracene–maleimide Diels–Alder cycloaddition (Reprinted with permission from ref.[41]. Copyright 2006 Wiley-VCH) and (c) RAFT–HDA concept (Reprinted with permission from ref.[44]. Copyright 2008 Royal Society of Chemistry).*

starpolymers synthesized by click approaches.[43,51] In order to extract more useful data out of size-exclusion chromatography measurements, it is possible to deconvolute the data via peak splitting and arrive at corrected M_n and PDI values as well as the relative ratios of different products in the sample.[43,51,52] Interestingly, in two reported examples of the synthesis of graft copolymer structures, a reduction in molecular weight of the graft product with respect to the linear precursors was observed.[53,54] These examples nicely show the difficulties that can occur through the above-mentioned effects.

One characteristic that all click reactions share is that the transformations from the separate reactive functional groups to the final adduct may be verified by a number of different spectroscopic techniques, depending upon the nature of each reaction.

NMR spectroscopy has been widely used across all forms of click chemistry in determining the success of the conjugation reaction. For example, the CuAAC may be monitored by the shift of the azido methylene protons as well as the terminal alkyne protons to higher ppm values after the triazole formation.[28,30,32,35,47] Futhermore, the conversion of anthracene and maleimide into the Diels–Alder adduct[41,42] as well as the conversion of dithioesters and dienes into the HDA adduct[43,44,55] may also be monitored via NMR spectroscopy.

FT-IR spectroscopy is a particularly useful technique for the monitoring of the CuAAC in that the disappearance of the azide signal (~2100 cm^{-1}) can be used as a indicator of the success of the conversion.[24,25,34,36,56] UV-vis spectroscopy has also been conveniently utilized in determining the progress of click reactions. The appearance of an absorbance signal of the 1,2,3-triazole ring ($\lambda_{max} = 258$ nm) clearly displays the success of the CuAAC.[56] In the anthracene–maleimide DA approach, the characteristic absorption pattern of the anthracene ($\lambda_{max}= 366$ nm) completely disappears after conversion to the conjugation adduct.[41,42] Furthermore, the transformation of the chromophoric RAFT thiocarbonyl group into the colourless 3,6-dihydro-2*H*-thiopyran ring in the RAFT–HDA concept can also be conveniently followed by UV–vis spectroscopy.[43]

Mass spectrometry has also been used to determine the success of click reactions between polymers and low molecular weight species. However, mass spectrometry of block structures can lead to spectra that are complex to the point of being devoid of any useful information. Nevertheless, click-functionalized polymers have been converted with low-molecular-weight species bearing the required complimentary click moiety and the relatively small shift in the mass spectrum peaks that results, very convincingly reveal the success and efficiency of the click reaction.[29,44]

6.3 Star Polymers

Star polymers can be subdivided into conventional star polymers with arms of similar chemical nature, miktoarm star polymers with three or more different arms and dendrimer-type star polymers with additional branching in each arm. While the CuAAC is predominant in all these approaches, other pathways have been explored additionally. The RAFT–HDA concept, which employs directly the thiocarbonyl group of the RAFT agent in combination with a diene, was as successfully employed as the maleimide-anthracene Diels–Alder approach.

6.3.1 Star polymers A_n

The simplest case of a spherical structure is a star polymer with several chemically identical arms.

CuAAC Chemistry Routes

For star polymer formation via CuAAC, a linear polymer chain, which carries either an alkyne or azide functionality, is coupled onto a core with multiple opposite functionalities. The core can theoretically be based upon either group. In reality, practical considerations play a role such as the type of reaction that has been used to generate the arm. An azide group can be generated at a chain end by the simple replacement of bromide groups – which are present in ATRP generated structures. Polymers bearing hydroxyl functionality such as PEG can be converted in a similar, facile approach via the tosylate and then into azides. It is therefore not surprising to find more reports using alkyne containing cores, which are then reacted with azide end-functional polymer chains (Scheme 6.9).

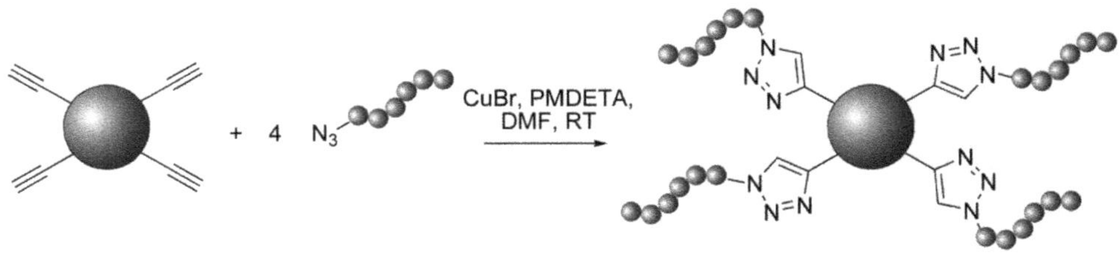

Scheme 6.9 *The synthesis of an A_4 star polymer via the CuAAC approach using typical reaction conditions.*

The success of a click reaction is expected to be a function of not only the length of the coupled polymer chain, but also the functionality of the core. A detailed study using ATRP-generated and azide-functionalized PS chains with molecular weights of 1400, 6800 and 18 100 g mol^{-1} confirmed that the yield is highly dependent upon the chain length of each arm. Di-, tri- and tetrafunctional alkyne containing cores were employed; the results that were obtained showed the yield drops from 90% (difunctional core) to 83% (tetrafunctional core) already with low molecular weight polystyrene (1400 g mol^{-1}). As expected, a more dramatic effect was observed when using higher molecular weight branches. A maximum yield of 80% was obtained when attempting to attach PS chains with a molecular weight of 18 100 g mol^{-1} to a difunctional core.[51] Similar results were obtained when coupling other polymers such as azide-functionalized PEG (M_n = 2600 g mol^{-1}) or PtBA (M_n = 6700 g mol^{-1}) to a trifunctional core. Conversions of not more than 87% were obtained, resulting in stars with an average number of arms of fewer than three.[57] An excess of the functionality belonging to the arm can sometimes enhance the completeness of the reaction. One such example has been observed in the case of the CuAAC between alkyne-functionalized PCL (M_n= 2200 g mol^{-1}) and azide-functionalized β-cyclodextrin. The formation of the seven-arm star was reported to be quantitative after using a 9-fold excess of the reactive linear precursor.[58] However, in another example, an excess of the linear precursor could not ensure the quantitative conversion in the synthesis of a three-arm star block copolymer (see Scheme 6.10).[59] Various ratios between PEG–azide (arm, M_n= 2000 g mol^{-1}) and alkyne (core) were tested showing that the maximum conversion of 85% could not be improved upon using alkyne–azide ratios of up to 4.5:1. In an alternative attempt reported by Monteiro *et al.*, the conversion of the binding sites could be improved (from 75 to 78%) by slowly feeding a solution of the core to the arm reaction mixture.[60]

The obtained star polymers described above required the use of a purification step to remove the copper catalyst, which was in most cases coordinated by PMDETA as the ligand. Synthetic strategies, which do not require the use of copper catalysts are therefore of potential interest. The predominant candidates as discussed below involve the Diels–Alder cycloaddition of anthracenes and maleimides as well as the RAFT-HDA concept.

The Diels–Alder reaction between maleimide and anthracene derivatives (Scheme 6.11) is carried out under the absence of any type of catalyst, but requires an extended reaction period at high temperatures.[52] Interestingly, this approach was observed to gain high conversions even for high molecular weight arms (PMMA, M_n = 8450 g mol^{-1}, 89%; PtBA, M_n = 10 600 g mol^{-1}, 93%).

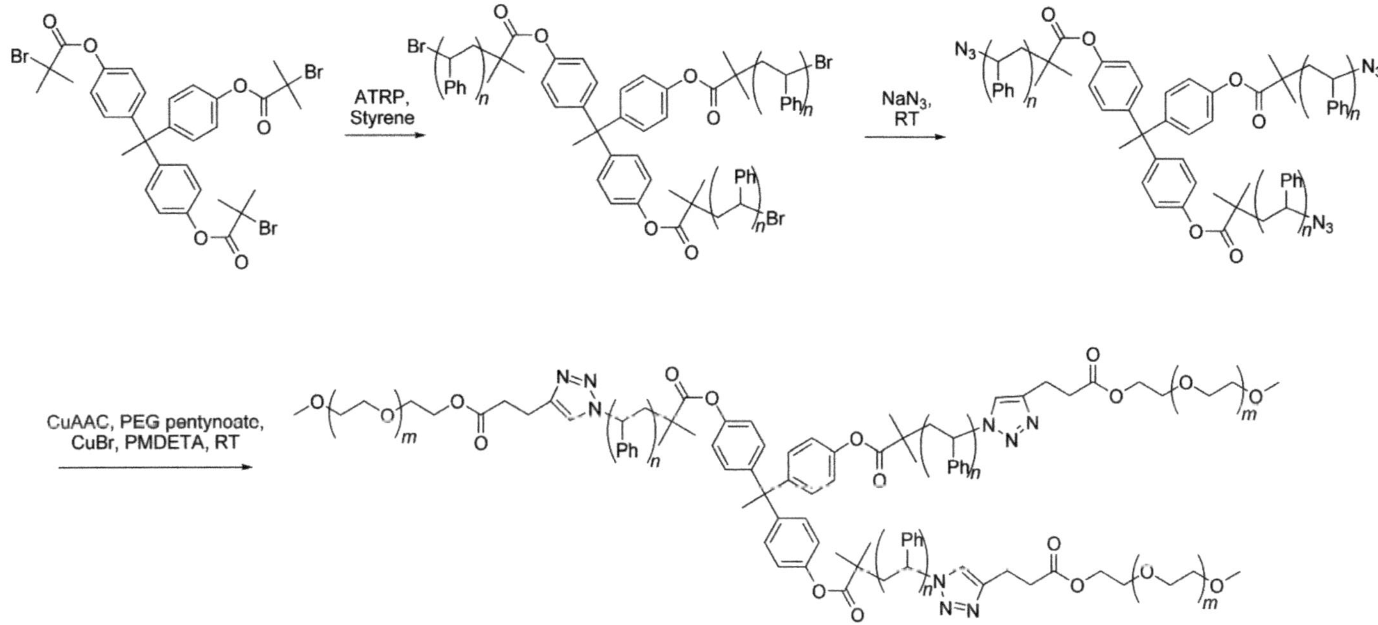

Scheme 6.10 *Synthesis of PS*-b-*PEG 3-arm start block copolymers by combination of core-first and coupling onto methods. For detailed reaction conditions please refer to Gao* et al.[59]

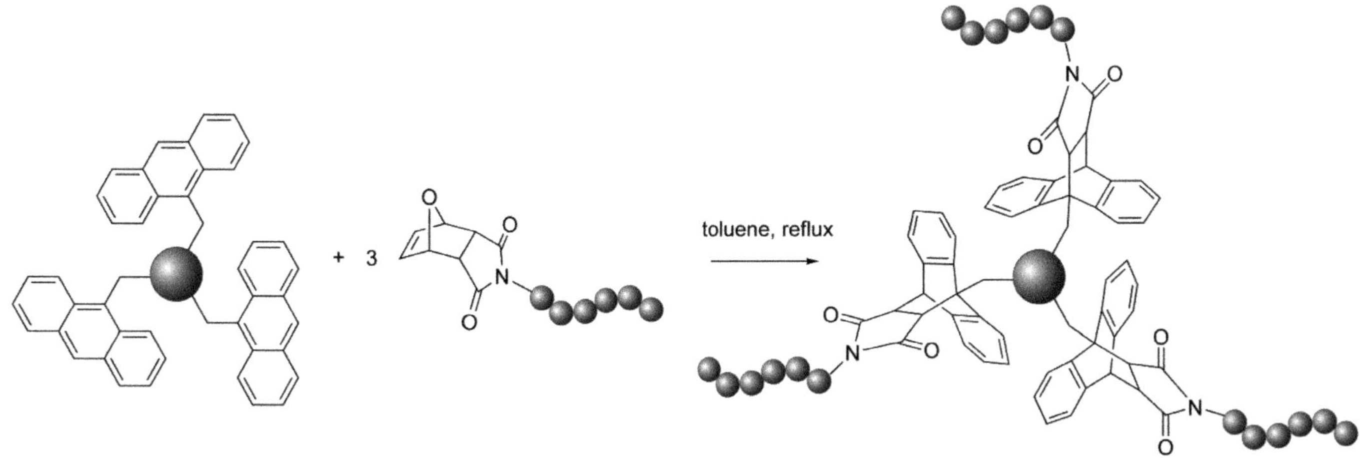

Scheme 6.11 *Synthesis of star polymers via a Diels–Alder reaction between furan-protected maleimide end-functionalized polymer and trianthracene core.*

The direct reaction between the RAFT groups and dienes, as described previously, can also be employed in the synthesis of star polymers by coupling RAFT-generated polymers onto multifunctional cores, which carry a multitude of diene functionalities. Similar to the CuAAC, the conversion between both functional groups was dependent upon the functionality of the coupling agent. While the reaction between PS ($M_n = 3600$ g mol^{-1}) synthesized using benzylpyridin-2-yldithioformate as RAFT agent and bis-diene functional core yielded 91% conversion, the tri- and tetrafunctional core resulting permitted a maximum conversion between arms and core of 86 and 81%, respectively. Interestingly, the type of the RAFT agent employed to generate the PS arms was found to have a pronounced influence on the reactivity, thus influencing the average number of arms. Replacing the pyridyl-Z-group of RAFT agent by diethoxyphosphoryl group to prepare a PS arm of similar molecular weight led to PS arms of lower HDA activity. As a result, the maximum conversion was reduced by more than 10% with conversions of 81, 77 and 65%.[43] In terms of efficiency, the RAFT HDA concept is therefore comparable to the CuAAC strategy. Importantly, both methods can be applied in a combined approach, which lends itself to independent control of two different binding sites.[55] The generation of three-arm stars with block structures in each arm by this technique is shown in Scheme 6.12.

A range of concepts were explored in the generation of star polymers with heteroarm structure (Scheme 6.13). These approaches typically involve the combination of a click reaction with other techniques such as RAFT,[61] ATRP,[60–64] NMP[62,65] or ROP.[63,65] Apart from one exception,[66] the CuAAC was the reaction of choice so far to synthesize these

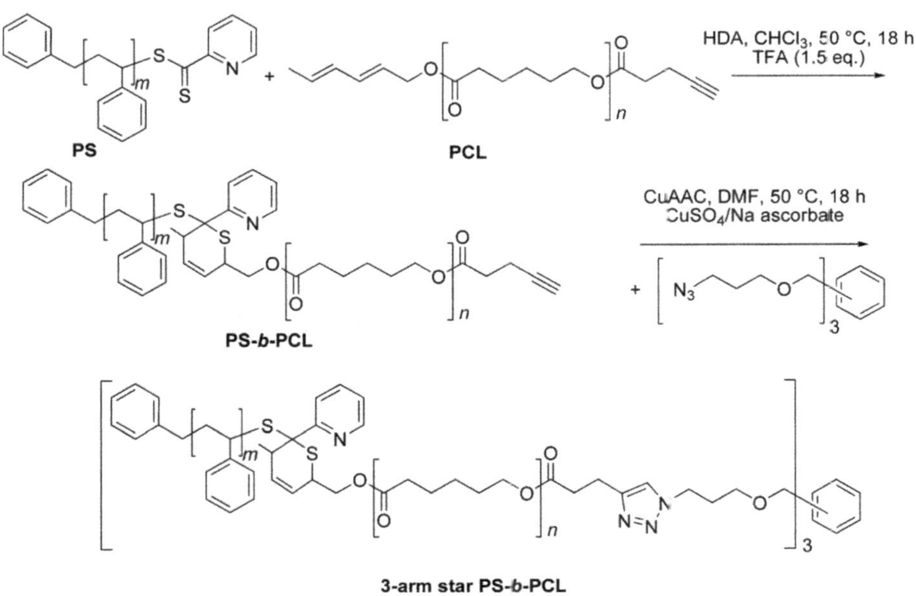

Scheme 6.12 *Arm-first strategy for the preparation of three-arm star PS-*b*-PCL via a consecutive combination of CuAAC and HDA chemistry. For detailed reaction conditions please refer to Barner-Kowollik* et al.[55]

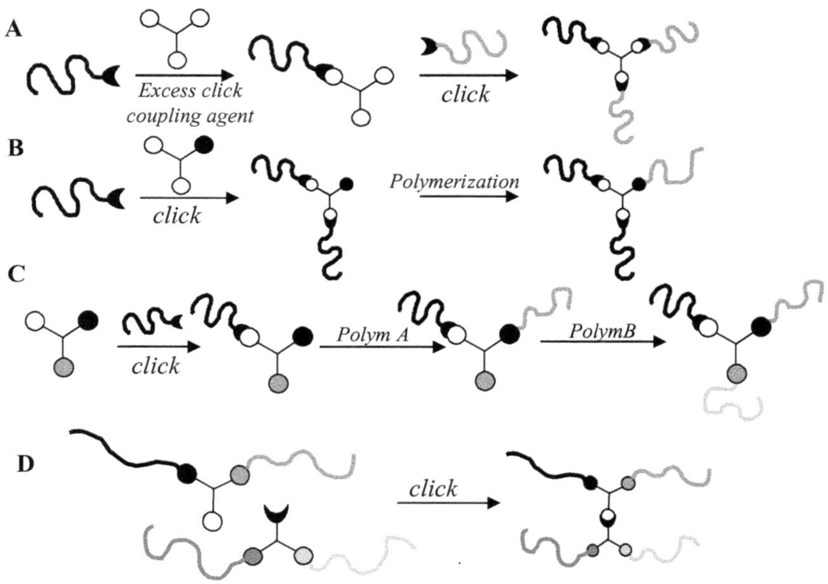

Scheme 6.13 Concepts to generate heteroarm star polymers: (A) AB_2; (B) AB_2; (C) ABC 3-miktoarm; (D) ABCD 4-miktoarm.

structures. A complete click approach to heteroarm star polymers was achieved by reacting the reactive polymer chain (azide-functionalized) with an excess of trialkyne-functionalized core. In a subsequent step, the remaining binding sites were coupled with further reactive polymer chains, resulting in AB_2 stars. A range of these stars with various combinations of PS, PMMA, PtBA and poly(methyl acrylate) (PMA) have been prepared using polymers with molecular weights of around 5000–7000 g mol^{-1} each.[60]

Most approaches to heteroarm stars, however, involve the synthesis of a multifunctional initiators that can undergo click reactions, while acting as initiators for a range of living polymerizations [Scheme 6.13(B, C)]. The click reaction was either carried out as an initial step[61,63,64] or as the final step after other arms had already been generated by the various polymerization techniques.[62]

An elegant approach is the use of a simultaneous click reaction with ROP and NMP. A multifunctional click-ROP–NMP compound was heated for 48 h at 125 °C in the presence of styrene, ϵ-caprolactone, azide-terminated PMMA (or PEG) and copper(I) bromide–PMDETA.[65]

Similar concepts were explored in the synthesis of ABCD 4-miktoarm star polymers. Two different trifunctional initiators were prepared – one carrying an azide group, the other containing an alkyne group.[66,67] Prior to the click reaction, various polymerizations were carried out, creating AB block copolymers carrying an azide group and CD block copolymers with an alkyne group between the two blocks. The two components were then combined – usually with a 10–30% excess of one block copolymer and an extended reaction period of 2 days. NMR studies were usually employed to calculate the conversion of the reaction, which was typically around 90%.

6.3.2 Dentritic Star Polymers

Star polymers with well-defined, branched arms were prepared using alternating ATRP and click reactions with an intermediate step replacing bromide with azide.[60,68,69] Divergent[68] and convergent[69] synthetic pathways were explored (Scheme 6.14). While the divergent approach involved only click reactions between polymers and a low-molecular-weight coupling agent, the convergent approach attempted in the final step to couple three-branched polymer chains to a trifunctional core. Not surprisingly, the reaction was reported to be sluggish, but fractionation allowed for the isolation of well-defined G_2 dendrimers.[69]

6.4 Graft Copolymers

Of the various macromolecular architectures, graft copolymers attract special interest owing to their unique material properties. Depending upon the chemical nature of the backbone and side chains, such structures are of potential application in tissue engineering, polymer-based biomaterials, nanotechnology and drug delivery vectors.

In principle, graft copolymers can be synthesized by three different routes: 'grafting from', 'grafting through' and 'grafting onto'. In particular, the efficiency of the 'grafting onto' methodology is often limited due to the steric repulsion of the reactive side chains. Grafting densities are therefore usually low and unreacted side chains are often left behind and have to be removed by further fractionation steps. To overcome these difficulties, highly efficient coupling strategies such as click reactions are of potential benefit. It is therefore not surprising that the first report about a CuAAC 'grafting onto' approach appeared already in 2005 soon after the first applications of click chemistry in polymer science, in which Emrick *et al.* grafted PEG chains and oligopeptides onto aliphatic polyesters.[70] Since then, a number of publications have appeared reporting the use of click strategies for the 'grafting onto' approach, a summary of which is presented in the following.

The two principal designs for the construction of graft copolymers via CuAAC 'grafting onto' chemistry are depicted in Figure 6.3. Either the macromolecular scaffold is equipped with azide functions (Table 6.1), which are converted with polymers carrying the alkyne moiety or alkyne-functionalized polymeric backbones (Table 6.2) are combined with azido-terminated polymers. The question of which linking strategy should be followed mostly depends upon which monomer family and which polymerization technique are chosen for the construction of the according building blocks. Finally, the use of protective group chemistry or post-polymerization functionalization allows for the synthesis of macromolecular scaffolds of varying chemical structure. Although there are a number of reports in the literature in which the CuAAC is used to graft low molecular weight reactants onto polymer backbones, the following section highlights the examples describing the construction of a variety of graft/brush copolymers. Reactive polymer chains, as counterparts in the 'grafting onto' methodology, are thoroughly discussed in the blocks section (see Section 6.2.1) and are therefore only briefly mentioned here.

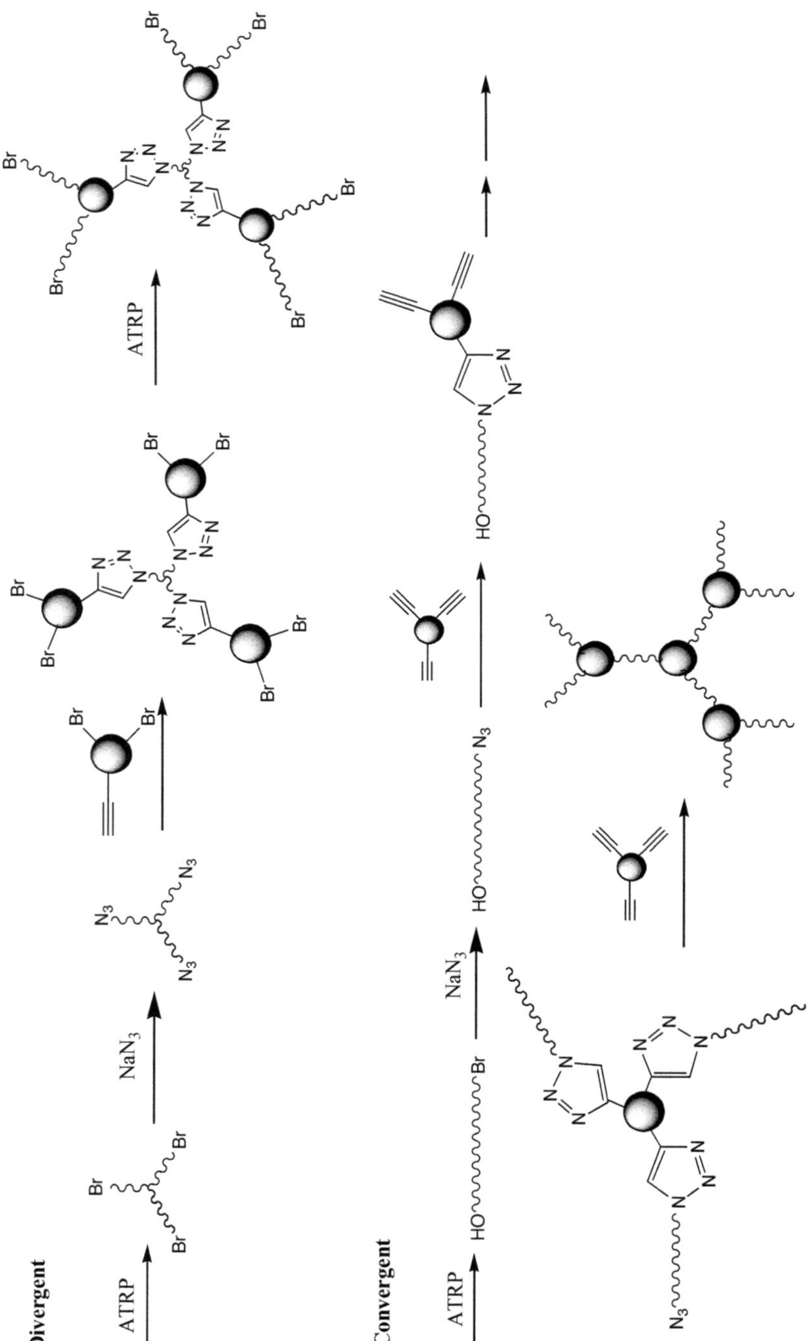

Scheme 6.14 *Divergent and convergent synthesis approach to dendritic star polymers by a combined ATRP/click approach.*

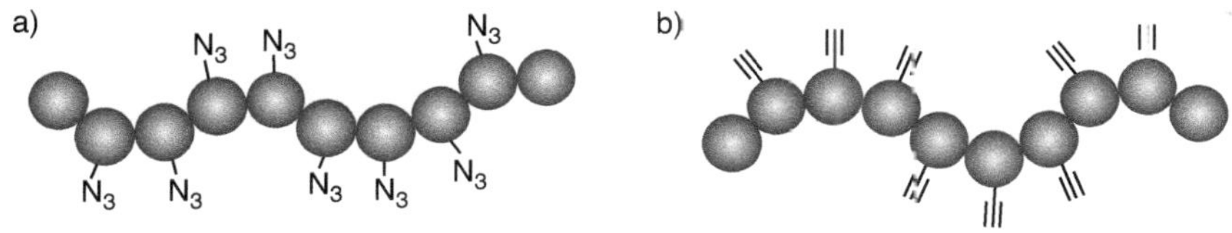

Figure 6.3 Azide- (a) and alkyne (b)-functionalized polymer chains as the two principal macromolecular scaffolds for the construction of graft copolymers *via* CuAAC chemistry.

6.4.1 'Grafting-to' Azide Main Chains

The main advantage of the use of azido polymer backbones is the nonrequirement of protective group chemistry in the case of radical polymerizations of vinylic azido monomers.

However, a recent study by Perrier *et al.* revealed that under typical radical polymerization conditions, electron-deficient monomers can undergo side reactions in the presence of organic azides, which result in a significant loss of orthogonality.[40] Nevertheless, short reaction times and low polymerization temperatures minimize the occurrence of this side reaction and well-defined materials are obtained. One such example was reported by Du Prez and coworkers, in which methyl methacrylate was copolymerized with 3-azidopropyl methacrylate via ATRP.[30] Subsequently, PEAA chains obtained after ATRP with an alkyne-containing initiator were attached to the macromolecular precursor using a copper(I) bromide–PMDETA system. GPC analysis revealed the successful coupling reaction. However, excessive PEEA needed to be removed via preparative GPC. Similar results were found by Liu *et al.*, who copolymerized 3-azidopropyl methacrylate with both *tert*-butyl methacrylate and 2-(dimethylamino)ethyl methacrylate.[71] Successive CuAAC 'grafting-to' reactions with poly(*N*-isopropyl acrylamide) (PNIPAM) carrying an alkyne end-group were performed with copper(I) bromide under ligand-free conditions. Complete conversions of the alkynyl residues were achieved by using alkyne to azide ratios of <0.5:1.

The versatility of click strategies for the construction of graft copolymers was further shown by Matyjaszewski *et al.*, who used a combination of two consecutive click reactions for the synthesis of polymeric brushes with PEG side chains.[72] An epoxide ring containing copolymer of glycidyl methacrylate and methyl methacrylate was converted with sodium azide to obtain the corresponding 1-hydroxy-2-azido compound, which was further converted in a CuAAC reaction in the presence of PEG pentynoate and copper(I) bromide in a ligand-free or PMDETA environment. With an alkyne:azide ratio of 1:1 a maximum conversion of the reactive alkyne side chains of 75% was reached within 1 h, which could not be further increased after longer reaction times or a higher catalyst loading.

The use of ROP techniques for the preparation of aliphatic polyesters carrying azide substituents was reported by Jerome and coworkers.[73] Copolymers of α-chloro-ϵ-caprolactone and ϵ-caprolactone (CL) or lactide (LA) were transformed into the corresponding azide containing copolymer by substitution of the chloride. After the CuAAC with an alkyne-functionalized PEG using copper(I) iodide–triethylamine–THF, an amphiphilic graft copolymer with a grafting density of ~25% was obtained. In a more recent work, the same authors improved the synthesis protocol and prepared a polyester PEG graft copolymer with a grafting density of 40%.[74] Finally, this synthesis protocol led to the formation of tadpole-shaped PCL with two PEG grafted tails and eight-shaped PCL-*g*-PEG with

Table 6.1 *Azide-functionalized macromolecular scaffolds*

Polymer structure[a]	Polymer synthesis	Reference
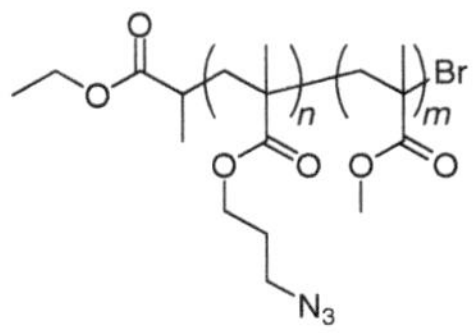	ATRP	30
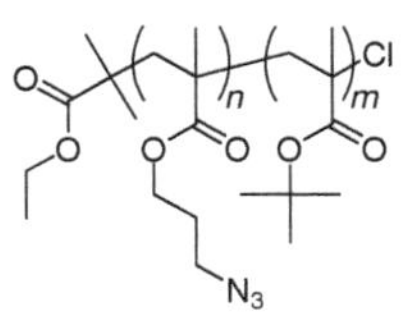	ATRP	71
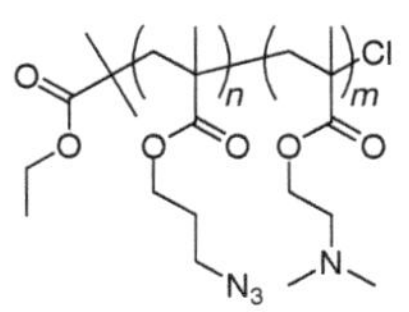	ATRP	71
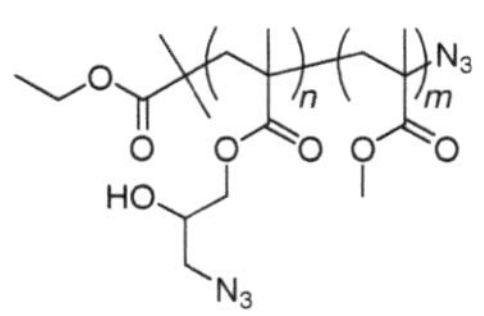	ATRP + 1 post-polymerization step	72
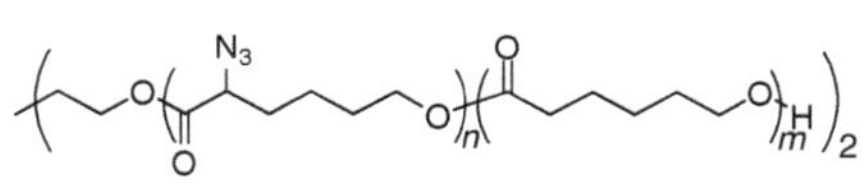	ROP + 1 post-polymerization step	74
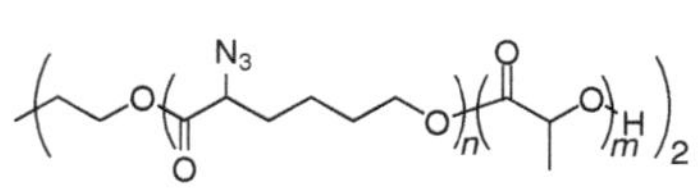	ROP + 1 post-polymerization step	73
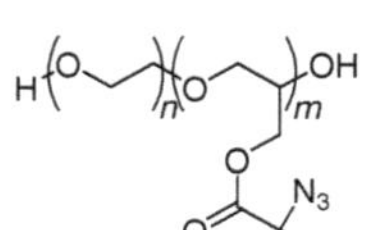	ROP + 3 consecutive post-polymerization steps	53

[a] All copolymer structures represent random copolymers.

Table 6.2 *Alkyne-functionalized macromolecular scaffolds*

Polymer structure[a]	Polymer synthesis	Reference
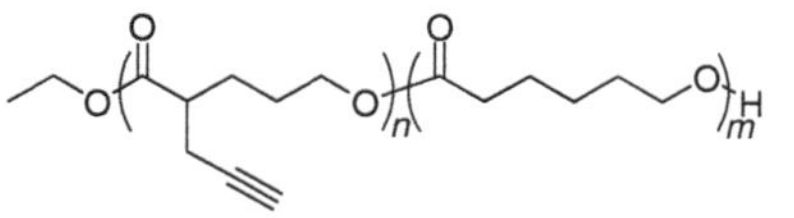	ROP	70
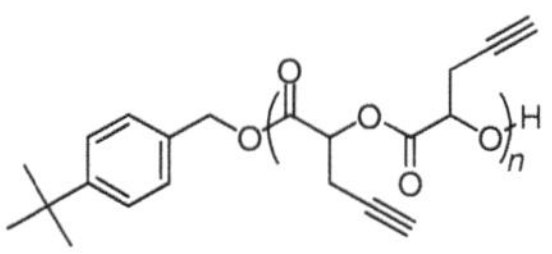	ROP	76
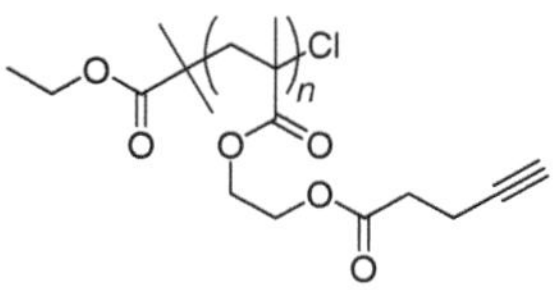	ATRP + 1 post-polymerization step	77
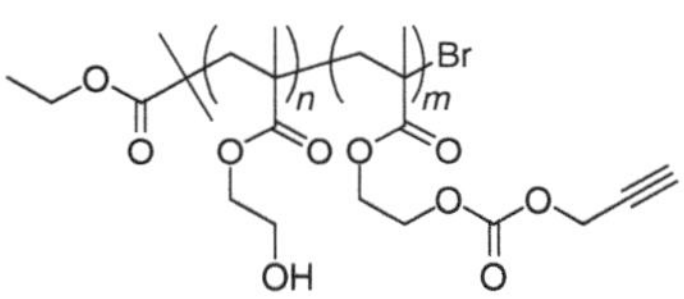	ATRP + 1 post-polymerization step	78
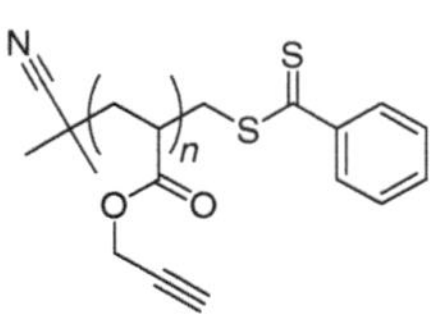	RAFT + 1 post-polymerization step	56

[a] All copolymer structures represent random copolymers.

potentially novel macroscopic and self-assembly properties.[54,75] The reverse strategy – a PEG-based main chain carrying azide groups – was presented by Huang and coworkers.[53] The polymer backbone was prepared via ROP and three consecutive post-polymerization modifications. CuAAC reactions between the main chain and alkyne-functionalized miktoarm star-shaped side chains were carried out in the presence of copper(I) bromide–PMDETA with a graft-efficiency of 63% (determined by NMR spectroscopy).

6.4.2 'Grafting-to' Alkyne Main Chains

Alternatively to the azide-functionalized main chains, polymer backbones carrying alkyne residues have been used for the construction of graft copolymers via CuAAC 'grafting onto' chemistry. ROP as well as controlled radical polymerization methods have been used for the synthesis of the macromolecular scaffolds; a summary is presented in Table 6.2.

In this case, the ROP provides an easy access to the desired alkyne main chains whereby routes via radical polymerization methods require the use of protective groups or

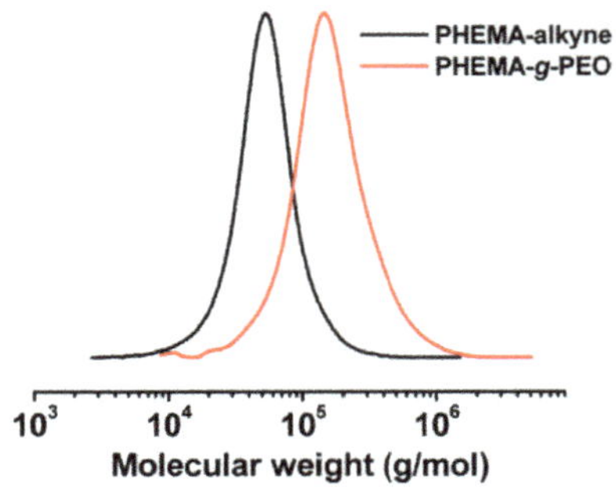

Figure 6.4 *GPC traces of PHEMA-g-PEG grafted copolymers synthesized by the 'grafting onto' method. For detailed reaction conditions please refer to Gao and Matyjaszewski.*[77] *Reprinted with permission from Gao, H. F., Matyjaszewski, K., (2007), Synthesis of molecular brushes by 'grafting onto' method: combination of ATRP and click reactions,* J. Am. Chem. Soc., ***129,*** *6633–6639. Copyright 2007 American Chemical Society.*

post-polymerization functionalization. Emrick and coworkers homo- and copolymerized α-propagyl-γ-valerolactone to obtain aliphatic polyesters with different degrees of alkyne content.[70] The obtained (co)polymers were subsequently used for the grafting of PEG and oligopeptide moieties. Interestingly, CuAAC reactions were performed in aqueous dispersion of the polyester, which solubilized in the aqueous environment as the reaction proceeded. The amphiphilic graft copolymers, with grafting densities of $>80\%$, were shown to be biocompatible. A similar approach was performed by Smith, Baker and coworkers who used a lactide-based monomer with two alkyne functions for the ring-opening (co)polymerization, resulting in polyglycolide homo and copolymers with pendant alkyne groups.[76] CuAAC reactions with an azide terminated PEG were performed with copper(II) sulfate in the presence of sodium ascorbate. The synthesis of polymer backbones via radical polymerization and their successive functionalization with alkyne groups was shown by Gao and Matyjaszewski.[77] The polymerization of 2-hydroxyethyl methacrylate (HEMA) via ATRP, followed by an esterification of the hydroxyl groups with 4-pentynoic acid yielded polymer main chains with high alkyne functionality. CuAAC reactions with different azide end-capped polymers, namely PEG (Figure 6.4), PS, poly(butyl acrylate) (PBA) and PBA-*b*-PS, were complete within 3 h and yielded the according graft copolymers with moderate graft densities ($<50\%$) for the bulkier polymers like PBA and PBA-*b*-PS and high densities (up to 88%) for the PEG side chains.

The use of poly(2-hydroxyethyl methacrylate) (PHEMA) as precursor for the alkyne functionalization was also reported by Hennink and coworkers.[78] In this case, propargyl alcohol was linked via a carbonate function onto the polymer backbone, which allowed for the hydrolytic degradation of the polymeric brushes obtained after CuAAC with PDMAEMA. A combination of RAFT polymerization and protective group chemistry was used by Stenzel, Barner-Kowollik and coworkers.[56] The polymerization of trimethylsilylpropargyl methacrylate and the successive deprotection of the obtained polymers led to well-defined polymers carrying alkyne groups on each monomer unit. CuAAC reactions with poly(vinyl acetate) (PVAc) chains obtained after RAFT polymerizations using an azide-functionalized RAFT agent led to PVAc brushes.

6.4.3 Non-CuAAC Routes

Besides the number of publications using CuAAC chemistry for the 'grafting-to' approach, there are two examples in which alternative synthetic routes were used. The Diels–Alder cycloaddition of anthracenes attached to a polymer backbone with maleimide-functionalized PEG as reactive polymer side chains was successfully used in the construction of graft copolymers.[79] The other example involved substitution reactions on poly(pentafluorostyrene) using amide-functionalized polymers as synthetic handles.[80]

6.5 Concluding Remarks

In summary, the construction of complex macromolecular architecture polymers via facile, rapid and orthogonal conjugation chemistries is one of the key driving forces in modern polymer science. The current chapter has highlighted, in detail, access routes to block, star and comb (co)polymers via several click reactions including the CuAAC as well as Diels–Alder conjugation protocols in combination with living/controlled polymerization (including RAFT, NMP, ATRP and ROP) methodologies. The focus lies on synthetic advances that have opened novel and efficient approaches to these architectures as well as the monomer classes that have been employed. The chapter also provides selection guidelines in terms of the molecular scaffolds and end-group structures to arrive at specific complex polymers.

Many authors have described click chemistry as a technique that attempts to widen the synthetic toolbox that is made available to the organic chemist in the generation of such complex structures. However, from the way in which click chemistry has been used, it more or less is akin to a universal tool that may be used to great avail in a wide variety of circumstances in polymer science. It thus has functioned to narrow down the synthetic tool box to a few select reactions, which attempt to vastly improve the efficiency with which one may achieve complex macromolecular architectures.

In the introduction to this chapter, an analysis of the molecular weight ranges and types of polymers that have been used in click conjugation methodologies is provided. However, upon inspection of these figures, one may observe that there is a lack of the versatility that click chemistry is claimed to be able to offer polymer science. The time has come for the combination of click chemistry and polymer science to be taken to the next step to take advantage of its potential versatility and drive forward the development of well-defined complex macromolecular architectures in the pursuit of new, functional materials.

References

(1) N. Hadjichristidis, M. Pitsikalis, S. Pispas and H. Iatrou, (2001), Polymers with complex architecture by living anionic polymerization, *Chem. Rev.*, **101**, 3747–3792.
(2) G. Moad, E. Rizzardo and S. H. Thang, (2005), Living radical polymerization by the RAFT process, *Aust. J. Chem.*, **58**, 379–410.
(3) S. Perrier and P. Takolpuckdee, (2005), Macromolecular design via reversible addition–fragmentation chain transfer (RAFT)/xanthates (MADIX) polymerization, *J. Polym. Sci., Part A: Polym. Chem.*, **43**, 5347–5393.

(4) A. Favier and M. T. Charreyre, (2006), Experimental requirements for an efficient control of free-radical polymerizations via the reversible addition–fragmentation chain transfer (RAFT) process, *Macromol. Rapid Commun.*, **27**, 653–692.
(5) C. Barner-Kowollik, T. P. Davis, J. P. A. Heuts, M. H. Stenzel, P. Vana and M. Whittaker, (2003), RAFTing down under: tales of missing radicals, fancy architectures, and mysterious holes, *J. Polym. Sci., Part A: Polym. Chem.*, **41**, 365–375.
(6) C. J. Hawker, A. W. Bosman and E. Harth, (2001), New polymer synthesis by nitroxide mediated living radical polymerizations, *Chem. Rev.*, **101**, 3661–3688.
(7) K. Matyjaszewski and J. H. Xia, (2001), Atom transfer radical polymerization, *Chem. Rev.*, **101**, 2921–2990.
(8) L. Barner, T. P. Davis, M. H. Stenzel and C. Barner-Kowollik, (2007), Complex macromolecular architectures by reversible addition fragmentation chain transfer chemistry: theory and practice, *Macromol. Rapid Commun.*, **28**, 539–559.
(9) K. Matyjaszewski, (2005), Macromolecular engineering: from rational design through precise macromolecular synthesis and processing to targeted macroscopic material properties, *Prog. Polym. Sci.*, **30**, 858–875.
(10) M. H. Stenzel, L. Cummins, G. E. Roberts, T. R. Davis, P. Vana and C. Barner-Kowollik, (2003), Xanthate mediated living polymerization of vinyl acetate: a systematic variation in MADIX/RAFT agent structure, *Macromol. Chem. Phys.*, **204**, 1160–1168.
(11) E. E. Malmstrom and C. J. Hawker, (1998), Macromolecular engineering via 'living' free radical polymerizations, *Macromol. Chem. Phys.*, **199**, 923–935.
(12) G. A. McConnell, A. P. Gast, J. S. Huang and S. D. Smith, (1993), Disorder–order transitions in soft-sphere polymer micelles, *Phys. Rev. Lett.*, **71**, 2102–2105.
(13) G. A. McConnell and A. P. Gast, (1997), Melting of ordered arrays and shape transitions in highly concentrated diblock copolymer solutions, *Macromolecules*, **30**, 435–444.
(14) S. Forster and M. Antonietti, (1998), Amphiphilic block copolymers in structure-controlled nanomaterial hybrids, *Adv. Mater.*, **10**, 195±.
(15) L. Leibler, (1980), Theory of microphase separation in block copolymers, *Macromolecules*, **13**, 1602–1617.
(16) F. S. Bates, (1991), Polymer–polymer phase-behavior, *Science*, **251**, 898–905.
(17) G. H. Fredrickson and F. S. Bates, (1996), Dynamics of block copolymers: theory and experiment, *Annu. Rev. Mater. Sci.*, **26**, 501–550.
(18) A. Harada and K. Kataoka, (1999), Chain length recognition: core–shell supramolecular assembly from oppositely charged block copolymers, *Science*, **283**, 65–67.
(19) A. Choucair, C. Lavigueur and A. Eisenberg, (2004), Polystyrene-b-poly(acrylic acid) vesicle size control using solution properties and hydrophilic block length, *Langmuir*, **20**, 3894–3900.
(20) H. C. Kolb, M. G. Finn and K. B. Sharpless, Click chemistry: diverse chemical function from a few good reactions, *Angew. Chem. Int. Edn*, **40**, 2004 (2001).
(21) P. Wu, A. K. Feldman, A. K. Nugent, C. J. Hawker, A. Scheel, B. Voit, J. Pyun, J. M. J. Frechet, K. B. Sharpless and V. V. Fokin, (2004), Efficiency and fidelity in a click-chemistry route to triazole dendrimers by the copper(I)-catalyzed ligation of azides and alkynes, *Angew. Chem. Int. Edn*, **43**, 3928–3932.
(22) B. Helms, J. L. Mynar, C. J. Hawker and J. M. J. Frechet, (2004), Dendronized linear polymers via 'click chemistry', *J. Am. Chem. Soc.*, **126**, 15020–15021.
(23) D. D. Diaz, S. Punna, P. Holzer, A. K. McPherson, K. B. Sharpless, V. V. Fokin and M. G. Finn, (2004), Click chemistry in materials synthesis. 1. Adhesive polymers from copper-catalyzed azide–alkyne cycloaddition, *J. Polym. Sci., Part A: Polym. Chem.*, **42**, 4392–4403.
(24) J. A. Opsteen and J. C. M. van Hest, (2005), Modular synthesis of block copolymers via cycloaddition of terminal azide and alkyne functionalized polymers, *Chem. Commun.*, 57–59.
(25) D. Quemener, T. P. Davis, C. Barner-Kowollik and M. H. Stenzel, (2006), RAFT and click chemistry: A versatile approach to well-defined block copolymers, *Chem. Commun.*, 5051–5053.
(26) K. Tsuda, T. Ishizone, A. Hirao and S. Nakahama, (1993), Protection and polymerization of functional monomers. 19. Synthesis of well-defimned poly(ethynylstyrene)s by means of anionic living polymerization of (trimethylsilyl)ethynylstyrenes, *Macromolecules*, **26**, 6985–6991.

(27) W. Agut, D. Taton and S. Lecommandoux, (2007), A versatile synthetic approach to polypeptide based rod–coil block copolymers by click chemistry, *Macromolecules*, **40**, 5653–5661.
(28) W. Lin, Q. Fu, Y. Zhang and J. Huang, (2008), One-pot synthesis of ABC type triblock copolymers via a combination of 'click chemistry' and atom transfer nitroxide radical coupling chemistry, *Macromolecules*, **41**, 4127–4135.
(29) J. A. Opsteen and J. C. M. Van Hest, (2007), Modular synthesis of ABC type block copolymers by 'click' chemistry, *J. Polym. Sci., Part A: Polym. Chem.*, **45**, 2913–2924.
(30) W. Van Camp, V. Germonpre, L. Mespouille, P. Dubois, E. J. Goethals and F. E. Du Prez, (2007), New poly(acrylic acid) containing segmented copolymer structures by combination of 'click' chemistry and atom transfer radical polymerization, *React. Funct. Polym.*, **67**, 1168–1180.
(31) S. R. S. Ting, A. M. Granville, D. Quemener, T. P. Davis, M. H. Stenzel and C. Barner-Kowollik, (2007), RAFT chemistry and Huisgen 1,3-dipolar cycloaddition: a route to block copolymers of vinyl acetate and 6-O-methacryloyl mannose? *Aust. J. Chem.*, **60**, 405–409.
(32) N. V. Tsarevsky, B. S. Sumerlin and K. Matyjaszewski, (2005), Step-growth 'click' coupling of telechelic polymers prepared by atom transfer radical polymerization, *Macromolecules*, **38**, 3558–3561.
(33) N. J. Nasrullah, A. Vora and D. C. Webster, (2008), Block copolymer synthesis by a combination of ATRP and RAFT via click chemsitry using a high throughput approach, *Polym. Prepr.*, **49**, 422–423.
(34) R. Ranjan and W. J. Brittain, (2007), Combination of living radical polymerization and click chemistry for surface modification, *Macromolecules*, **40**, 6217–6223.
(35) W. Agut, R. Agnaou, S. Lecommandoux and D. Taton, (2008), Synthesis of block copolypeptides by click chemistry, *Macromol. Rapid Commun.*, **29**, 1147–1155.
(36) X. H. He, L. Y. Liang, M. R. Xie, Y. Q. Zhang, S. L. Lin and D. Y. Yan, (2007), Synthesis of novel linear PEO-b-PS-b-PCL triblock copolymers by the combination of ATRP, ROP, and a click reaction, *Macromol. Chem. Phys.*, **208**, 1797–1802.
(37) G. Mantovani, V. Ladmiral, L. Tao and D. M. Haddleton, (2005), One-pot tandem living radical polymerisation-Huisgens cycloaddition process ("click") catalysed by *N*-alkyl-2-pyridylmethanimine/Cu(I)Br complexes, *Chem. Commun.*, 2089–2091.
(38) J. W. Lee, J. H. Kim, B. K. Kim, J. H. Kim, W. S. Shin and S. H. Jin, (2006), Convergent synthesis of PAMAM dendrimers using click chemistry of azide-functionalized PAMAM dendrons, *Tetrahedron*, **62**, 9193–9200.
(39) Y. Li, J. W. Yang and B. C. Benicewicz, (2007), Well-controlled polymerization of 2-azidoethyl methacrylate at near room temperature and click functionalization, *J. Polym. Sci., Part A: Polym. Chem.*, **45**, 4300–4308.
(40) V. Ladmiral, T. M. Legge, Y. Zhao and S. Perrier, (2008), 'Click' chemistry and radical polymerization: potential loss of orthogonality, *Macromolecules*, 10.1021/ma8010262.
(41) H. Durmaz, B. Colakoclu, U. Tunca and G. Hizal, (2006), Preparation of block copolymers via diels alder reaction of maleimide- and anthracene-end functionalized polymers, *J. Polym. Sci., Part A: Polym. Chem.*, **44**, 1667–1675.
(42) H. Durmaz, A. Dag, O. Altintas, T. Erdogan, G. Hizal and U. Tunca, (2007), One-pot synthesis of ABC type triblock copolymers via *in situ* click [3+2] and Diels–Alder [4+2] reactions, *Macromolecules*, **40**, 191–198.
(43) A. J. Inglis, S. Sinnwell, T. P. Davis, C. Barner-Kowollik and M. H. Stenzel, (2008), Reversible addition fragmentation chain transfer (RAFT) and hetero-Diels–Alder chemistry as a convenient conjugation tool for access to complex macromolecular designs, *Macromolecules*, **41**, 4120–4126.
(44) S. Sinnwell, A. J. Inglis, T. P. Davis, M. H. Stenzel and C. Barner-Kowollik, (2008), An atom-efficient conjugation approach to well-defined block copolymers using RAFT chemistry and hetero Diels–Alder cycloaddition, *Chem. Commun.*, 2052–2054.
(45) P. L. Golas, N. V. Tsarevsky, B. S. Sumerlin, L. M. Walker and K. Matyjaszewski, (2007), Multisegmented block copolymers by 'click' coupling of polymers prepared by ATRP, *Aust. J. Chem.*, **60**, 400–404.

(46) J. F. Lutz, H. G. Borner and K. Weichenhan, (2007), 'Click' bioconjugation of a well-defined synthetic polymer and a protein transduction domain, *Aust. J. Chem.*, **60**, 410–413.
(47) L. Mespouille, M. Vachaudez, F. Suriano, P. Gerbaux, O. Coulembier, P. Degee, R. Flammang and P. Dubois, (2007), One-pot synthesis of well-defined amphiphilic and adaptative block copolymers via versatile combination of 'click' chemistry and ATRP, *Macromol. Rapid Commun.*, **28**, 2151–2158.
(48) S. Hilf, N. Hanik and A. F. M. Kilbinger, (2008), A 'click' approach to ROMP block copolymers, *J. Polym. Sci., Part A: Polym. Chem.*, **46**, 2913–2921.
(49) M. Jeffries-EL and M. Mitchell, (2008), Efficient synthesis of poly(3-hexylthiophene) block copolymers using 'click' chemistry, *Polym. Prepr.*, **49**, 168–169.
(50) R. Sachsenhofer and W. H. Binder, (2008), Novel synthesis of poly(isobutylene-block-ethyleneoxide), *Polym. Prepr.*, **49**, 193–194.
(51) H. F. Gao and K. Matyjaszewski, (2006), Synthesis of star polymers by a combination of ATRP and the 'click' coupling method, *Macromolecules*, **39**, 4960–4965.
(52) A. Dag, H. Durmaz, G. Hizal and U. Tunca, (2008), Preparation of 3-arm star polymers (A(3)) via Diels–Alder click reaction, *J. Polym. Sci., Part A: Polym. Chem.*, **46**, 302–313.
(53) X. L. Luo, G. W. Wang, X. C. Pang and J. L. Huang, (2008), Synthesis of a novel kind of amphiphilic graft copolymer with miktoarm star-shaped side chains, *Macromolecules*, **41**, 2315–2317.
(54) H. Y. Li, R. Riva, R. Jerome and P. Lecomte, (2007), Combination of ring-opening polymerization and 'click' chemistry for the synthesis of an amphiphilic tadpole-shaped poly(epsilon-caprolactone) grafted by PEO, *Macromolecules*, **40**, 824–831.
(55) S. Sinnwell, A. J. Inglis, M. H. Stenzel and C. Barner-Kowollik, (2008), Access to three-arm star block copolymers by a consecutive combination of the copper(I)-catalyzed azide–alkyne cycloaddition and the RAFT hetero Diels–Alder concept, *Macromol. Rapid Commun.*, **29**, 1090–1096.
(56) D. Quemener, M. Le Hellaye, C. Bissett, T. P. Davis, C. Barner-Kowollik and M. H. Stenzel, (2008), Graft block copolymers of propargyl methacrylate and vinyl acetate via a combination of RAFT/MADIX and click chemistry: Reaction analysis, *J. Polym. Sci., Part A: Polym. Chem.*, **46**, 155–173.
(57) O. Altintas, B. Yankul, G. Hizal and U. Tunca, (2006), A(3)-type star polymers via click chemistry, *J. Polym. Sci., Part A: Polym. Chem.*, **44**, 6458–6465.
(58) R. Hoogenboom, B. C. Moore and U. S. Schubert, (2006), Synthesis of star-shaped poly(epsilon-caprolactone) via 'click' chemistry and 'supramolecular click' chemistry, *Chem. Commun.*, 4010–4012.
(59) H. F. Gao, K. Min and K. Matyjaszewski, (2007), Synthesis of 3-arm star block copolymers by combination of 'core-first' and 'coupling-onto' methods using ATRP and click reactions, *Macromol. Chem. Phys.*, **208**, 1370–1378.
(60) M. R. Whittaker, C. N. Urbani and M. J. Monteiro, (2006), Synthesis of 3-miktoarm stars and 1st generation mikto dendritic copolymers by 'living' radical polymerization and 'click' chemistry, *J. Am. Chem. Soc.*, **128**, 11360–11361.
(61) J. Zhu, X. L. Zhu, E. T. Kang and K. G. Neoh, (2007), Design and synthesis of star polymers with hetero-arms by the combination of controlled radical polymerizations and click chemistry, *Polymer*, **48**, 6992–6999.
(62) O. Altintas, G. Hizal and U. Tunca, (2006), ABC-type hetero-arm star terpolymers through 'click' chemistry, *J. Polym. Sci., Part A: Polym. Chem.*, **44**, 5699–5707.
(63) G. H. Deng, D. Y. Ma and Z. Z. Xu, (2007), Synthesis of ABC-type miktoarm star polymers by 'click' chemistry, ATRP and ROP, *Eur. Polym. J.*, **43**, 1179–1187.
(64) H. Durmaz, F. Karatas, U. Tunca and G. Hizal, (2006), Preparation of ABC miktoarm star terpolymer containing poly(ethylene glycol), polystyrene, and poly(tert-butylacrylate) arms by combining Diels–Alder reaction, atom transfer radical, and stable free radical polymerization routes, *J. Polym. Sci., Part A: Polym. Chem.*, **44**, 499–509.
(65) O. Altintas, B. Yankul, G. Hizal and U. Tunca, (2007), One-pot preparation of 3-miktoarm star terpolyrmers via click [3 + 2] reaction, *J. Polym. Sci., Part A: Polym. Chem.*, **45**, 3588–3598.

(66) G. W. Wang, X. L. Luo, C. Liu and J. L. Huang, (2008), Synthesis of ABCD 4-miktoarm star-shaped quarterpolymers by combination of the 'click' chemistry with multiple polymerization mechanism, *J. Polym. Sci., Part A: Polym. Chem.*, **46**, 2154–2166.
(67) O. Altintas, G. Hizal and U. Tunca, (2008), ABCD 4-miktoarm star quarterpolymers using click [3 + 2] reaction strategy, *J. Polym. Sci., Part A: Polym. Chem.*, **46**, 1218–1228.
(68) Q. C. Liu, P. Zhao and Y. M. Chen, (2007), Divergent synthesis of dendrimer-like macromolecules through a combination of atom transfer radical polymerization and click reaction, *J. Polym. Sci., Part A: Polym. Chem.*, **45**, 3330–3341.
(69) C. N. Urbani, C. A. Bell, M. R. Whittaker and M. J. Monteiro, (2008), Convergent synthesis of second generation AB-type miktoarm dendrimers using 'click' chemistry catalyzed by copper wire, *Macromolecules*, **41**, 1057–1060.
(70) B. Parrish, R. B. Breitenkamp and T. Emrick, (2005), PEG- and peptide-grafted aliphatic polyesters by click chemistry, *J. Am. Chem. Soc.*, **127**, 7404–7410.
(71) J. Y. Zhang, Y. M. Zhou, Z. Y. Zhu, Z. S. Ge and S. Y. Liu, (2008), Polyion complex micelles possessing thermoresponsive coronas and their covalent core stabilization via 'click' chemistry, *Macromolecules*, **41**, 1444–1454.
(72) N. V. Tsarevsky, S. A. Bencherif and K. Matyjaszewski, (2007), Graft copolymers by a combination of ATRP and two different consecutive click reactions, *Macromolecules*, **40**, 4439–4445.
(73) R. Riva, P. Schmeits, F. Stoffelbach, C. Jerome, R. Jerome and P. Lecomte, (2005), Combination of ring-opening polymerization and 'click'' chemistry towards functionalization of aliphatic polyesters, *Chem. Commun.*, 5334–5336.
(74) R. Riva, S. Schmeits, C. Jerome, R. Jerome and P. Lecomte, (2007), Combination of ring-opening polymerization and 'click chemistry': toward functionalization and grafting of poly(epsilon-caprolactone), *Macromolecules*, **40**, 796–803.
(75) H. Y. Li, R. Riva, H. R. Kricheldorf, R. Jerome and P. Lecomte, (2008), Synthesis of eight- and star-shaped poly(epsilon-caprolactone)s and their amphiphilic derivatives, *Chem. Eur. J.*, **14**, 358–368.
(76) X. Jiang, E. B. Vogel, M. R. Smith and G. L. Baker, (2008), 'Clickable' polyglycolides: tunable synthons for thermoresponsive, degradable polymers, *Macromolecules*, **41**, 1937–1944.
(77) H. F. Gao and K. Matyjaszewski, (2007), Synthesis of molecular brushes by 'grafting onto' method: combination of ATRP and click reactions, *J. Am. Chem. Soc*, **129**, 6633–6639.
(78) X. Jiang, M. C. Lok and W. E. Hennink, (2007), Degradable-brushed pHEMA-pDMAEMA synthesized via ATRP and click chemistry for gene delivery, *Bioconjugate Chem.*, **18**, 2077–2084.
(79) B. Gacal, H. Durmaz, M. A. Tasdelen, G. Hizal, U. Tunca, Y. Yagci and A. L. Demirel, (2006), Anthracene–maleimide-based Diels–Alder 'click chemistry' as a novel route to graft copolymers, *Macromolecules*, **39**, 5330–5336.
(80) C. Ott, R. Hoogenboom and U. S. Schubert, (2008), Post-modification of poly(pentafluorostyrene): a versatile 'click' method to create well-defined multifunctional graft copolymers, *Chem. Commun.*, 3516–3518.

7

Click Chemistry on Supramolecular Materials

Wolfgang H. Binder and Robert Sachsenhofer

7.1 Introduction

A plethora of materials has been generated in the past decades, often built from molecules in highly defined configurations and conformations. Additionally, many modern functional materials rely on defined arrangement of molecular aggregates, in which the arrangement of molecules dictates the use of the underlying material,[1] thus putting supramolecular structure and ordering in the limelight.[2] Thus the use of optoelectronically active materials is strongly influenced by their arrangement in crystals or semicrystals, controlling band-overlap or charge-transport. As examples, the ordering of sexithiophenes in solar-cell devices strongly influences their ability to harvest photons and convert them into excitons; the conjugation length of oligo-(phenylene-vinylenes) strongly influences their absorption spectrum and thus their use in organic-light emitting diods; push–pull liquid crystalline molecules are ordered into liquid-crystalline phases via dipole–dipole interactions, which can be switched by external electrical fields from one liquid crystalline phase into another, thus changing the reflection of light as required in LCDs. Similarly, materials for use in biochemical applications are strongly influenced by noncovalent bonds acting through space, making hydrogen bonds or dipolar interactions the main directing forces for the spatial arrangement of biochemical receptors (Figure 7.1).

These examples demonstrate the close proximity of material science and supramolecular chemistry,[3] which are connected via the proper spatial and orientational positioning of intermolecular forces and interaction within molecular building blocks. Thus, often a molecular (= functional) scaffold needs to be oriented in space via appropriately affixed

Click Chemistry for Biotechnology and Materials Science Edited by Joerg Lahann

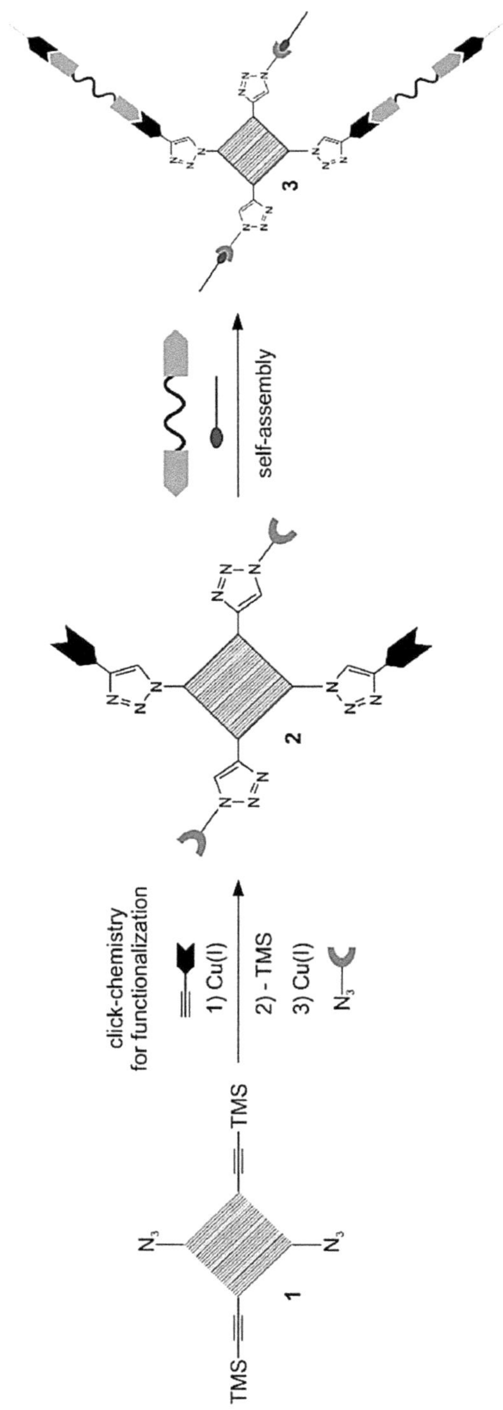

***Figure 7.1** Concept for the integration of azide–alkyne click chemistry into supramolecular science: highly efficient functionalization of the scaffold (1) is achieved with two different supramolecular receptors, furnishing the supramolecule (2), which assembled into aggregates (3).*

supramolecular interactions, which must be properly fixed and arranged around the molecule of interest. This often requires considerable synthetic force, since the molecules used are multifunctional, and thus multistep pathways are necessary – with all the disadvantages of modern synthetic organic chemistry. Thus, as most of the mentioned structures require tedious synthetic pathways, the approach to a specific structure is often limited by long-step syntheses and purification issues, often hampered by incomplete and insufficient chemical reactions. Moreover, as the molecular weights approach the limit of oligomeric and polymeric structures, the defined functionalization of such materials becomes problematic, as purification of incompletely reacted starting materials from their final products is difficult due to similar chemical structures or comparable size.

Universal chemical reactions, which are able to link many molecular species without protecting groups, offer high yields and provide an inherent insensitivity to chemical structure and solvents, are an important step forward in supramolecular material science. Moreover, as the energy of assembly in many supramolecular structures is close to the thermal energy, such a universal reaction in the ideal case would be of catalytic nature, not requiring strong acceleration by temperature increases, thus keeping supramolecular assemblies in place during reaction. Click chemistry is a proponent set of reactions, able to act as universal chemistry within supramolecular (material) science. According to the definition of Sharpless *et al.*,[4] a 'click reaction' is defined by a gain of thermodynamic enthalpy of at least 20 kcal mol^{-1}, thus opening way to a high-yielding and thus nearly substrate-insensitive reaction. This type of reaction was found most of all in the azide–alkyne click reaction,[4–6] which represents a metal-catalyzed variant of the Huisgen 1,3-dipolar cycloaddition reaction[7,8] between CC triple, CN triple bonds[9] and alkyl-/aryl-/sulfonyl azides. The relevant outcomes of this reaction are (a) tetrazoles,[5,10] (b) 1,2,3-triazoles[11–14] or (c) 1,2-oxazoles, respectively. Briefly, the basic process of the Huisgen 1,3-dipolar cycloaddition[2,10,11] generates 1,4- and 1,5-triazoles, respectively (Scheme 7.1). The main metal salts used to accelerate this (at room-temperature rather slow) reaction are copper (I) salts [Cu(I)Br, Cu(I)I, in amounts of approximately 0.25–2 mol% with respect to the azide or alkyne substrate], aqueous regenerative systems [i.e. Cu(II) salts–ascorbic acid] as well as various copper clusters (Cu–Cu-oxide nanoparticles, sized 7–10 nm[15] or ~4 nm[16]), metallic $Cu^{(O)}$clusters[16–18] and copper–charcoal.[19] Recently, the use of a Cu(I)-free variant using the ring-strain of substituted 1,1,-difluoro-cyclooctynes to promote the dipolar cycloaddition process has been described, enabling mild reactions on living (cellular) systems.[20] Besides copper, other metals employed include Ru complexes[21] {[CpRuCl(PPh_3), [Cp∗$RuCl_2$]$_2$, Cp∗RuCl(NBD) and Cp∗RuCl(COD) favouring 1,5-addition [i.e. with $Ru(OAc)_2(PPh_3)_2$]}, and Au(I),[22] Ni, Pd[23] and Pt salts, although with much less catalytic activity.[24]

The mechanism of the reaction is different from that of a purely thermal 1,3-dipolar cycloaddition. According to Sharpless *et al.*,[11] modified by Finn *et al.*[25,26] by computational methods,[27,28] and finally revised by Bock *et al.*,[29] the metal-catalyzed reaction involves: (a) an up to 10^5th-rate acceleration and an absolute 1,4-regioselectivity of the Cu(I)-catalyzed process; (b) a kinetic feature of the reaction indicating at least second-order kinetics with respect to the concentration of the copper species,[26] thus involving at least two copper centers within the catalytic cycle, presumably linking two acetylenes via a μ-bridge;[30] (c) a significant autoacceleration if multiple triazoles are formed,[31] revealing intermolecular ligands effects; and (d) a significant rate-reduction with strongly increasing amount of copper. A basic feature, however is the formation of a copper-acetylide, resulting in the

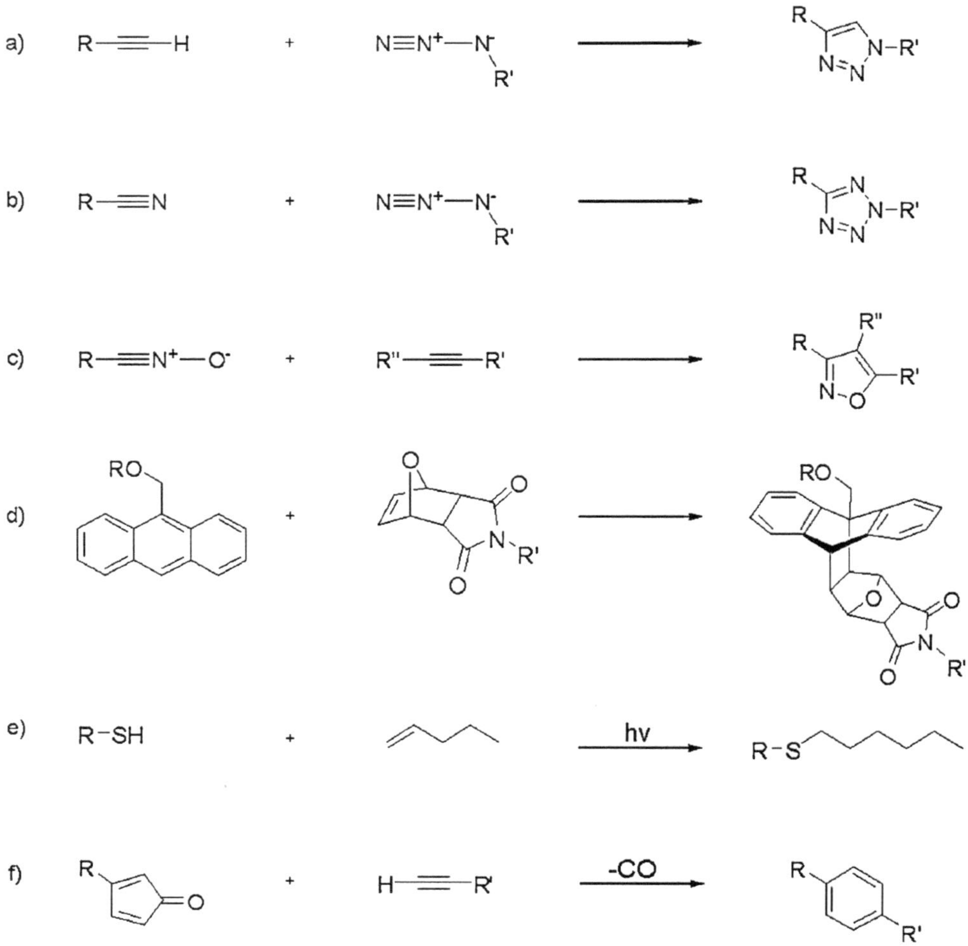

Scheme 7.1

lowering of the pK_a-value of the Cu-acetylide by up to 9.8 units as calculated[28] via DFT-calculations.

Thus a relatively complex supramolecular assembly (**3**) can be generated from the supramolecule with two different receptor structures (**2**) (Figure 7.1) (representing different supramolecular interactions) in a two-step procedure, using the central azide- or alkyne-modified starting molecule (**1**), as nearly all functional groups are compatible with this process, except those that are (a) self-reactive or (b) able to yield stable complexes with the [Cu(I)-metal] under catalyst deactivation. The main interfering functional groups are terminal azides and alkynes,[32] strongly activated cyanides,[5,6,10] free (= accessible) thiol-moieties (R-SH) via the Staudinger reaction as well as strained or electronically activated alkenes.[8,33] However, the possibility to use free-thiols prior to an azide–alkyne click reaction has been demonstrated on polymers[34] and surfaces,[35] thus enabling the use of free thiols despite the often interfering azide–amine reduction by the free thio-moiety. Most known solvents and biphasic reaction systems (mixtures of water–alcohol to water–toluene) can be applied with excellent results. Cocatalytic systems[31] often used include amino-bases[17,36] [mono- and multivalent triazoles[36], but also phosphines such as tris(carboxyethyl)phosphine (TCPE)]. Many reviews descriebe the azide–alkyne click reaction in general,[4,29] for application in polymer chemistry,[37–43] dendrimers,[40,44] carbohydrate chemistry,[45–47] materials-chemistry[37,38] and organic chemistry,[42,48] as well as for peptides[49] and drug discovery.[50] The following review focuses on the use of the azide–alkyne click reaction for the synthesis and assembly of multifunctional molecules in supramolecular (material) science.

7.2 Click Reactions on Rotaxanes, Cyclodextrines and Macrocycles

Rotaxanes, cyclodextrines and macrocycles are among the 'oldest' supramoelcular molecules developed. They are highly defined, and there has been intense research during recent decades. The azide–alkyne click reaction has promoted research in this area, as many synthetic approaches have become simpler and more effective.

7.2.1 Click with Rotaxanes

The basic approach to rotaxanes mediated via click reactions is shown in Figure 7.2.[51–57] Thus interlocked structures by (a) stoppering-reactions, (b) 'click polymerization or (c) macrocyclization have been achieved. A recent review has focused on this topic, describing rotaxanes and catenanes via click chemistry.[52] Two classical 'click stoppering' approaches have been described by Sauvage *et al.*[54] and Stoddart *et al.*[55] recently (see Figures 7.3 and 7.4).

Thus the assembly of the molecules is driven first by noncovalent interactions generating the complex (**4**), and subsequently the 'stoppers' fix the corresponding rotaxanes (**5**) in yields above 80%. An interesting example of 'click stoppering' by Sauvage *et al.*[54] (see Figure 7.4), generated the Cu-complexed molecule (**7**) via the click reaction. As the copper(I)-species is directly linked to the central rotaxane-core, thus acting not only as scaffold-forming metal, but also as catalyst in the subsequent click reaction, Leigh *et al.*[56] have formulated a highly interesting mechanism for the catalytic cycle as shown in Figure 7.4. Thus the copper-species interlocks molecule (**8**) and forms the catalytic species to add the propargylic alcohol (**6**) and the azido alcohol (**9**) to furnish the species (**10**) in catalytic amounts. Thus only a small amount of copper is needed to effect the efficient formation of the rotaxane (**7**). A similar formation of [3]-rotaxanes has been described furnishing the interlocked molecule (**11**) (see Figure 7.5).

One of the oldest examples of rotaxane formation was described before the discovery of Cu(I) catalysis by Steinke *et al.*,[57,58] focusing on the reaction of cucurbiturils into rotaxanes via 'click polymerization'. This approach has recently been revived,[53] generating pseudo-polyrotaxanes (**12**) threading cucurbit[7]uril and β-cyclodextrine in an alternating fashion onto the respective polymer, which is generated by a polyaddition process. As the click reaction is highly moderate in its reaction condition, the simple synthesis of such molecules is only possible via this strategy (Figure 7.6).

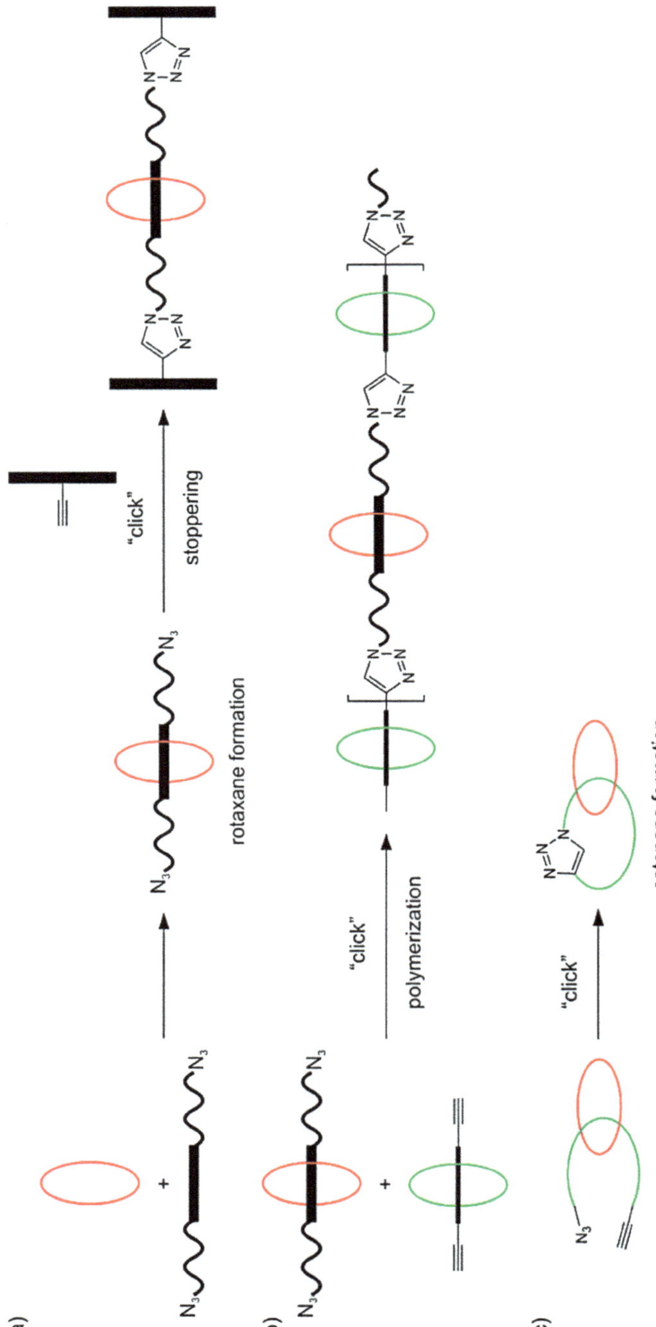

Figure 7.2 *Concept for the generation of rotaxanes via azide–alkyne click reactions.*

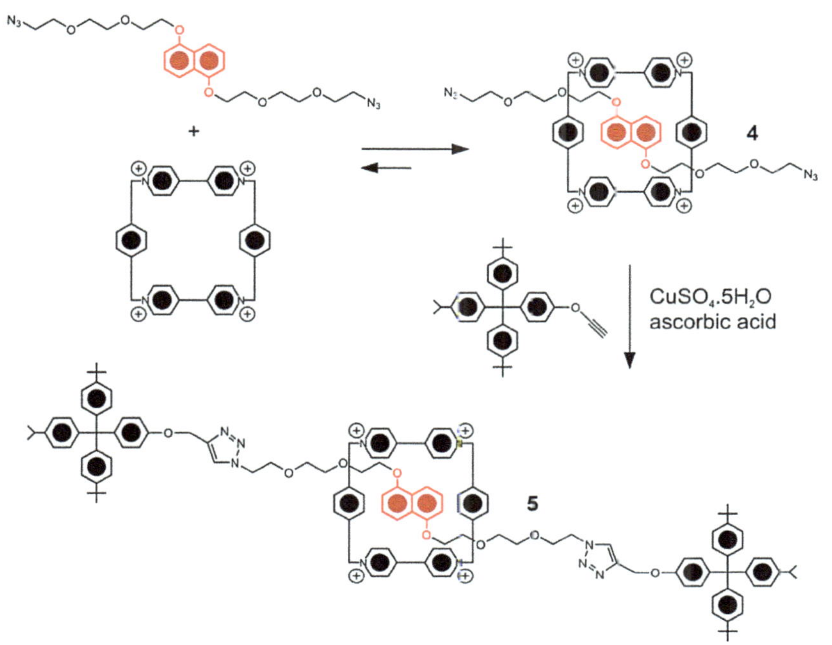

Figure 7.3 Formation of rotaxanes via click stoppering.

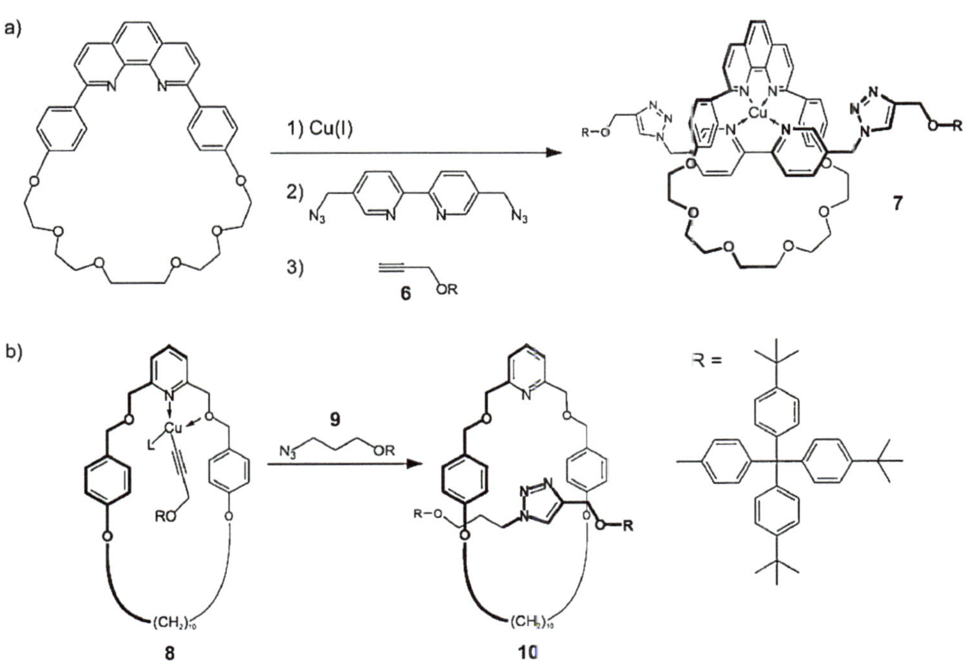

Figure 7.4 Formation and mechanistic considerations for the generation of roatxane (7).

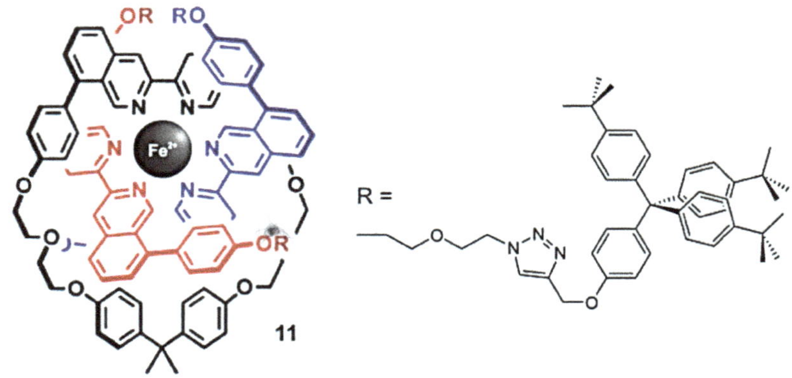

11

Figure 7.5 Formation of interlocked structures (**11**).

7.2.2 Click on Cyclodextrines

As the azide–alkyne click chemistry has been strongly applied to carbohydrates[45,46,59] due to their multifunctional nature and the excellent compatibility of the azide–alkyne click reaction with multihydroxy moieties; also cyclodextrines as well as the logically related calixarenes[60] have been functionalized via this reaction. Cyclodextrines are important supramolecular molecules, with a highly defined hydrophobic cavity and a hydrophilic exterior, allowing the complexation of hydrophobic guest within their interior, controlled by substituents on the outer rim. The functionalization of the outer rim via their 6′-hydroxy-moieties thus is an important point, often hampered by steric constraints or insufficient reactions that can nicely be accomplished vie the azide–alkyne click reaction, as shown in Figure 7.7.[53,61–69] The transformation of the 6′-hydroxylmoieties into azido-moieties can be achieved via direct reactions, thus opening the possibility for the

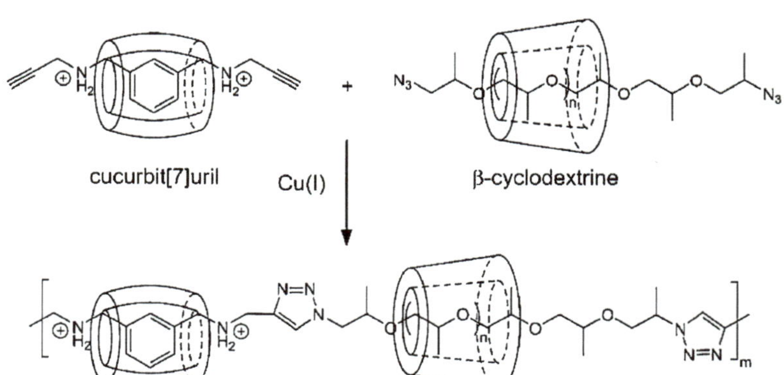

12

Figure 7.6 Formation of alternating copolymers (**12**) consisting of rotaxanes made from b-cyclodextrin and cucurbit[7]uril.

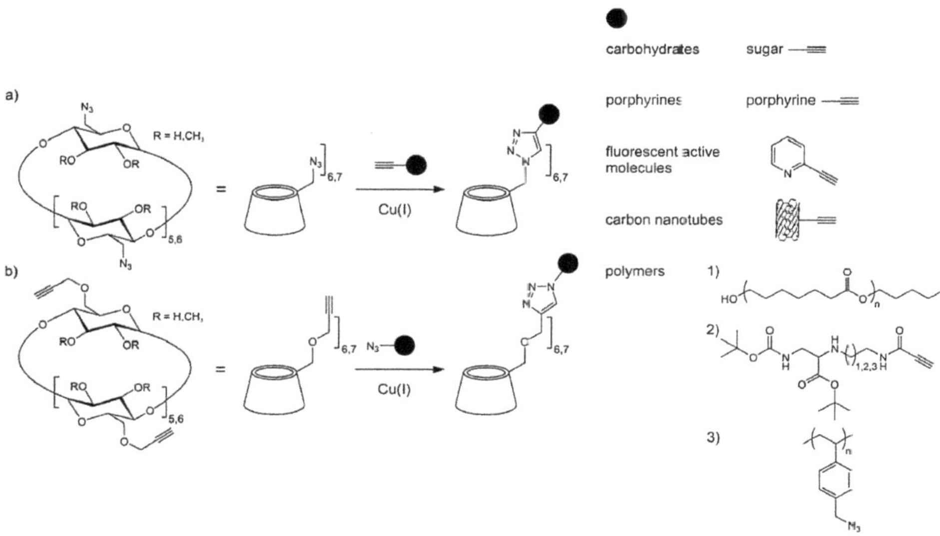

Figure 7.7 Generation of modified cyclodextrines by azide–alkyne click reaction.

attachment of other carbohydrates,[66,68] porphyrines,[64] other cyclodextrines,[63] fluorescent-labels,[67] carbon-nanotubes[65] and a variety of polymers.[62,69]

An excellent example for the use of the azide–alkyne click reaction has been recently described by Liu *et al.*[64] by attaching 6′-mono-azido-cyclodextrine moieties onto all four edges of porphyrine-molecules, thus generating brick-type structures as shown in Figure 7.8. The resulting structures [either the methoxylated structures (**13**) or the hydroxylated structures (**14**)] can self-assemble into highly regular structures using porphyrines with four phenyl groups, able to insert into the cyclodextrine cavities. Thus highly organized nanostructures can be built from these easily available associates within a one-step synthetic procedure.

Hoogenboom *et al.*[69] have attached seven poly(caprolactones) to the 6-positions of a β-cyclodextrin-derivative [see Figure 7.9(a)], generating a polymer-modifed cyclodextrin (star-shaped polymer **15**). As the steric demand of such polymeric chains is fairly large, the result is important and could not have been achieved by other method.

Another modification of cyclodextrines with polymers has been reported by attaching cationic polyimine polymers to a β-cyclodextrine-derivative.[62] The resulting star-polymer (**16**) can self-assemble into aggregates of about 50–100 nm and complex DNA due to its highly poly(cationic) nature. The molecules are highly efficient, nontoxic delivery agents for nucleic acids into HELA cells due to their complex formation and masking of the DNA within their assemblate structure.

Modification of carbon-nanotubes has been reported with cyclodextrines. Similar to the method employed previously with PS polymers,[70] an alkyne moiety was introduced via a nitrene addition onto the carbon nanotube backbone. Subsequently, the monofunctionalized cyclodextrine was attached to the carbon nanotubes, leading to uniformly labeled CD nanotubes (**17**), visible using TEM methods (Figure 7.10).

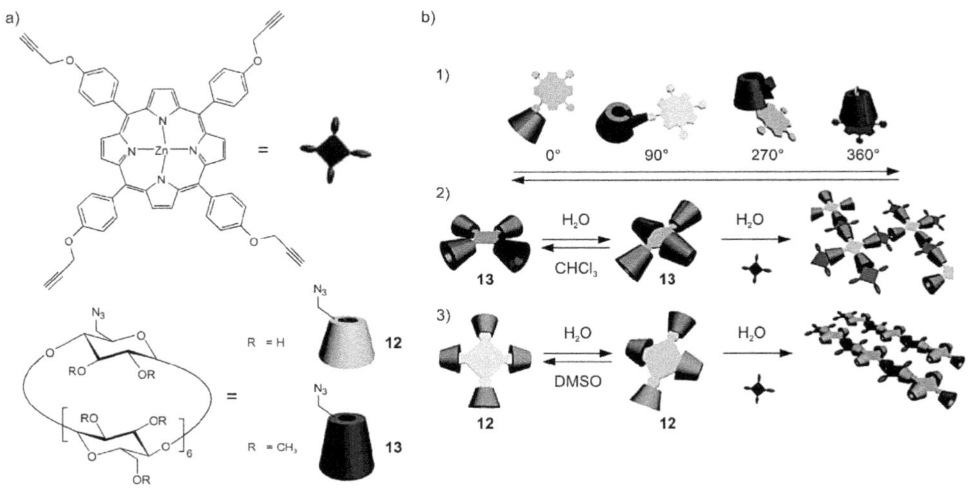

Figure 7.8 Assembly of cyclodextrine-modifed porphyrines **12** and **13** into regular grids via key/lock-type interactions. Reprinted with permission from Y. Liu et al., (2008), Complexation-induced transition of nanorod to network aggregates: alternate porphyrin and cyclodextrin arrays, J. Am. Chem. Soc., **130** (2), 600–605. Copyright 2008 American Chemical Society.

7.2.3 Click on Macrocycles

Macrocycles are important supramolecular structures – early examples include crown-ethers or cryptands; later examples are often related to membrane-spanning channels,[71,72] defined amphiphilic macrocycles[73] or cyclic polymers.[74] The engineering of such structures is difficult, since supramolecular interactions and defined rigid/flexible segments are

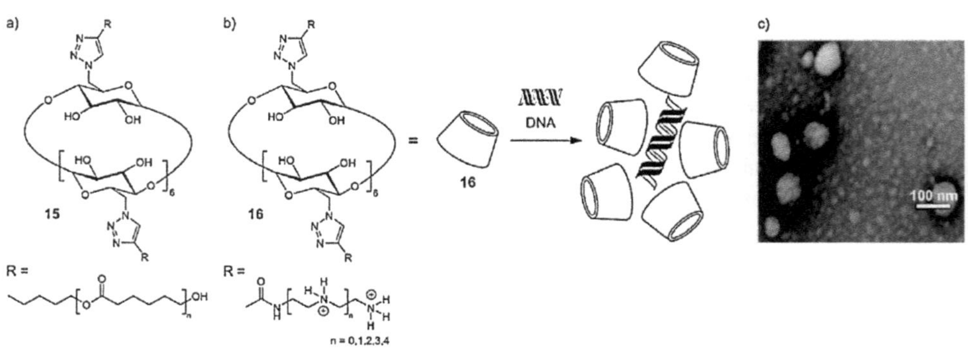

Figure 7.9 (a) Generation of DNA–cyclodextrine complexes via attachment of cationic polymers onto the outer rim, yielding cyclodextrine **16**. (b) TEM-micrograph of the complexes formed between DNA and **16**. Reprinted with permission from S. Srinivasachari et al., (2008), Polycationic beta-cyclodextrin click clusters: monodisperse and versatile scaffolds for nucleic acid delivery, J. Am. Chem. Soc., **130**, 4618–4627. Copyright 2008 American Chemical Society.

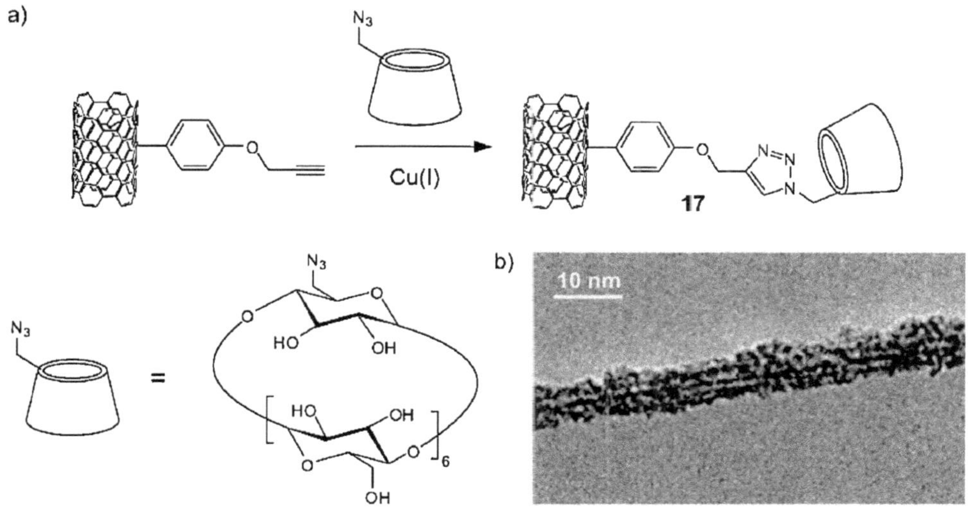

Figure 7.10 (a) Attachment of cyclodextrines onto carbon-nanotubes. (b) TEM-micrograph of the formed-CD/nanotube structure. Reprinted with kind permission from Z. Guo et al., (2008), Covalently β-cyclodextrin modified single-walled carbon nanotubes: a novel artificial receptor synthesized by click chemistry, J. Nanopart. Res., **10** (6), 1077–1083. Copyright 2008 Springer Science and Business Media.

difficult to be introduced into large cycles due to the poor efficiency of the cyclization reaction, whose efficiency declines with increasing chain length and steric constraints[75] (cyclization efficiency $\sim 1/N^{3/2}$ (N = chain length of the linear structure).[76] As the topic has been reviewed recently, only selected examples are provided in this review.

Cyclic polymers have been generated via the α,ω-end group functionalization PS (**18**),[77] PNIPAM (**20**, **22**)[78,79] and poly(ε-caprolactones)[80] with azide and alkyne groups. After macrocyclization, this strategy yields the corresponding cyclic polymers **20**, **22** in yields of ~80% efficiency (see Figure 7.11). starting from their linear precursor-structures **19** or **21**.

This method is a highly efficient method to generate cyclic polymers with higher molecular weights from their linear counterparts. As the properties of cyclic polymers are very different from those of the corresponding linear structures in terms of chain-conformation, crystallization and supramolecular ordering, this simple approach to cyclic polymers will definitely be a landmark for further investigations.

Haridas *et al.*[81] have described the synthesis of macrocycles **23** and **24**, starting from the open-bisacetylene via ring-closure reaction. (Figure 7.12). The generated macrocycles (triazolophanes) display a nonclassical hydrogen bonding system, as solvent molecules (i.e. acetonitrile) can be embedded in the interior of the cycle via these nonclassical hydrogen bonds. Currently, the macrocycle is investigated for its ionophoric properties due to this hydrogen-bonding ability.[81]

Another example of such nonclassical hydrogen bonds for ionophoric abilities has been reported by Flood *et al.*,[82] generating macrocycle **25** with four triazole-rings in its cavity. A dynamic equilibrium is observed upon addition of chlorine ions, which can be complexed to the interior, furnishing the ioniphore **26** with an association constant of $K_{\text{assn}} \approx 130\,000\ \text{M}^{-1}$.

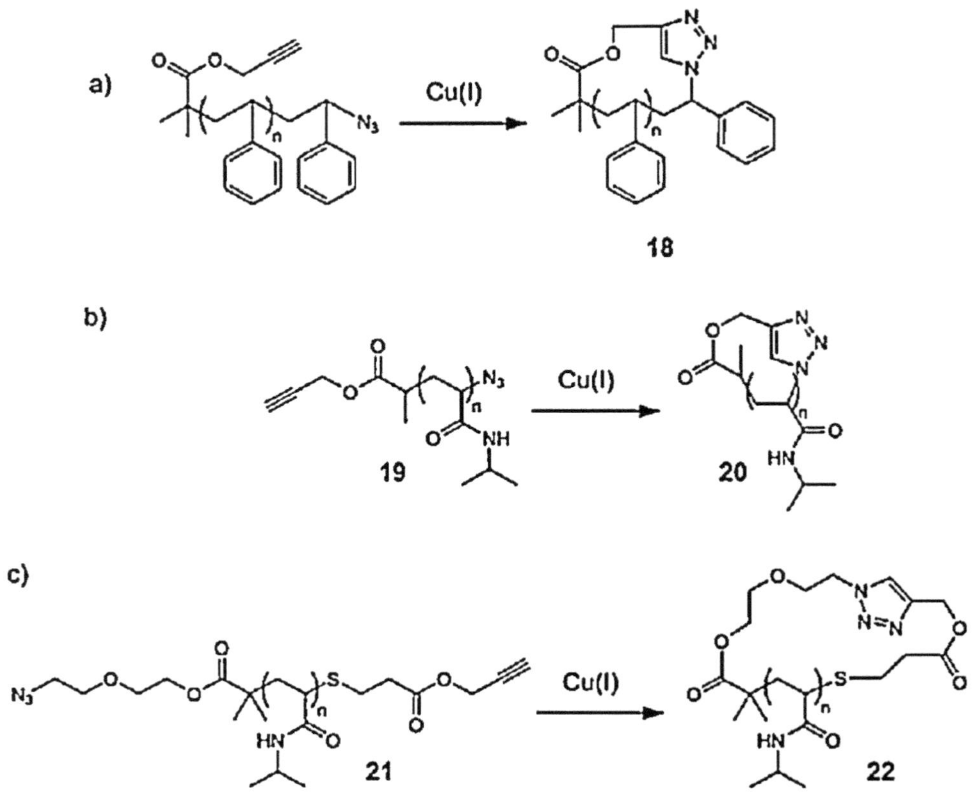

Figure 7.11 *Macrocyclization reactions of α,ω-modified polymers: (a) formation of cyclic-poly(styrene) (PS)* ***18****; (b, c) macrocyclic poly(*N*-isopropylacrylamides)* ***20*** *and* ***22****.*

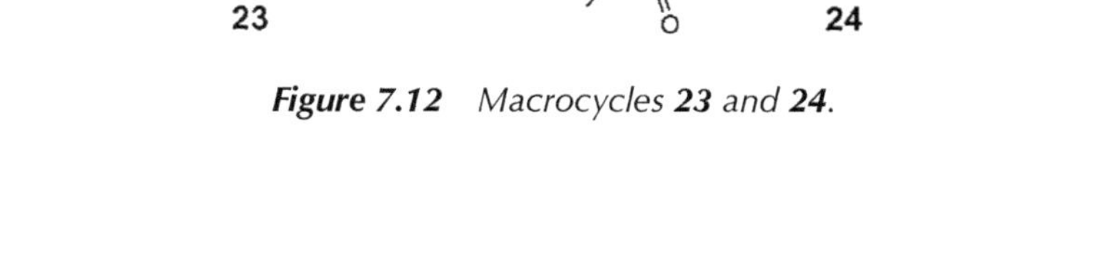

Figure 7.12 *Macrocycles* ***23*** *and* ***24****.*

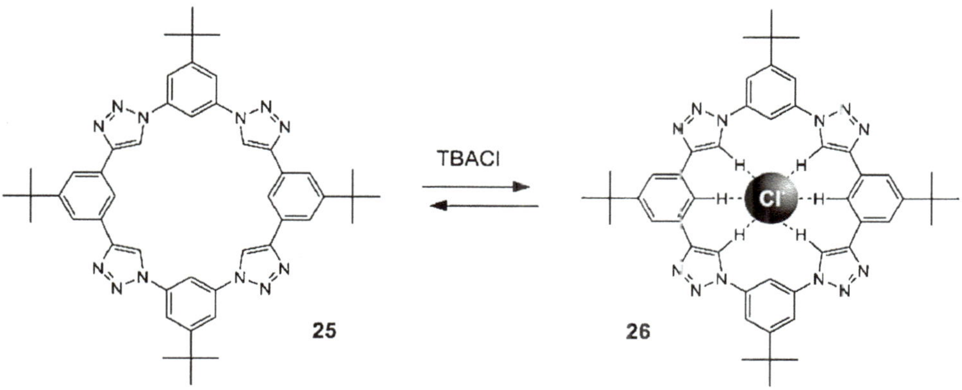

Figure 7.13 *Ionophoric activity of the macrocycle* ***25*** *via its chlorine-form* ***26****.*

Again, such structures are difficult to make via conventional pathways, thus opening new supramolecular macrocycles via the azide–alkyne click reaction (Figure 7.13).

Cyclic peptides for membrane-spanning,[72] helical peptides[72,83] or as binding domains for SH2 domains[84] have been prepared by the azide–alkyne click reaction. Thus the starting peptide **27** has been reacted with Cu(I)–ascorbate system in aqueous media, generating the SH2-binding macrocycle **28** as well as the dimeric macrocycle **29**. Both were found to exhibit increased affinities towards the Sh2-domain in sub-μM concentrations (Figure 7.14).

Cyclic carbohydrate molecules have been generated via click reactions (see Figure 7.15). Thus the dimerization of the trisachharide **30** furnishes the macrocycle **31** in nearly 80% yield.[68] As slightly different strategy has been used relying a combination of ring closing metathesis (RCM) and the azide–alkyne click reaction:[85] thus the macrocycles **32**, **33** and **34** containing the hexo- and pentopyranoses within their ring structure have been prepared in yields between 73 and 95% (Figure 7.15).

7.3 Click Reactions on DNA

It is unquestionable that DNA and RNA represent some of the best-studied supramolecular systems, as several types of supramolecular interactions are present and can be used for

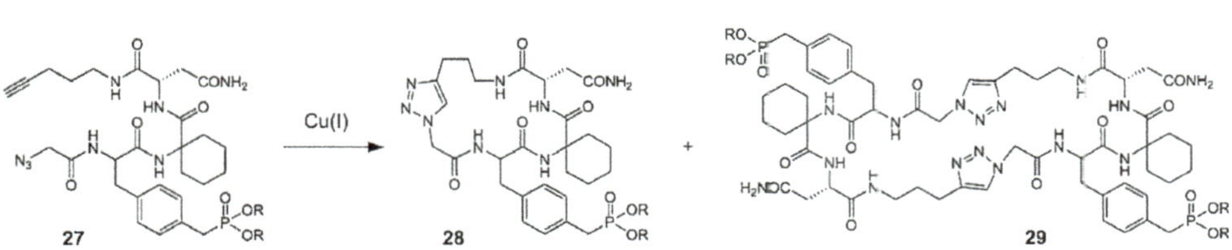

Figure 7.14 *Dimerization of the peptide* ***27*** *into monomeric cycle* ***28*** *and the dimer* ***29****.*

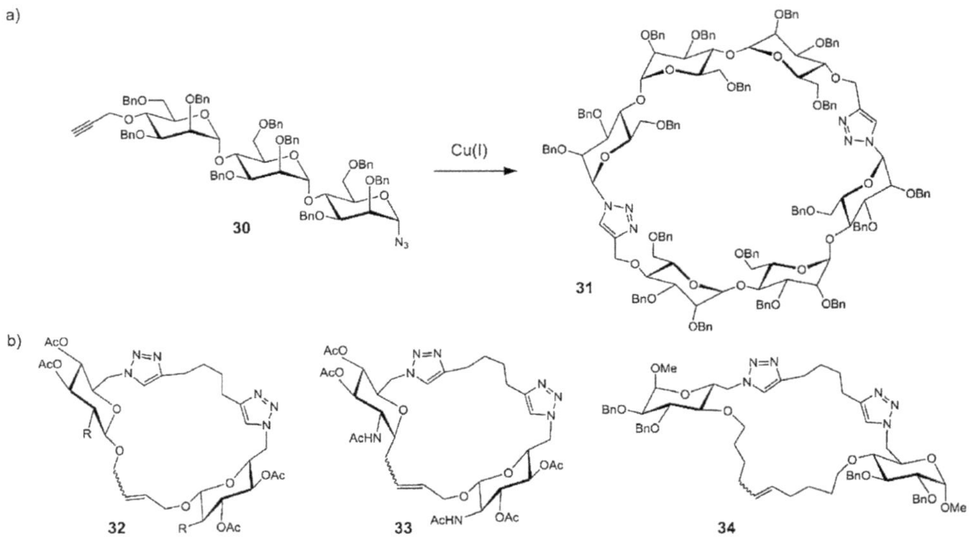

Figure 7.15 *Formation of macrocyclic carbohydrates. (a) Dimerization of* ***30*** *yielding* ***31****. (b) Various carbohydrate containing macrocycles formed by a combination of RCM and click.*

scaffolding. The hybridization of DNA and the subsequent PCR and other detection methods are the most prominent, followed by triplex formation (via Hogsteen-base-paring), cyclic DNA and cross-linking. As evident by the large number of publications, the azide–alkyne click reaction has had a strong impact on DNA modification and the subsequent use of DNA in supramolecular recognition processes (see Figure 7.16):[86–102] thus labeling of DNA with terminal azide–alkyne moieties in the side chain or at the chain end can be either achieved via chemical synthesis (phosphoamidite-method),[90,95] the cellular DNA-polymerases[87] or PCR,[90,91,93,94,97,99,100] as well as via chemical labeling of the respective end groups.[96,98,102] Thus unnatural nucleosides or nucleotides **35a–f** are required as shown in Figure 7.16, displaying purine and pyrimidine bases with attached alkynes or azides,[86,90,91,93,94,97,99,100] which subsequently can be incorporated into the DNA. It has been demonstrated that the structure of the incorporated nucleoside has a pronounced impact on the efficiency of the click reaction[93] (Figure 7.16). Thus both nucleosidic structures **35b** and **35c** were incorporated in the DNA via PCR and their click reaction in single- and double-stranded DNA subsequently investigated. It could be demonstrated that nucleoside **35b** is more efficient than nucleoside **35c**, presumably due to steric effects, since the alkyne moiety in dsDNA of nucleoside **35b** is sterically less hindered within the major groove of the DNA molecule.[88,93]

The method of side chain modification of DNA has been intensely investigated towards DNA metallization by Carell *et al.*,[99] as shown in Figure 7.17. Thus the side chain of DNA was modified with aldehyde residues via the click reaction, and subsequently metalized with Ag ions via the Tollens reaction by reaction with the pendant aldehydes. It was demonstrated that the attachment of poly-hydroxylated dendrimers yields a higher

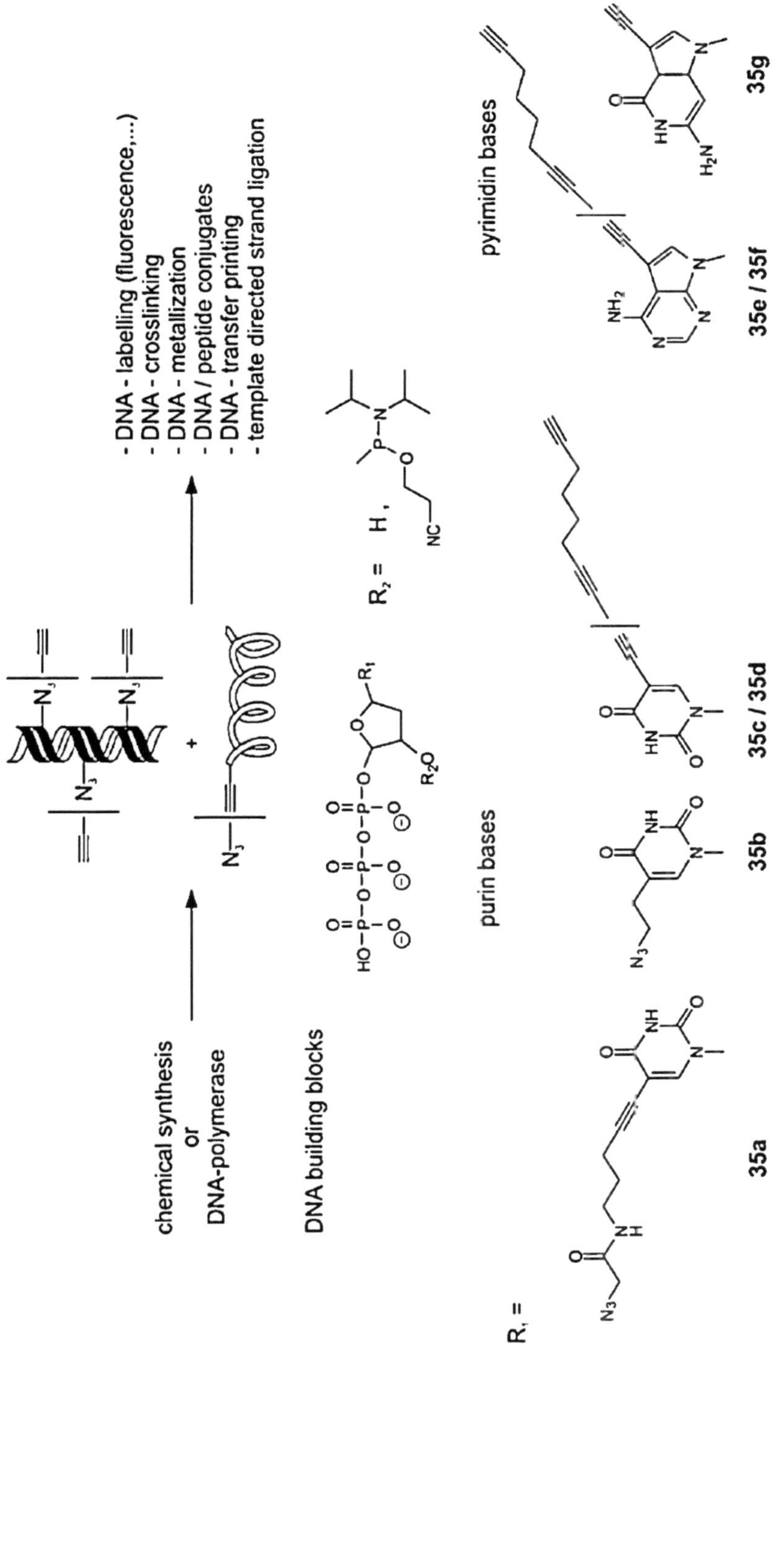

Figure 7.16 Strategy for the incorporation of click concepts with DNA.

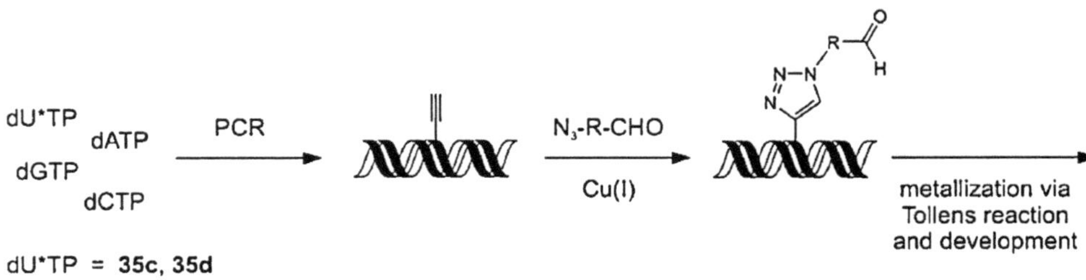

Figure 7.17 *Formation of alkyne modified DNA via PCR using the modified nucleotides **35c, d** and their metallization via attached aldehydes using the Tollens reaction.*

density of aldehyde moieties on the DNA strand during the Tollens reaction, followed by an easier metallization reaction of the DNA due to the increased presence of reducing moieties.

An enormous improvement of assaying methods of DNA has been developed by incorporating nucleotide **35c** directly into DNA within living cells [see Figure 7.18(a)].[87,92] After incorporation of **36c** by the DNA polymerases within the cells, the DNA is visualized by addition of a fluorescein derivative, attached to the labeled DNA via the click reaction. Thus whole amounts of large tissues or organ explants can be labeled via this method in minutes, which represents an enormous improvement over the conventionally used techniques. Moreover, the labeling can be achieved in live cells, thus enabling the assaying of *in-situ*-gene activity.

A slightly different approach for DNA detection *in vitro* used a process derived from photography for 'naked-eye'-DNA-detection [Figure 7.18(b)].[97] Thus DNA was labeled with alkyne moieties as described before, using PCR and the nucleosides **35b, c** and a pinacyanol dye was linked to the DNA via the azide–alkyne click reaction. Subsequent spotting generated strongly diluted DNA, which – after incubation with Ag^{+}-ions – generated dark-spots, which allow the selective detection of DNA after conventional hybridization down to 600 fmol by the naked eye.

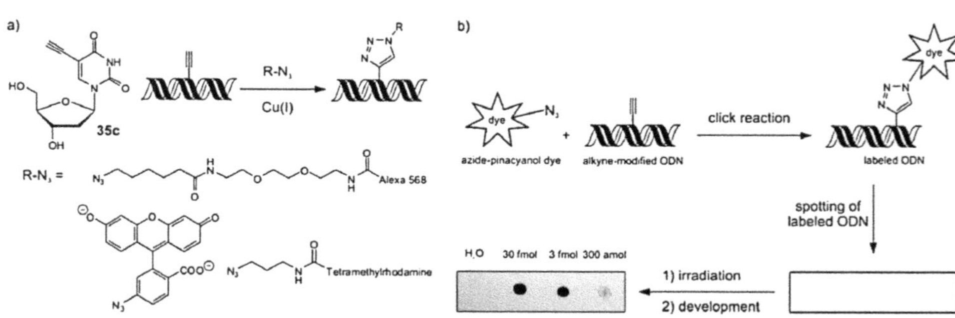

Figure 7.18 *(a) DNA-assaying in vivo by incorporation of **35c** and subsequent attachment of fluorescein-dye. (b) DNA-assaying for the 'naked-eye'. Reprinted with permission from D.M. Hammond* et al., *(2007), DNA photography: an ultrasensitive DNA-detection method based on photographic techniques,* Ang. Chem. Int. Ed., ***46** (22), 4184–4187. Copyright 2007 Wiley-VCH.*

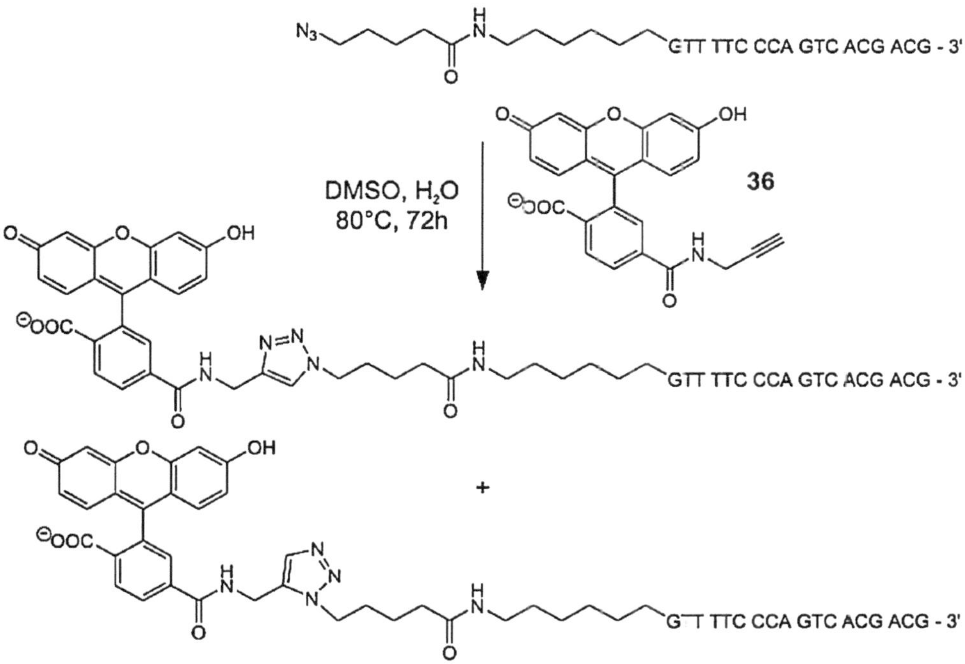

Figure 7.19 *DNA labeling at the end using the fluoresceine-molecule **36**.*

The classical attachment of fluorescent dyes to the end of DNA has been reported as one of the first examples for DNA labeling (see Figure 7.19).[102] MALDI-TOF methods were used to verify the completeness of the ligation-reaction with the fluorescent dye **36**, and the labeled DNA was directly used without further purification for the DNA sequencing via capillary electrophoresis due to the high efficiency of the azide–alkyne click reaction.

DNA cross-linking has been use as demonstrated the proximity effect of the azide–alkyne click reaction in adjacent DNA-strands (see Figure 7.20).[88] Thus exactly one nucleotide with an azide–alkyne moiety was incorporated into the DNA strands and subsequently hybridized. After hybridization, the addition of Cu(I) salts led to the cross-linking of the adjacent moieties under triazol formation. Nucleotides with longer side chains (i.e. octydiynyl **37**, **38** vs ethinyl **35c**) gave better cross-linking for steric reasons, as already described by Carell *et al.*[93] for the attachment of other groups to alkyne labeled DNA.

The transfer-printing of DNA onto azide-terminated glass surfaces is an important method to attach DNA in a simple process to surfaces for use in biochip-technology (Figure 7.21).[94] A method reported previously on polymeric surfaces via microcontact printing[103] or AFM tips[104] has thus been transferred to DNA: alkyne-labeled DNA (side chain- or end group-labeled) was stamped via a dendri-stamp onto a glass surface and modified with azido-moieties. The dendri-stamp presents multiple functional groups in order to improve the adhesion process by multivalent binding effects. Because of the

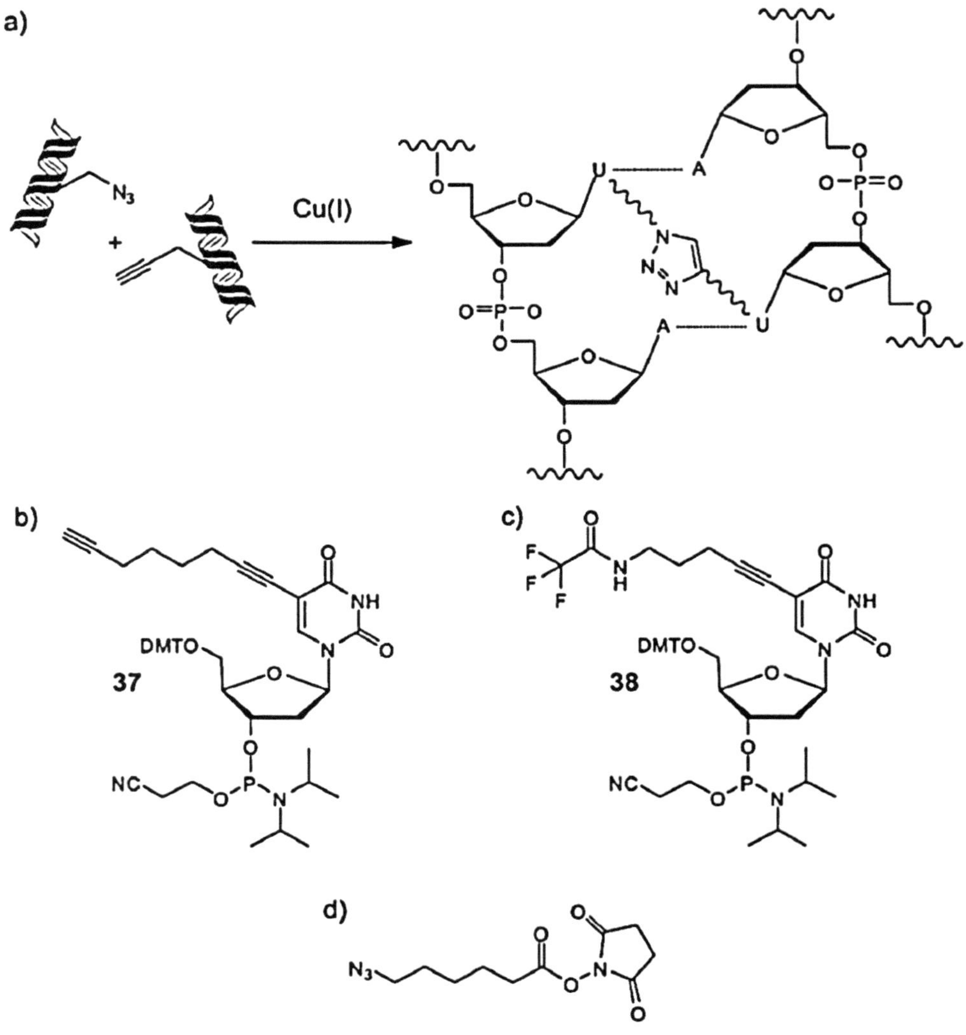

Figure 7.20 *(a) Crosslinking of DNA using the click reaction. (b) Incorporated **37** and **38** used for DNA-cross-linking.*

multiply-present reactive groups between the DNA (alkynes) and the surface (azides), no Cu(I) ions were required, just a contact time of approximately 1 h while applying a load of 120 g. Subsequent hybridization experiments demonstrated the effectiveness of the method for DNA recognition and assaying methods.

A combined method of DNA labeling and methylation has been described using the azide–alkyne click reaction [see Figure 7.22(a)].[101] An alkyne-labeled nucleoside **39** able to methylate DNA was incubated with DNA and a methyltransferase. During methyltransferase reaction, the azide–alkyne click reaction took place, thus demonstrating the

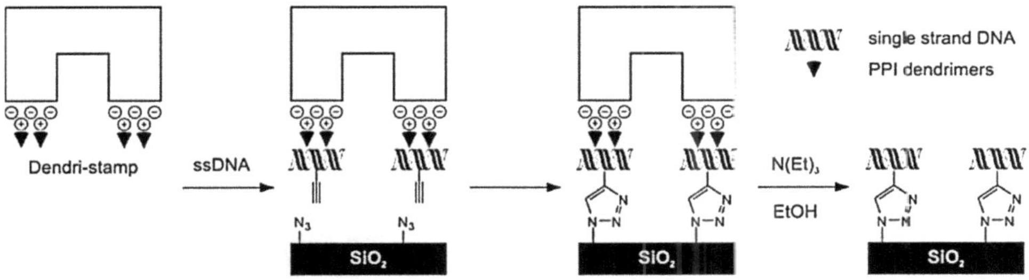

Figure 7.21 *Attachment of DNA to surfaces via stamping-methods.*

possibility to conduct both reactions within the active center of the methyltransferase enzyme.

A template-directed ligation process has been described, linking two DNA-fragments to yield a cyclic-structure (Figure 7.22).[96] Thus labeled DNA (3′-azide and a 5′-alkyne) was hybridized and incubated with Cu(I) ions, furnishing either a template-directed ligation into linear structures, or a nontemplated cyclization yielding cyclic DNA. The latter is very difficult to achieve by other methods and thus represents the first example of such a reaction to cyclic DNA.

Finally, DNA can be used for the assembly of nanoparticles, if corresponding additional supramolecular interactions are affixed to it (Figure 7.23).[91] Thus alkyne moieties were incorporated via PCR-methods, yielding DNA with a high density of alkyne moietes in the major groove. Au nanoparticles, equipped with high densities of azide moieties, can be assembled on the DNA strand, and subsequently fixed covalently after addition of Cu(I) salts via the triazole-linkages. Thus a stable adhesion and binding of nanoparticles can be effected on the DNA without the use of a reduction process directly.

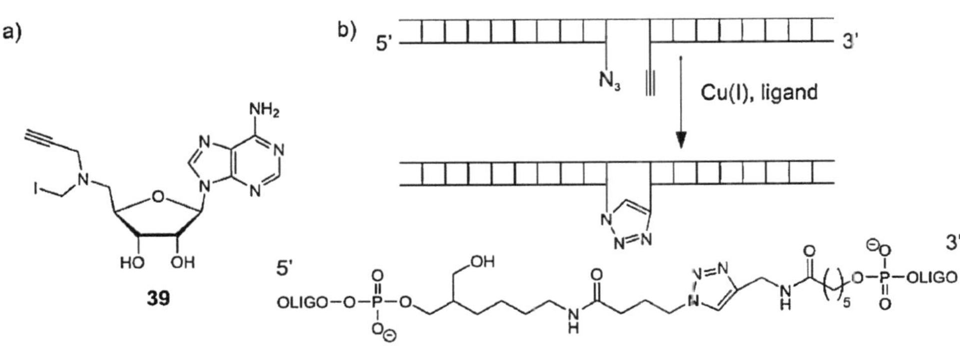

Figure 7.22 *(a) Chemical structure of* ***39*** *used for DNA-methylation. (b) Formation of cyclic DNA.*

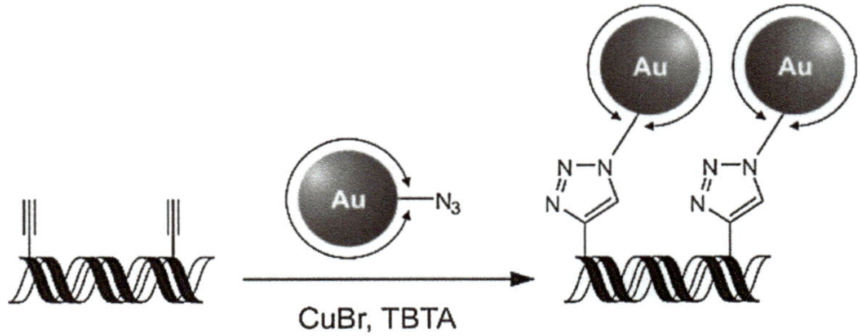

Figure 7.23 *DNA-metallization by use of alkine-modified-DNA and subsequent nanoparticle-attachment. Reproduced with permission from M. Fischler, U. Simon, H. Nir* et al.*, (2007), Formation of bimetallic Ag–Au nanowires by metallization of artificial DNA duplexes,* Small*, **3** (6), 1049–1055; G. A. Burley, J. Gierlich, M. R. Mofid* et al.*, (2006), Directed DNA metallization,* J. Am. Chem. Soc.*, **128** (5), 1398–1399.*

7.4 Click Reactions on Supramolecular Polymers

Supramolecular polymers[105–107] are an increasingly important class of polymeric materials, where noncovalent bonds mediate the adhesion of oligomers or polymers. As supramolecular interactions in polymers can be tuned very efficiently (i.e. hydrogen-bonds ranging from ~7 to ~40 kJ/mol),[107] the molecular interaction between the chains can be tuned in high precision. Thus polymeric materials with highly dynamic properties can be generated, allowing the generation of dynamics polymers (so called 'dynamers'),[108] self-healing materials,[109] microphase-separated polymer blends[110] and gels[111] with tunable properties, or new biomaterials[112] and nanocomposites.[105]

As the topic of the azide–alkyne click reaction has had enormous impact on polymer chemistry and synthetic macromolecular chemistry,[43,113] it is not surprising that the field of supramolecular polymer chemistry has been strongly influenced by this reaction as it allows the affixation of supramolecular interactions at specific sites of a polymer chain. As one of the most prosperous combinations, many living polymerization methods have been combined with the azide–alkyne click reaction (for a recent reviews see Binder and Sachsenhofer[43,113]).

We were the first to exploit the use of the azide–alkyne click chemistry for the attachment of supramolecular entities onto the backbone of polymers prepared by living polymerization methods.[35,38,114–123] One of the first examples concerned the combination of ROMP with click chemistry,[114,116,117,123] thus achieving a controllable density of supramolecular entities in homopolymers **40**,[114] statistical copolymers[117] **41** and blockcopolymers **42**.[116,123] As shown in Figure 7.24(a–c), the possibility to 'first-click-then-ROMP' or 'first-ROMP-then-click' proved useful for the synthesis of an enormous number of different ROMP polymers with nearly any thinkable architecture. The method represents a universal scaffold for the attachment of many supramolecular entities, e.g. **43** (Hamilton-receptor–barbituric acid interaction), since the click chemistry is nearly substrate insensitive and allows the easy attachment of even complex supramolecular entities.

The scaffolds can be used to take advantage of two highly defined supramolecular interactions upon spreading as films, as shown in Figure 7.24 (d–f):[116,117]one the one hand

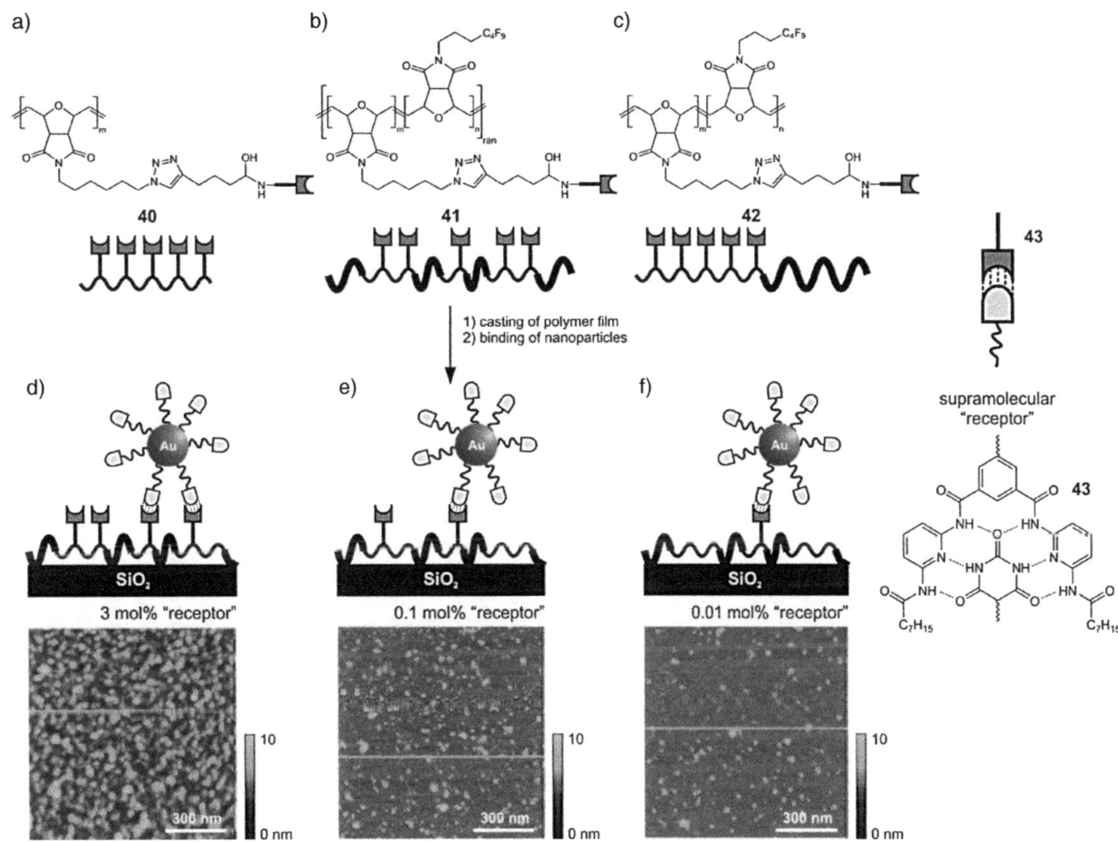

Figure 7.24 *Synthesis of polymers via ROMP/click methods, attaching supramolecular moiety **43** to the side chain, thus controlling the density of the supramolecular interaction. (a) Homopolymer **40**; (b) statistical copolymer **41**; (c) block-copolymer **42**. (d–f) The polymers can be cast into films, used for the attachment of nanoparticles via the selective hydrogen bond **43**.*

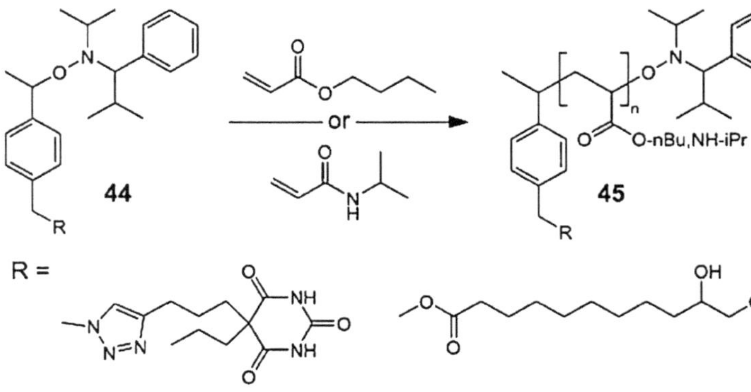

Figure 7.25 *Combining nitroxide-mediated polymerization (NMP) with click chemistry for the attachment of the supramolecular structures to the Haker-type-initiator* ***44****, yielding the monofunctional supramolecular polymers* ***45****. Reprinted with permission from W. Binder* et al.*, (2007), Magnetic and temperature-sensitive release gels from supramolecular polymers,* Adv. Func. Mat.*,* ***17*** *(8), 1317–1326. Copyright 2007 Wiley-VCH.*

microphase separation of block copolymers takes place if immiscible blocks exist and the supramoelcular entities are affixed to different blocks of the blockcopolymers, and on the other hand hydrogen-bonding interactions that can be presented by the polymer are useful to attach nanoparticles to the surface via supramolecular recognition. Thus statistically distributed hydrogen bonds shown in Figure 7.24(e) yield the correspondingly statistically distributed nanoparticles on the polymer film,[117] whereas the block copolymers yield controlled aggregates of the nanoparticles on the polymeric surface.[116] Thus the density and distribution of the nanoparticles can be controlled by use of the underlying polymeric scaffold. Without azide–alkyne click reaction it is nearly impossible to modify the density of such interactions on a polymeric chain without enormous synthetic effort.

Another combination of living polymerization and azide–alkyne click reactions has been reported by us, combing nitroxide-mediated polymerization (NMP) with click chemistry (Figure 7.25).[124] Thus the supramolecular entities (hydrogen bonds) have been affixed by use of a modified Hawker-type-nitroxide initiator **44**. Subsequent NMP of *n*-butylacrylate or *N*-isopropylacrylamide furnished the correspondingly mono-functional polymer chains **45**, as proven by MALDI-TOF-analysis. The method was further extended to the grafting-from reaction of NIPAM from iron-oxide-nanoparticle surfaces. The telechelic PNIPAM used was then incorporated into supramolecular gels, achieving an additional element of thermoresponsiveness into the material.[125]

The combination of living carbocationic polymerization of poly(isobutene) with the azide–alkyne click reaction allows the generation of star[111] and block copolymers[126,127] functionalized with hydrogen bonding end groups (Figure 7.26). Thus the three-arm star-poly(isobutylene) **46** was prepared with the respective multiple hydrogen bonds affixed to its end group moieties. MALDI-TOF and NMR-spectroscopy have been used to prove the generated structures, which in turn can be combined into highly temperature-sensitive amphiphilic gels by mixing with their . Superparamagnetic iron-oxide nanoparticles can be incorporated into these gels, yielding responsive materials with two-sensitivities: (a) those

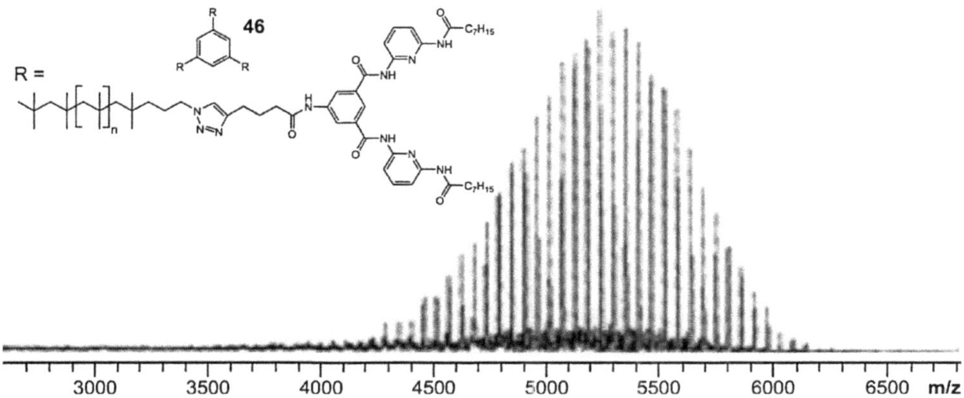

Figure 7.26 Synthesis of three-arm-star polyisobutylene (PIB) with three attached supramolecular hydrogen bonding receptors **46**. Shown is the corresponding MALDI spectrum, demonstrating the effectiveness of the synthetic method.

induced by the reversibility of the hydrogen bond and another (b) induced by an oszillating magnetic field, heating up the iron-oxide nanoparticles and stimulating the breakup of the gel. Moreover, the gel is self-healing as it assumes its original shape after mechanical deformation, as proven by rheological experiments.

The binding of nanoparticles to surfaces made from self-assembled monolayers (SAMs) has been achieved by the use of hydrogen-bonding systems between surface-modifed nanoparticles and SAMs with the matching interaction [see Figure 7.27(a, b)].[35,119] Thus a controlled density of hydrogen bonds (multiple-hydrogen bonds) was attached to mixed self-assembled monolayers via the azide–alkyne click reaction. As the molar ratio of the mixed SAM could be adjusted perfectly, surfaces with a defined density of molecular 'stickiness'[128] were prepared.[35] Thus a large variety of nanoparticles, surface modified with a similar strategy[119,129,130] (i.e. ligands modified via the azide–alkyne click reaction) could be deposited selectively onto the SAM-surface. Thus CdSe,[119] iron-oxide[124,129,131] and Au nanoparticles[35,116,117] were bound to the respective surfaces, allowing control of the layer thickness, morphology and density of the underlying layers.

A fine example of supramolecular poymer organization has been described by Hecht *et al.*,[132] taking advantage of selective chain-folding (see Figure 7.28). Thus pyridine units have been linked by triazoles, generating the helical structure **47**. Upon addition of metal ions, the helically folded chains are transformed into gels due to the bridging of the chains into networks.

The functionalization of styrene polymers with a supramolecular metal complex (iridium complexes) has been described using the click reaction (Figure 7.29). As these metallo-supramolecular structures may be important for the light-harvesting and charge-transfer in solar cells, these systems represent another contribution towards chain-organization via the polymeric backbones.

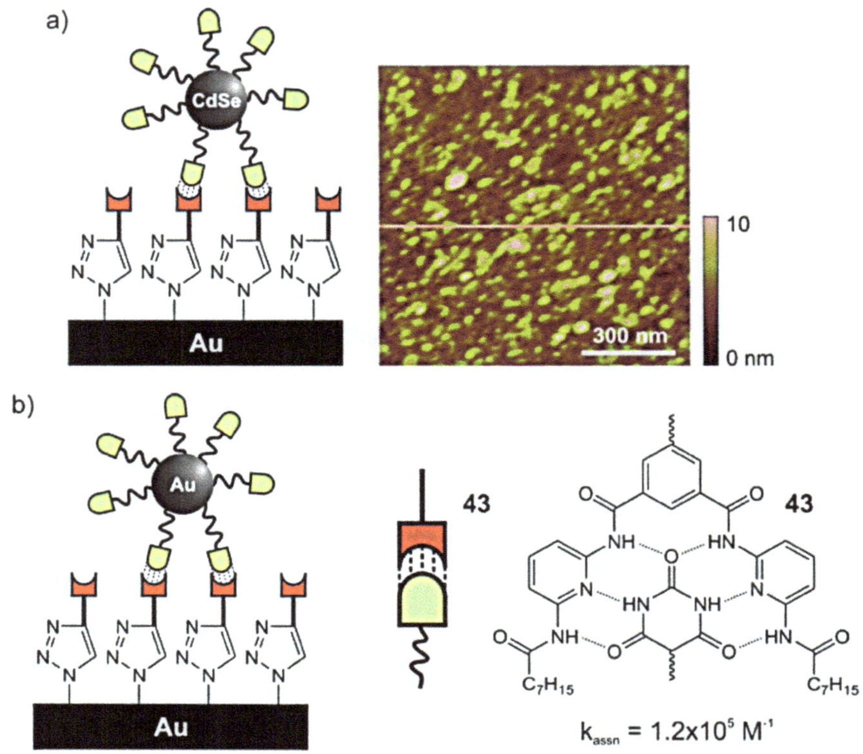

Figure 7.27 *(a) Modifications of surfaces via the azide–alkyne click reaction yielding controllable densities of supramolecular interactions using the Hamilton-receptor* ***43****. (b) CdSe and Au nanoparticles binding via the Hamilton receptor* ***43****. Reprinted with permission from W. H. Binder, R. Sachsenhofer, C. J. Straif* et al., *(2007), Surface-modified nanoparticles via thermal and Cu(I)-mediated click chemistry: generation of luminescent CdSe nanoparticles with polar ligands guiding supramolecular recognition,* J. Mater. Chem.*,* ***17*** *(20), 2125–2132. Copyright 2007 Royal Society of Chemistry.*

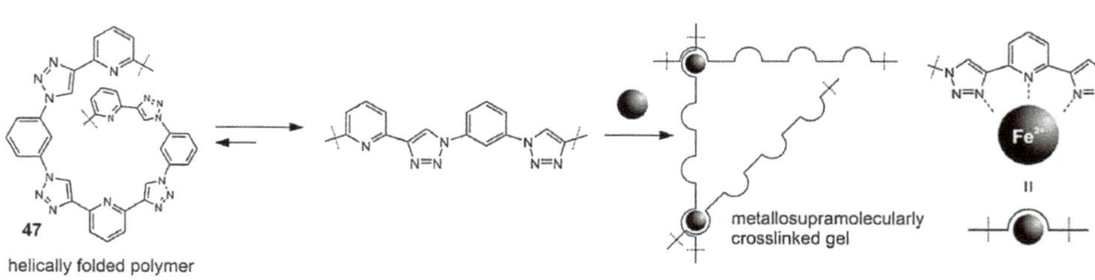

Figure 7.28 *Reversible folding of helical polymers* ***47*** *into gels by addition of iron-(II)-salts.*

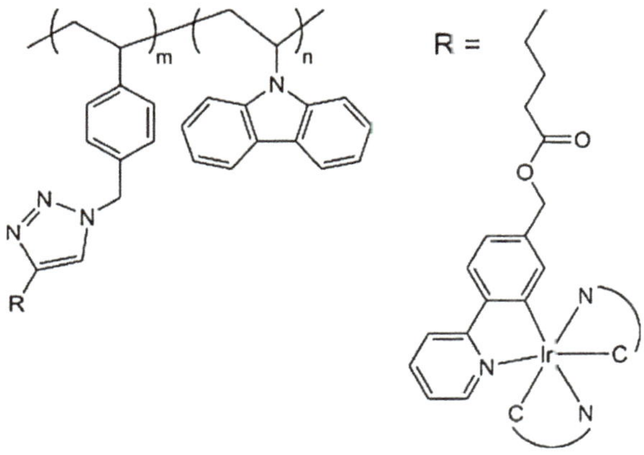

Figure 7.29 *Formation of block copolymers with attached iridium complexes.*

7.5 Click Reactions on Membranes

Biological membranes are highly organized assemblates of lipid molecules, being present either as closed lipid bilayer structures (called liposomes, vesicles) or as (artificial) monolayer systems (Langmuir layers).[133] Usually, biological membranes display a variety of physical effects (liquid crystalline transition temperatures, mixing/demixing effects),[134] which are highly dynamic in nature and reflect the lability of biological membranes, whose stability is limited by temperature, pressure and mechanical deformation. Another type of membrane consists of polymers, called polymersomes, where phase-separation phenomena between polymer chains (microphase-separation) or differential solubility (selective solubility) of polymer chains generate closed membranes.[135] These membranes represent a small fraction of the overall phase structures and are kinetically labile structures. However, when compared with their lipid counterparts, their stability is significantly higher and the membrane thickness scales with the length of the polymer chains.[136] In both cases (polymersomes and liposomes), modifications of the outer surface are important and crucial to effect molecular recognition at the outside of the membranes, thus studying, e.g., membrane-binding processes, membrane transport,[71] nanoparticle–membrane interaction[126,137] or encapsulation and triggered release.[138] Modification of polymersomes or liposomes thus is an important point for studying such processes, but often hampered by the inhereint lability of the underlying structures, especially with liposomes. As the azide–alkyne click reaction works under relatively mild reaction conditions (low temperature with high efficiency), it is a useful alternative to other methods such as thiol addition, disulfide reactions or *N*-hydroxysuccinimide additions to effect the modification of polymersomal–liposomal structures.[82,139–143]

We have described the self-assembly of hydrophilic and hydrophobic nanoparticles into liposomal[137] and polymersomal membranes[126] (see Figure 7.30). One of the best systems for this purpose proved to be a diblock copolymer made by linking PEO and PIB chains via a click reaction.[126] These blockcopolymers can be assembled into polymersomes, if the

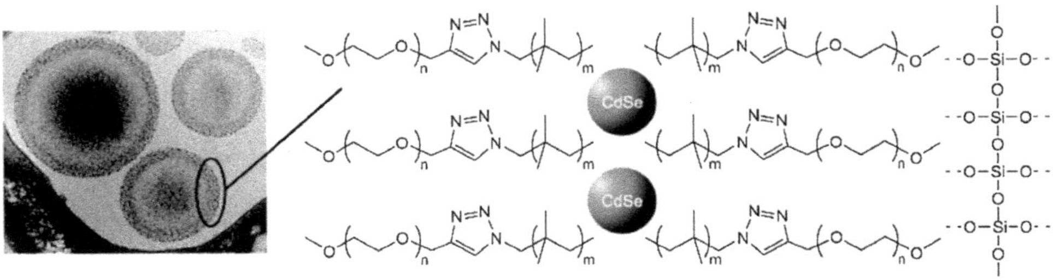

Figure 7.30 Incorporation of CdSe nanoparticles into polymersome membranes made from PEO–PIB blockcopolymers and silicification of their outer shell via sol–gel-processes.

ratio between the PEO and the PIB-block is appropriate. The simple testing of different systems can be achieved via generation of a library of different telechelic PIB-N_3/PEO-alkyne systems, clicked together in a simple manner. Thus the formation of the respective polymersomes could be acheived, whose membrane was subsequently used to incorporate hydrophobic nanoparticles into their hydrophobic interior. Furthermore, the outer shell was stabilized by sol–gel processes, yielding stable capsules with the embedded nanoparticles as a highly organized, supramolecular system.

The direct modification of polymersomal outer layers has been achieved via two different routes [see Figure 7.31(a, b)].[82,140] The use of a blockcopolymer (PS-PEG) with alkyne end groups allowed the generation of polymersomes with pendant alkyne moieties in multiple fashion.[82] The Cu(I)-mediated azide–alkyne click reaction can be subsequently conducted, enabling the fixation of azide-modified candida-lipase (CalB) onto the surface of the polymersome.

The analogous pathway has been described, using an endlabeled PS-*b*-PAA diblock-copolymer, which – after assembly into polymersomes – presents multiple azido-moieties [Figure 7.31(b)].[140] These were used for the reaction with either a fluorescence label (dansyl-dye) or the attachment of green-fluorescent protein onto the surface of the polymersome. A similar reaction pathway, reporting on the attachment of dendritic moieties onto the surface of PBD-PEO-polymersomes, is reported [Figure 7.31(c)].[143] Again, an appropriately functionalized PBD–PEO–N_3 diblock copolymer forms the scaffold for the polymersome, which is then decorated with azido moieties for further reaction with alkyne-modified dendrimers. The click reactions were carried out in aqueous systems, taking advantage of the regenerative system ($CuSO_4$–sodium ascorbate), thus nearly working under physiological conditions.

With liposomes, two examples of a direct azide–alkyne click reaction have been described (see Figure 7.32).[142,144] Using a DOPE alkyne [Figure 7.32(a)], the outer surface of a liposomal membrane was decorated with about 50% of alkyne moieties, embedded into a membrane consisting of DOPC.[142] The resulting liposomes were then incubated with an oxazole dye, which was attached covalently to the outer surface under Cu(I) catalysis. The presence of the oxazole dye was proven by FRET-measurements between a dye already present within the membrane.

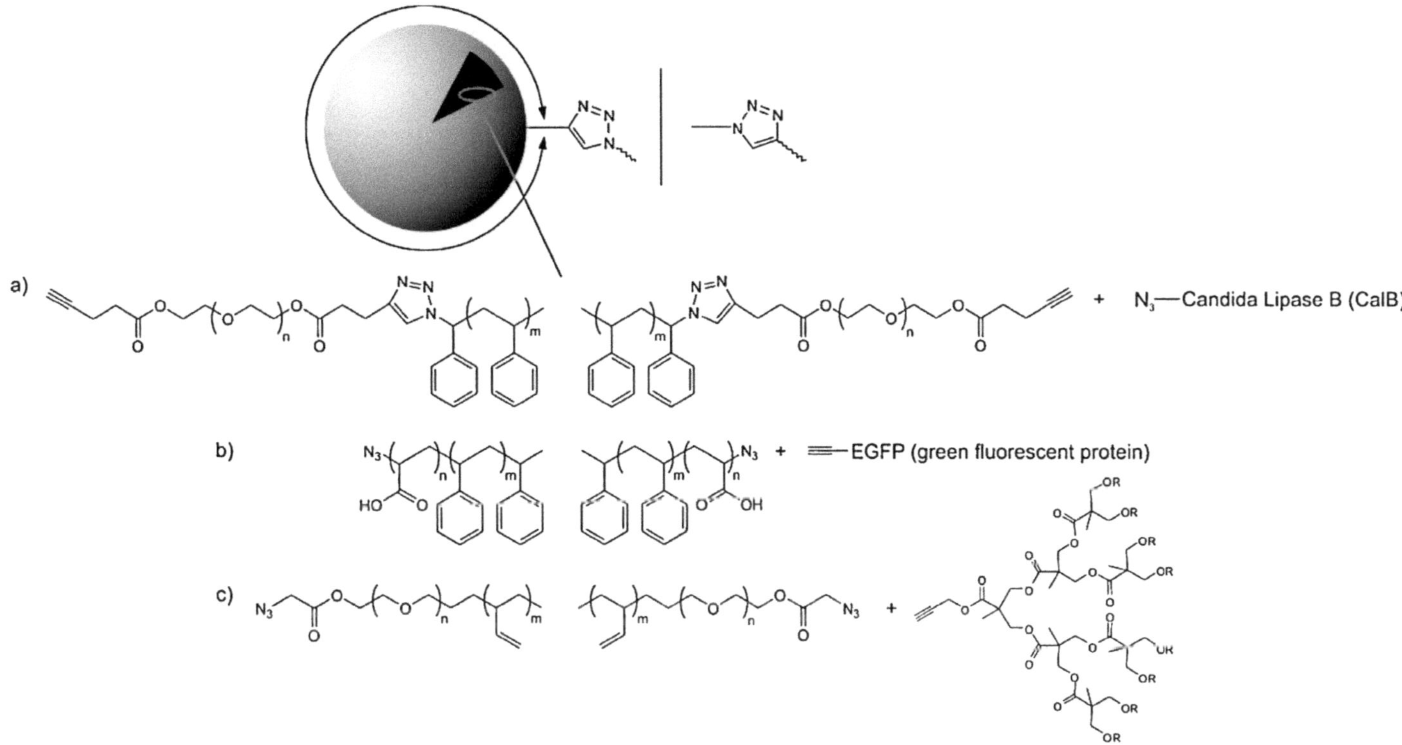

Figure 7.31 *Modification of polymersomal surfaces via the azide–alkyne click reaction: (a) modification of azide-modified polymersomes (polymer: PEG) with candida lipase (CalB); (b) PS polymersomes modified with green fluorescent protein (EGFP); (c) PS–PEO blockcopolymers within polymersomes and their modification with dendrimers.*

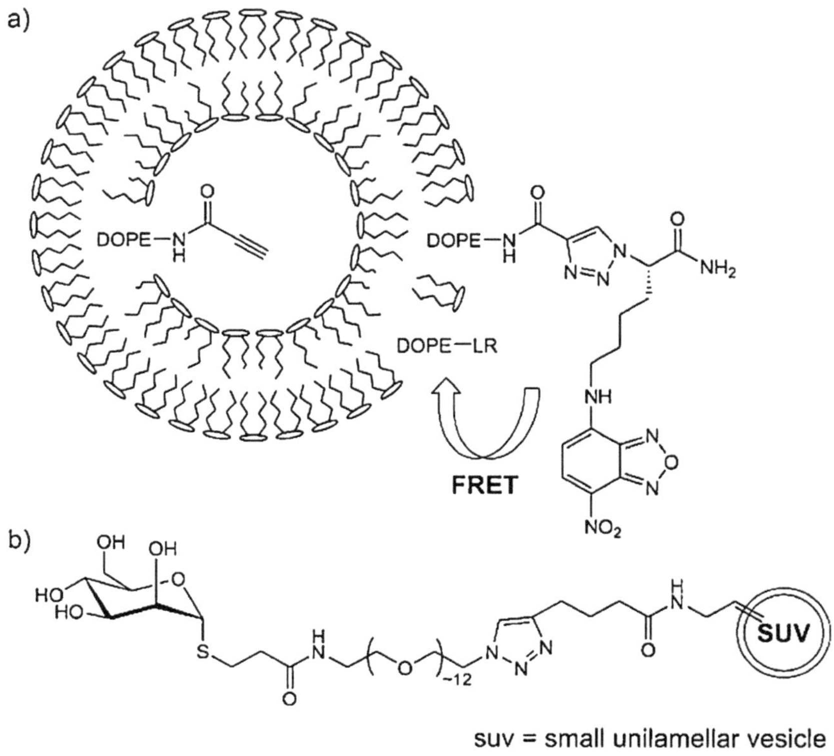

Figure 7.32 *Azide–alkyne click reactions using liposomes (a) using DOPE lipid; (b) using a small unilamellar vesicle (SUV) attaching a carbohydrate ligand to the outer surface.*

A similar strategy for labeling the outer surface of liposomes has been used relying on a glycerol-anchored lipid with a terminal alkyne moiety added to a conventional liposome-forming lipid mixture in 5–10 mol% [Figure 7.32(b)].[144] After liposome formation, the attachment of a mannose conjugate bearing a terminal azide moiety was investigated. It turned out that the efficiency of the reaction was strongly enhanced, if an appropriate bathophenanthrolinedisulfonic acid ligand for complexing copper ions was present. Only under these conditions could a complete surface functionalization be achieved, as proven via subsequent agglutination assays, which allow for a quantification of the attached mannose units to the liposomal surface.

Finally, it should be mentioned that the Cu(I)-mediated reactions are not useful for living (i.e. cellular) systems, owing to the toxicity of the Cu ions, which inhibit cell growth. Bertozzi *et al.*[20] have therefore developed a copper-free variant of the azide–alkyne click reaction, which relies on the use of highly strained substituted cyclooctynes, whose release of ring strain promotes the dipolar cycloaddition process without the use of Cu species. This method is now the method of choice for the labeling of cellular surfaces via incorporation of artificial amino-acids into membrane proteins.

7.6 Click Reactions on Dendrimers

The usefulness of the azide–alkyne click reaction is well demonstrated in the build-up of larger polymeric structures, in particular the generation of dendrimers. As dendrimers are important scaffolds for assembly into higher-ordered supramolecular structures, the value of the azide–alkyne click reaction is high. A recent review has been focussing on this combination, especially on the use of high-yielding, high-energy reactions for this purpose.[44] Besides the synthesis and functionalization of dendrimers,[38,41,44,46,145–158] hyperbranched polymers[44,158–161] can also be prepared using this methodology. Briefly, dendrimers can be generated by convergent or divergent methods, using the azide–alkyne click reaction as internal bond for the synthesis. This can lead to hyperbranched polymers either in one step or via sequential reaction. Additionally, whole dendron structures may be assembled via the azide–alkyne click reaction, using appropriately functionalized dendrons. Another issue concerns the generation of surface-modified dendrimers, which generates dendrimers with a high density of outer azide–alkyne moieties, which subsequently are then reacted with the appropriate functional groups, thus attaching a large number of these moieties onto the outer shell of the respective dendrimer. A large variety of different dendrimers, such as PAMAM-type dendrimers,[147,162] benzyl-type,[155] PS/PMDETA dendrimers,[152] triazole-containing dendrimers (in each generation)[149,163] and polyester-type dendrimers have been prepared via convergent methods, where the buildup of the central structure has been achieved by linking azide–alkynes.[148] Dendron attachment (i.e. divergent synthetic methodologies) to the side chain of poly(vinylacetylenes)[164] and inorganic ruthenium oligomers have been described.[165] Moreover, the generation of hyperbranched polymers in a one step-procedure, given that the starting material is present sufficiently pure and sterically not too crowded. This strategy has been used by several authors, generating medium-branched hyperbranched polymers in good yields.[159,166]

The surface of a large variety of different dendrimers can be modified generating, e.g., ferrocenyl-modified triazolyl-silane dendrimers,[146] dendritic peptides,[150] surface-modified polybenzyl and Boltorn dendrimers,[158] PEG-modified carbamate-dendrimers[167] and carbohydrate modified Boltorn dendrimers.[151] All these surfaces are more or less designed to act as recognition or organization sites for some supramolecular activity on the dendritic surface, whether it is a pure steric effect of organization or a defined key/lock-recognition.

The strategy to run multiple azide–alkyne click chemistry has been also transferred to the synthesis of polymer-brushes[168,169] or cross-linked capsules.[170] Similar to dendrimers, these structures display a high density of functional groups at their surface, thus requiring highly efficient linking-reaction for their functionalization.

7.7 Click Reactions on Gels and Networks

Gels and networks[62,79,120,171–184] have been formed additionally via azide–alkyne click reactions. This strategy has been proven useful as a simple cross-linking strategy, but also for the formation of highly sensitive gel and network structures not accessible by other methods.[174] As gels and networks are often either highly defined structures (e.g. fibers, organized by supramolecular interactions between small and medium-sized organic molecules or block-copolymeric micelles linked into gels) or relatively rough-organized systems (weakly cross-linked gels, networks formed by covalent cross-linking), their structural definition is sometimes vague. This makes the following rather an assembly of gels, where the main structural definition has been achieved by use of the azide–alkyne click reaction, or where this reaction is a main structural element of the final material generated (see Figure 7.33). Thus multuivalent azide **48** and alkynes **49** (Figure 7.33) can be directly reacted, generating networks with a high level of cross-linking density due to the high efficiency of the azide–alkyne click reaction. By appropriate choice of the corresponding building block (examples of small monomer, oligomeric or polymeric) azides–alkynes as given in Figure 7.33(b, c), the corresponding properties of the networks, such as swelling character, hydrophobic/hydrophilic properties, density or functionality, can be nicely controlled.

Highly dense networks have been used extensively as scaffolds for synthetic reactions. Thus a large number of investigations have been carried out using highly cross-linked resins and support, mostly in the field of organic chemistry.[13,29,72,185] In these cases, the highly cross-linked Rink, Wang or Merrifield resins serve as a (porous) solid phase, presenting terminal azido- or alkyne moieties able to attach substrates via the azide–alkyne click reaction.

As supramolecularily preorganized molecules often tend to disintegrate upon thermal treatment, the azide–alkyne click reactions represent an important step towards stable networks of defined cross-linking density, thus 'freezing-in' a specific supramolecular structure.[175,179] Thus block copolymer micelles can be easily cross-linked using the azide–alkyne click reaction after assembly of the block copolymers (BCPs) into the respective micelles, yielding the well-known cross-linked BCP-micelles,[154,157,186–188] with a highly defined degree of cross-linking within their core- or corona-structure.

Besides the work of Wooley *et al.*[157,186,187], who used the azide–alkyne click reaction for the cross-linking of BPC-micellar core, a highly innovative example for cross-linking the shell of a BCP micelle has been described by Meier *et al.*[180] (Figure 7.34). Thus a diblockcopolymer **50** generated via ROMP, whose one block was modified via a highly cationic moiety via the azide–alkyne click reaction. The BCP was able to incorporate DNA, generating particles sized 20–120 nm, able to deliver DNA. The particles display a highly dendritic structure on their outside, thus presenting a high cationic charge to the outside of the carrier nanoparticle.

An approach to highly sophisticated and smart networks has been descriebd by Turro *et al.*[175,181] (see Figure 7.35). Thus telechelic macromolecules (P^{t}BuA, PMA) were prepared via ATRP methods, and finally equipped with terminal azido–alkin-moieties. Because of the presence of a photocleavable linker (*o*-nitrobenzyl-unit; Figure 7.35) or internal double bonds, the corresponding networks can be cleaved either by UV irradiation or via ozonolysis. As the initial chain length of the polymers is highly defined, the density of the networks has been adjusted with high precision.

Other examples of defined networks with relatively controllable network densities have recently been reported, derived from PEGs,[171] polyvinylacohols,[177] hyaluronic acids[182] or cross-linked hydrophilic polymer beads.[183]

A fine example of a liquid crystalline polymer via click reaction has been reported by Grubbs *et al.* by using an endfunctionalized bitelechelic ROMP-polymer **51** with pendant

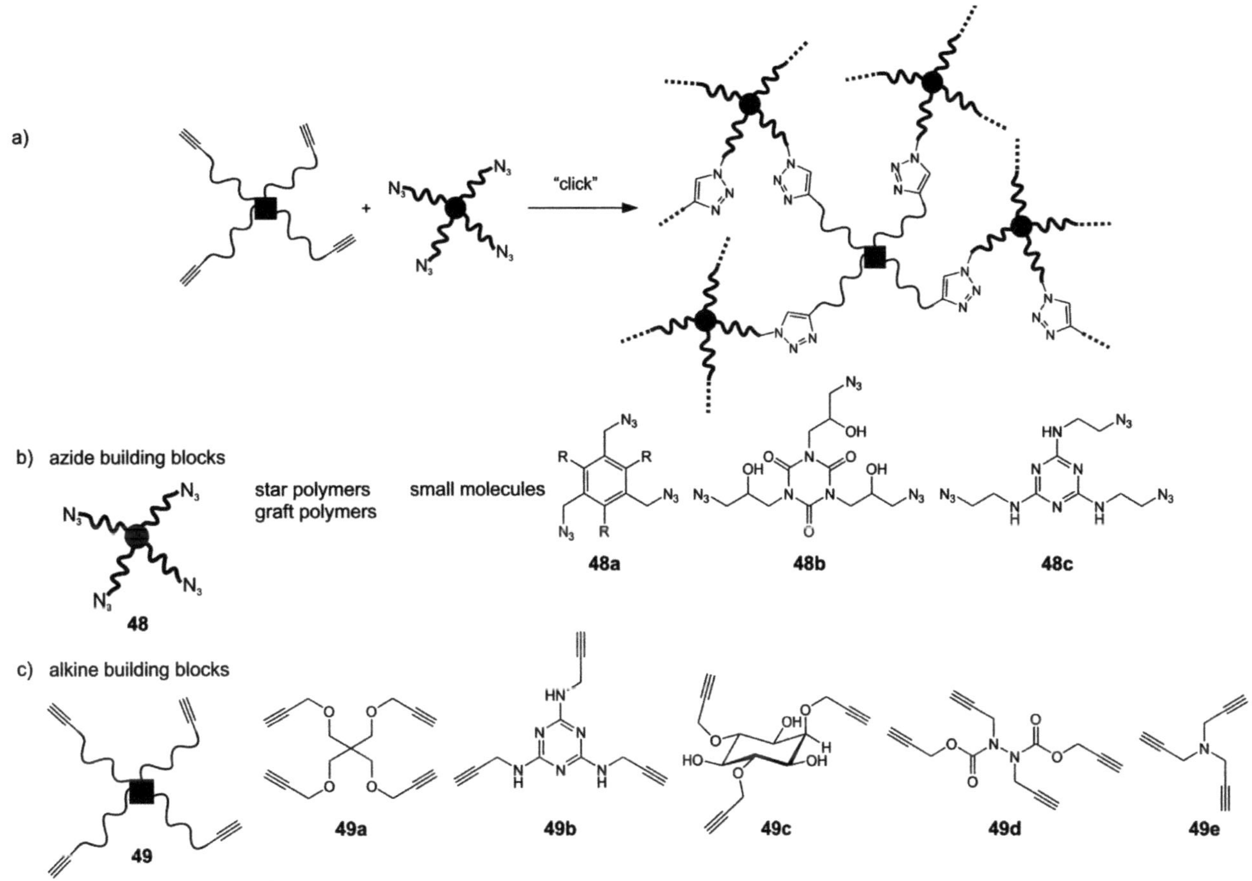

Figure 7.33 *Formation of networks using the azide–alkyne click reaction via multivalent building blocks* ***48*** *and* ***49*** *(selected examples).*

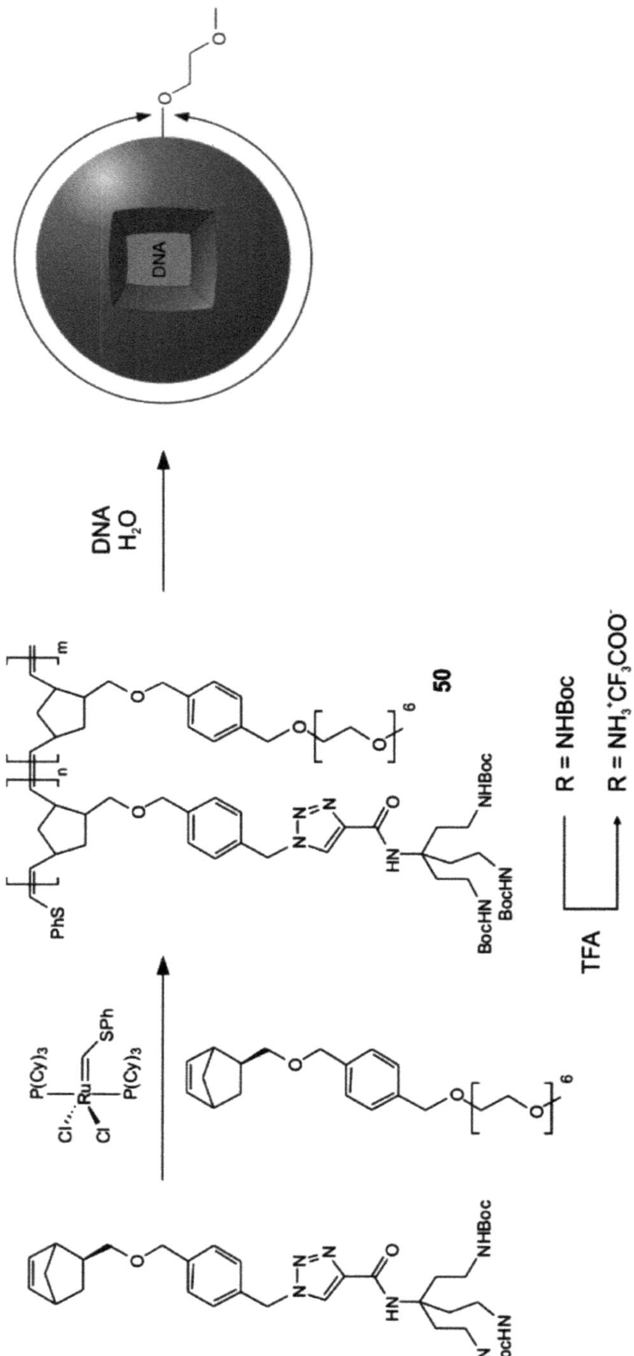

Figure 7.34 *Formation of DNA block copolymeric micelles using the ROMP polymer* ***50****.*

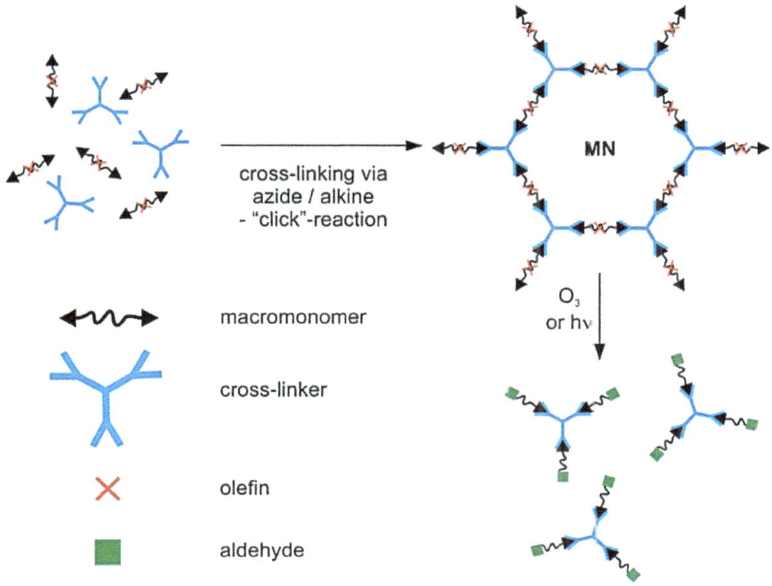

Figure 7.35 *Formation of photo- and ozone-cleavable networks via the azide–alkyne click reaction. Reprinted with permission from Y. Xia* et al., *(2008), Well-defined liquid crystal gels from telechelic polymers,* J. Am. Chem. Soc., ***130*** *(5), 1735–1740. Copyright 2008 American Chemical Society.*

liquid crystalline moieties (see Figure 7.36).[178] The final structure was cross-linked with a trivalent alkyne moiety, thus generating a highly defined network **52**, allowing study of the influence of network density on the orientation and nature of the liquid crystallinity.

Supramolecular gels are often generated from the interplay between hydrogen-bonding systems and hydrophobic interaction, often well balanced by solvent effects. The stabilization of such structures is difficult, but can be achieved using the azide–alkyne click reaction. Thus Finn *et al.*[184] and Diaz *et al.*[179] have reported on gels formed from amidic-bond networks, subsequently stabilized by the azide–alkyne click reaction (Figure 7.37)

We have reported on the generation of supramolecular gels built from multiple hydrogen bonds, attached to star-like PIB or PEG-poylmer (see Figure 7.38).[111,125] Thus trivalent star-PIBs **53** were prepared by a combination of living cationic polymerization and azide–alkyne click chemistry, being able to control the chain length of the (hydrophobic) PIB polymer. Upon assembly with matching hydrogen bonds **43** (supplied via end group-modified PEGs **54**), gel formation was observed, resulting in highly thermoreversible gels.[111] Furthermore, suprerparamagnetic nanoparticles or PNIPAM[111] could be incorporated into the gel, enabling strong thermoreversibility. The nanoparticles are located selectively within the hydrophobic cavities provided by the PIB-polymer, thus leading to a microphase-induced segregation of the nanoparticles.

The most picturesque example of network formation via the azide–alkyne click reaction has been provided by Finn *et al.*[172,173,189] in a series of publications (see Figure 7.39). The simplicity correlates to the effectiveness in the buildup of mechanically highly stable thermosets. Thus small molecules, multivalent in their azide–alkyne structures, were

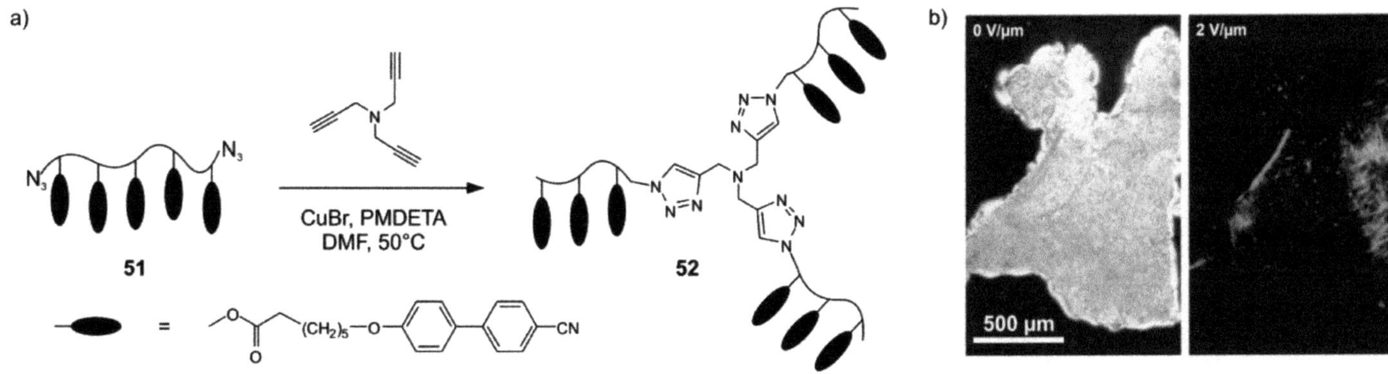

Figure 7.36 *Formation of cross-linked networks from liquid crystalline polymers. (a) Synthesis the network via cross-linking of **51** into the network **52**. (b) TEM micrograph of the cross-linked liquid crystalline gel.*

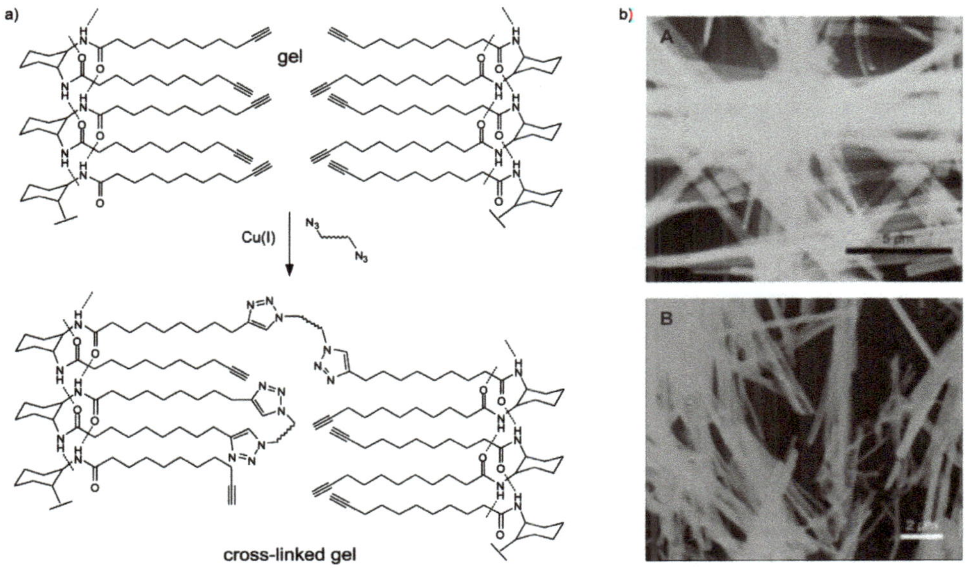

Figure 7.37 Crosslinking of transient fibers, formed via amid-bond-assembly (organic gelators). (b) TEM micrograph of the fibers before (top) and after cross-linking (bottom), indicating the preservance of the fiber's-integrity. Reprinted with permission from D. D. Diaz et al., (2006), Click chemistry in a supramolecular environment: stabilization of organogels by copper(I)-catalyzed azide–alkyne [3 + 2] cycloaddition, J. Am. Chem. Soc., **128** (18), 6056–6057. Copyright (2006) American Chemical Society.

mixed, and subsequently cross-linked between metal plates, i.e. metallic copper. Because of the high efficiency of the azide–alkyne click reaction, enormous hardness and glue strength could be achieved, representing the first directly applicable example of the click reaction. It furthermore shows, that often the most simple approach may be the most effective.

7.8 Click Reactions on Self-assembled Monolayers

Surfaces and interfaces often are not counted to supramolecular chemistry. However, as self-assembled monolayers or nanoparticle surfaces are highly organized structures, they are included in this review chapter. The chemistry on surfaces is as manifold as the chemistry on polymers or other materials; therefore the present data cannot be discussed in full detail, as they would represent a chapter on their own. An interesting aspect of the azide–alkyne click reaction lies in the fact that a reduced or enforced distance between the reaction partners leads to a strongly enhanced reaction rate. This effect has been demonstrated in the azide–alkyne click reaction within the pocket of enzymes (activity-based protein profiling, ABPP)[14,32,190] by direct microcontact printing[103,191] or via AFM-tips,[192] thus opening the chance for a sufficiently complete reaction at an interface.

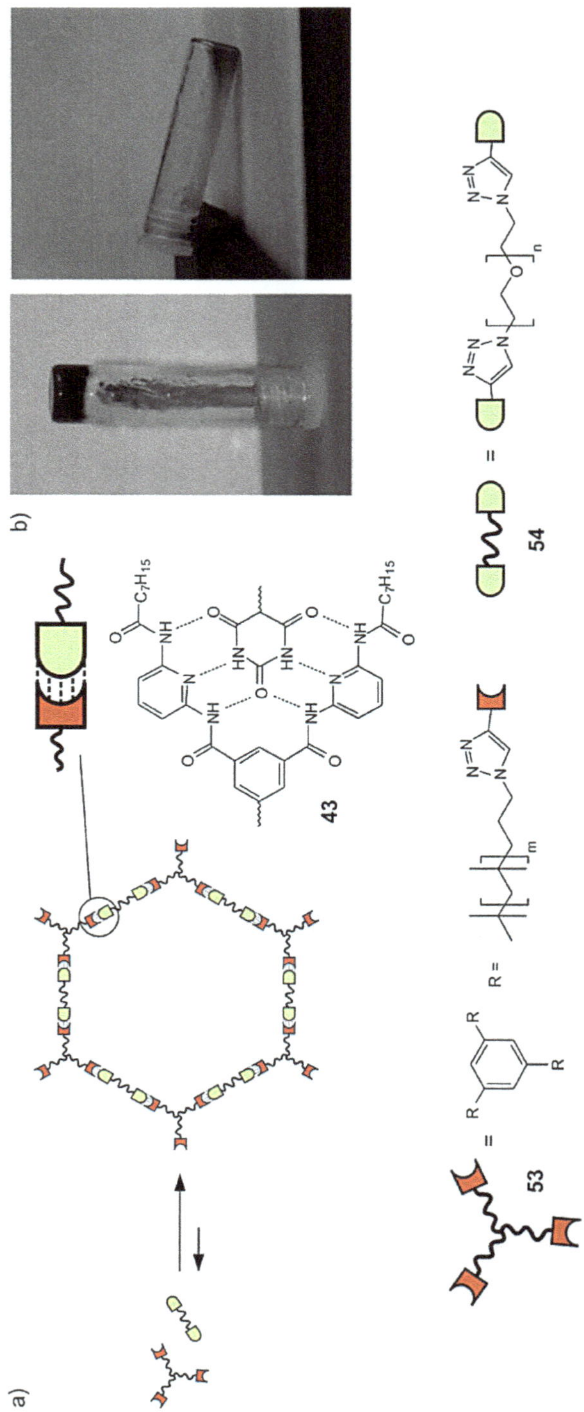

Figure 7.38 *(a) Formation of supramolecular gels by noncovalent interactions (hydrogen bonds) using a three-arm star PIB and a PEG-polymer, equipped with hydrogen bonds* ***43****. (b) Thermoreversibility of the gel by heating from room temperature (left) to 40 °C (right).*

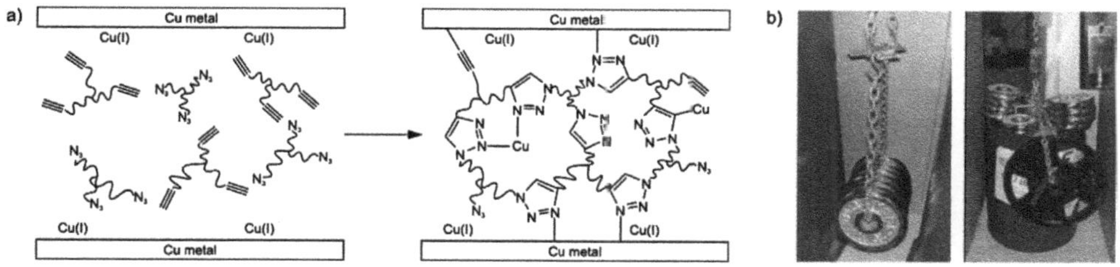

Figure 7.39 *Thermoset-formation by cross-linking via the azide–alkyne click reaction. (a) Linking of Cu plates by small-molecule-network-formation (b) Load tests demonstrating the effectiveness of the final thermoset. Reprinted with permission from D. D. Diaz* et al.*, (2004), Click chemistry in materials synthesis. 1. Adhesive polymers from copper-catalyzed azide–alkyne cycloaddition,* J. Polym. Sci., Part A: Polym. Chem.*,* **42** *(17), 4392–4403. Copyright (2004) John Wiley and Sons, Inc.*

Table 7.1 gives an overview on the most relevant click reactions on surfaces. One of the most important aspects is the generation of the corresponding azide–alkyne modifed surfaces, which is a prerequisite for the subsequent azide–alkyne click reaction.

In the case of SAMs, the use of appropriately azide-[35,193–197] or alkyne-functionalized[156,198–202] surfaces by direct ligand-adsorption has been described. Alternatively, *in-situ*-generation of terminal azides by bromide–azide-exchange directly on the ω-bromoalkyl-functional monolayer can be effected,[196,203] eliminating the pressing instability of ω-azido-1-thioalkanes prior to the SAM-formation process. Thus a large variety of click reactions on SAMs,[35,118,156,169,193–196,198–203] polymeric surfaces[191,204–210] or Langmuir–Blodget layers (LbL layers),[160,206] has been reported .

Similar to SAMs, the surface of nanoparticles can be modified with the azide–alkyne click reaction.[118,121,122,124,211–214] Thus a large variety of nanoparticles (Au,[118,211,212,214] CdSe,[121] Fe_2O_3[122,124,213] and SiO_2[197]) as well as viruses[215] and Au nanorods[216] have been surface-functionalized with this method. Mostly, the attached ligands serve as recognition sites to direct the location of such nanosized objects onto materials via defined or nonspecific interactions. Selected examples of such recognition processes rely, e.g., on hydrogen bonding moieties, which allow the corresponding nanoparticles to be directed to a SAM surface,[35,119]a polymeric surface,[116,117] a liquid–liquid-interface[217] or a block-copolymeric phase or interface.[218] The interested reader is referred to the references for further reading.[35,116,117,119,217–219]

Finally, an important point has been observed upon comparing the Cu(I)-catalyzed reaction with the uncatalyzed, purely thermal, click reaction on CdSe nanoparticles.[121] Since copper ions interfere with the fluorescence properties in semiconductive nanoparticles, the use of Cu(I)-species is not advantageous for their surface modification. Thus without the use of the Cu(I) catalyst, the photoluminescence of the final, surface-modified CdSe nanoparticles remains nearly unchanged, whereas under Cu(I) catalysis a significant drop in the quantum yields is observed. Therefore, the purely thermal azide–alkyne reaction may be sometimes advantageous over the metal-catalyzed click process. Compared with conventional surface-modification methods, the azide–alkyne-methodology enables an elegant, fast and efficient approach to functionalized nanoparticles in a simple mode.

Table 7.1 *Azide–alkyne click reactions on surfaces, nanoparticles and interfaces*

Entry	Polymer/substrate	Surface	Catalyst/conditions	Reference
1	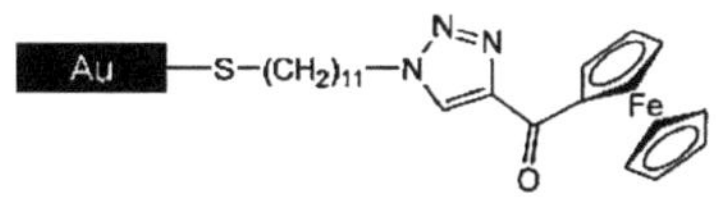	SAM on Au	$CuSO_4 \cdot 5H_2O$/sodium ascorbate/H_2O/EtOH	193
2	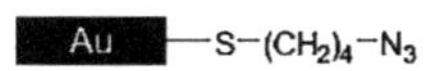	SAM on Au	$CuSO_4 \cdot 5H_2O$/sodium ascorbate and $Cu(Ph_3)_3Br$/H_2O/EtOH	35
3	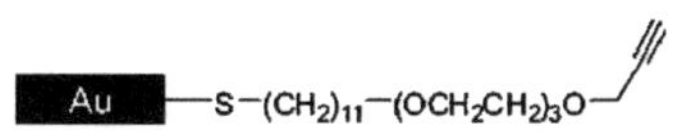	SAM on Au	$CuSO_4 \cdot 5H_2O$/sodium ascorbate/H_2O/EtOH	198
4	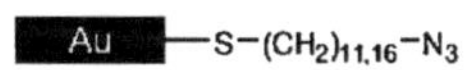	SAM on Au	$CuSO_4 \cdot 5H_2O$/sodium ascorbate/H_2O/EtOH and DMSO/H_2O	194
5	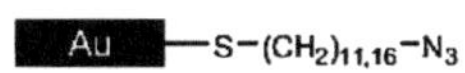	SAM on Au	TBTA $CuBF_4$/hydroquinone/DMSO/H_2O	195
6	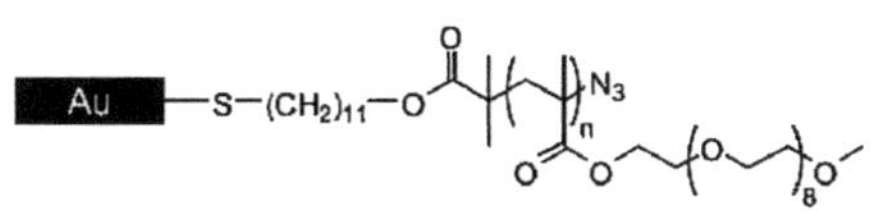	SAM on Au	$CuSO_4 \cdot 5H_2O$/sodium ascorbate/r.t.	205
7	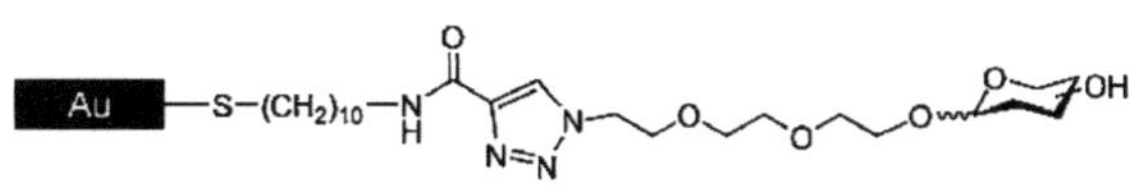	SAM on Au	$CuSO_4$/sodium ascorbate/H_2O:EtOH = 1:1	201
8	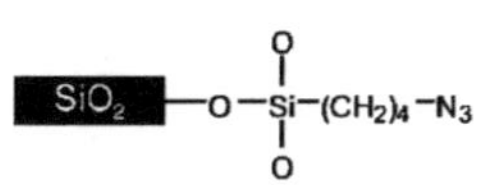	SAM on SiO_2	Thermal/70 °C/neat	196
9	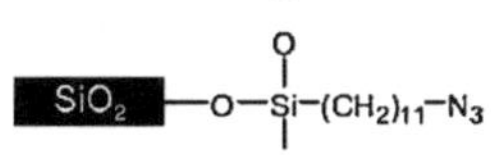	SAM on SiO_2	$CuSO_4 \cdot 5H_2O$/sodium ascorbate	203

10	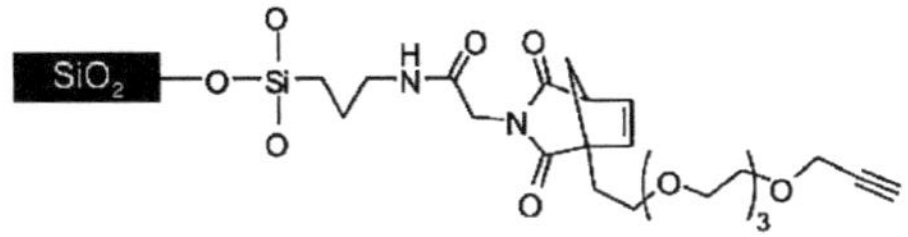	SAM on SiO_2	$CuSO_4 \cdot 5H_2O$/TBTA/TCEP/PBS-buffer/^tBuOH/4 °C	202
11	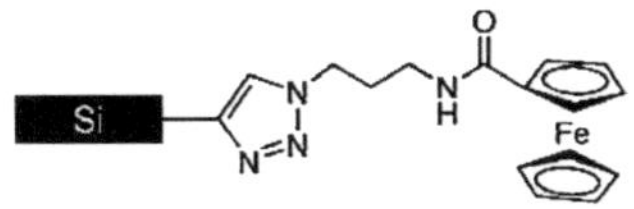	SAM on Si	$CuSO_4 \cdot 5H_2O$/sodium ascorbate/DMF	199
12	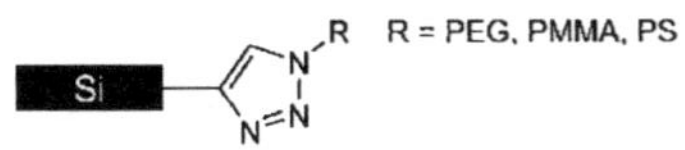	SAM on Si	$Cu(PPh_3)_3Br$/DIPEA/THF	169
13	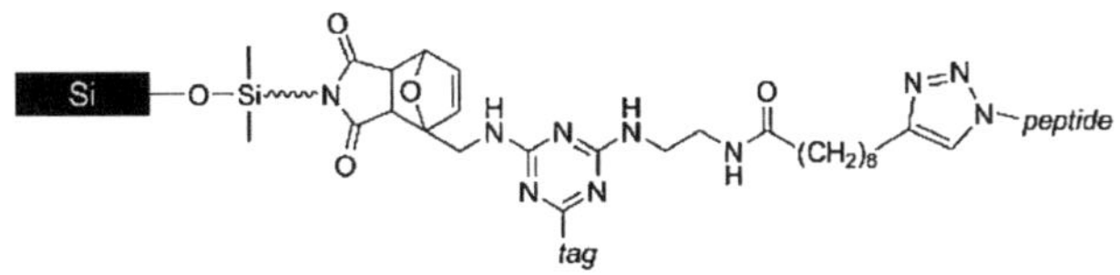	Porous Si	$CuSO_4$/ascorbic acid,MeCN/tris-buffer/pH = 8.0/r.t.	200,220
14	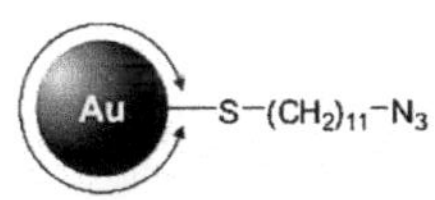	SAM on Au nanoparticles	Dioxane/hexane/r.t.	211
15	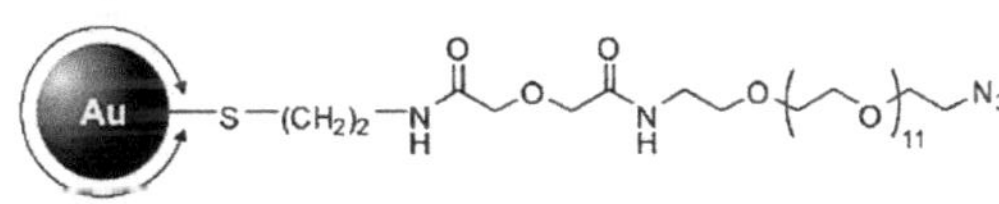	SAM on Au nanoparticles	Cu(I)/r.t.	212
16	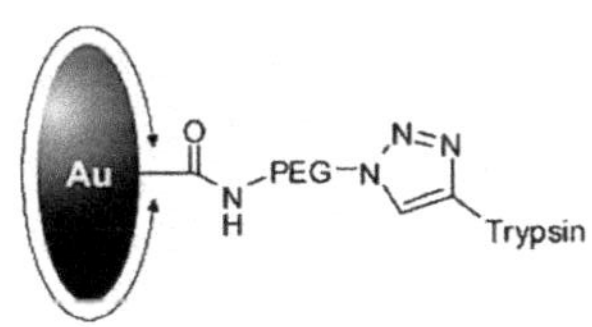	SAM on Au nanorods	$CuSO_4$/ascorbic acid/4 °C	216

(Continued)

Table 7.1 *Azide–alkyne click reactions on surfaces, nanoparticles and interfaces (Continued)*

Entry	Polymer/substrate	Surface	Catalyst/conditions	Reference
17	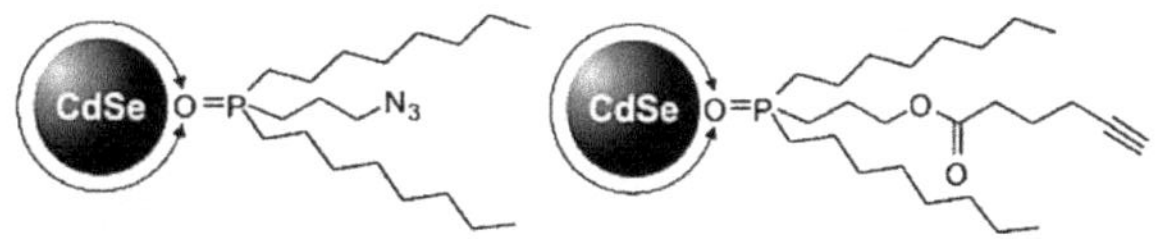	SAM on CdSe nanoparticles	CuBr/TBTA/DIPEA or ΔT	118,121
18	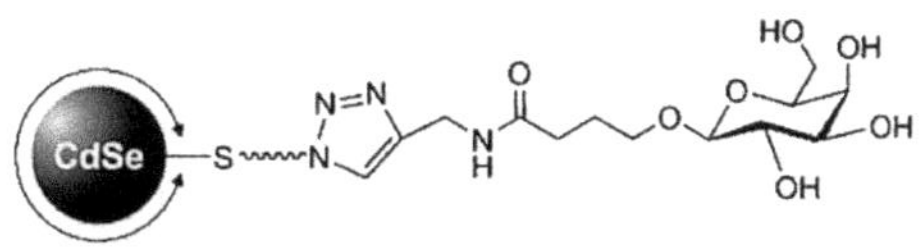	SAM on CdSe nanoparticles	$CuSO_4$/sodium ascorbate/$^tBuOH:H_2O=1:1$	221
19	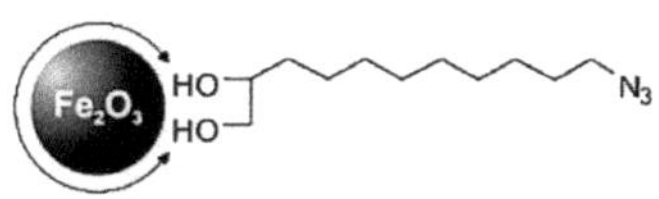	SAM on Fe_2O_3 nanoparticles	ΔT/toluene	122
20	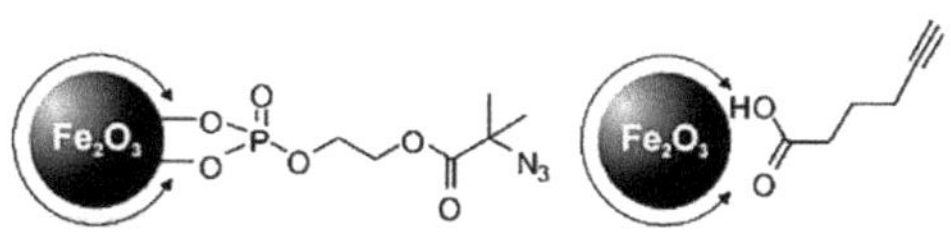	SAM on Fe_2O_3 nanoparticles	$CuSO_4$	213
21	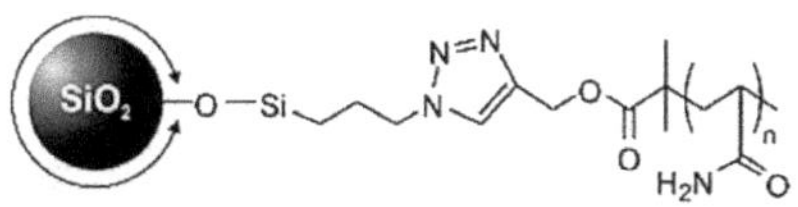	SAM on SiO_2 nanoparticles	$CuSO_4 \cdot 5H_2O$/sodium ascorbate/DMSO/ 50 °C	197
22	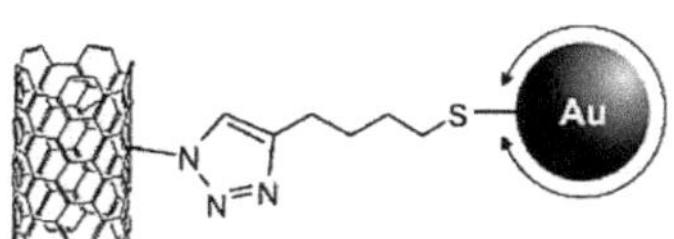	SWNT-nanocomposites	Cu(I)	214
23	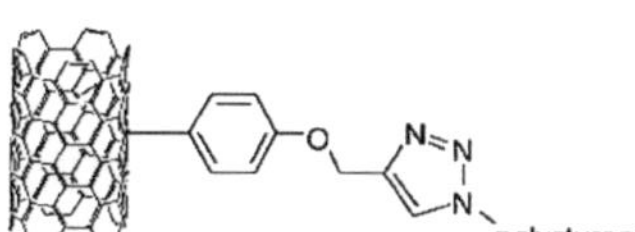	SWNT-nanocomposites	Cu(I)/DMF	70

24	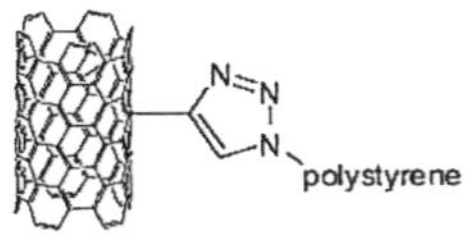	SWNT-nanocomposites	Cu(I)	222
25	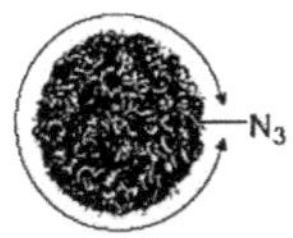	Surface-functionalized micelles	$CuSO_4 \cdot 5H_2O$/sodium ascorbate/H_2O/r.t.	187
26	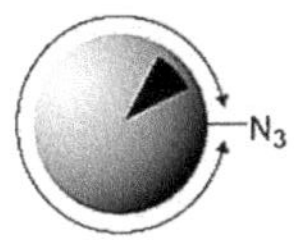	Surface-functionalized polymersomes	$CuSO_4 \cdot 5H_2O$/sodium ascorbate/TBTA	140
27	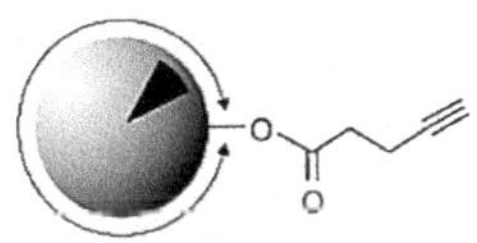	Surface-functionalized polymersomes	$CuSO_4 \cdot 5H_2O$/sodium ascorbate/bathophenanthroline/4 °C	82
28	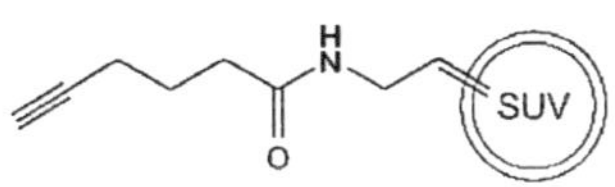	Surface-functionalized liposomes	$CuSO_4 \cdot 5H_2O$/sodium ascorbate/H_2O	141
29	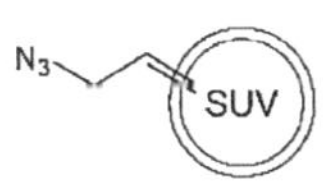	Surface-functionalized liposomes	$CuSO_4$/sodium ascorbate/HEPES-buffer/pH = 6.5	223
30	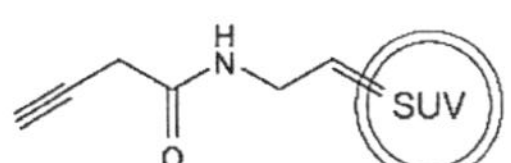	Surface-functionalized liposomes	$CuBr/H_2O$	142

(Continued)

Table 7.1 *Azide–alkyne click reactions on surfaces, nanoparticles and interfaces (Continued)*

Entry	Polymer/substrate	Surface	Catalyst/conditions	Reference
31	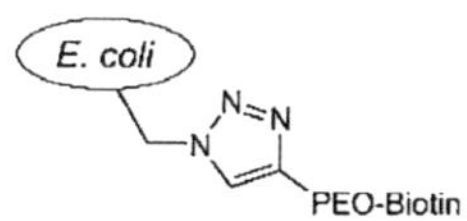	Cell surface	$CuSO_4$/TCEP	224
32	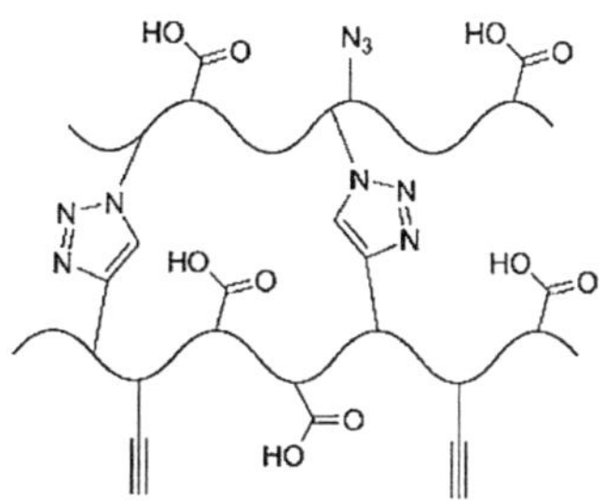	Responsive polymer capsules	$CuSO_4 \cdot 5H_2O$/sodium ascorbate	206
33	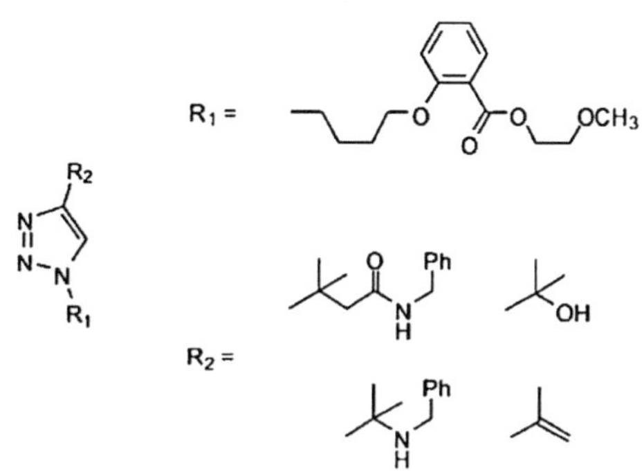	pH sensitive releasing systems	$CuSO_4 \cdot 5H_2O$/sodium ascorbate/$^tBuOH:H_2O$ = 1:1	225

34	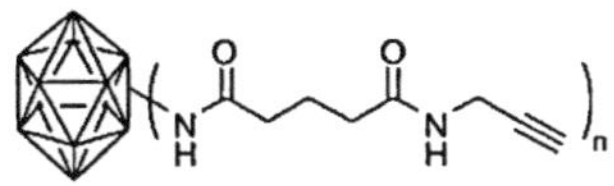	Surface-functionalized bionanoparticle	CuBr/PCDS	215
35	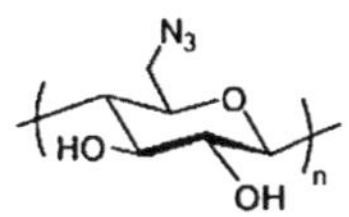	Polysaccharides	$CuSO_4{\cdot}5H_2O$/sodium ascorbate/DMSO/r.t.	208
36	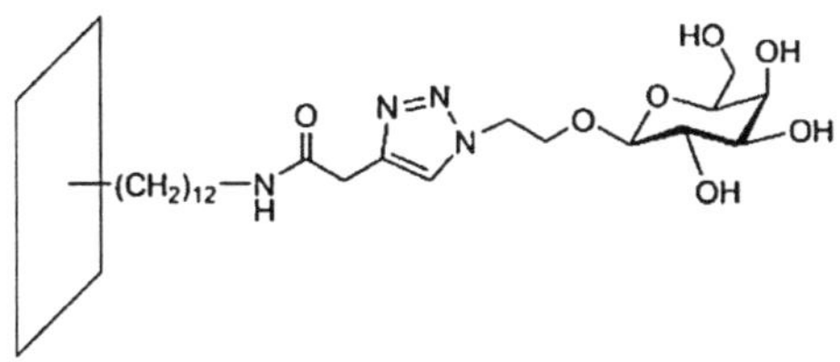	Sugar arrays in microtiter plate	CuI/DIPEA/toluene	226
37	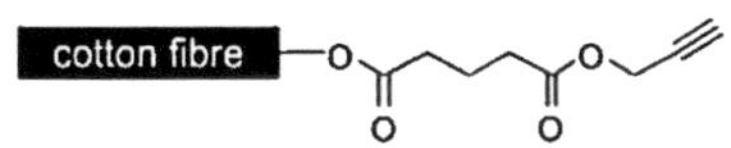	Surface-functionalized cotton fibers	CuBr/*N*-(*n*-propyl)-2-pyridylmethanimine/toluene/70 °C	227
38	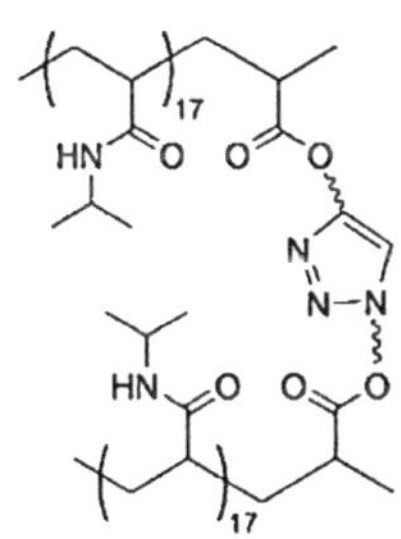	Layer by layer (LbL)	$CuSO_4{\cdot}5H_2O$/sodium ascorbate/H_2O	160

(Continued)

Table 7.1 *Azide–alkyne click reactions on surfaces, nanoparticles and interfaces (Continued)*

Entry	Polymer/substrate	Surface	Catalyst/conditions	Reference
39	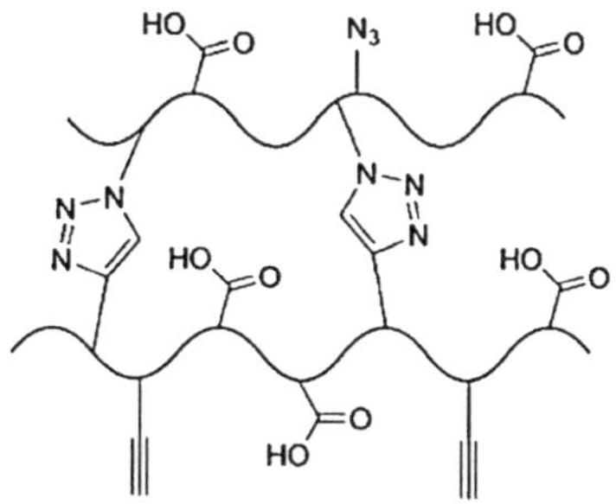	Layer by layer (LbL)	$CuSO_4 \cdot 5H_2O$/sodium ascorbate	204
40	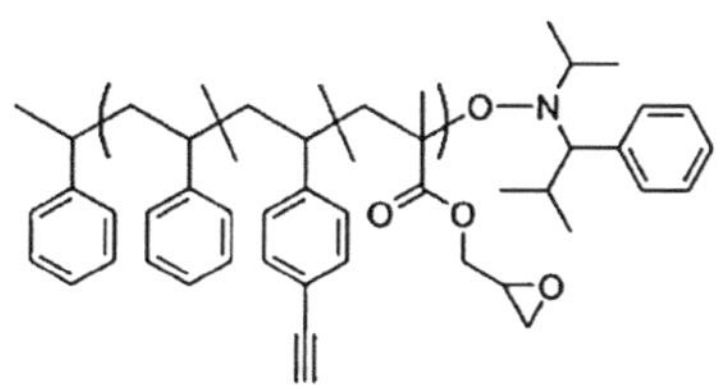	Polymer film	$CuSO_4$/sodium ascorbate/DIPEA/ H_2O/THF/r.t.	228
41	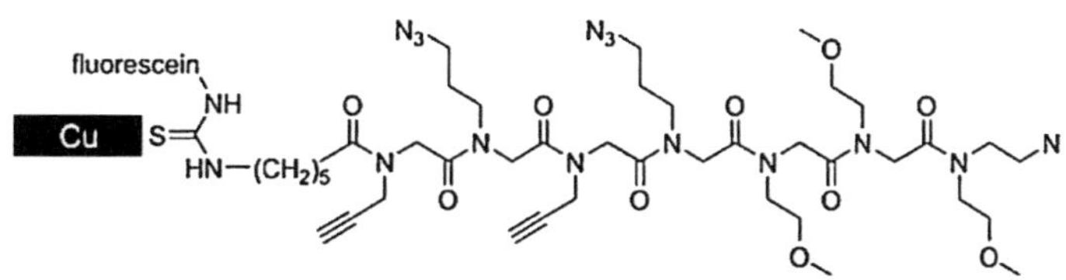	Polymer film	TEA.HCl/45 °C	229
42	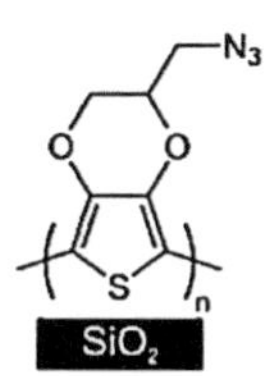	Conductive polymers	$CuSO_4$/sodium ascorbate/DMF/r.t.	230

43	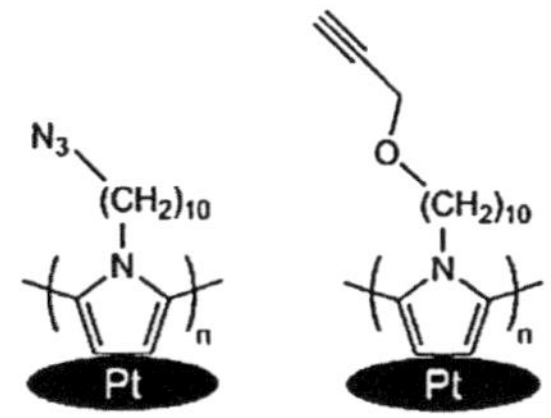	Conductive polymers	$CuCl_2$/sodium ascorbate/$^{t}BuOH:H_2O=2:1$/r.t.	209
44	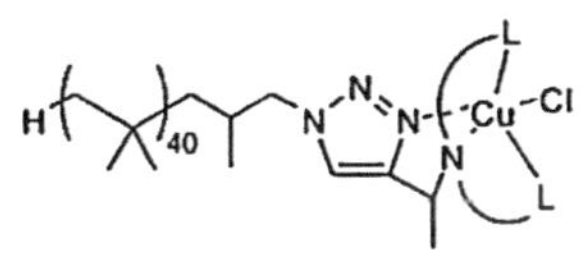	Self-separating homogeneous Cu(I) catalysts	CuCl/heptanes/EtOH	207
45	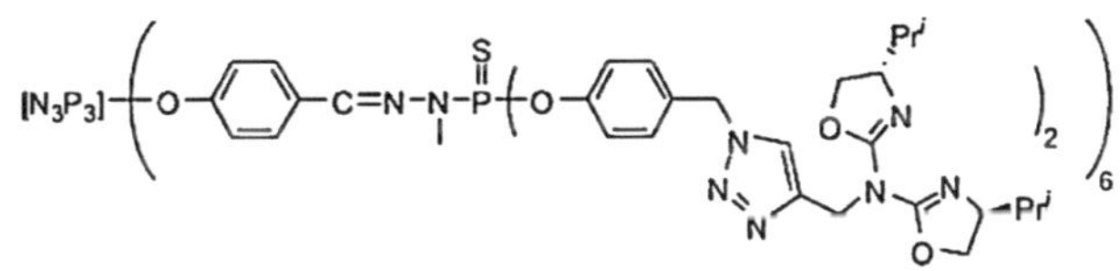		$CuI/N(C_2H_5)_3$/THF	153

References

(1) F. Mohammad, (2007), *Speciality Polymers: Materials and Applications*, International Publishing House, Kent.
(2) J. W. Steed and J. L. Atwood, (2000), *Supramolecular Chemistry*, John Wiley & Sons Ltd, Chichester.
(3) J.-M. Lehn, (2002), Supramolecular chemistry and self-assembly special feature: toward complex matter: supramolecular chemistry and self-organization, *Proc. Natl Acad. Sci. USA*, **99** (8), 4763–4768.
(4) H. C. Kolb, M. G. Finn, and K. B. Sharpless, (2001), Click chemistry: diverse chemical function from a few good reactions, *Angew. Chem., Int. Edn*, **40** (11), 2004–2021.
(5) Z. P. Demko and K. B. Sharpless, (2001), Preparation of 5-substituted 1H-tetrazoles from nitriles in water, *J. Org. Chem.*, **66** (24), 7945–7950.
(6) Z. P. Demko and K. B. Sharpless, (2001), An intramolecular [2 + 3] cycloaddition route to fused 5-heterosubstituted tetrazoles, *Org. Lett.*, **3** (25), 4091–4094.
(7) R. Huisgen, (1989), Kinetics and reaction mechanisms: selected examples from the experience of forty years, *Pure Appl. Chem.*, **61** (4), 613–628.
(8) R. Huisgen, G. Szeimies and L. Möbius, (1967), *Chem. Ber.*, **100**, 2494.
(9) V. Aureggi and G. Sedelmeier, (2007), 1,3-Dipolar cycloaddition: click chemistry for the synthesis of 5-substituted tetrazoles from organoaluminum azides and nitriles, *Angew. Chem., Int. Edn*, **46** (44), 8440–8444.
(10) Z. P. Demko and K. B. Sharpless, (2002), A click chemistry approach to tetrazoles by huisgen 1,3-dipolar cycloaddition: synthesis of 5-sulfonyl tetrazoles from azides and sulfonyl cyanides, *Angew. Chem., Int. Edn*, **41** (12), 2110–2113.
(11) V. V. Rostovtsev, L. G. Green, V. V. Fokin *et al.*, (2002), A stepwise Huisgen cycloaddition process: copper(I)-catalyzed regioselective ligation of azides and terminal alkynes, *Angew. Chem., Int. Edn*, **41** (14), 2596–2599.
(12) M. Meldal and C. W. Tornoe, (2001), Peptidotriazoles: copper(I)-catalyzed 1,3-dipolar cycloadditions on solid-phase, *Proceedings of the Second International and the Seventeenth American Peptide Symposium*, 263–264.
(13) C. W. Tornoe, C. Christensen, and M. Meldal, (2002), Peptidotriazoles on solid phase: [1,2,3]-triazoles by regiospecific copper(I)-catalyzed 1,3-dipolar cycloadditions of terminal alkynes to azides, *J. Org. Chem.*, **67** (9), 3057–3064.
(14) W. G. Lewis, L. G. Green, F. Grynszpan *et al.*, (2002), Click chemistry *in situ*: acetylcholinesterase as a reaction vessel for the selective assembly of a femtomolar inhibitor from an array of building blocks, *Angew. Chem., Int. Edn*, **41** (6), 1053–1057.
(15) G. Molteni, C. I. Bianchi, G. Marinoni *et al.*, (2006), *New. J. Chem.*, **30**, 1137.
(16) L. D. Pachón, J. H. van Maarseveen, and G. Rothenberg, (2005), Click chemistry: copper clusters catalyse the cycloaddition of azides with terminal alkynes, *Adv. Synth. Catal.*, **347** (6), 811–815.
(17) H. A. Orgueira, Demosthenes Fokas, Yuko Isome *et al.*, (2005), Regioselective synthesis of [1,2,3]-triazoles catalyzed by Cu(I) generated *in situ* from Cu(0) nanosize activated powder and amine hydrochloride salts, *Tetrahedron Lett.*, **46** (16), 2911–2914.
(18) M. B. Thathagar, J. Beckers, and G. Rothenberg, (2002), Copper-catalyzed Suzuki cross-coupling using mixed nanocluster catalysts, *J. Am. Chem. Soc.*, **124** (40), 11858–11859.
(19) B. H. Lipshutz and B. R. Taft, (2006), Heterogeneous copper-in-charcoal-catalyzed click chemistry, *Angew. Chem., Int. Edn*, **45** (48), 8235–8238.
(20) J. M. Baskin, J. A. Prescher, S. T. Laughlin *et al.*, (2007), From the cover: copper-free click chemistry for dynamic *in vivo* imaging, *Proc. Natl Acad. Sci. USA*, **104** (43), 16793–16797; X. Ning, J. Guo, M. A. Wolfert *et al.*, (2008), Visualizing metabolically labeled glycoconjugates of living cells by copper-free and fast Huisgen cycloadditions, *Angew. Chem. Int. Edn*, **47** (12), 2253–2255.
(21) L. Zhang, X. Chen, P. Xue *et al.*, (2005), Ruthenium-catalyzed cycloaddition of alkynes and organic azides, *J. Am. Chem. Soc.*, **127** (46), 15998–15999; S. Oppilliart, G. Mousseau, L.

Zhang *et al.*, (2007), 1-Protected 5-amido 1,2,3-triazoles via ruthenium-catalyzed [3 + 2] cycloaddition of azides and ynamides, *Tetrahedron*, **63** (34), 8094–8098.
(22) D. V. Partyka, J. B. Updegraff, M. Zeller *et al.*, (2007), Carbon-gold bond formation through [3 + 2] cycloaddition reactions of gold(I) azides and terminal alkynes, *Organometallics*, **26** (1), 183–186.
(23) C. Chowdhury, S. B. Mandal, and B. Achari, (2005), Palladium-copper catalysed heteroannulation of acetylenic compounds: an expeditious synthesis of isoindoline fused with triazoles, *Tetrahedron Lett.*, **46** (49), 8531–8534.
(24) P. L. Golas, N. V. Tsarevsky, B. S. Sumerlin *et al.*, (2006), Catalyst performance in click coupling reactions of polymers prepared by ATRP: ligand and metal effects, *Macromolecules*, **39** (19), 6451–6457.
(25) V. O. Rodionov, V. V. Fokin, and M. G. Finn, (2005), Mechanism of the ligand-free Cu(I)-catalyzed azide–alkyne cycloaddition reaction, *Angew. Chem. Int. Edn*, **44** (15), 2210–2215.
(26) S. Punna, J. Kuzelka, Q. Wang *et al.*, (2005), Head-to-tail peptide cyclodimerization by copper-catalyzed azide–alkyne cycloaddition, *Angew. Chem. Int. Edn*, **44** (15), 2215–2220.
(27) G. Molteni and A. Ponti, (2003), Arylazide cycloaddition to methyl propiolate: dft-based quantitative prediction of regioselectivity, *Chem. Eur. J.*, **9** (12), 2770–2774.
(28) F. Himo, T. Lovell, R. Hilgraf *et al.*, (2005), Copper(I)-catalyzed synthesis of azoles. DFT study predicts unprecedented reactivity and intermediates, *J. Am. Chem. Soc.*, **127** (1), 210–216.
(29) V. D. Bock, H. Hiemstra, and J. H. van Maarseveen, (2006), Cu(I)-catalyzed alkyne–azide click cycloadditions from a mechanistic and synthetic perspective, *Eur. J. Org. Chem.*, **2006** (1), 51–68.
(30) B. F. Straub, (2007), my-Acetylide and my-alkenylidene ligands in click triazole syntheses, *Chem. Commun.*, 3868–3870.
(31) J.-C. Meng, V. V. Fokin, and M. G. Finn, (2005), Kinetic resolution by copper-catalyzed azide–alkyne cycloaddition, *Tetrahedron Lett.*, **46** (27), 4543–4546.
(32) V. P. Mocharla, B. Colasson, L. V. Lee *et al.*, (2005), *In situ* click chemistry: enzyme-generated inhibitors of carbonic anhydrase II, *Angew. Chem., Int. Edn*, **44** (1), 116–120.
(33) P. Scheiner, J. H. Schomaker, S. Deming *et al.*, (1965), The addition of aryl azides to norbornene. A kinetic investigation, *J. Am. Chem. Soc.*, **87** (2), 306–311.
(34) S. Hiki and K. Kataoka, (2007), A facile synthesis of azido-terminated heterobifunctional poly(ethylene glycol)s for 'click' conjugation, *Bioconjugate Chem.*, **18** (6), 2191–2196.
(35) R. Zirbs, F. Kienberger, P. Hinterdorfer *et al.*, (2005), Directed assembly of Au nanoparticles onto planar surfaces via multiple hydrogen bonds, *Langmuir*, **21** (18), 8414–8421.
(36) W. G. Lewis, F. G. Magallon, V. V. Fokin *et al.*, (2004), Discovery and characterization of catalysts for azide–alkyne cycloaddition by fluorescence quenching, *J. Am. Chem. Soc.*, **126** (30), 9152–9153.
(37) H. Nandivada, X. Jiang, and J. Lahann, (2007), Click chemistry: versatility and control in the hands of materials scientists, *Adv. Mater.*, **19** (17), 2197–2208.
(38) W. H. Binder and R. Sachsenhofer, (2007), Click chemistry in polymer and materials science, *Macromol. Rapid Commun.*, **28** (1), 15–54.
(39) J.-F. Lutz, (2007), 1,3-Dipolar cycloadditions of azides and alkynes: a universal ligation tool in polymer and materials science, *Angew. Chem., Int. Edn*, **46** (7), 1018–1025; D. Fournier, R. Hoogenboom, and U. S. Schubert, (2007), Clicking polymers: a straightforward approach to novel macromolecular architectures, *Chem. Soc. Rev.*, **36** (8), 1369–1380.
(40) L. Barner, T. P. Davis, M. H. Stenzel *et al.*, (2007), Complex macromolecular architectures by reversible addition fragmentation chain transfer chemistry: theory and practice, *Macromol. Rapid Commun.*, **28** (5), 539–559.
(41) R. A. Evans, (2007), The rise of azide/alkyne 1,3-dipolar-click-cycloaddition and its application to polymer science and surface modification, *Aust. J. Chem.*, **60** (6), 384–395.
(42) W. H. Binder and C. Kluger, (2006), Azide/alkyne-click reactions: applications in material science and organic synthesis, *Curr. Org. Chem.*, **10**, 1791–1815.
(43) W. H. Binder and R. Sachsenhofer, (2008), Click-chemistry in polymer and material science: an update, *Macromol. Rapid Commun.*, **29** (12–13), 952–981.

(44) B. Voit, (2007), The potential of cycloaddition reactions in the synthesis of dendritic polymers, *New J. Chem.*, **31** (7), 1139–1151.
(45) S. Dedola, S. A. Nepogodiev, and R. A. Field, (2007), Recent applications of the CuI-catalysed Huisgen azide–alkyne 1,3-dipolar cycloaddition reaction in carbohydrate chemistry, *Org. Biomol. Chem.*, **5** (7), 1006–1017.
(46) A. Dondoni, (2007), Triazole: the keystone in glycosylated molecular architectures constructed by a click reaction, *Chem. Asian J.*, **2** (6), 700–708.
(47) S. G. Spain, M. I. Gibson, and N. R. Cameron, (2007), Recent advances in the synthesis of well-defined glycopolymers, *J. Polym. Sci., Part A: Polym. Chem.*, **45** (11), 2059–2072.
(48) M. V. Gil, M. J. Arévalo, and Ó. López, (2007), Click chemistry – what's in a name? Triazole synthesis and beyond, *Synthesis*, **11**, 1589–1620.
(49) Y. L. Angell and K. Burgess, (2007), Peptidomimetics via copper-catalyzed azide–alkyne cycloadditions, *Chem. Soc. Rev.*, **36** (10), 1674–1689.
(50) H. C. Kolb and K. B. Sharpless, (2003), The growing impact of click chemistry on drug discovery, *Drug Discov. Today*, **8** (24), 1128–1137.
(51) A. I. Prikhodko, F. Durola, and J.-P. Sauvage, (2008), Iron(II)-templated synthesis of [3]rotaxanes by passing two threads through the same ring, *J. Am. Chem. Soc.*, **130** (2), 448–449; J.-S. Marois, K. Cantin, A. Desmarais *et al.*, (2008), [3]Rotaxane–porphyrin conjugate as a novel supramolecular host for fullerenes, *Org. Lett.*, **10** (1), 33–36; P. Mobian, J.-P. Collin, and J.-P. Sauvage, (2006), Efficient synthesis of a labile copper(I)–rotaxane complex using click chemistry, *Tetrahedron Lett.*, **47** (28), 4907–4909; W. R. Dichtel, O. S. Miljanic, J. M. Spruell *et al.*, (2006), Efficient templated synthesis of donor–acceptor rotaxanes using click chemistry, *J. Am. Chem. Soc.*, **128** (32), 10388–10390.
(52) O. S. Miljanic, W. R. Dichtel, I. Aprahamian *et al.*, (2007), Rotaxanes and catenanes by click chemistry, *QSAR Combin. Sci.*, **26** (11–12), 1165–1174.
(53) T. Ooya, D. Inoue, H. S. Choi *et al.*, (2006), pH-Responsive movement of cucurbit[7]uril in a diblock polypseudorotaxane containing dimethyl beta-cyclodextrin and cucurbit[7]uril, *Org. Lett.*, **8** (15), 3159–3162.
(54) P. Mobian, J.-P. Collin, and J.-P. Sauvage, (2006), Efficient synthesis of a labile copper(I)–rotaxane complex using click chemistry, *Tetrahedron Lett.*, **47** (28), 4907–4909.
(55) W. R. Dichtel, O. S. Miljanic, J. M. Spruell *et al.*, (2006), Efficient templated synthesis of donor–acceptor rotaxanes using click chemistry, *J. Am. Chem. Soc.*, **128** (32), 10388–10390.
(56) V. Aucagne, K. D. Haenni, D. A. Leigh *et al.*, (2006), Catalytic click rotaxanes: a substoichiometric metal-template pathway to mechanically interlocked architectures, *J. Am. Chem. Soc.*, **128** (7), 2186–2187.
(57) D. Tuncel and J. H. G. Steinke, (2002), The synthesis of [2], [3] and [4]rotaxanes and semirotaxanes, *Chem. Commun.*, **5**, 496–497.
(58) T. C. Krasia and J. H. G. Steinke, (2002), Formation of oligotriazoles catalysed by cucurbituril, *Chem. Commun.*, **1**, 22–23.
(59) J. M. Langenhahn and J. S. Thorson, (2005), Recent carbohydrate-based chemoselective ligation applications, *Curr. Org. Synth.*, **2** (1), 59–81.
(60) E. H. Ryu and Y. Zhao, (2005), Efficient synthesis of water-soluble calixarenes using click chemistry, *Org. Lett.*, **7** (6), 1035–1037.
(61) Y. Zhang, Z. Guo, J. Ye *et al.*, (2008), Preparation of novel β-cyclodextrin chiral stationary phase based on click chemistry, *J. Chromatogr. A*, **1191** (1–2), 188–192; A. Binello, B. Robaldo, A. Barge *et al.*, (2008), Synthesis of cyclodextrin-based polymers and their use as debittering agents, *J. Appl. Polym. Sci.*, **107** (4), 2549–2557; X.-M. Liu, H.-T. Lee, R. A. Reinhardt *et al.*, (2007), Novel biomineral-binding cyclodextrins for controlled drug delivery in the oral cavity, *J. Controlled Release*, **122** (1), 54–62.
(62) S. Srinivasachari, K. M. Fichter, and T. M. Reineke, (2008), Polycationic beta-cyclodextrin click clusters: monodisperse and versatile scaffolds for nucleic acid delivery, *J. Am. Chem. Soc.*, **130**, 4618–4627.
(63) M. Mourer, F. Hapiot, E. Monflier *et al.*, (2008), Click chemistry as an efficient tool to access β-cyclodextrin dimers, *Tetrahedron*, **64** (30–31), 7159–7163.

(64) Y. Liu, C.-F. Ke, H.-Y. Zhang *et al.*, (2008), Complexation-induced transition of nanorod to network aggregates: alternate porphyrin and cyclodextrin arrays, *J. Am. Chem. Soc.*, **130** (2), 600–605.
(65) Z. Guo, L. Liang, J.-J. Liang *et al.*, (2008), Covalently β-cyclodextrin modified single-walled carbon nanotubes: a novel artificial receptor synthesized by click chemistry, *J. Nanopart. Res.*, **10** (6), 1077–1083.
(66) M. Ortega-Munoz, J. Morales-Sanfrutos, F. Perez-Balderas *et al.*, (2007), Click multivalent neoglycoconjugates as synthetic activators in cell adhesion and stimulation of monocyte/machrophage cell lines, *Org. Biomol. Chem.*, **5** (14), 2291–2301.
(67) O. David, S. Maisonneuve, and J. Xie, (2007), Generation of new fluorophore by click chemistry: synthesis and properties of β-cyclodextrin substituted by 2-pyridyl triazole, *Tetrahedron Lett.*, **48** (37), 6527–6530.
(68) K. D. Bodine, D. Y. Gin, and M. S. Gin, (2004), Synthesis of readily modifiable cyclodextrin analogues via cyclodimerization of an alkynyl-azido trisaccharide, *J. Am. Chem. Soc.*, **126** (6), 1638–1639.
(69) R. Hoogenboom, B. C. Moore, and U. S. Schubert, (2006), Synthesis of star-shaped poly(e-caprolactone) viaclick chemistry and supramolecular click chemistry, *Chem. Commun.*, **38**, 4010–4012.
(70) H. Li, F. Cheng, A. M. Duft *et al.*, (2005), Functionalization of single-walled carbon nanotubes with well-defined polystyrene by click coupling, *J. Am. Chem. Soc.*, **127** (41), 14518–14524.
(71) W. H. Binder, (2008), Polymer-Induced transient pores in lipid membranes, *Angew. Chem. Int. Edn*, **47** (17), 3092–3095.
(72) W. S. Horne, C. D. Stout, and M. R. Ghadiri, (2003), A heterocyclic peptide nanotube, *J. Am. Chem. Soc.*, **125** (31), 9372–9376.
(73) W.-Y. Yang, J.-H. Ahn, Y.-S. Yoo *et al.*, (2005), Supramolecular barrels from amphiphilic rigid-flexible macrocycles, *Nat. Mater.*, **4** (5), 399–402.
(74) A. Deffieux and R. Borsali, (2007), in *Macromolecular Engineering*, edited by K. Matyjaszewski, Y. Gnanou, and L. Leibler, Wiley-VCH, Weinheim, vol. 2, pp. 875–908.
(75) R. E. Looper, D. Pizzirani, and S. L. Schreiber, (2006), Macrocycloadditions leading to conformationally restricted small molecules, *Org. Lett.*, **8** (10), 2063–2066.
(76) A. M. Rubio, M. Pita, and J. J. Freire, (2002), Cyclization kinetics of nondiluted bond fluctuation chains, *Macromolecules*, **35** (14), 5681–5687.
(77) B. A. Laurent and S. M. Grayson, (2006), An efficient route to well-defined macrocyclic polymers via click cyclization, *J. Am. Chem. Soc.*, **128** (13), 4238–4239.
(78) X. P. Qiu, F. Tanaka, and F. M. Winnik, (2007), Temperature-induced phase transition of well-defined cyclic poly(*N*-isopropylacrylamide)s in aqueous solution, *Macromolecules*, **40** (20), 7069–7071.
(79) J. Xu, J. Ye, and S. Liu, (2007), Synthesis of well-defined cyclic poly(*N*-isopropylacrylamide) via click chemistry and its unique thermal phase transition behavior, *Macromolecules*, **40** (25), 9103–9110.
(80) H. Li, R. Riva, R. Jerome *et al.*, (2007), Combination of ring-opening polymerization and click chemistry for the synthesis of an amphiphilic tadpole-shaped poly(epsilon-caprolactone) grafted by PEO, *Macromolecules*, **40** (4), 824–831.
(81) V. Haridas, Kashmiri Lal, Yogesh K. Sharma *et al.*, (2008), Design, synthesis, and self-assembling properties of novel triazolophanes, *Org. Lett.*, **10** (8), 1645–1647.
(82) S. F. M. van Dongen, M. N. Sanne, S. Jeroen *et al.*, (2008), A block copolymer for functionalisation of polymersome surfaces, *Macromol. Rapid Commun.*, in press.
(83) W. S. Horne, M. K. Yadav, C. D. Stout *et al.*, (2004), Heterocyclic peptide backbone modifications in an alpha-helical coiled coil, *J. Am. Chem. Soc.*, **126** (47), 15366–15367.
(84) W. J. Choi, Z.-D. Shi, K. M. Worthy *et al.*, (2006), Application of azide–alkyne cycloaddition click chemistry for the synthesis of Grb2 SH2 domain-binding macrocycles, *Bioorg. Med. Chem. Lett.*, **16** (20), 5265–5269.
(85) S. Dorner and B. Westermann, (2005), A short route for the synthesis of sweet macrocycles via a click-dimerization-ring-closing metathesis approach, *Chem. Commun.*, **22**, 2852–2854.

(86) F. Seela, V. R. Sirivolu, and P. Chittepu, (2008), Modification of DNA with octadiynyl side chains: synthesis, base pairing, and formation of fluorescent coumarin dye conjugates of four nucleobases by the alkyne–azide click reaction, *Bioconjugate Chem.*, **19** (1), 211–224; F. Seela and V. R. Sirivolu, (2006), DNA containing side chains with terminal triple bonds: base-pair stability and functionalization of alkynylated pyrimidines and 7-deazapurines, *Chem. Biodiversity*, **3** (5), 509–514.
(87) A. Salic and T. J. Mitchison, (2008), A chemical method for fast and sensitive detection of DNA synthesis *in vivo*, *Proc. Natl Acad. Sci. USA*, **105** (7), 2415–2420.
(88) P. Kocalka, A. H. El-Sagheer, and T. Brown, (2008), Rapid and efficient DNA strand cross-linking by click chemistry, *ChemBioChem*, **9** (8), 1280–1285.
(89) K. Ahmed, N. Shankaraiah, V. Devaiah *et al.*, (2008), Synthesis of 1,2,3-triazole-linked pyrrolobenzodiazepine conjugates employing click chemistry: DNA-binding affinity and anticancer activity, *Bioorg. Med. Chem. Lett.*, **18** (4), 1468–1473; S. Srinivasachari, Y. Liu, L. E. Prevette *et al.*, (2007), Effects of trehalose click polymer length on pDNA complex stability and delivery efficacy, *Biomaterials*, **28** (18), 2885–2898; A. D. Moorhouse, A. Mafalda Santos, M. Gunaratnam *et al.*, (2006), Stabilization of G-quadruplex DNA by highly selective ligands via click chemistry, *J. Am. Chem. Soc.*, **128** (50), 15972–15973.
(90) P. M. E. Gramlich, S. Warncke, J. Gierlich *et al.*, (2008), Click click click: single to triple modification of DNA, *Angew. Chem. Int. Edn*, **47** (18), 3442–3444.
(91) M. Fischler, A. Sologubenko, J. Mayer *et al.*, (2008), Chain-like assembly of gold nanoparticles on artificial DNA templates via click chemistry, *Chem. Commun.*, **2**, 169–171.
(92) A. Doerr, (2008), DNA synthesis lights up, *Nat. Meth.*, **5** (4), 286.
(93) C. T. Wirges, P. M. E. Gramlich, K. Gutsmiedl *et al.*, (2007), Pronounced effect of DNA hybridization on click reaction efficiency, *QSAR Combin. Sci.*, **26** (11–12), 1159–1164.
(94) D. I. Rozkiewicz, J. Gierlich, G. A. Burley *et al.*, (2007), Transfer printing of DNA by click chemistry, *ChemBioChem*, **8** (16), 1997–2002.
(95) F. Morvan, A. Meyer, A. Jochum *et al.*, (2007), Fucosylated pentaerythrityl phosphodiester oligomers (PePOs): automated synthesis of DNA-based glycoclusters and binding to *Pseudomonas aeruginosa* Lectin (PA–IIL), *Bioconjugate Chem.*, **18** (5), 1637–1643.
(96) R. Kumar, A. El-Sagheer, J. Tumpane *et al.*, (2007), Template-directed oligonucleotide strand ligation, covalent intramolecular DNA circularization and catenation using click chemistry, *J. Am. Chem. Soc.*, **129** (21), 6859–6864.
(97) D. M. Hammond, A. Manetto, J. Gierlich *et al.*, (2007), DNA photography: an ultrasensitive DNA-detection method based on photographic techniques, *Angew. Chem. Int. Edn*, **46** (22), 4184–4187.
(98) K. Gogoi, M. V. Mane, S. S. Kunte *et al.*, (2007), A versatile method for the preparation of conjugates of peptides with DNA/PNA/analog by employing chemo-selective click reaction in water, *Nucleic Acid. Res.*, **35** (21), e139/131–e139/137.
(99) M. Fischler, U. Simon, H. Nir *et al.*, (2007), Formation of bimetallic Ag–Au nanowires by metallization of artificial DNA duplexes, *Small*, **3** (6), 1049–1055; G. A. Burley, J. Gierlich, M. R. Mofid *et al.*, (2006), Directed DNA metallization, *J. Am. Chem. Soc.*, **128** (5), 1398–1399.
(100) J. Gierlich, G. A. Burley, P. M. E. Gramlich *et al.*, (2006), Click chemistry as a reliable method for the high-density postsynthetic functionalization of alkyne-modified DNA, *Org. Lett.*, **8** (17), 3639–3642.
(101) R. L. Weller and S. R. Rajski, (2005), DNA methyltransferase-moderated click chemistry, *Org. Lett.*, **7** (11), 2141–2144.
(102) T. S. Seo, Z. Li, H. Ruparel *et al.*, (2003), Click chemistry to construct fluorescent oligonucleotides for DNA sequencing, *J. Org. Chem.*, **68** (2), 609–612.
(103) D. I. Rozkiewicz, D. Janczewski, W. Verboom *et al.*, (2006), Click chemistry by microcontact printing, *Angew. Chem., Int. Edn*, **45** (32), 5292–5296.
(104) S. Bakbak, P. J. Leech, B. E. Carson *et al.*, (2006), 1,3-Dipolar cycloaddition for the generation of nanostructured semiconductors by heated probe tips, *Macromolecules*, **39** (20), 6793–6795.
(105) H. Xu, S. Srivastava, and V. Rotello, (2007), Nanocomposites based on hydrogen bonds, *Adv. Polym. Sci. Hydrogen Bonded Polym.*, 179–198.

(106) G. ten Brinke, J. Ruokolainen, and O. Ikkala, (2007), Supramolecular materials based on hydrogen-bonded polymers, *Adv. Polym. Sci. Hydrogen Bonded Polym.*, 113–177; W. Marcus, (2007), Side-chain functionalized supramolecular polymers, *Polym. Int.*, **56** (4), 453–460; L. Bouteiller, (2007), Assembly via hydrogen bonds of low molar mass compounds into supramolecular polymers, *Adv. Polym. Sci. Hydrogen Bonded Polym.*, 79–112.

(107) W. Binder and R. Zirbs, (2007), Supramolecular polymers and networks with hydrogen bonds in the main- and side-chain, *Adv. Polym. Sci. Hydrogen Bonded Polym.*, 1–78.

(108) J.-M. Lehn, (2005), Dynamers: dynamic molecular and supramolecular polymers, *Progr. Polym. Sci.*, **30** (8–9), 814–831.

(109) P. Cordier, F. Tournilhac, C. Soulie-Ziakovic *et al.*, (2008), Self-healing and thermoreversible rubber from supramolecular assembly, *Nature*, **451**, 977–980.

(110) W. H. Binder, S. Bernstorff, C. Kluger *et al.*, (2005), Tunable materials from hydrogen-bonded pseudo block copolymers, *Adv. Mater.*, **17** (23), 2824–2828; M. J. Kunz, G. Hayn, R. Saf *et al.*, (2004), Hydrogen-bonded supramolecular poly(ether ketone)s, *J. Polym. Sci., Part A: Polym. Chem.*, **42** (3), 661–674; W. H. Binder, M. J. Kunz, and E. Ingolic, (2004), Supramolecular poly(ether ketone)-polyisobutylene pseudo-block copolymers, *J. Polym. Sci., Part A: Polym. Chem.*, **42** (1), 162–172.

(111) W. H. Binder, L. Petraru, T. Roth *et al.*, (2007), Magnetic and temperature-sensitive release gels from supramolecular polymers, *Adv. Funct. Mater.*, **17** (8), 1317–1326.

(112) Patricia Y. W. Dankers, Martin C. Harmsen, Linda A. Brouwer *et al.*, (2005), A modular and supramolecular approach to bioactive scaffolds for tissue engineering, *Nat. Mater.*, **4** (7), 568–574.

(113) W. H. Binder and R. Sachsenhofer, (2007), Click chemistry in polymer and materials science, *Macromol. Rapid Commun.*, **28** (1), 15–54; W. H. Binder and C. Kluger, (2006), Azide/alkyne–click reactions: applications in material science and organic synthesis, *Curr. Org. Chem.*, **10** (14), 1791.

(114) W. H. Binder and C. Kluger, (2004), Combining ring-opening metathesis polymerization (ROMP) with Sharpless-type click reactions: an easy method for the preparation of side chain functionalized poly(oxynorbornenes), *Macromolecules*, **37** (25), 9321–9330.

(115) W. H. Binder, D. Machl, and C. Kluger, (2004), Connecting polymeric fragments by Sharpless-type click-reactions, *Polym. Prepr.*, **45** (2), 692–693; W. H. Binder and R. Sachsenhofer, (2008), Polymersome/silica capsules by click-chemistry, *Macromol. Rapid Commun.*, in press.

(116) W. H. Binder, C. Kluger, C. J. Straif *et al.*, (2005), Directed nanoparticle binding onto microphase-separated block copolymer thin films, *Macromolecules*, **38** (23), 9405–9410.

(117) W. H. Binder, C. Kluger, M. Josipovic *et al.*, (2006), Directing supramolecular nanoparticle binding onto polymer films: film formation and influence of receptor density on binding densities, *Macromolecules*, **39** (23), 8092–8101.

(118) W. H. Binder, L. Petraru, R. Sachenshofer *et al.*, (2006), Synthesis of surface-modified nanoparticles via cycloaddition-reactions, *Monatsh. Chem.*, **137** (7), 835–841.

(119) W. H. Binder, R. Sachsenhofer, C. J. Straif *et al.*, (2007), Surface-modified nanoparticles via thermal and Cu(i)-mediated click chemistry: generation of luminescent CdSe nanoparticles with polar ligands guiding supramolecular recognition, *J. Mater. Chem.*, **17** (20), 2125–2132.

(120) W. H. Binder, L. Petraru, T. Roth *et al.*, (2007), Magnetic and temperature-sensitive release gels from supramolecular polymers, *Adv. Funct. Mater.*, **17** (8), 1317–1326.

(121) W. H. Binder, R. Sachsenhofer, C. J. Straif *et al.*, (2007), Surface-modified nanoparticles via thermal and Cu(i)-mediated click chemistry: generation of luminescent CdSe nanoparticles with polar ligands guiding supramolecular recognition, *J. Mater. Chem.*, **17** (20), 2125–2132.

(122) W. H. Binder and H. C. Weinstabl, (2007), Surface-modified superparamagnetic iron-oxide nanoparticles, *Monatsh. Chem.*, **138** (4), 315–320.

(123) C. Kluger and W. H. Binder, (2007), Functionalized poly(oxanorbornene)-block-copolymers: preparation via ROMP/click-methodology, *J. Polym. Sci., Part A: Polym. Chem.*, **45** (3), 485–499.

(124) W. H. Binder, D. Gloger, H. Weinstabl *et al.*, (2007), Telechelic poly(*N*-isopropylacrylamides) via nitroxide-mediated controlled polymerization and click chemistry: livingness and grafting-from methodology, *Macromolecules*, **40** (9), 3097–3107.

(125) W. H. Binder, L. Petraru, H. Weinstabl *et al.*, (2007), Hard and soft capsules: from branched polymers to controlled release via gels, *Macromol. Symp.*, **254** (1), 62–66.
(126) W. H. Binder and R. Sachsenhofer, (2008), Polymersome/silica capsules by click-chemistry, *Macromol. Rapid Commun.*, **29** (12–13), 1097–1103.
(127) R. Sachsenhofer, W. H. Binder, D. Farnik *et al.*, (2007), Polymersome-embedded nanoparticles, *Macromol. Symp.*, **254** (1), 375–377.
(128) S. K. Yang and M. Weck, (2008), Modular covalent multifunctionalization of copolymers, *Macromolecules*, **41** (2), 346–351; X.-Y. Wang, A. Kimyonok, and M. Weck, (2006), Functionalization of polymers with phosphorescent iridium complexes via click chemistry, *Chem. Commun.*, 3933–3935.
(129) W. H. Binder, H. Weinstabl, and R. Sachsenhofer, (2008), Superparamagnetic ironoxide nanoparticles via ligand exchange reactions: organic 1,2-diols as versatile building blocks for surfaceengineering, *J. Nanomater.*, in press; W. H. Binder and H. C. Weinstabl, (2007), Surface-modified superparamagnetic iron-oxide nanoparticles, *Monatsh. Chem./Chemical Mthly*, **138** (4), 315–320; W. H. Binder, L. Petraru, R. Sachenshofer *et al.*, (2006), Synthesis of surface-modified nanoparticles via cycloaddition-reactions, *Monatsh. Chem./Chem. Mthly*, **137** (7), 835–841.
(130) L. Petraru and W. H. Binder, (2005), Azide/alkyne – functionalized oligomeric silsesquioxanes, *Polym. Prepr.*, **46** (2), 841.
(131) H. Weinstabl and W. H. Binder, (2006), Magnetic nanoparticles with supramolecular recognition sites, *Polym. Prepr.*, **47** (2), 866.
(132) R. M. Meudtner and S. Hecht, (2008), Helicity inversion in responsive foldamers induced by achiral halide ion guests, *Angew. Chem. Int. Edn*, **47** (26), 4926–4930.
(133) H. Ringsdorf, B. Schlarb, and J. Venzmer, (1988), Molekulare Architektur und Funktion von polymeren orientierten Systemen – Modelle für das Studium von Organisation, Oberflächenerkennung und Dynamik bei Biomembranen, *Angew. Chem.*, **100** (1), 117–162; T. Kunitake, (2000), Self-assemblies of biomembrane mimics, in *Physical Chemistry of Biological Interfaces*, A. Baszkin, W. Norde (Eds) (Marcel Dekker, New York), pp. 283–305.
(134) W. H. Binder, V. Barragan, and F. M. Menger, (2003), Domains and rafts in lipid membranes, *Angew. Chem., Int. Edn*, **42** (47), 5802–5827.
(135) W. Meier, (2000), Polymer nanocapsules, *Chem. Soc. Rev.*, **29** (5), 295–303; D. E. Discher and A. Eisenberg, (2002), Polymer vesicles, *Science*, **297** (5583), 967–973.
(136) H. Bermudez, D. A. Hammer, and D. E. Discher, (2004), Effect of bilayer thickness on membrane bending rigidity, *Langmuir*, **20** (3), 540–543.
(137) W. H. Binder, R. Sachsenhofer, D. Farnik *et al.*, (2007), Guiding the location of nanoparticles into vesicular structures: a morphological study, *Phys. Chem. Chem. Phys.*, **9** (48), 6435–6441.
(138) M. Yotaro, (2007), Thermally responsive polymer vesicles, *Angew. Chem. Int. Edn*, **46** (9), 1370–1372; A. Napoli, M. Valentini, N. Tirelli *et al.*, (2004), Oxidation-responsive polymeric vesicles, *Nat. Mater.*, **3** (3), 183–189.
(139) W. H. Binder and R. Sachsenhofer, (2008), Polymersome/silica capsules by click-chemistry, *Macromol. Rapid. Commun.*, **29** (12–13), 1097–1103.
(140) J. A. Opsteen, R. P. Brinkhuis, R. L. M. Teeuwen *et al.*, (2007), Clickable polymersomes, *Chem. Commun.*, **30**, 3136–3138.
(141) F. SaidHassane, B. Frisch, and F. Schuber, (2006), Targeted liposomes: convenient coupling of ligands to preformed vesicles using click chemistry, *Bioconjugate Chem.*, **17** (3), 849–854.
(142) S. Cavalli, A. R. Tipton, M. Overhand *et al.*, (2006), The chemical modification of liposome surfaces via a copper-mediated [3 + 2] azide-alkyne cycloaddition monitored by a colorimetric assay, *Chem. Commun.*, **30**, 3193–3195.
(143) B. Li, A. L. Martin, and E. R. Gillies, (2007), Multivalent polymer vesicles via surface functionalization, *Chem. Commun.*, **48**, 5217–5219.
(144) F. Said Hassane, B. Frisch, and F. Schuber, (2006), Targeted liposomes: convenient coupling of ligands to preformed vesicles using click chemistry, *Bioconjugate Chem.*, **17** (3), 849–854.

(145) E. Fernandez-Megia, J. Correa, I. Rodriguez-Meizoso *et al.*, (2006), A click approach to unprotected glycodendrimers, *Macromolecules*, **39** (6), 2113–2120; S. Svenson and D. A. Tomalia, (2005), Dendrimers in biomedical applications – reflections on the field, *Adv. Drug Deliv. Rev.*, **57** (15), 2106–2129; P. Wu, A. K. Feldman, A. K. Nugent *et al.*, (2004), Efficiency and fidelity in a click-chemistry route to triazole dendrimers by the copper(I)-catalyzed ligation of azides and alkynes, *Angew. Chem.*, **116** (30), 4018–4022; J. Lenoble, N. Maringa, S. Campidelli *et al.*, (2006), Liquid-crystalline fullerodendrimers which display columnar phases, *Org. Lett.*, **8** (9), 1851–1854; J. W. Lee, J. H. Kim, and B.-K. Kim, (2006), Synthesis of azide-functionalized PAMAM dendrons at the focal point and their application for synthesis of PAMAM-like dendrimers, *Tetrahedron Lett.*, **47** (16), 2683–2686; E. Fernandez-Megia, J. Correa, and R. Riguera, (2006), Clickable PEG-dendritic block copolymers, *Biomacromolecules*, **7** (11), 3104–3111; S. Campidelli, J. Lenoble, J. Barbera *et al.*, (2005), Supramolecular fullerene materials: dendritic liquid-crystalline fulleropyrrolidines, *Macromolecules*, **38** (19), 7915–7925; M. J. Joralemon, A. K. Nugent, J. B. Matson *et al.*, (2004), Clicking together dendritic macromolecular divergently, *PMSE Prepr.*, **91**, 195; S. Campidelli, E. Vazquez, D. Milic *et al.*, (2004), Liquid-crystalline fullerene-ferrocene dyads, *J. Mater. Chem.*, **14** (8), 1266–1272.

(146) C. Ornelas, J. Ruiz Aranzaes, E. Cloutet *et al.*, (2007), Click assembly of 1,2,3-triazole-linked dendrimers, including ferrocenyl dendrimers, which sense both oxo anions and metal cations, *Angew. Chem., Int. Edn*, **46** (6), 872–877.

(147) J. W. Lee, B. K. Kim, H. J. Kim *et al.*, (2006), Convergent synthesis of symmetrical and unsymmetrical PAMAM dendrimers, *Macromolecules*, **39** (6), 2418–2422; J. W. Lee, B. K. Kim, J. H. Kim *et al.*, (2006), Facile approach for diblock codendrimers by fusion between Frechet dendrons and PAMAM Dendrons, *J. Org. Chem.*, **71** (13), 4988–4991; J. W. Lee, J. H. Kim, H. J. Kim *et al.*, (2007), Synthesis of symmetrical and unsymmetrical PAMAM dendrimers by fusion between azide- and alkyne-functionalized PAMAM dendrons, *Bioconjugate Chem.*, **18** (2), 579–584.

(148) Q. Liu, P. Zhao, and Y. Chen, (2007), Divergent synthesis of dendrimer-like macromolecules through a combination of atom transfer radical polymerization and click reaction, *J. Polym. Sci., Part A: Polym. Chem.*, **45** (15), 3330–3341; P. Antoni, D. Nystrom, C. J. Hawker *et al.*, (2007), A chemoselective approach for the accelerated synthesis of well-defined dendritic architectures, *Chem. Commun.*, **22**, 2249–2251.

(149) P. Wu, A. K. Feldman, A. K. Nugent *et al.*, (2004), Efficiency and fidelity in a click-chemistry route to triazole dendrimers by the copper(I)-catalyzed ligation of azides and alkynes, *Angew. Chem., Int. Edn*, **43** (30), 3928–3932.

(150) D. T. S. Rijkers, G. W. van Esse, R. Merkx *et al.*, (2005), Efficient microwave-assisted synthesis of multivalent dendrimeric peptides using cycloaddition reaction (click) chemistry, *Chem. Commun.*, **36**, 4581–4583.

(151) P. Wu, M. Malkoch, J. N. Hunt *et al.*, (2005), Multivalent, bifunctional dendrimers prepared by click chemistry, *Chem. Commun.*, **46**, 5775–5777.

(152) C. N. Urbani, C. A. Bell, D. E. Lonsdale *et al.*, (2007), Reactive alkyne and azide solid supports to increase purity of novel polymeric stars and dendrimers via the click reaction, *Macromolecules*, **40** (19), 7056–7059; C. N. Urbani, C. A. Bell, D. Lonsdale *et al.*, (2008), Self-assembly of amphiphilic polymeric dendrimers synthesized with selective degradable linkages, *Macromolecules*, **41** (1), 76–86.

(153) A. Gissibl, C. Padie, M. Hager *et al.*, (2007), Synthesis and application of phosphorus dendrimer immobilized azabis(oxazolines), *Org. Lett.*, **9** (15), 2895–2898.

(154) R. K. O'Reilly, M. J. Joralemon, C. J. Hawker *et al.*, (2005), Synthesis and applications of click-functionalized dendrimers as crosslinkers for nanoparticle delivery agents, *PMSE Prepr.*, **46** (1), 92–93.

(155) J. W. Lee, J. H. Kim, B.-K. Kim *et al.*, (2006), Synthesis of Frechet type dendritic benzyl propargyl ether and Frechet type triazole dendrimer, *Tetrahedron*, **62** (5), 894–900.

(156) R. Vestberg, M. Malkoch, M. Kade *et al.*, (2007), Role of architecture and molecular weight in the formation of tailor-made ultrathin multilayers using dendritic macromolecules and click chemistry, *J. Polym. Sci., Part A: Polym. Chem.*, **45** (14), 2835–2846.

(157) R. K. O'Reilly, M. J. Joralemon, C. J. Hawker *et al.*, (2007), Preparation of orthogonally-functionalized core click cross-linked nanoparticles, *New J. Chem.*, **31** (5), 718–724.
(158) M. Malkoch, K. Schleicher, E. Drockenmuller *et al.*, (2005), Structurally diverse dendritic libraries: a highly efficient functionalization approach using click chemistry, *Macromolecules*, **38** (9), 3663–3678.
(159) A. P. Vogt, S. R. Gondi, and B. S. Sumerlin, (2007), Hyperbranched polymers via RAFT copolymerization of an acryloyl trithiocarbonate, *Aust. J. Chem.*, **60** (6), 396–399; A. J. Scheel, H. Komber, and B. I. Voit, (2004), Novel hyperbranched poly([1,2,3]-triazole)s derived from AB2 monomers by a 1,3-dipolar cycloaddition, *Macromol. Rapid Commun.*, **25** (12), 1175–1180; M. Smet, K. Metten, and W. Dehaen, (2004), Synthesis of new AB2 monomers for polymerization to hyperbranched polymers by 1,3-dipolar cycloaddition, *Collect. Czech. Chem. Commun.*, **69**, 1097–1108.
(160) D. E. Bergbreiter and B. S. Chance, (2007), Click-based covalent layer-by-layer assembly on polyethylene using water-soluble polymeric reagents, *Macromolecules*, **40** (15), 5337–5343.
(161) G. W. Goodall and W. Hayes, (2006), Advances in cycloaddition polymerizations, *Chem. Soc. Rev.*, **35** (3), 280–312.
(162) J. W. Lee, J. H. Kim, B.-K. Kim *et al.*, (2006), Convergent synthesis of PAMAM dendrimers using click chemistry of azide-functionalized PAMAM dendrons, *Tetrahedron*, **62**, 9193–9200.
(163) M. J. Joralemon, R. K. O'Reilly, J. B. Matson *et al.*, (2005), Dendrimers clicked together divergently, *Macromolecules*, **38** (13), 5436–5443.
(164) B. Helms, J. L. Mynar, C. J. Hawker *et al.*, (2004), Dendronized linear polymers via click chemistry, *J. Am. Chem. Soc.*, **126** (46), 15020–15021.
(165) W. Z. Chen, P. E. Fanwick, and T. Ren, (2007), Dendronized diruthenium compounds via the copper(I)-catalyzed click reaction, *Inorg. Chem.*, **46** (9), 3429–3431.
(166) A. Qin, J. W. Y. Lam, C. K. W. Jim *et al.*, (2008), Hyperbranched polytriazoles: click polymerization, regioisomeric structure, light emission, and fluorescent patterning, *Macromolecules*, **41** (11), 3808–3822.
(167) A. Gopin, S. Ebner, B. Attali *et al.*, (2006), Enzymatic activation of second-generation dendritic prodrugs: conjugation of self-immolative dendrimers with poly(ethylene glycol) via click chemistry, *Bioconjugate Chem.*, **17**, 1432–1440.
(168) H. Gao and K. Matyjaszewski, (2007), Synthesis of molecular brushes by grafting onto method: combination of ATRP and click reactions, *J. Am. Chem. Soc.*, **129** (20), 6633–6639; X. Jiang, M. C. Lok, and W. E. Hennink, (2007), Degradable-brushed pHEMA-pDMAEMA synthesized via ATRP and click chemistry for gene delivery, *Bioconjugate Chem.*, **18** (6), 2077–2084.
(169) R. V. Ostaci, D. Damiron, S. Capponi *et al.*, (2008), Polymer brushes grafted to passivated silicon substrates using click chemistry, *Langmuir*, **24** (6), 2732–2739.
(170) Bruno G. De Geest, Wim Van Camp, Filip E. Du Prez *et al.*, (2008), Biodegradable microcapsules designed via click chemistry, *Chem. Commun.*, 190–192; D. D. Evanoff Jr, S. E. Hayes, Y. Ying *et al.*, (2007), Functionalization of crystalline colloidal arrays through click chemistry, *Adv. Mater.*, **19** (21), 3507–3512.
(171) M. Malkoch, R. Vestberg, N. Gupta *et al.*, (2006), Synthesis of well-defined hydrogel networks using click chemistry, *Chem. Commun.*, **26**, 2774–2776.
(172) D. D. Diaz, S. Punna, P. Holzer *et al.*, (2004), Click chemistry in materials synthesis. 1. Adhesive polymers from copper-catalyzed azide–alkyne cycloaddition, *J. Polym. Sci., Part A: Polym. Chem.*, **42** (17), 4392–4403.
(173) Yi Liu, David D. Diaz, Adrian A. Accurso *et al.*, (2007), Click chemistry in materials synthesis. III. Metal-adhesive polymers from Cu(I)-catalyzed azide–alkyne cycloaddition, *J. Polym. Sci., Part A: Polym. Chem.*, **45** (22), 5182–5189; N. Le Baut, D. D. Diaz, Sreenivas Punna *et al.*, (2007), Study of high glass transition temperature thermosets made from the copper(I)-catalyzed azide–alkyne cycloaddition reaction, *Polymer*, **48** (1), 239–244.
(174) D. D. Diaz, K. Rajagopal, E. Strable *et al.*, (2006), Click chemistry in a supramolecular environment: stabilization of organogels by copper(I)-catalyzed azide–alkyne [3 + 2] cycloaddition, *J. Am. Chem. Soc.*, **128** (18), 6056–6057.

(175) J. A. Johnson, D. R. Lewis, D. D. Diaz *et al.*, (2006), Synthesis of degradable model networks via ATRP and click chemistry, *J. Am. Chem. Soc.*, **128** (20), 6564–6565.

(176) P. Screenivas, D. D. Diaz, L. Chunmei *et al.*, (2004), Click chemistry in polymer synthesis, *PMSE Prepr.*, **45** (1), 778–779; A. R. Katritzky, N. K. Meher, S. Hanci *et al.*, (2008), Preparation and characterization of 1,2,3-triazole-cured polymers from endcapped azides and alkynes, *J. Polym. Sci., Part A: Polym. Chem.*, **46** (1), 238–256; J. Rao, Y. Zhang, J. Zhang *et al.*, (2008), Facile preparation of well-defined AB2 Y-shaped miktoarm star polypeptide copolymer via the combination of ring-opening polymerization and click chemistry, *Biomacromolecules*; R. M. Meudtner and S. Hecht, (2008) Responsive backbones based on alternating triazole-pyridine/benzene copolymers: from helically folding polymers to metallosupramolecularly crosslinked gels, *Macromol. Rapid. Commun.*, **29** (4), 347–351; H. M. Konig, T. Gorelik, U. Kolb *et al.*, (2007), Supramolecular PEG-co-Oligo(*p*-benzamide)s prepared on a peptide synthesizer, *J. Am. Chem. Soc.*, **129** (3), 704–708.

(177) D. A. Ossipov and J. Hilborn, (2006), Poly(vinyl alcohol)-based hydrogels formed by click chemistry, *Macromolecules*, **39** (5), 1709–1718.

(178) Y. Xia, R. Verduzco, R. H. Grubbs *et al.*, (2008), Well-defined liquid crystal gels from telechelic polymers, *J. Am. Chem. Soc.*, in press.

(179) D. D. Diaz, J. J. M. Tellado, D. G. Velazquez *et al.*, (2008), Polymer thermoreversible gels from organogelators enabled by click chemistry, *Tetrahedron Lett.*, **49** (8), 1340–1343.

(180) T. J. Wigglesworth, F. Teixeira, F. Axthelm *et al.*, (2008), Dendronised block copolymers as potential vectors for gene transfection, *Org. Biomol. Chem.*, **6** (11), 1905–1911.

(181) J. A. Johnson, M. G. Finn, J. T. Koberstein *et al.*, (2007), Synthesis of photocleavable linear macromonomers by ATRP and star macromonomers by a tandem ATRP-click reaction: precursors to photodegradable model networks, *Macromolecules*, **40** (10), 3589–3598.

(182) V. Crescenzi, L. Cornelio, C. DiMeo *et al.*, (2007), Novel hydrogels via click chemistry: synthesis and potential biomedical applications, *Biomacromolecules*, **8** (6), 1844–1850.

(183) G. Chen, L. Tao, G. Mantovani *et al.*, (2007), A modular click approach to glycosylated polymeric beads: design, synthesis and preliminary lectin recognition studies, *Macromolecules*, **40** (21), 7513–7520.

(184) D. D. Diaz, K. Rajagopal, E. Strable *et al.*, (2006), Click chemistry in a supramolecular environment: stabilization of organogels by copper(I)-catalyzed azide–alkyne [3 + 2] cycloaddition, *J. Am. Chem. Soc.*, **128** (18), 6056–6057.

(185) P. R. Loaiza, S. Lober, H. Hubner *et al.*, (2006), Click chemistry on solid phase: parallel synthesis of *N*-benzyltriazole carboxamides as super-potent G-protein coupled receptor ligands, *J. Comb. Chem.*, **8** (2), 252–261; K. Ruck-Braun, T. H. E. Freysoldt, and F. Wierschem, (2005), 1,3-Dipolar cycloaddition on solid supports: nitrone approach towards isoxazolidines and isoxazolines and subsequent transformations, *Chem. Soc. Rev.*, **34** (6), 507–516; K. Harju, M. Vahermo, I. Mutikainen *et al.*, (2003), Solid-phase synthesis of 1,2,3-triazoles via 1,3-dipolar cycloaddition, *J. Comb. Chem.*, **5** (6), 826–833; J. Nielsen, (2002), Combinatorial synthesis of natural products, *Curr. Opin. Chem. Biol.*, **6** (3), 297–305.

(186) R. K. O'Reilly, Maisie J. Joralemon, Craig J. Hawker *et al.*, (2006), Fluorogenic 1,3-dipolar cycloaddition within the hydrophobic core of a shell cross-linked nanoparticle, *Chem. Eur. J.*, **12** (26), 6776–6786.

(187) Rachel K. O'Reilly, Maisie J. Joralemon, Craig J. Hawker *et al*, (2006), Facile syntheses of surface-functionalized micelles and shell cross-linked nanoparticles, *J. Polym. Sci., Part A: Polym. Chem.*, **44**, 5203–5217; R. K. O'Reilly, M. J. Joralemon, K. L. Wooley *et al.*, (2005), Functionalization of micelles and shell cross-linked nanoparticles using click chemistry, Chem. Mater., **17** (24), 5976–5988.

(188) R. K. O'Reilly, M. J. Joralemon, A. K. Nugent *et al.*, (2004), A novel approach to regioselectively functionalized amphiphilic block copolymers and nanoparticles, *PMSE Prepr.*, **45** (2), 292.

(189) Chunmei Li and M. G. Finn, (2006), Click chemistry in materials synthesis. II. Acid-swellable crosslinked polymers made by copper-catalyzed azide–alkyne cycloaddition, *J. Polym. Sci., Part A: Polym. Chem.*, **44**, 5513–5518.

(190) L. V. Lee, M. L. Mitchell, S. J. Huang *et al.*, (2003), A potent and highly selective inhibitor of human alpha-1,3-fucosyltransferase via click chemistry, *J. Am. Chem. Soc.*, **125** (32), 9588–9589; J. Wang, G. Sui, V. P. Mocharla *et al.*, (2006), Integrated microfluidics for parallel screening of an *in situ* click chemistry library, *Angew. Chem., Int. Edn*, **45** (32), 5276–5281.

(191) H. Nandivada, H.-Y. Chen, L. Bondarenko *et al.*, (2006), Reactive polymer coatings that click, *Angew. Chem., Int. Edn*, **45** (20), 3360–3363.

(192) D. A. Long, K. Unal, R. C. Pratt *et al.*, (2007), Localized click chemistry through dip-pen nanolithography, *Adv. Mater.*, **19** (24), 4471–4473.

(193) J. P. Collman, N. K. Devaraj, and C. E. D. Chidsey, (2004), Clicking functionality onto electrode surfaces, *Langmuir*, **20** (4), 1051–1053.

(194) J. P. Collman, N. K. Devaraj, T. P. A. Eberspacher *et al.*, (2006), Mixed azide-terminated monolayers: a platform for modifying electrode surfaces, *Langmuir*, **22** (6), 2457–2464.

(195) N. K. Devaraj, R. A. Decreau, W. Ebina *et al.*, (2006), Rate of interfacial electron transfer through the 1,2,3-triazole linkage, *J. Phys. Chem. B*, **110** (32), 15955–15962.

(196) T. Lummerstorfer and H. Hoffmann, (2004), Click chemistry on surfaces: 1,3-dipolar cycloaddition reactions of azide-terminated monolayers on silica, *J. Phys. Chem. B*, **108** (13), 3963–3966.

(197) R. Ranjan and W. J. Brittain, (2007), Combination of living radical polymerization and click chemistry for surface modification, *Macromolecules*, **40** (17), 6217–6223.

(198) J. K. Lee, Y. S. Chi, and I. S. Choi, (2004), Reactivity of acetylenyl-terminated self-assembled monolayers on gold: triazole formation, *Langmuir*, **20** (10), 3844–3847.

(199) R. D. Rohde, H. D. Agnew, W. S. Yeo *et al.*, (2006), A non-oxidative approach toward chemically and electrochemically functionalizing Si(111), *J. Am. Chem. Soc.*, **128** (29), 9518–9525.

(200) J.-C. Meng, C. Averbuj, W. G. Lewis *et al.*, (2004), Cleavable linkers for porous silicon-based mass spectrometry, *Angew. Chem., Int. Edn*, **43** (10), 1255–1260.

(201) Y. Zhang, S. Luo, Y. Tang *et al.*, (2006), Carbohydrate–protein interactions by clicked carbohydrate self-assembled monolayers, *Anal. Chem.*, **78** (6), 2001–2008.

(202) X. L. Sun, C. L. Stabler, C. S. Cazalis *et al.*, (2006), Carbohydrate and protein immobilization onto solid surfaces by sequential Diels–Alder and azide–alkyne cycloadditions, *Bioconjugate Chem.*, **17** (1), 52–57.

(203) S. Prakash, T. M. Long, J. C. Selby *et al.*, (2007), Click modification of silica surfaces and glass microfluidic channels, *Anal. Chem.*, **79** (4), 1661–1667.

(204) G. K. Such, J. F. Quinn, A. Quinn *et al.*, (2006), Assembly of ultrathin polymer multilayer films by click chemistry, *J. Am. Chem. Soc.*, **128** (29), 9318–9319.

(205) B. S. Lee, J. K. Lee, W. J. Kim *et al.*, (2007), Surface-initiated, atom transfer radical polymerization of oligo(ethylene glycol) methyl ether methacrylate and subsequent click chemistry for bioconjugation, *Biomacromolecules*, **8** (2), 744–749.

(206) G. K. Such, E. Tjipto, A. Postma *et al.*, (2007), Ultrathin, responsive polymer click capsules, *Nano Lett.*, **7** (6), 1706–1710.

(207) D. E. Bergbreiter, P. N. Hamilton, and N. M. Koshti, (2007), Self-separating homogeneous copper (I) Catalysts, *J. Am. Chem. Soc.*, **129** (35), 10666–10667.

(208) T. Liebert, C. Hänsch, and T. Heinze, (2006), Click chemistry with polysaccharides, *Macromol. Rapid Commun.*, **27** (3), 208–213.

(209) Y. Li, W. Zhang, J. Chang *et al.*, (2008), Click on conducting polymer coated electrodes: a versatile platform for the modification of electrode surfaces, *Macromol. Chem. Phys.*, **209** (3), 322–329.

(210) S. Fleischmann, H. Komber, and B. Voit, (2008), Diblock copolymers as scaffolds for efficient functionalization via click chemistry, *Macromolecules*, **41** (14), 5255–5264.

(211) D. A. Fleming, C. J. Thode, and M. E. Williams, (2006), Triazole cycloaddition as a general route for functionalization of Au nanoparticles, *Chem. Mater.*, **18** (9), 2327–2334.

(212) J. L. Brennan, N. S. Hatzakis, T. R. Tshikhudo *et al.*, (2006), Bionanoconjugation via click chemistry: the creation of functional hybrids of lipases and gold nanoparticles, *Bioconjugate Chem.*, **17** (6), 1373–1375.

(213) M. A. White, J. A. Johnson, J. T. Koberstein *et al.*, (2006), Toward the syntheses of universal ligands for metal oxide surfaces: controlling surface functionality through click chemistry, *J. Am. Chem. Soc.*, **128** (35), 11356–11357.
(214) R. Voggu, P. Suguna, S. Chandrasekaran *et al.*, (2007), Assembling covalently linked nanocrystals and nanotubes through click chemistry, *Chem. Phys. Lett.*, **443** (1–3), 118–121.
(215) Q. Zeng, T. Li, B. Cash *et al.*, (2007), Chemoselective derivatization of a bionanoparticle by click reaction and ATRP reaction, *Chem. Commun.*, **14**, 1453–1455.
(216) A. Gole and C. J. Murphy, (2008), Azide-derivatized gold nanorods: functional materials for click chemistry, *Langmuir*, **24** (1), 266–272.
(217) W. H. Binder, (2005), Supramolecular assembly of nanoparticles at liquid–liquid interfaces, *Angew. Chem., Int. Edn*, **44** (33), 5172–5175.
(218) A. Haryono and W. H. Binder, (2006), Controlled arrangement of nanoparticle arrays in block-copolymer domains, *Small*, **2** (5), 600–611.
(219) H. Xu, S. Srivastava, and V. Rotello, (2007), Nanocomposites based on hydrogen bonds, *Adv. Polym. Sci.*, 179–198; S. Kinge, M. Crego-Calama, and D. N. Reinhoudt, (2008) Self-assembling nanoparticles at surfaces and interfaces, *ChemPhysChem*, **9** (1), 20–42.
(220) J.-C. Meng, G. Siuzdak, and M. G. Finn, (2004), Affinity mass spectrometry from a tailored porous silicon surface, *Chem. Commun.*, **18**, 2108–2109.
(221) C.-T. Chen, Y. S. Munot, S. B. Salunke *et al.*, (2008), A triantennary dendritic galactoside-capped nanohybrid with a ZnS/CdSe nanoparticle core as a hydrophilic, fluorescent, multivalent probe for metastatic lung cancer cells, *Adv. Funct. Mater.*, **18** (4), 527–540.
(222) Y. Gao H. Xiong, H. M. Li, (2007), Non-isothermal crystallization kinetics of syndiotactic polystyrene – polystyrene functionalized SWNTs nanocomposites, *ePolymers*, 416–426.
(223) H.-J. Musiol, S. Dong, M. Kaiser *et al.*, (2005), Toward Semisynthetic lipoproteins by convergent strategies based on click and ligation chemistry, *ChemBioChem*, **6** (4), 625–628.
(224) A. J. Link, M. K. S. Vink, and D. A. Tirrell, (2004), Presentation and detection of azide functionality in bacterial cell surface proteins, *J. Am. Chem. Soc.*, **126** (34), 10598–10602; A. J. Link and D. A. Tirrell, (2003), Cell surface labeling of *Escherichia coli* via copper(I)-catalyzed [3 + 2] cycloaddition, *J. Am. Chem. Soc.*, **125** (37), 11164–11165.
(225) P. Bertrand and J. P. Gesson, (2007), Click chemistry with *O*-dimethylpropargylcarbamate for preparation of pH-sensitive functional groups. A Case Study, *J. Org. Chem.*, **72** (9), 3596–3599.
(226) F. Fazio, M. C. Bryan, O. Blixt *et al.*, (2002), Synthesis of sugar arrays in microtiter plate, *J. Am. Chem. Soc.*, **124** (48), 14397–14402.
(227) G. Chen, L. Tao, G. Mantovani *et al.*, (2007), Synthesis of azide/alkyne-terminal polymers and application for surface functionalisation through a [2 + 3] Huisgen cycloaddition process, click chemistry, *Soft Matter*, **3** (6), 732–739.
(228) S. Fleischmann, K. Hinrichs, U. Oertel *et al.*, (2008), Modification of polymer surfaces by click chemistry, *Macromol. Rapid Commun.*, **29** (12–13), 1177–1185.
(229) N. H. Shah and K. Kirshenbaum, (2008), Direct generation of polymer films on copper surfaces through azide–alkyne cycloaddition reactions between peptidomimetic oligomers, *Macromol. Rapid Commun.*, **29** (12–13), 1134–1139.
(230) A. E. Daugaard, S. Hvilsted, T. S. Hansen *et al.*, (2008), Conductive polymer functionalization by click chemistry, *Macromolecules*.

8

Dendrimer Synthesis and Functionalization by Click Chemistry for Biomedical Applications

Daniel Q. McNerny, Douglas G. Mullen, Istvan J. Majoros, Mark M. Banaszak Holl and James R. Baker Jr

8.1 Introduction

Dendrimer-based platforms have achieved increasing attention for use in biomedical applications including, but not limited to, targeted drug delivery, imaging and transfection. Interest in dendrimers can be attributed to their unique branching structure that results in exceptionally high degrees of monodispersity as compared with other polymeric materials. Furthermore, well-defined terminal groups allow the conjugation of multiple functional molecules. Dendrimer systems with multiple copies of targeting ligands enhance the interaction of targeting molecules with cell membrane receptors due to multivalent binding.[1] While the concept of preferential targeting is not limited to dendrimer platforms, the ability to create dendrimers that mimic the size and shape of human proteins makes the technology an ideal choice for many therapeutic and diagnostic applications. The dendrimer's nanometer size enables efficient diffusion across the vascular endothelium and directs internalization into cancer cells, and facilitates rapid renal clearance of these molecules from the blood stream.

The most widely used dendrimers in biomedical applications are poly(amidoamine) (PAMAM) dendrimers. The polyamide backbone helps the macromolecule maintain water solubility and minimizes immunogenicity. PAMAM dendrimers exhibit little toxicity if the surface amines have been neutralized or modified (Figure 8.1).[2–7]

Click Chemistry for Biotechnology and Materials Science Edited by Joerg Lahann

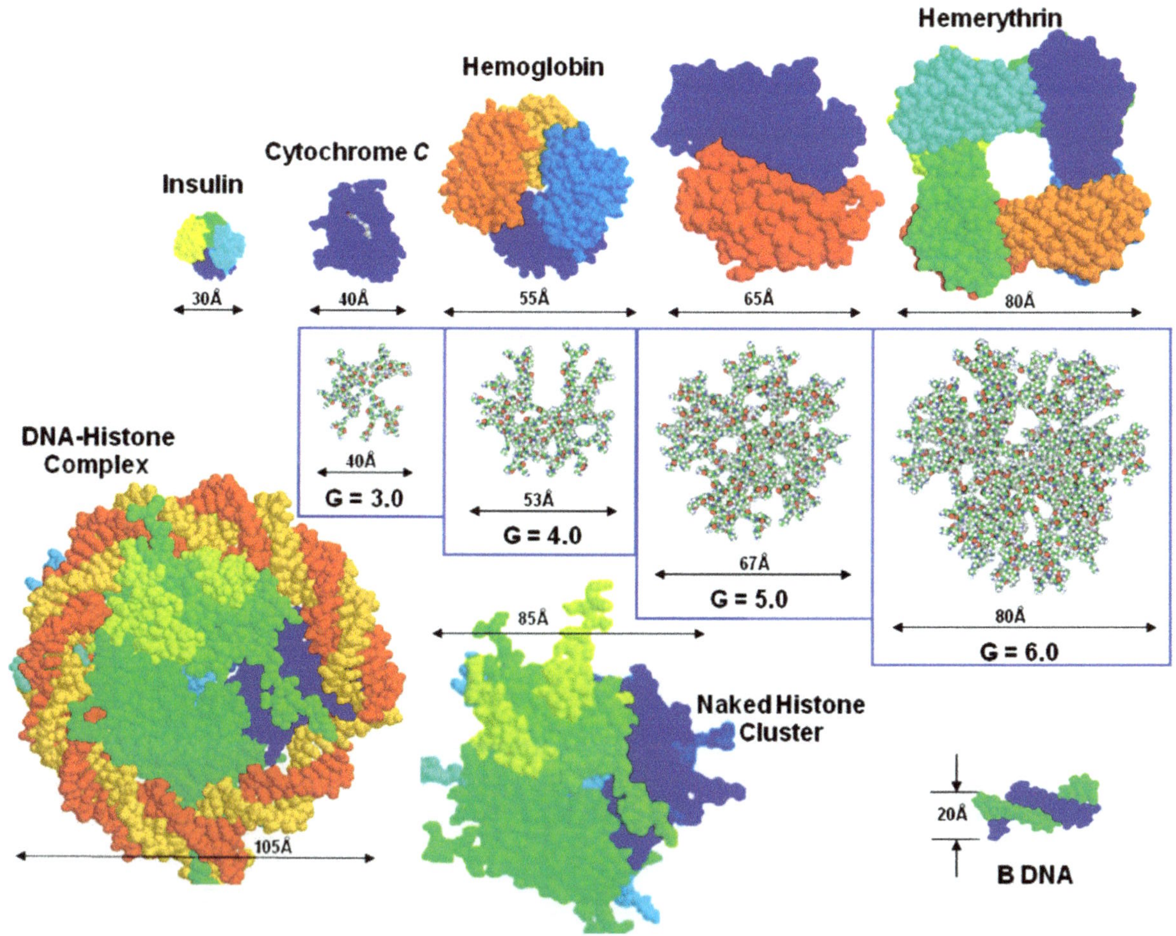

Figure 8.1 Comparison of absolute size of PAMAM dendrimers and proteins.

The PAMAM platform has been successfully used as a scaffold for the attachment of targeting molecules including antibodies,[8–12] peptides,[13] T-antigens[14–16] and folic acid.[17–26] The targeting ligands anchor the dendrimers to locations where receptors are expressed on cell surfaces. Dendrimers have also been used to carry and solubilize therapeutic molecules, such as methotrexate.[19,22,23,27,28] Targeted dendrimer–drug conjugates deliver a higher dose specifically to tumor cells while avoiding normal cells, thus avoiding the systemic toxicity of current therapeutics. PAMAM dendrimers with folate and methotrexate have been shown to specifically bind to carcinoma KB cells expressing high levels of the folate receptor.[23] Targeted chemotherapy with dendrimers showed 10 times the efficacy and decreased toxicity compared with standard chemotherapy with free methotrexate during *in vivo* studies. Phase I clinical trials for this cancer therapy are planned (Figure 8.2).

One can imagine dendrimer scaffolds being modified to create a vast library of targeted therapeutics, each tailored to address the properties and overexpressed receptors of an individual target cell. Dendrimer treatment could be personalized after a screening to determine which therapeutic would have the greatest effect.

Several dendrimer-based products are under development for the treatment of a variety of diseases. Starpharma is testing a topical polylysine dendrimer-based microbicide, VivaGel™, for the prevention of HIV transmission and other sexually transmitted diseases during intercourse. SuperFect®, developed by Qiagen, is an activated dendrimer used for gene transfection in a broad range of cell lines. Dendrimers have also been used as diagnostic tools. Gadomer-17, a polylysine dendrimer functionalized with gadolinium chelates, from Schering AG is used as MRI contrast agent. The US Army Research Laboratory developed Alert Ticket™ as a dendrimer-based diagnostic for anthrax detection. Stratus® CS, has been commercialized by Dade Behring for the rapid diagnosis of heart attacks by acting as a biosensor for cardiac markers.

Despite the promise and successes of dendrimer platforms, the field's ability to provide materials for biomedical applications has been slowed by synthetic challenges in producing mono-dispersed bulk dendrimers. PAMAM dendrimer synthesis can lead to undesired side reactions causing defect structures, and these defect structures propagate as the dendrimer grows with each additional generation.[29,30] Therefore, although PAMAM dendrimers maintain a very low PDI, maintaining batch-to-batch reproducibility can be challenging. When these polymers are conjugated with multiple targeting and therapeutic molecules, the heterogeneity of the conjugates is magnified. Unfortunately, this makes the characterization of these materials difficult and the material itself must be reproducible and consistent if one hopes to eventual administer these agents to human subjects. Other dendrimers based on different subunit structures may avoid side reactions and ease reproducibility concerns, but are poor choices for many biomedical applications because of poor solubility at physiological conditions and/or cytotoxicity. In addition, creating multifunctional dendrimers for therapeutic applications currently involves multiple, step-wise conjugations that are time-consuming and hard to reproduce. New synthetic approaches will be needed to produce consistent materials. In addition, when producing patient-specific drugs ('personalized medicine') some method of parallel synthesis must be achieved to produce many combinations of drug and targeting ligands that allow for individualized therapeutics.

'Click chemistry' is an approach that could enhance the synthesis of dendrimeric structures. There have been several examples involving the synthesis and conjugation of

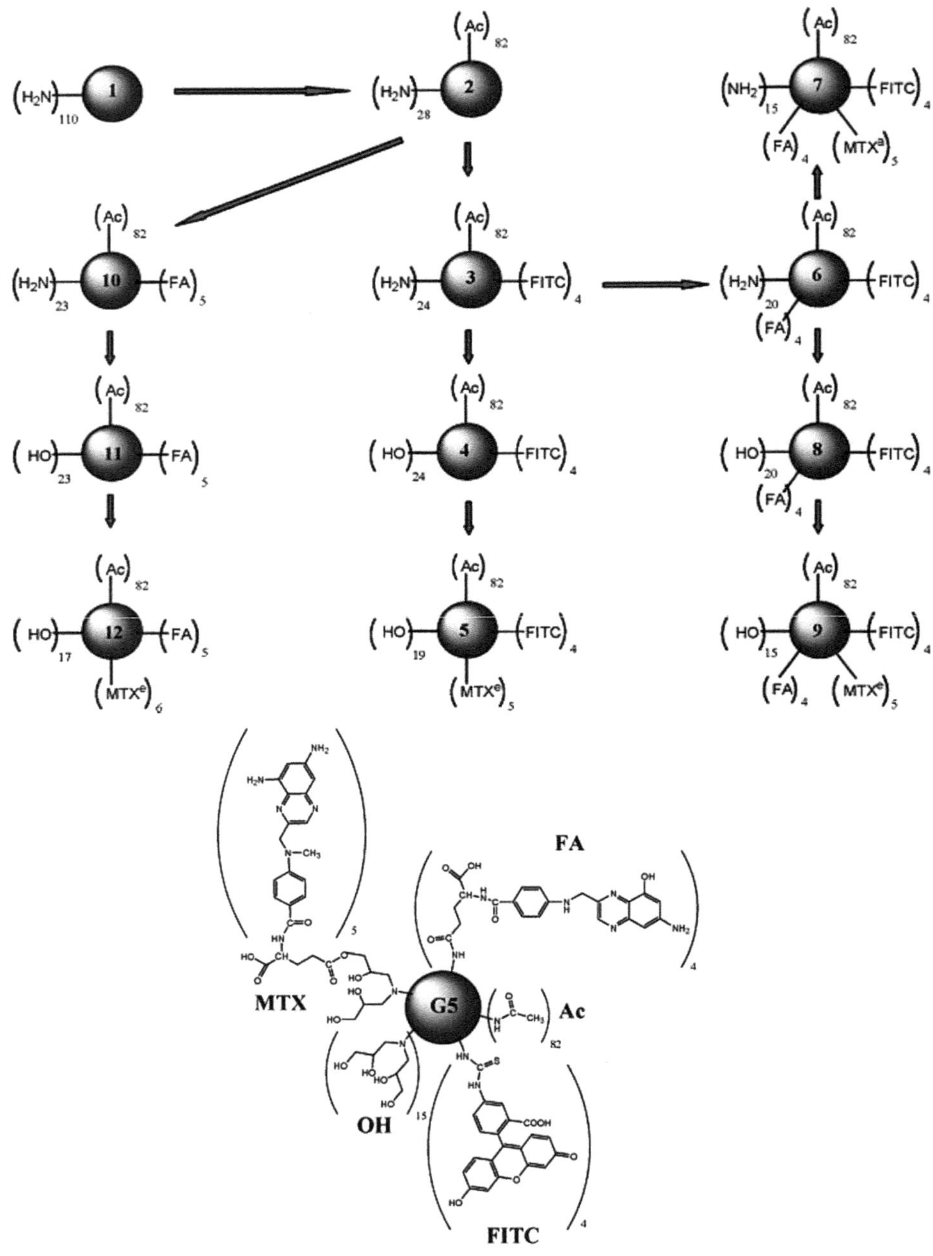

Figure 8.2 *Multifunctional device on a single dendrimer. Reprinted with permission from ref.*[22]*. Copyright 2005 American Chemical Society.*

dendrimers using click chemistry since Sharpless and co-workers popularized the copper(I)-catalyzed 1,3-dipolar cycloaddition.[31] Researchers hope to use click chemistry to obtain greater control over the synthesis of the dendrimer platforms by minimizing defect structures and reducing the need for purification. Others use click chemistry as a means to control the conjugation of desired functionalities.[32–39] The orthoganality of click chemistry provides a means to avoid incompatible conjugation reactions, reduce product heterogeneity and potentially add specific numbers of functional moieties.

8.2 Dendrimer Synthesis

Dendrimers have been synthesized by two methods: divergent and convergent.[40] These methods both create dendritic macromolecules through repeated growth and activation reactions, and each additional cycle adds an additional dendrimer 'generation'. Divergent synthesis proceeds radially from the core. Convergent synthesis proceeds from the surface inward, where molecular building blocks form dendrons which are then joined together at their unique focal points. Convergent synthesis typically yields dendrimers of higher purity and reproducibility compared with divergent methods; however, there is greater difficulty in producing larger generation dendrimers using convergent methods due to increasing steric constraints on the synthesis.[41]

Traditional synthetic approaches to dendrimer synthesis, whether convergent or divergent, are inefficient in their time and materials requirements (Figure 8.3). Some of the dendrimers that are commercially available may not be of appropriate quality for biomedical purposes, including PAMAM, phosphorous-based, polypropylenimine, polylysine, polyester and 2,2-bis(methylol)propionic acid (bis-MPA) dendrimers. A highly-trained synthetic chemist may require months to prepare dendrimers in significant quantities and quality adequate for bioconjugation. New synthetic methods which accelerate and simplify dendrimer production may be required in order to increase the amount and type of dendrimers necessary for many pharmaceutical applications. Attractive methods must produce dendrimer in good yields, minimize defect population distribution and display tolerance of functional groups and reaction conditions while cutting back on reaction and purification time. To accomplish these goals, various click chemistry reactions have been used to synthesize dendrimers by both divergent and convergent methods.

8.2.1 Divergent Synthesis

Click chemistry has been used on multiple occasions to divergently synthesize dendrimers. The copper(I)-catalyzed azide–alkyne cycloaddition (CuAAC) is used during the growth step, where the dendrimer is reacted with an alkyne or azide monomer unit.[43] These reactions have been performed at quantitative yields with minimal workup. Unfortunately, dendrimers divergently synthesized via the CuAAC have yet to be tested in biological systems. A possible issue with these molecules is the large number of heteroaromatics in these dendrimers, which may prohibit the material from being soluble in many physiological conditions.

An exciting development by Hawker and colleagues employs thiol-ene click chemistry to divergently synthesize dendrimers.[44] The thiol–alkene reaction is thought to be a robust,

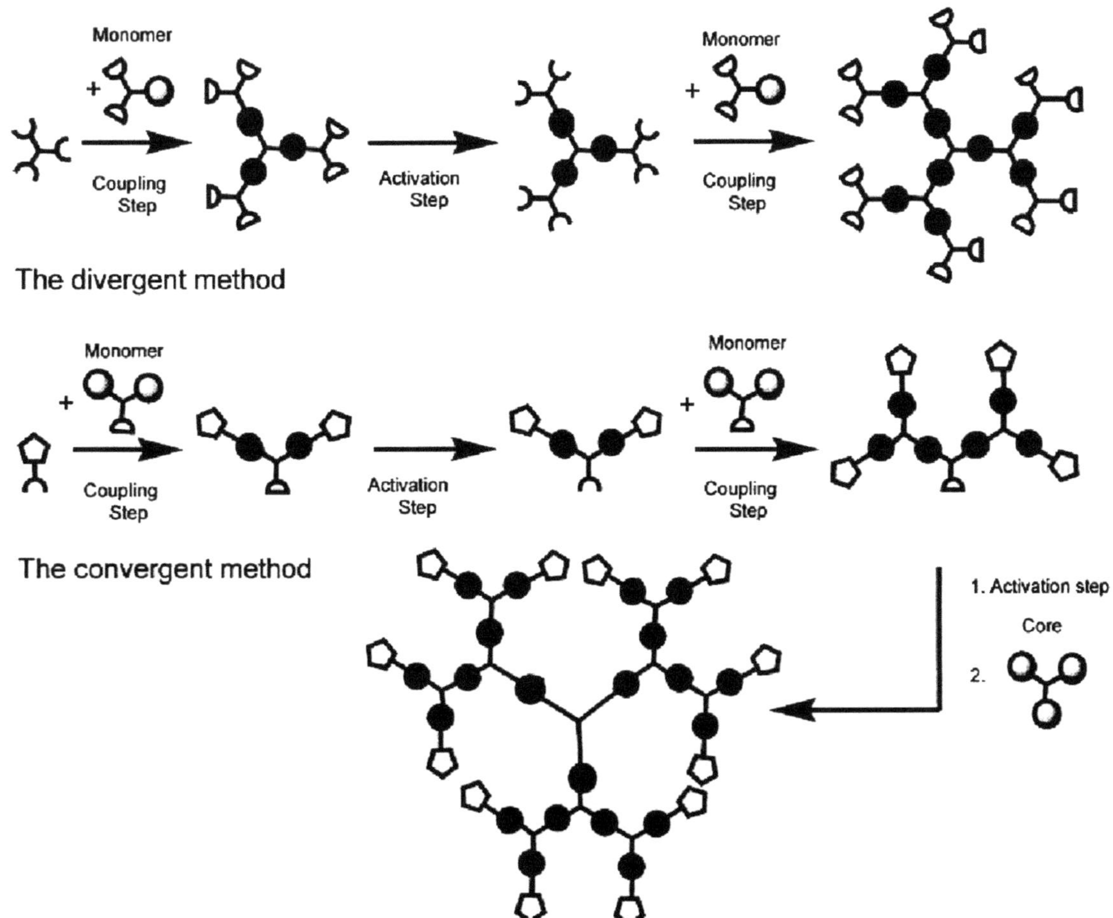

Figure 8.3 *Divergent and convergent dendrimer synthesis pathways. Reprinted with permission from ref.[42]. Copyright 2005 John Wiley and Sons, Inc.*

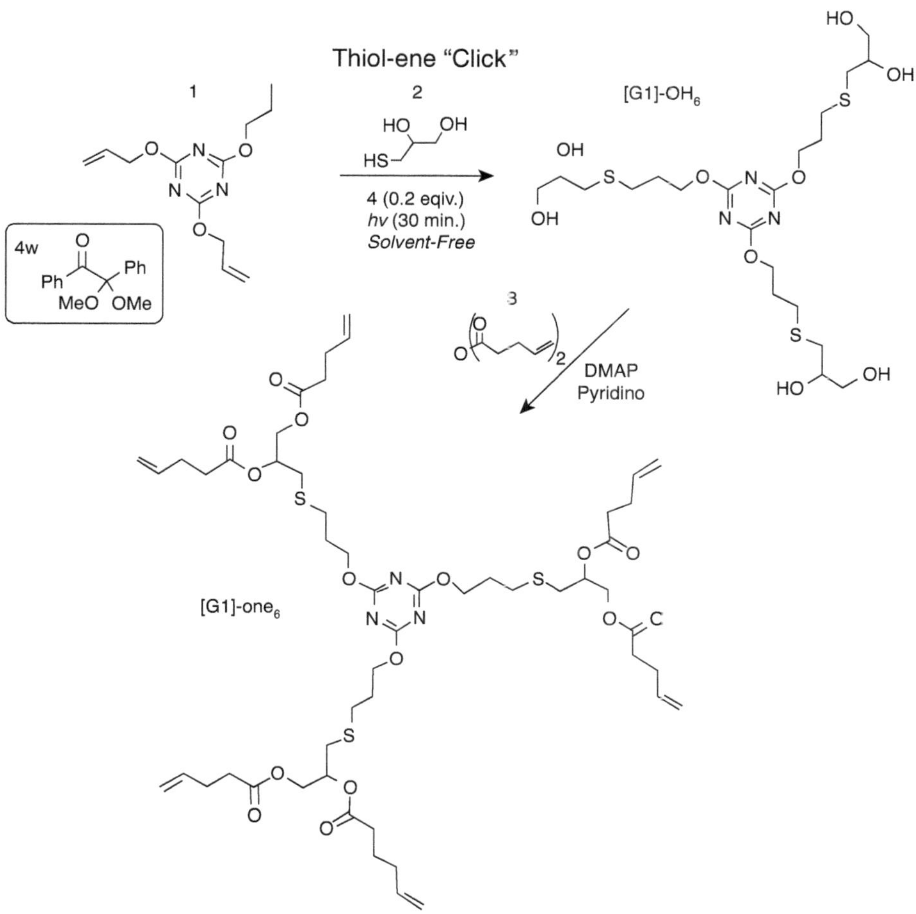

Figure 8.4 *Dendrimer synthesized using thiol-ene chemistry. Reprinted with permission from ref.[44]. Copyright 2008 American Chemical Society.*

clean reaction that avoids the use of a metal catalyst as with some other 'click' approaches. In addition, this chemistry can be performed without a solvent while maintaining the specificity properties of click chemistry. Generation 4 poly(thio-ether) dendrimers were constructed using this approach and functionalized with various surface groups using this chemistry. No biological studies have yet been published for this material, but the approach may open the door for the creation of a larger number of dendritic platforms for biomedical applications (Figure 8.4).

8.2.2 Convergent Synthesis

Convergent synthesis of dendrimers has been achieved for both symmetrical[32,34,45–50] and asymmetrical dendrimers.[33,45,49,50] A click reaction can be employed as the final step to combine dendrons with each other or to a core after traditional approaches are used to synthesize the dendrons.[33,45–52] CuAAC[48,50] and Diels–Alder[53] reactions have both been employed to make dendrimers. In 2007, Haridas reported the design of peptide dendrimers with varied cores using CuAAC.[34] The group proposes conjugating biologically relevant molecules to the amino groups of the dendrimer. Wysozogrodzka applied the convergent method to polyglycerol dendrons of varying generation to create core-shell structures.[32] This approach successfully solubilized different hydrophobic cores and the hydrophobic dye Nile Red. Wysozogrodzka showed the amount of Nile Red solubilized depended on core size and generation of the polyglycerol dendrons (Figure 8.5).

The application of click chemistry to convergent synthesis also has the potential to be used for a mix-and-match approach to creating multifunctional macromolecules based on dendrimers. Uniquely functionalized dendrons can be selected from a library and combined to create functionalized devices for specific applications. Proof-of-concept reactions have been performed by Lee and co-workers on methyl ester-terminated, half-generation PAMAM dendrons to create symmetrical and unsymmetrical dendrimers.[49–51] This methodology has the potential to be adapted to functionalized full-generation dendrons that are soluble in physiological conditions. The same group has successfully coupled dendrons with different backbones to create diblock co-dendron devices.[51] Wu and co-workers synthesized unsymmetrical bis-MPA dendrimers, which were then functionalized to create a bifunctional, targeted device.[33]

In general, studies have indicated that the yield of the click reaction decreases with increasing dendron generation. It is proposed that steric effects, the dendron branches backfolding over the focal point, cause the decreased yield (Figure 8.6).[45,48–50]

8.3 Dendrimer Functionalization

The orthoganality and efficiency of click chemistry makes it an attractive means to functionalize dendritic macromolecules. Click chemistry has repeatedly been shown to be an efficient way to functionalize dendrimers with either carbohydrates,[35] peptides[36] or solubilizing linkers[37] for biomedical applications. This has provided increased potency and multivalent targeting because of the dendritic platform.

Wu and co-workers successfully produced a multifunctional bis-MPA dendron with targeted and imaging units using the 1,3-dipolar cycloaddition.[33] The dendron-bound mannose was reported to show a 240-fold greater potency in hemagglutination *vs* monomeric mannose. Dijkgraaf showed that radiolabeled tetrameric RGD dendrimers had enhanced uptake in both *in vitro* binding assays and $\alpha_V\beta_3$ integrin expressing tumors *in vivo* when compared with dimeric and monomeric RGD controls. This increased avidity was accomplished without altering the specificity of binding.[36] In 2006, Fernandez-Megia surface conjugated unprotected carbohydrates to produce PEG-lyated glycodendrimers for potential applications in exploring carbohydrate–receptor interactions or for targeted drug delivery (Figures 8.7 and 8.8).

Urbani recently combined atom transfer radical polymerization and click chemistry to design third-generation polymeric dendrimers with a peripheral generational layer that could be selectively cleaved off from the second-generation via basic hydrolysis of the dendrimers ester groups.[39] This platform can be used for the applications requiring a slow and controlled release of its peripheral layer (Figure 8.9).

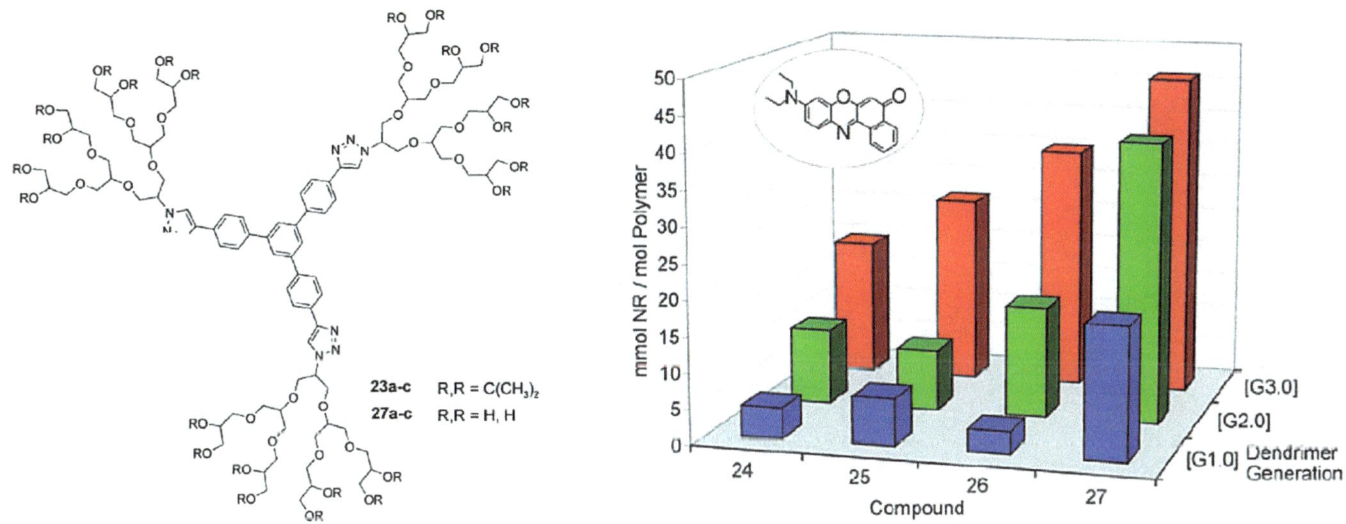

Figure 8.5 *Convergent synthesis of polyglycerol dendrimers and effect of generation on Nile Red solubilization. Reprinted with permission from ref.[32]. Copyright 2008 Wiley-VCH.*

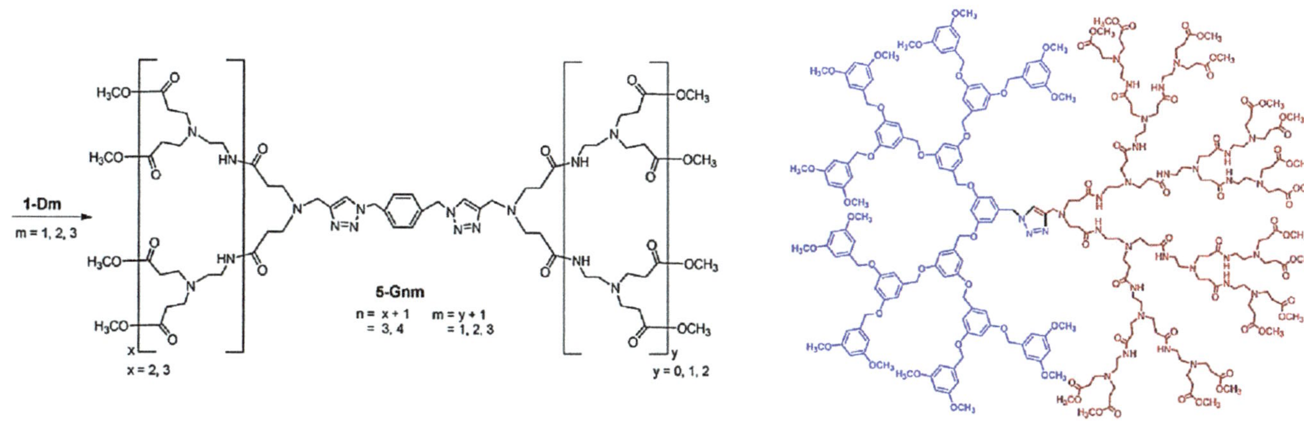

Figure 8.6 Convergent synthesis of asymmetrical dendrimers. Reprinted with permission from ref.[51]. Copyright 2006 American Chemical Society.

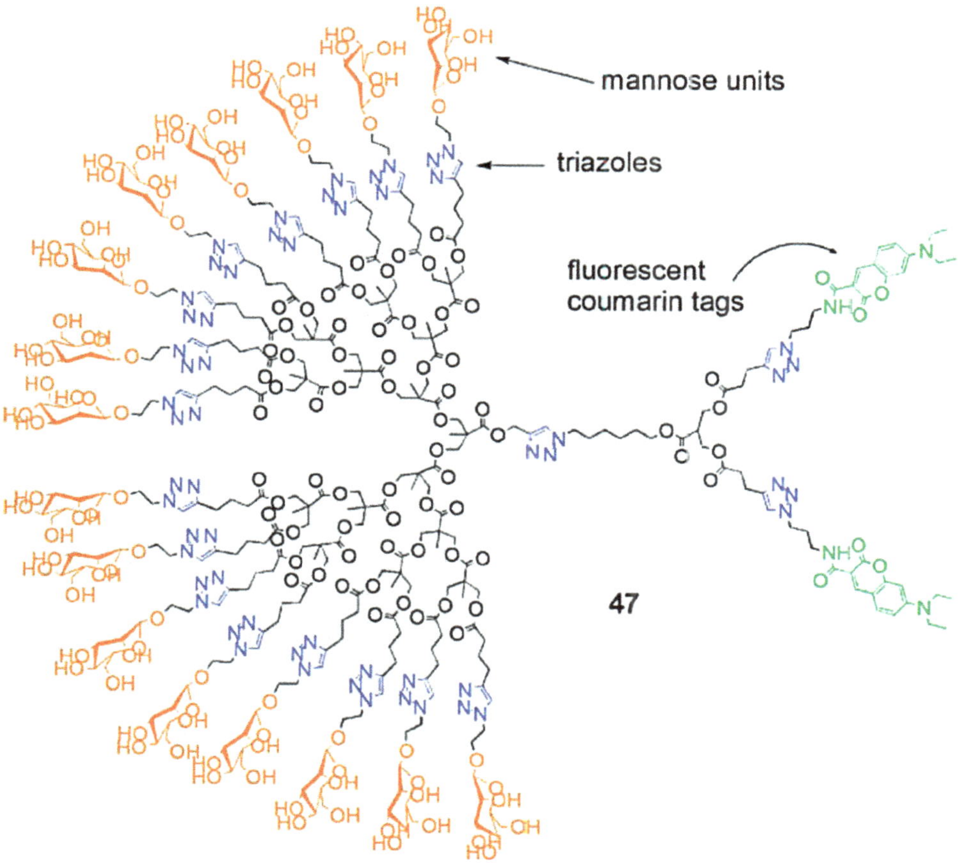

Figure 8.7 Bifunctional bis-MPA dendrimer with 16 mannose units and 2 coumarin units. Reprinted with permission from ref.[33]. Copyright 2005 The Royal Society of Chemistry.

CuAAC reactions have been used as an efficient means to functionalize solid supports or surfaces with dendritic materials. For example, Ortega-Munoz and co-workers conjugated glyco-dendrimers to a silica core to create a bio-selective affinity chromatography matrix for potential applications for the inmobilization of other biomolecules.[38] Pohl showed the functionalization of cellulose surfaces with propargyl–PAMAM dendrons via CuAAC.[54] A model enzyme, glucose oxidase, was covalently attached to the dendron and protein attachment was shown. These biofunctionalized surfaces have the potential to be used for various sensor, catalytic and delivery applications.

Click reactions have also been exploited for their orthoganality and neighboring group tolerance. When stoichiochemistry is critical, click chemistry can be used as a valuable tool to specifically couple single functionalities. The most obvious example is having a unique 'clickable' focal point on a dendron (Figure 8.10).

Work has been done by the Weck group to apply unique reactive sites to dendrimers as well. Polyamide dendrimers have been designed with a single azide or alkyne moiety on

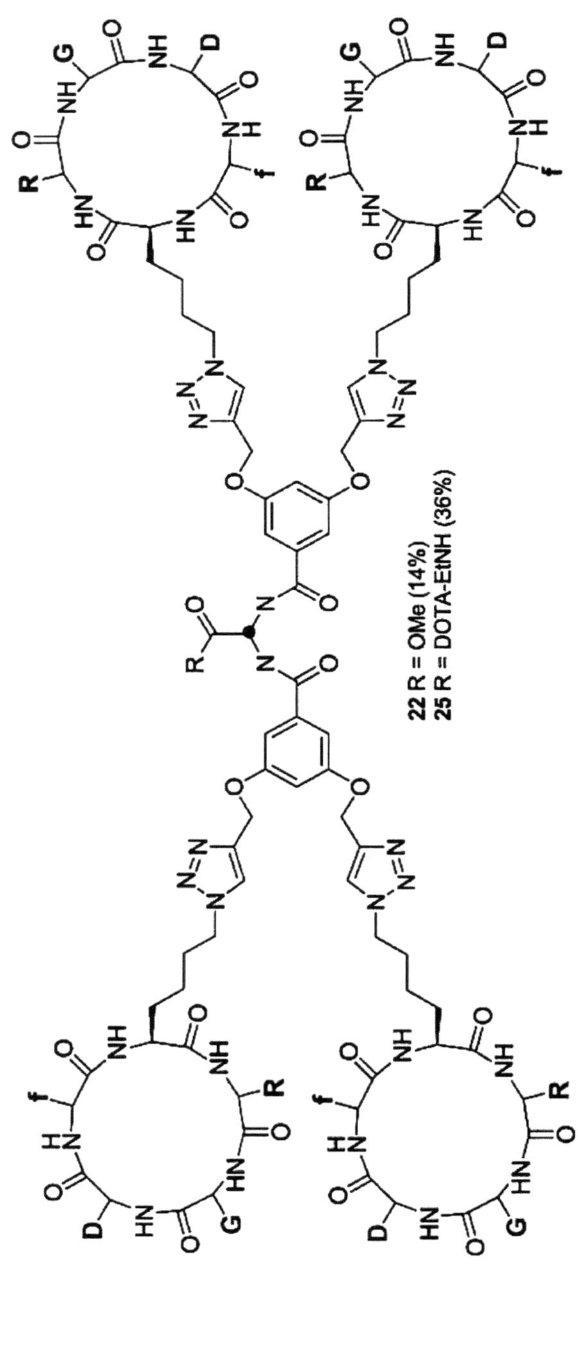

Figure 8.8 *Tetrameric RGD-dendrimer synthesized via CuAAC. Reprinted with permission from ref.*[36]*. Copyright 2007 The Royal Society of Chemistry.*

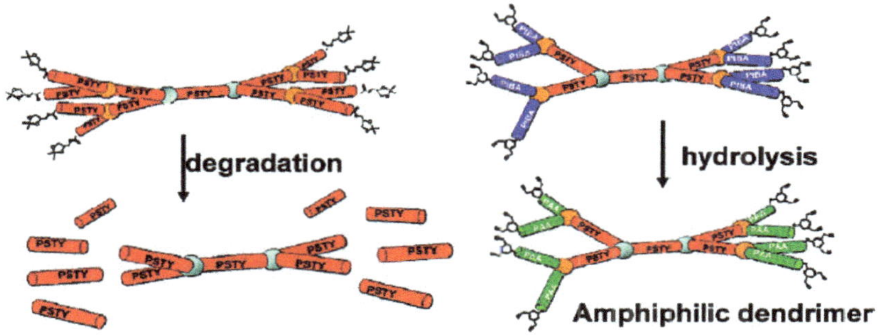

Figure 8.9 *Generation 3 polymeric dendrimer with cleavable periphery. Reprinted with permission from ref.[39]. Copyright 2008 American Chemical Society.*

the surface.[55,56] One can imagine using similar scaffolds when product distributions must be avoided or to create dumbbell structures without fear of forming insoluble networks (Figure 8.11).

8.4 Conclusions and Future Directions

The unique branched structure of dendrimers offers many benefits, from multivalent binding to increased load capacity. Yet standard techniques in dendrimer chemistry are laborious and the costs can be prohibitive. The inability of commercial suppliers to reproducibility

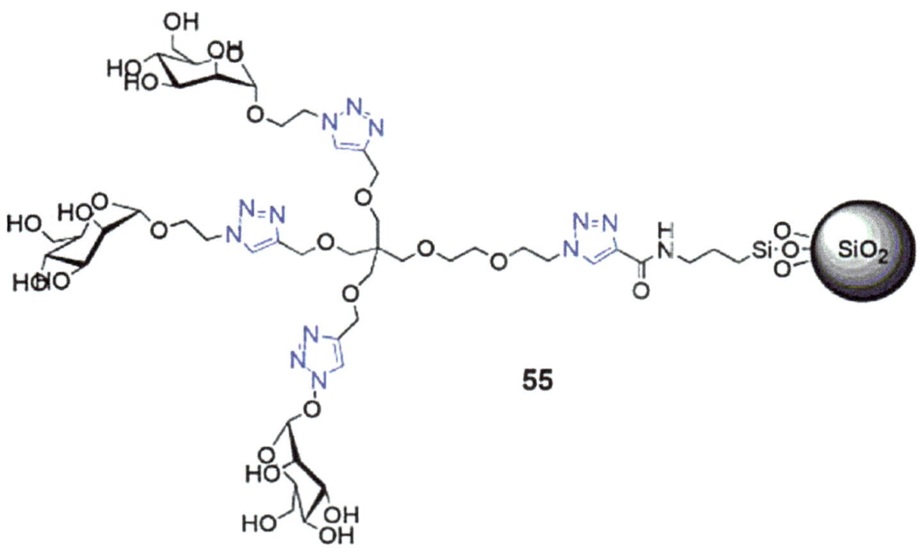

55

Figure 8.10 *Glyco-dendrimer conjugated to a silica core for affinity chromatography. Reprinted with permission from ref.[38]. Copyright 2006 Wiley-VCH.*

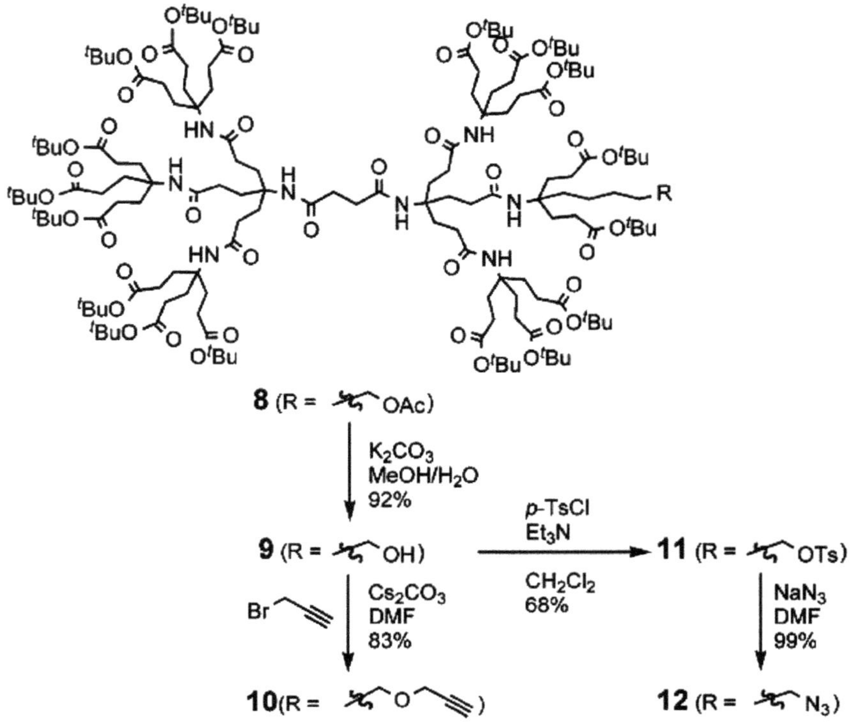

Figure 8.11 Polyamide dendrimer with single orthogonal surface reactive site. Reprinted with permission from ref.[56]. Copyright 2007 American Chemical Society.

minimize defect structures in highly biocompatible PAMAM has slowed their progress in reaching therapeutic trials. Many alternative dendrimers either suffer from the same problems or lack biocompatible and nontoxic properties. For dendrimers to become reliable and affordable scaffolds for medical use as targeted therapeutics, detection agents or diagnostics, their production must become more efficient and reproducible on a large scale. Expanding the use of click chemistry in dendrimer science will most likely continue to facilitate these goals.

Specifically, click chemistry has been applied to address dendrimer synthetic concerns. New shorter-term developments will probably come via functionalization of dendrimers, where reaction orthoganality and simpler purification will speed up and increase the production of new devices. Longer-term solutions needed include using click reactions to create methods of dendrimer synthesis. Click-type reactions have been applied to dendrimer synthesis, but the majority have used repeated triazole-based click reactions which typically produce dendrimers that are not soluble in physiological conditions. The use of the thiol-ene reaction shows that click methodologies can be developed that maintain biological applicability. Dendrimer synthesis will dramatically improve as additional clean, efficient reactions are developed to create non-immunogenetic, biofunctional scaffolds.

References

(1) Hong, S., Leroueil, P. R., Majoros, I. J., Orr, B. G., Baker, J. R. Jr and Banaszak Holl, M. M., (2007), The binding avidity of a nanoparticle-based multivalent targeted drug delivery platform, *Chem. Biol.*, **14**, 107–115.

(2) Majoros, I. J., Keszler, B., Woehler, S., Bull, T. and Baker, J. R., (2003), Acetylation of poly(amidoamine) dendrimers, *Macromolecules*, **36**, 5526–5529.

(3) Hong, S. P., Bielinska, A. U., Mecke, A., Keszler, B., Beals, J. L., Shi, X. Y., Balogh, L., Orr, B. G., Baker, J. R. and Holl, M. M. B., (2004), Interaction of poly(amidoamine) dendrimers with supported lipid bilayers and cells: Hole formation and the relation to transport, *Bioconjugate Chem.*, **15**, 774–782.

(4) Lee, C. C., MacKay, J. A., Frechet, J. M. J. and Szoka, F. C., (2005), Designing dendrimers for biological applications, *Nat. Biotechnol.*, **23**, 1517–1526.

(5) Svenson, S. and Tomalia, D. A., (2005), Commentary – dendrimers in biomedical applications – reflections on the field, *Adv. Drug Deliv. Rev.*, **57**, 2106–2129.

(6) Hong, S. P., Leroueil, P. R., Janus, E. K., Peters, J. L., Kober, M. M., Islam, M. T., Orr, B. G., Baker, J. R. and Holl, M. M. B., (2006), Interaction of polycationic polymers with supported lipid bilayers and cells: nanoscale hole formation and enhanced membrane permeability, *Bioconjugate Chem.*, **17**, 728–734.

(7) Leroueil, P. R., Hong, S. Y., Mecke, A., Baker, J. R., Orr, B. G. and Holl, M. M. B., (2007), Nanoparticle interaction with biological membranes: does nanotechnology present a Janus face?, *Acc. Chem. Res.*, **40**, 335–342.

(8) Thomas, T. P., Patri, A. K., Myc, A., Myaing, M. T., Ye, J. Y., Norris, T. B. and Baker, J. R. Jr, (2004), *In vitro* targeting of synthesized antibody-conjugated dendrimer nanoparticles, *Biomacromolecules*, **5**, 2269–2274.

(9) Patri, A. K., Myc, A., Beals, J., Thomas, T. P., Bander, N. H. and Baker, J. R., (2004), Synthesis and in vitro testing of J591 antibody–dendrimer conjugates for targeted prostate cancer therapy, *Bioconjugate Chem.*, **15**, 1174–1181.

(10) Shukla, R., Thomas, T. P., Peters, J. L., Desai, A. M., Kukowska-Latallo, J., Patri, A. K., Kotlyar, A. and Baker, J. R. Jr, (2006), HER2 specific tumor targeting with dendrimer conjugated anti-HER2 mAb, *Bioconjugate Chem.*, **17**, 1109–1115.

(11) Wu, G., Barth, R. F., Yang, W. L., Chatterjee, M., Tjarks, W., Ciesielski, M. J. and Fenstermaker, R. A., (2004), Site-specific conjugation of boron-containing dendrimers to anti-EGF receptor monoclonal antibody cetuximab (IMC-C225) and its evaluation as a potential delivery agent for neutron capture therapy, *Bioconjugate Chem.*, **15**, 185–194.

(12) Wu, G., Barth, R. F., Yang, W. L., Kawabata, S., Zhang, L. W. and Green-Church, K., (2006), Targeted delivery of methotrexate to epidermal growth factor receptor-positive brain tumors by means of cetuximab (IMC-C225) dendrimer bioconjugates, *Mol. Cancer Ther.*, **5**, 52–59.

(13) Shukla, R., Thomas, T. P., Peters, J., Kotlyar, A., Myc, A. and Baker Jr, J. R., (2005), Tumor angiogenic vasculature targeting with PAMAM dendrimer–RGD conjugates, *Chem. Commun. (Camb).*, **46**, 5739–5741.

(14) Sheng, K. C., Kalkanidis, M., Pouniotis, D. S., Esparon, S., Tang, C. K., Apostolopoulos, V. and Pietersz, G. A., (2008), Delivery of antigen using a novel mannosylated dendrimer potentiates immunogenicity *in vitro* and *in vivo*, *Eur. J. Immunol.*, **38**, 424–436.

(15) Baek, M. G. and Roy, R., (2002), Synthesis and protein binding properties of T-antigen containing GlycoPAMAM dendrimers, *Bioorg. Med. Chem.*, **10**, 11–17.

(16) Taite, L. J. and West, J. L., (2006), Poly(ethylene glycol)–lysine dendrimers for targeted delivery of nitric oxide, *J. Biomater. Sci. – Polym. Edn*, **17**, 1159–1172.

(17) Kono, K., Liu, M. and Frechet, J. M., (1999), Design of dendritic macromolecules containing folate or methotrexate residues, *Bioconjug.Chem.*, **10**, 1115–1121.

(18) Shukla, S., Wu, G., Chatterjee, M., Yang, W., Sekido, M., Diop, L. A., Muller, R., Sudimack, J. J., Lee, R. J., Barth, R. F. and Tjarks, W., (2003), Synthesis and biological evaluation of folate receptor-targeted boronated PAMAM dendrimers as potential agents for neutron capture therapy, *Bioconjug. Chem.*, **14**, 158–167.

(19) Majoros, I. J., Myc, A., Thomas, T., Mehta, C. B. and Baker, J. R. Jr, (2006), PAMAM dendrimer-based multifunctional conjugate for cancer therapy: synthesis, characterization, and functionality, *Biomacromolecules*, **7**, 572–579.
(20) Thomas, T. P., Majoros, I. J., Kotlyar, A., Kukowska-Latallo, J. F., Bielinska, A., Myc, A. and Baker, J. R. Jr, (2005), Targeting and inhibition of cell growth by an engineered dendritic nanodevice, *J. Med. Chem.*, **48**, 3729–3735.
(21) Myc, A., Douce, T. B., Ahuja, N., Kotlyar, A., Kukowska-Latallo, J., Thomas, T. P. and Baker, J. R., (2008), Preclinical antitumor efficacy evaluation of dendrimer-based methotrexate conjugates, *Anti-Cancer Drugs*, **19**, 143–149.
(22) Majoros, I. J., Thomas, T. P., Mehta, C. B. and Baker, J. R. Jr, (2005), Poly(amidoamine) dendrimer-based multifunctional engineered nanodevice for cancer therapy, *J. Med. Chem.*, **48**, 5892–5899.
(23) Kukowska-Latallo, J. F., Candido, K. A., Cao, Z., Nigavekar, S. S., Majoros, I. J., Thomas, T. P., Balogh, L. P., Khan, M. K. and Baker, J. R. Jr, (2005), Nanoparticle targeting of anticancer drug improves therapeutic response in animal model of human epithelial cancer, *Cancer Res.*, **65**, 5317–5324.
(24) Myc, A., Majoros, I. J., Thomas, T. P. and Baker, J. R. Jr, (2007), Dendrimer-based targeted delivery of an apoptotic sensor in cancer cells, *Biomacromolecules*, **8**, 13–18.
(25) Myc, A., Patri, A. K. and Baker, J. R., (2007), Dendrimer-based BH3 conjugate that targets human carcinoma cells, *Biomacromolecules*, **8**, 2986–2989.
(26) Landmark, K. J., DiMaggio, S., Ward, J., Kelly, C., Vogt, S., Hong, S., Kotlyar, A., Myc, A., Thomas, T. P., Penner-Hahn, J. E., Baker, J. R., Holl, M. M. B. and Orr, B. G., (2008), Synthesis, characterization, and in vitro testing of superparamagnetic iron oxide nanoparticles targeted using folic acid-conjugated dendrimers, *ACS Nano.*, **2**, 773–783.
(27) Patri, A. K., Kukowska-Latallo, J. F. and Baker, J. R. Jr, (2005), Targeted drug delivery with dendrimers: Comparison of the release kinetics of covalently conjugated drug and non-covalent drug inclusion complex, *Adv. Drug Deliv. Rev.*, **57**, 2203–2214.
(28) Patri, A. K., Majoros, I. J. and Baker, J. R., (2002), Dendritic polymer macromolecular carriers for drug delivery, *Curr. Opin. Chem. Biol.*, **6**, 466–471.
(29) Tomalia, D. A., Baker, H., Dewald, J., Hall, M., Kallos, G., Martin, S., Roeck, J., Ryder, J. and Smith, P., (1985), A new class of polymers: starburst-dendritic macromolecules, *Polym. J.*, **17**, 117–132.
(30) Giordanengo, R., Mazarin, M., Wu, J., Peng, L. and Charles, L., (2007), Propagation of structural deviations of poly(amidoamine) fan-shape dendrimers (generations 0–3) characterized by MALDI and electrospray mass spectrometry, *Int. J. Mass Spectrom.*, **266**, 62–75.
(31) Kolb, H. C., Finn, M. G. and Sharpless, K. B., (2001), Click chemistry: diverse chemical function from a few good reactions, *Angew. Chem. Int. Edn Engl.*, **40**, 2004–2021.
(32) Wyszogrodzka, M. and Haag, R., (2008), A convergent approach to biocompatible polyglycerol 'click' dendrons for the synthesis of modular core-shell architectures and their transport behavior, *Chemistry*, **14**, 9202–9214.
(33) Wu, P., Malkoch, M., Hunt, J. N., Vestberg, R., Kaltgrad, E., Finn, M. G., Fokin, V. V., Sharpless, K. B. and Hawker, C. J., (2005), Multivalent, bifunctional dendrimers prepared by click chemistry, *Chem. Commun. (Camb).*, **46**, 5775–5777.
(34) Haridas, V., Lal, K. and Sharma, Y. K., (2007), Design and synthesis of triazole-based peptide dendrimers, *Tetrahedron Lett.*, **48**, 4719.
(35) Fernandez-Megia, E., Correa, J., Rodriguez-Meizoso, I. and Riguera, R., (2006), A click approach to unprotected glycodendrimers, *Macromolecules*, **39**, 2113–2120.
(36) Dijkgraaf, I., Rijnders, A. Y., Soede, A., Dechesne, A. C., van Esse, G. W., Brouwer, A. J., Corstens, F. H., Boerman, O. C., Rijkers, D. T. and Liskamp, R. M., (2007), Synthesis of DOTA-conjugated multivalent cyclic-RGD peptide dendrimers via 1,3-dipolar cycloaddition and their biological evaluation: Implications for tumor targeting and tumor imaging purposes, *Org. Biomol. Chem.*, **5**, 935–944.
(37) Gopin, A., Ebner, S., Attali, B. and Shabat, D., (2006), Enzymatic activation of second-generation dendritic prodrugs: conjugation of self-immolative dendrimers with poly(ethylene glycol) via click chemistry, *Bioconjugate Chem.*, **17**, 1432–1440.

(38) Ortega-Muñoz, M., Lopez-Jaramillo J., Hernandez-Mateo F., Santoyo-Gonzalez, F., (2006), Synthesis of glyco-silicas by cu(I)-catalyzed ldquoClick-chemistryrdquo and their applications in affinity chromatography, *Adv. Synth. Catal.*, **348**, 2410–2420.
(39) Urbani, C. N., Bell, C. A., Lonsdale, D., Whittaker, M. R. and Monteiro, M. J., (2008), Self-assembly of amphiphilic polymeric dendrimers synthesized with selective degradable linkages, *Macromolecules*, **41**, 76–86.
(40) Frechet, J. M. J. and Tomalia, D.A., (2001), *Dendrimers and Other Dendritic Polymers*. John Wiley & Sons Ltd, Chichester.
(41) Svenson, S. and Tomalia, D. A., (2005), Dendrimers in biomedical applications – reflections on the field, *Adv. Drug Deliv. Rev.*, **57**, 2106–2129.
(42) Adi Dahan, M. P., (2005), Dendrons and dendritic catalysts immobilized on solid support: synthesis and dendritic effects in catalysis, *J. Polym. Sci. Pt A: Polym. Chem.*, **43**, 235–262.
(43) Joralemon, M. J., O'Reilly, R. K., Matson, J. B., Nugent, A. K., Hawker, C. J. and Wooley, K. L., (2005), Dendrimers clicked together divergently, *Macromolecules*, **38**, 5436–5443.
(44) Killops, K. L., Campos, L. M. and Hawker, C. J., (2008), Robust, efficient, and orthogonal synthesis of dendrimers via thiol-ene 'click' chemistry, *J. Am. Chem. Soc.*, **130**, 5062–5064.
(45) Lee, J. W. and Kim, B. K., (2006), Synthesis of symmetric and unsymmetric triazole dendrimers via dipolar cycloaddition reaction, *Synthesis-Stuttgart*, 615–618.
(46) Lee, J. W., Kim, H. J., Han, S. C., Kim, J. H. and Jin, S. H., (2008), Designing poly(amido amine) dendrimers containing core diversities by click chemistry of the propargyl focal point poly(amido amine) dendrons, *J. Polym. Sci. Pt a – Polym. Chem.*, **46**, 1083–1097.
(47) Lee, J. W., Kim, J. H. and Kim, B. K., (2006), Synthesis of azide-functionalized PAMAM dendrons at the focal point and their application for synthesis of PAMAM-like dendrimers, *Tetrahedron Lett.*, **47**, 2683–2686.
(48) Lee, J. W., Kim, J. H., Kim, B. K., Kim, J. H., Shin, W. S., Jin, S. H. and Kim, M., (2006), Convergent synthesis of PAMAM-like dendrimers from azide-functionalized PAMAM, *Bull. Korean Chem. Soc.*, **27**, 1795–1800.
(49) Lee, J. W., Kim, J. H., Kim, H. J., Han, S. C., Kim, J. H., Shin, W. S. and Jin, S. H., (2007), Synthesis of symmetrical and unsymmetrical PAMAM dendrimers by fusion between azide- and alkyne-functionalized PAMAM dendrons, *Bioconjug. Chem.*, **18**, 579–584.
(50) Lee, J. W., Kim, B., Kim, H. J., Han, S. C., Shin, W. S. and Jin, S., (2006), Convergent synthesis of symmetrical and unsymmetrical PAMAM dendrimers, *Macromolecules*, **39**, 2418–2422.
(51) Lee, J. W., Kim, B. K., Kim, J. H., Shin, W. S. and Jin, S. H., (2006), Facile approach for diblock codendrimers by fusion between frechet dendrons and PAMAM dendrons, *J. Org. Chem.*, **71**, 4988–4991.
(52) Lee, J. W., Kim, J. H., Kim, B. K., Shin, W. S. and Jin, S. H., (2005), Synthesis of Frechet type dendritic benzyl propargyl ether and frechet type triazole dendrimer, *Tetrahedron*, **62**, 894–900.
(53) Szalai, M. L., McGrath, D. V., Wheeler, D. R., Zifer, T. and McElhanon, J. R., (2007), Dendrimers based on thermally reversible Furan–Maleimide Diels–Alder adducts, *Macromolecules*, **40**, 818–823.
(54) Pohl, M., Michaelis, N., Meister, F. and Heinze, T., (2009), Biofunctional surfaces based on dendronized cellulose, *Biomacromolecules*, **10**, 382–389.
(55) Goyal, P., Yoon, K. and Weck, M., (2007), Multifunctionalization of dendrimers through orthogonal transformations, *Chemistry*, **13**, 8801–8810.
(56) Yoon, K., Goyal, P. and Weck, M., (2007), Monofunctionalization of dendrimers with use of microwave-assisted 1,3-dipolar cycloadditions, *Org. Lett.*, **9**, 2051–2054.

9

Reversible Diels–Alder Cycloaddition for the Design of Multifunctional Network Polymers

Amy M. Peterson and Giuseppe R. Palmese

9.1 Introduction

Click chemistries such as the Diels–Alder reaction have been used to prepare many novel and difficult to achieve multifunctional polymer networks.[1–9] Such chemistries are popular because they offer high yields and minimal side reactions under mild reaction conditions. The Diels–Alder reaction is a cycloaddition of a diene and a dienophile to form a substituted cyclohexene.[10–12] In some cases, the reaction is thermally reversible. This reversibility makes the Diels–Alder reaction particularly desirable for the development of multifunctional polymer networks.[13]

The Diels–Alder reaction, shown in its general form in Figure 9.1, was discovered by Otto Diels and Kurt Alder in the 1920s. A substituted cyclohexene is formed in this reaction through the [4 + 2] cycloaddition of a diene and a dienophile. This reaction proceeds through unsymmetrical transition states.[14] While the simplest dienophile, ethylene, reacts poorly with dienes, there are many dienes and dienophiles that undergo the Diels–Alder cycloaddition readily. Conjugated dienes react as long as they can achieve a *cisoid* geometry and cyclic dienes tend to be more reactive than open chain dienes.[11] The Diels–Alder reaction is reversible when the reactants are stable molecules or when one can be consumed in a side reaction. The reverse reaction is known as the retro-Diels–Alder reaction. The chemical structures of well-known dienes, dienophiles and adducts as well as the

Click Chemistry for Biotechnology and Materials Science Edited by Joerg Lahann

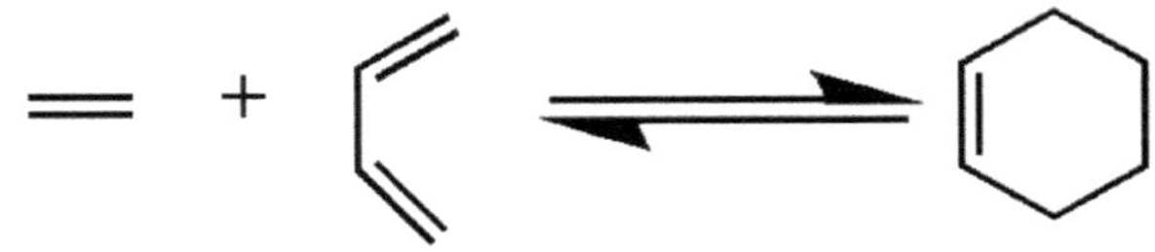

Figure 9.1 Generalized Diels–Alder reaction of a dienophile with a diene.

temperatures at which they undergo the forward and reverse Diels–Alder reactions are shown in Table 9.1.

Much of the literature has focused on the reaction of furan (diene) and maleimide (dienophile), presumably because of the mild conditions under which the retro Diels–Alder reaction occurs.[15–54] However, the reactions of maleimide with cyclopentadienone,[55] fulvene,[56] pyrone,[57,58] and anthracene,[4–6,49,59–62] of benzene with cyclopentadienone[63–65] and of quinone with cyclopentadiene[66–69] have been investigated as well. Cyclopentadiene has also been shown to react reversibly with itself via the Diels–Alder reaction, acting as both the diene and the dienophile.[70–80] Additionally, poly(vinyl chloride) demonstrates reversible crosslinking when thermal degradation results in the formation of polyenes in the polymer chain.[81–84]

Table 9.1 Some well known thermoreversible Diels–Alder reactions

Diene	Dienophile	Adduct	Forward reaction upper limit - reverse reaction lower limit	References
Furan 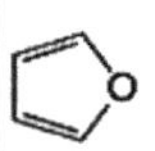	Maleimide 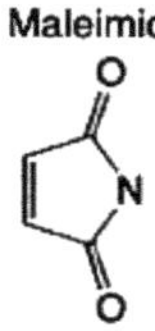	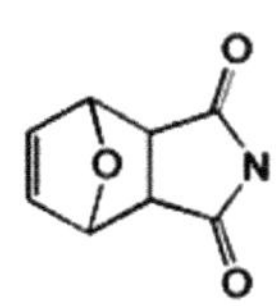	60-90 °C	17, 34, 37, 39, 40, 47, 48
Anthracene 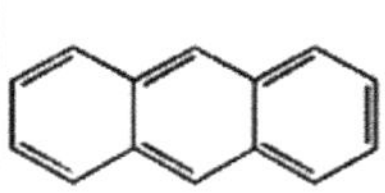	Maleimide 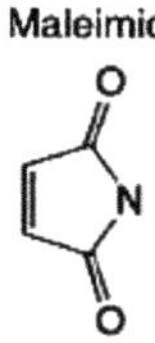	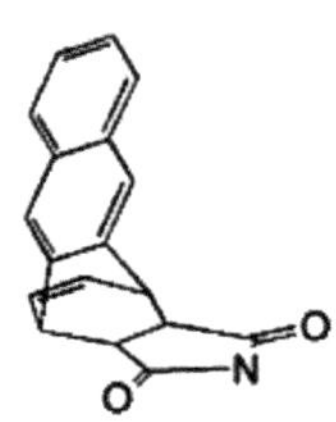	250-300 °C	60
Fulvene 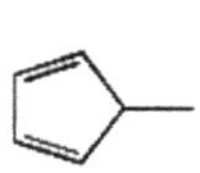	Maleimide 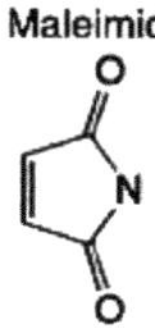	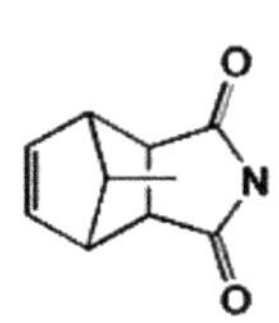	30-100 °C	56
Cyclopentadiene 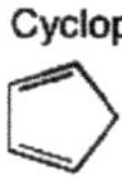	Cyclopentadiene 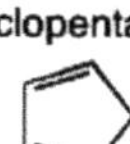		80-180 °C	8, 72, 78, 79

Since the discovery of the Diels–Alder reaction over 80 years ago, many groups have investigated important aspects of this reaction's behavior such as reaction mechanisms, as well as thermodynamic and kinetic concerns. As a whole these inquiries have elucidated factors that influence reactivity and reversibility, including the chemical nature of the diene–dienophile selected, steric considerations, reaction conditions, solvents and diffusion limitations. The following is not meant to be a comprehensive summary of these works but rather one that highlights important aspects that need to be considered when polymer networks containing Diels–Alder function linkages are designed.

The Diels–Alder adduct exhibits *endo* and *exo* isomers with different properties. The major product of the Diels–Alder reaction is typically the *endo* adduct. This preference has been explained with secondary orbital interactions, inductive or charge-transfer interactions and overlap of π orbitals.[11] Solvent choice for the reaction of furan and maleimide in solution does not affect the isomer produced, although reaction temperature does. At 90 °C, the *exo* product, with a melting temperature of 162 °C, is achieved. Performing the reaction at 25 °C produces the *endo* isomer, with a melting temperature of 131 °C.[85] Herndon *et al.* discovered that *exo*-dicyclopentadiene is less reactive than the *endo* isomer by an order of magnitude.[71]

Neukam and Grimme found that the retro-Diels–Alder reaction was exothermic if it combined the formation of an aromatic group with the release of ring strain from the adduct.[84] They also showed that the cycloreversion of the cyclopentadiene and 5-cyanocyclopentadiene adduct can be accelerated by deprotonation. Loss of basicity was also proposed to significantly increase the rate of cycloreversion.[86] Additionally, Brand and Klapper found that substituted furans undergo Diels–Alder reaction with normal electron demand and that the reactivity of forward and reverse reactions can be controlled by changing the donor/acceptor character of the substituent.[87]

The first mention of Diels–Alder reaction order was made by Kwart and King.[27] They reported that the forward reaction of a diene and a dienophile generally followed second-order kinetics. This finding was later corroborated,[44,51,52,66,67] although pseudo-first-order conditions can be approximated with an excess of either diene or dienophile.[9,69,88,89] Liu and Hsieh found the activation energy of the forward Diels–Alder reaction between furan and maleimide groups to be 67 kJ mol^{-1} in one case[51] and 32.1 kJ mol^{-1} in another.[52] In other studies the reverse reaction was found to follow a first-order rate law with an activation energy of 21 kJ mol^{-1}.[42] There have been numerous investigations of the forward and reverse reaction rates and it is thought that for each system an equilibrium is established over a range of temperatures so that the equilibrium concentration of Diels–Alder linkages can be controlled by setting a temperature in this range. However, comprehensive experimental studies have not been conducted to establish the quantitative relationship between temperature and equilibrium composition in polymer systems.

The influence of surfaces on the Diels–Alder reaction has also been studied. In the case of quinone immobilized on a gold surface reacting with cyclopentadiene, the reaction was found to follow a pseudo-first-order rate law, but the rate constant is not linearly related to diene concentration.[69] Gawalt and Mrksich propose that this behavior is the result of electrochemical oxidation, which creates two forms of quinone, with one that is two orders of magnitude more reactive than the other. Production of the highly reactive form is the rate-determining step in this reaction. Kwon and Mrksich studied the effect of steric hindrance on reaction rate for the reaction of cyclopentadiene with benzoquinone contained within a self-assembled monolayer (SAM).[68] The rate constant decreased significantly when the quinone was positioned below the monolayer interface. When the quinone was positioned above the other groups of the SAM, a rate constant between 0.18 and 0.22 $\text{M}^{-1}\ \text{s}^{-1}$ was found, while the rate for quinone below the monolayer interface was approximately 0.03 $\text{M}^{-1}\ \text{s}^{-1}$.

Although click chemistries are often selected because of the dearth of side reactions, there are a few secondary reactions in which typical Diels–Alder dienes and dienophiles participate. Maleimides react readily with amines, alcohols and thiols through Michael addition.[46,91] Maleimides can also homopolymerize.[92] Furfuryl groups have been found to be unstable and can ring open or form dihydro- or tetrahydrofuran rings.[13,80] The Diels–Alder adduct may degrade via aromatization to form an irreversible linkage.[13,29] Vinyl addition of some dienes and dienophiles has been shown.[76] Additionally, steric hindrance of reactants can affect the extent of the Diels–Alder reaction.[15] Some researchers have applied these side reactions in the preparation of novel polymer networks.[46,91,92] The focus of this chapter is the application of reversible Diels–Alder chemistries to polymer networks to provide additional functionality. Polymer network design considerations are presented first, followed by examples of polymer networks with Diels–Alder linkages providing additional functionality are presented. We begin the discussion of Diels–Alder functional systems with linear polymers, which, although not network polymers, have been investigated thoroughly and can provide important information about the behavior of the Diels–Alder reaction in a polymer system. A number of reversibly crosslinked systems are discussed including remoldable crosslinked materials, thermally removable encapsulants, reversibly crosslinked polymer-solvent gels, remendable materials, recyclable thermosets, and smart materials.

9.2 Design of Polymer Networks

A polymer network is defined as a polymeric system in which the mer units are connected such that numerous paths exist through the macromolecule.[93] Crosslinked networks are desirable because of their mechanical properties, thermal stability and insolubility in solvents. Once crosslinked, a polymer network cannot melt. This chapter focuses on polymer networks given additional functionality with the incorporation of reversible Diels–Alder linkages because of the ubiquity of crosslinked systems and because of the novel characteristics of polymer networks containing such bonds. In our discussion of networks, we will also consider branched systems such as dendrimers.

When crosslinks are formed with Diels–Alder bonds, the resulting material is a crosslinked material at ambient conditions but can be remolded and remended at increased temperatures.[16,23,94–96] One interesting result of incorporating thermoreversible linkages is that chains within the system can diffuse and stress can relax, allowing for creep of the crosslinked network.[97]

In the design of networks containing reversible linkages, the concentration, placement and type of reversible bonds are very important considerations. One method for making such crosslinked networks using diene and dienophile reactive groups is to form polymer chains with pendant functional groups (diene or dienophile) that crosslink via reaction with difunctional molecules[16,28,77] or other polymer chains[72,73,78,79] with the complementary functionality (dienophile or diene). The other method relies on the reaction of multifunctional ($f \geq 3$) monomers to form a network.[70] If the first method is used, gelation generally occurs at significantly lower conversion than the second method because the polymer chains have more functional groups than monomers. The statistical approaches of Flory[98,99] and Stockmayer[100,101] state that, as the number of functional groups increases, the percolation conversion decreases according to the following equation:

$$p_c = \frac{1}{(f-1)^{1/2}} \tag{9.1}$$

Equation (9.1) assumes a stoichiometric mixture of dienes and dienophiles with a difunctional unit consisting of one type of functional group (A) and a branching unit consisting of the other type of functional group (B). Additionally, equal reactivity of A and B, reactivity independent of conversion, and no intramolecular reaction prior to gelation are assumed. Based on Equation (9.1), gelation of a system with branching units of $f = 4$ (such as those of Chen *et al.*[21–23]) is predicted to occur at $p_c = 0.577$. However, the critical conversion of a furan-modified polystyrene, $f = 94.2$ (Goussé *et al.*[18]), reacting with a bismaleimide, is 0.104.

As shown in Table 9.1, the choice of diene and dienophile affects the temperature at which the Diels–Alder adduct forms and breaks or forms and degrades. Desired properties and applications dictate the use of various Diels–Alder reactants. Additionally, and as mentioned in the previous section, equilibrium composition is influenced by temperature over a considerable range, so mechanical behavior would be influenced not only by typical molecular mobility effects but also by changes in polymer structure like crosslink density. An example of these new factors that must be considered in the design of polymer networks using reversible linkages is the influence of the retro-Diels–Alder reaction on viscoelastic behavior. The time-temperature superposition principle states that time is equivalent to temperature for viscoelastic materials like polymers. A phenomenon that occurs over a long time scale at a low temperature is equivalent to the phenomenon that occurs over a short time scale at a high temperature. As a result, behavior can be determined at a wider range of temperature and frequencies than can be measured. One way to relate time and temperature is through the Williams–Landel–Ferry (WLF) equation, which is given below.[90,102]

$$\log_{10} a_T = \frac{-C_1(T - T_0)}{C_2 + (T - T_0)} \tag{9.2}$$

In the WLF equation, a_T is the shift factor and is positive if the curve is to the left of the reference and is negative if the curve is to the right of the reference. T is the measured temperature and T_0 is the reference temperature, while C_1 and C_2 are constants. In many cases, $C_1 = 17.4$ and $C_2 = 51.6$. Gotsmann *et al.* found that the indentation kinetics of a polymer comprised of a trisfuran and bismaleimide deviate from the WLF equation.[50] Deviation from WLF behavior is expected in polymers containing Diels–Alder linkages because the number of bonds, and therefore the inherent viscoelastic behavior of the material, is temperature dependant.

9.3 Application of Diels–Alder Linkages to Polymer Systems

Diels–Alder bonds have been applied to many types of networks to form materials designed for a variety of uses, ranging from biomedical and encapsulant[24–26,39] technologies to recyclable[19,53] and remendable materials.[94,95] The following sections are organized according to the type of polymer developed and potential applications. Taken as a whole they provide a comprehensive review of polymer architectures that can be created to provide multifunctionality based on the thermoreversible nature of the Diels–Alder linkage.

9.3.1 Molecular Weight Control of Linear Polymers

Although linear polymers are not considered to be polymer networks, a significant portion of the literature on reversible Diels–Alder linkages is concerned with noncrosslinked systems.[12] Additionally, much of the early work in Diels–Alder-based polymers was concerned with linear polymers.[55,56,70,76,104,105] Diels–Alder bonds are desirable in linear polymers for control of chain length,[74] control of processing viscosity[87] and for improved recyclability.[105] Ladder polymers have also been prepared using Diels–Alder bonds.[106]

The first mention of Diels–Alder polymers appears in the patent literature. Upson prepared copolymers of *p*-xylene bis(5-cyclopentadiene) and cyclopentadiene.[70] Cure conditions dictated the physical properties of the resulting polymer. Fusible polymers were obtained with low temperature cure short, high-temperature cure cycles. Insoluble polymers were probably formed as a result of vinyl addition.[76] Kraiman prepared permanently bonded linear polymers via the reaction of bismaleimide with cyclopentadienone[55] and alpha pyrone.[55] In both cases, polymerization proceeded through the Diels–Alder reaction, the product of which condensed a smaller group (carbon monoxide and carbon dioxide, respectively) and then homopolymerized. Similar polymers include the copolymer of bismaleimide and thiophene dioxide, which emitted sulfur dioxide,[103] the copolymer of bismaleimide and cyclopentadienone, which emitted carbon monoxide,[104] and the linear polymer of bismaleimide and bispyrone, which emitted carbon dioxide.[57]

Figure 9.2 is a schematic showing the thermally reversible behavior of a linear polymer based on Diels–Alder linkages. The first example of reversibly bonded polymers appears in work from Stille and Plummer.[76] Since cyclopentadiene can act as a diene and a dienophile, dicyclopentadiene was homopolymerized and also copolymerized with *p*-benzoquinone and a bismaleimide. At increased reflux temperatures, inherent viscosity decreased, indicating the occurrence of the retro-Diels–Alder reaction. Another copolymer formed from the reaction of cyclopentadiene with maleimide shows chain extension at 80 °C.[74] Reaction equilibrium was shifted towards the reactants at higher temperature, while lower temperatures decreased the reaction rate. However, even when polymerized at 80 °C these opposing phenomena prevented the formation of high-molecular-weight polymers.

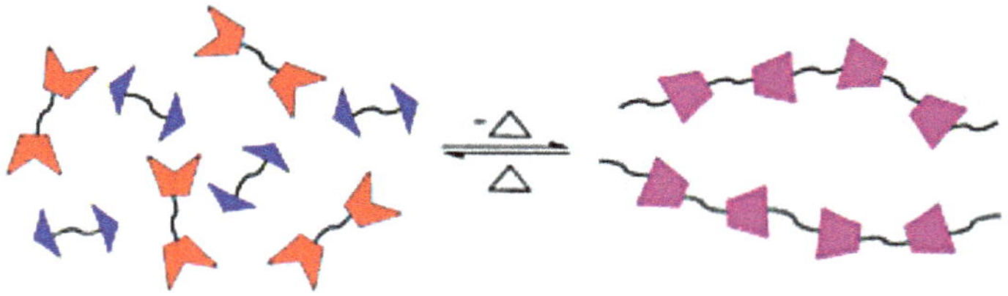

Figure 9.2 Schematic of a reversibly forming linear polymer of a bisdiene and a bisdienophile. Dienes are shown as notched trapezoids, dienophiles are shown as triangles and Diels–Alder adducts are shown as trapezoids.

A similar effect was shown in the polymerization of bisfuran with bismaleimide. Kuramoto *et al.* observed that, as the reaction temperature increased from 25 to 60 °C, reaction rate increased, but chain length decreased.[34] Additionally, spectroscopic analysis showed that after heating the polymerized system at 90 °C for 2 h, only 20% of the initial maleimide was recovered. This suggests that either some of the Diels–Alder adducts could not be reversed at this temperature as a result of the chemical equilibrium or that secondary maleimide reactions occurred. Teramoto *et al.* reported maximum molecular weights for polymerization temperatures of 70 and 55 °C for two bismaleimides.[53] Molecular weight was shown to decrease for reaction temperatures above 90 °C and gel permeation chromatography demonstrated that the retro-Diels–Alder reaction proceeded rapidly above 100 °C. A copolymer of bismaleimide and bisfulvene also displayed reversibility of Diels–Alder adducts.[56] When heated above 60 °C the polymer demonstrated bond reversal through changes in viscosity. Temperature-controlled viscosity and molecular weight were achieved for linear polymer systems by Brand and Klapper.[87] Polymerization of bisanthracene and bismaleimide was reported by Grigoras and Colotin.[60] The product was an oligomer, perhaps due to steric effects or equilibrium with the retro-Diels–Alder reaction at the cure temperature of 120 °C. However, thermogravimetric analysis indicates that the retro-Diels–Alder reaction occurs between 250 and 300 °C in these systems.

Siloxane-modified bisfurans have been prepared and polymerized with bismaleimides for application in electronics or coatings.[29] To improve thermal stability of the material, the Diels–Alder adduct was aromatized by refluxing the solid product in acetic anhydride for 2 h. Aromatization of the Diels–Alder adduct was also used to prepare thermally stable polyimides[31] and linear polymers bearing pendant cyano groups[107] that could be further stabilized by crosslinking through said cyano groups.

Ladder polymers have been synthesized via Diels–Alder reaction. Blatter and Schlueter formed a soluble ladder polymer from a bisdiene and a bisdienophile.[108,109] No mention was made as to the reversible nature of this material. Kintzel *et al.* also prepared a ladder polymer system.[106]

Monomers containing both diene and dienophile functionality are of interest because a 1:1 stoichiometric ratio is always achieved. Mikroyannidis prepared AB monomers from furfuryl-substitution of maleimic acids[31–33] Goussé and Gandini questioned the success of Mikroyannidis based on limited polymer characterization and prepared 2-furfurylmaleimide themselves.[35] The polymerization product following reaction at 90–180 °C for 15 min to several hours was a brown solid insoluble in most solvents. No mention of reversibility was made by either group; however, cure temperatures may have degraded the maleimide[17] or caused vinyl addition polymerization.[76] Crosslinking through a secondary reaction is apparent in the product achieved by Goussé and Gandini due to its insolubility.

Grafting is also possible using the Diels–Alder reaction. Jones *et al.*[59] and Vargas *et al.*[61] reported poly(ethylene terephthalate) (PET) with anthracene groups. Reaction of these anthracenes with substituted maleimides provided hydrophobic and hydrophilic materials, depending on the type of maleimide used. This process can be applied to thin films since the Diels–Alder reaction occurs below the melting temperature of the polymer.

Compounds with potential biological applications have been prepared through diene modification of poly(ethylene glycol) (PEG) derivatives and subsequent reaction with maleimides.[111,112] Substitution of the maleimide can provide novel PEG derivates. Block copolymers were developed based on maleimide- and anthracene functionalization of PEG, poly(methyl methacrylate) (PMMA), polystyrene (PS) and poly(*tert* butyl acrylate) (PtBA).[4] Controlled synthesis was achieved by protecting maleimide groups with furan. When block copolymer synthesis was desired, the system was heated to 110 °C, cleaving the furan and allowing for reaction between maleimide and anthracene terminated blocks. A similar strategy was also used by Durmaz *et al.* to graft PEG onto PS chains.[49]

9.3.2 Remoldable Crosslinked Materials

Thermosets are desirable for many applications because of their mechanical properties, thermal stability, and resistance to solvents. However, once cast, crosslinked systems cannot be reformed. The desire to create remoldable crosslinked networks was the catalyst for developing polymers with thermoreversible crosslinks. One way to do this is shown schematically in Figure 9.3, in which linear polymers with pendant dienes/dienophiles crosslink through reaction with bisdienes/bisdienophiles.

Craven was the first to report on a reversibly crosslinking network with chains bearing furans crosslinked via reaction with bismaleimides. Upon heating to 120–140 °C the material was capable of being reshaped. Although high reversal temperatures were used, Craven warns, '"Reversed" does not mean that the crosslinked polymer product can be completely converted to its original components, but that it becomes a more plastic material, capable of being formed and shaped'.[16] This observation may be the result of secondary crosslinking, perhaps through homopolymerization of maleimides or aromatization of Diels–Alder adducts.

In another example, anthracene-functionalized PET was found to reversibly crosslink in the presence of bismaleimides, even when the PET copolymer contained just 2 mol% of the anthracene unit.[59] Crosslinking was partially (27%, according to ^{1}H NMR) reversible upon heating to 250 °C for 7 h. Reversible Diels–Alder linkages have also been applied to elastomeric siloxane-based polymers, with the motivation of developing a recyclable material for tires. In this system, pendant furans react with bismaleimides at room temperature and adducts break apart at 80 °C.[40]

When cyclopentadiene, which can act as diene and dienophile, was used as the pendant functional group, linear polymers crosslinked via reaction with each other. Remolding

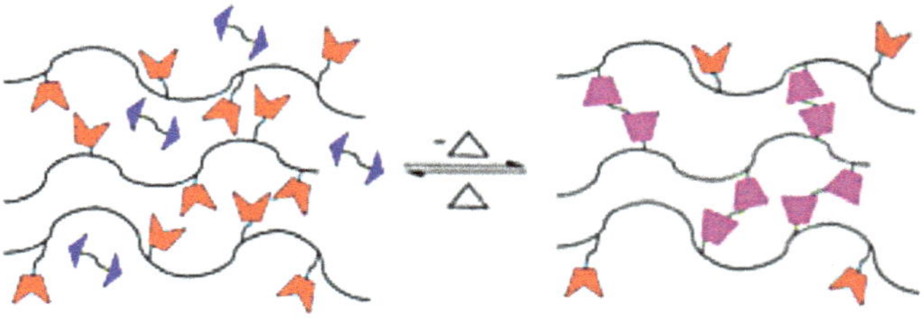

Figure 9.3 Schematic of a reversibly crosslinking system comprised of a diene-functionalized linear polymer with a bisdienophile. Dienes are shown as notched trapezoids, dienophiles are shown as triangles and Diels–Alder adducts are shown as trapezoids.

was possible in one system when heated to 150–400 °C, although the preferred temperature range was 200–300 °C.[77] Kennedy and Castner also observed gelation of a cyclopentadienylated polymer. When heated to 215 °C, the system was insoluble in hexachlorobutadiene; however, when heated to 215 °C in the presence of maleic anhydride, the polymer was soluble, demonstrating that crosslinking occurred through pendant cyclopentadienes.[72,73] This material behaved as an elastomer and could be shaped by a mold when cured at 170 °C.[78] A cyclopentadienylated polyphosphazene demonstrated crosslinking at 80 °C and crosslink decoupling at 160 °C.[79] However, degradation associated with a loss of crosslink reversibility as well as a mass loss of 15–35% was observed when the polymer was heated to 320 °C.

9.3.3 Thermally Removable Encapsulants

Encapsulants are used in the electronics industry to protect components from the environment. Thermosets are often used for their high strength, durability and resistance to heat and chemical degradation. However, it is sometimes desirable for these encapsulants to be removed. As a result, thermally removable encapsulants have been designed using the Diels–Alder chemistry.[39,54] These materials are similar in many ways to remoldable materials, except that the complete removal of the encapsulant is necessary. Small *et al.* reported on a thermally removable encapsulant from at least one bismaleimide and at least one trisfuran or tetrafuran cured below 90 °C.[39] Such encapsulants could be easily removed by heating above 90 °C, preferably in a polar solvent.

Thermally removable epoxies and polyurethanes were developed by Loy *et al.*[25,26] The epoxy-based system was prepared by reacting bismaleimide with furfuryl glycidyl ether to form a diepoxy, which reacted further with a diamine. Below 90 °C, the polymer is a thermoset, but above 90 °C the reverse Diels–Alder reaction dominates and adducts break apart. Two methods were proposed for the synthesis of polyurethanes: hydroxyl-bearing furan reacted with bismaleimide to form a diol, which was cured with a diisocyanate; and isocyanate-bearing furan reacted with bismaleimide to form a diisocyanate, which was cured with a diol.

Brock *et al.* developed a 'smart' encapsulant fluid capable of flowing into a system, polymerizing upon thermal stimulus and being removed upon further thermal stimulus.[54] The original encapsulant fluid contained a multifuran with $f \geq 3$ and a multimaleimide with $f \geq 2$. Either the furans or the maleimides were protected with maleimides or furans, respectively, to prevent polymerization before it is desired. Upon heating, reactive groups were unprotected and capable of reacting with other monomers. Gelation was observed at room temperature in 3–6 h. The crosslinked network could then be removed by heating above 120 °C.

9.3.4 Reversibly Crosslinked Polymer–Solvent Gels

Because of the high crystallinity of many bismaleimides, a number of polymer networks have been prepared in solution, resulting in polymer–solvent gels. Such gels could be used for biomedical applications[30] or as stimuli-responsive gels,[52] for example. Since the polymer networks are thermoreversible, gels become liquid upon heating and reform gels upon cooling. The first mention of swelled networks based on Diels–Alder chemistry in the literature was a copolymer of polystyrene and *N*-chloromethylmaleimide crosslinked with a bisfuran and a dicyclopentadiene.[28] When crosslinked with the bisfuran at 80 °C, gelation occurred in 15 min, and with dicyclopentadiene at 270 °C, gelation was observed within 5 min.

Thermally reversible hydrogels were prepared by modifying poly(*N*-acetylethylenimine) with maleimide and furan.[30] Hydrogels were synthesized at room temperature in the dark in methanol. Swelling of these gels was controlled by the degree of substitution as well as the temperature.

Canary and Stevens compared a copolymer of polystyrene and *N*-chloromethylmaleimide crosslinked with difurfuryl adipate and a polymer containing both furan and maleimide groups.[80] Gels of both were prepared with acetophenone as the solvent. The gel crosslinked with difurfyl adipate became liquid after 2.5 min at 150 °C, while the gel crosslinked with the polymer liquefied in 15 s under the same conditions. However, these polymers, particularly the furan-modified one, are rather unstable and the authors question the practicality of using furfuryl-based polymer networks.

A copolymer of styrene and furfuryl methacrylate was crosslinked using a bismaleimide. Gelation was observed after 12 days in toluene for a 12 wt% solution of copolymer[112] and after 100 h in chloroform for a 25 wt% solution of copolymer.[42] Solutions in toluene could be reverted to liquid form after 1 h at 110 °C. Further studies showed that unconsumed bismaleimide remains at the gel point.[44] Additionally, the highest crosslinking efficiency was observed when the ratio of maleimide to furan was 0.5.[45]

Reversibly crosslinking polyamide gels of maleimide-bearing polyamides and trisfurans were prepared in *N*,*N*-dimethylacetamide and evaluated for their thermally-responsive nature.[52] As the maleimide content of the polymer increased, the time necessary to liquefy the gel increased. By drying gels at room temperature under vacuum, reversibly crosslinked polyamides were obtained.

Hybrid organic–inorganic gels have also been prepared. In one case Diels–Alder linkages were applied to couple the organic polymer and the silica matrix.[43] Homogeneity of the gel was controlled by reaction temperature. The linear polymer, a furan-bearing polystyrene, could be extracted from the gel by heating to 130 °C. Another hybrid gel consisted of an interpenetrating network (IPN) of furan- and maleimide-functionalized poly(2-methyl-2-oxazoline) and silica gel.[37] The IPN increases modulus and solvent resistance. Reaction efficiency, the fraction of reactive sites used, was 47%.

9.3.5 Remendable Materials

More recent work in reversibly bonded networks has focused on their application for thermally remendable materials. The first report of a thermally remendable polymer based on the Diels–Alder reaction came from Chen *et al.*[21] The polymer consisted of a tetrafuran and a trismaleimide. Based on solid-state ^{13}C NMR, the bonds in this material were irreversible below 120 °C, showed 12% reversibility at 130 °C and showed 25% reversibility at 150 °C. Healing efficiency was evaluated by breaking a compact tension specimen, realigning the surfaces, heating at 120–150 °C for 2 h to break Diels–Alder bonds, cooling to room temperature to reform bonds across the crack surface, and retesting the specimen. Heating at 150 °C resulted in approximately 50% healing efficiency, while healing at 120 °C gave 41% healing efficiency. Although Chen *et al.* assert that the Diels–Alder reaction of furans

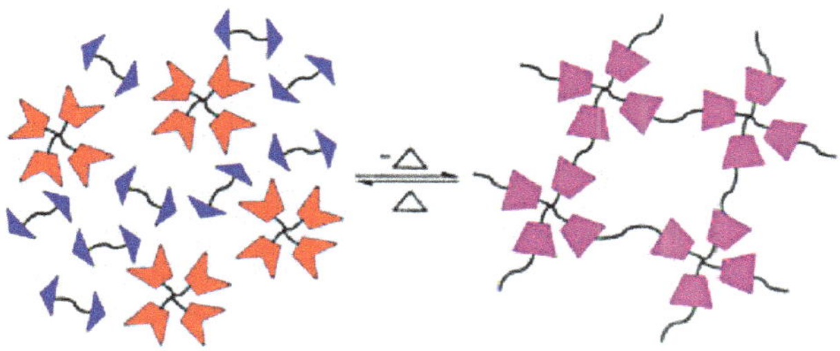

Figure 9.4 *Self-healing thermoset of a tetradiene and a bisdienophile. Dienes are shown as notched trapezoids, dienophiles are shown as triangles and Diels–Alder adducts are shown as trapezoids.*

and maleimides is irreversible below 120 °C, work from our group and others shows that the reaction equilibrium strongly favors the reverse reaction at temperatures above 90 °C.[113] Differences in healing efficiencies for healing temperatures above 90 °C are more likely a result of higher reactant mobility associated with higher temperatures.

Chen *et al.* also prepared remendable networks using two bismaleimides, 1,8-bis(maleimido)-3,6-dioxaoctane (2ME) and 1,8-bis(maleimido)-1-ethylpropane (2MEP) and a tetrafuran.[22] A schematic representation of such a system is shown in Figure 9.4. These polymers are hard, colorless and transparent at room temperature. Unlike their previous system, which was synthesized in a solvent that was evaporated off during cure, the new materials were synthesized in bulk, thanks to lower melting point bismaleimides. The polymer network synthesized with 2ME could be remolded at 160 °C; however, healing efficiency could not be evaluated because the material changed shape at the healing temperature. The polymer network prepared with 2MEP softened at 180 °C. When healed at 115 °C for 30 min, the material exhibited 80% healing efficiency for the first heal and 71% healing efficiency for the second heal.

Plaisted and Nemat-Nasser also investigated the mechanical and healing properties of the 2MEP4F system from Chen *et al.*[114] They found a fracture toughness of 0.71 MPa $m^{1/2}$. Interestingly, they discovered that healing at 85 °C for 30 min under reasonable pressure following by 30 min at 95 °C was sufficient for obtaining maximum healing. This research relied on a different geometry for mechanical testing, the double cleavage drilled compression specimen, which had the distinct advantage of a self-arresting crack so that fractured samples were in one piece following testing. The geometry significantly eased the realignment of opposing crack surfaces. The healing efficiencies reported for 4MEP4F were generally above 98%, significantly higher than that reported by Chen *et al.* for the same material.

Additional work by Wudl and Chen focused on remendable polymers based on a variety of multifurans and multimaleimides with $f \geq 3$. As with their other systems, these materials are hard and transparent. Above 120 °C, approximately 30% of the Diels–Alder adducts reverted to furans and maleimides, which reformed Diels–Alder adducts when cooled.[23] Although no mention was made of secondary reactions such as adduct aromatization or maleimide homopolymerization, it seems likely that such phenomena were at work in this system, since only 30% of the adducts broke apart at 120 °C, 30 °C above the temperature at which all adducts should have reverted to furans and maleimides.

One disadvantage of remendable materials using thermoreversible bonds, such as Diels–Alder adducts, is that the materials are not self-healing, that is, an external force is necessary to cause healing. This lack of autonomy is why these materials are characterized as remendable instead of self-healing.[115] In order to impart remendable materials with the ability to self-heal, arrays of conductive elements such as copper wires and coils were incorporated in fiber-reinforced composites of a trismaleimide tetrafuran polymer from the Wudl group.[116]

Liu and Hsieh prepared two thermosetting systems that exhibited healing. The first polymer consisted of epoxy-based trisfurans and trismaleimides that were mixed in acetone and cured at 50 °C for 12 h. The retro-Diels–Alder reaction of the material was observed after heating at 170 °C for 30 min. However, a cut in the material healed following 2 h at 120 °C and 12 h at 50 °C.[51] Polyamides with maleimide and furan functionality were also crosslinked. A cut in the surface of the material was healed after 3 h at 120 °C and 5 days at 50 °C.[20] The low temperature healing is curious since the reverse reaction does not occur below 60 °C. However, healing is possible at 50 °C, provided there is sufficient energy to bring furans and maleimides in contact with each other across the cut surface, because the reversible nature of the Diels–Alder reaction means that bonds are continually being formed and broken to keep the reaction at equilibrium.

Atom transfer radical polymerization was used to synthesize polymethacrylates with pendant furans and known molecular weights, which were then crosslinked with bismaleimide.[117] The Diels–Alder bonds were shown to be thermally mendable and fully reversible using Fourier Transform Infrared (FT-IR) spectroscopy and differential scanning calorimetry (DSC).

The most recent work in the area of remendable polymers based on the Diels–Alder reaction used dicyclopentadiene-based monomer. Upon heating of the monomer to 120 °C, the dicyclopentadiene opened via retro-Diels–Alder reaction, allowing the monomer to react with other cyclopentadienes. This formed a polymer backbone. Crosslinking of the system occurred through further reaction of backbone dicyclopentadiene with cyclopentadiene. The maximum healing efficiency achieved was 46%.[75]

9.3.6 Recyclable Thermosets

One disadvantage inherent in thermosetting polymers is the inability to reuse them. Once a thermoset has cured, it cannot be made to flow or mold to a different shape. However, if crosslinks are formed through reversible bonds, in theory the crosslinks can be removed generating a flowable material that can be recycled. Recyclable thermosets employing a number of reactions, including the Diels–Alder reaction of furan and maleimide, were proposed by Nakano *et al.*[118] Higher molecular weight monomer units were used because one condition of the material was that the properties of the base polymer not be compromised by recyclability. The polymer can be recycled by heating and filtering the system or by extracting monomer with solvent. Numerous other systems have been studied and are described below.

A random copolymer of styrene and furan-functionalized styrene was prepared and crosslinked with bismaleimide. Its recyclability was evaluated by heating at 130 °C in a solution of 2-methylfuran. 2-Methylfuran acted as a trap, capturing the bismaleimide and preventing re-crosslinking.[18]

Watanabe and Yoshie developed recyclable polymers from furan-terminated poly(ethylene adiapate) and bis- and tris-maleimides.[19] The copolymer of furan-terminated poly(ethylene adiapate) and bismaleimide was a linear polymer, while the copolymer containing trismaleimide was a crosslinked network. Heating of both to 145 °C for 20 min reversed the Diels–Alder linkages. The linear polymer was found to be recyclable at least four times and the crosslinked network was recyclable at least eight times.

Diels–Alder crosslinking materials have also been prepared from renewable resources. Reinecke and Ritter developed a dieneophile-modified unsaturated oligoester that formed branched networks when reacted with trisdienes and formed crosslinked networks when reacted with tetradienes.[119] Furfuryl amides and sorbic acid were both used successfully as dienes. No mention was made as to the reversibility of the network. Laita *et al.* were also interested in using the Diels–Alder reaction to develop novel polymers from renewable resources.[15] They prepared polyurethanes and acrylic copoylmers bearing furan groups. Crosslinking with bismaleimides occurred; however, the retro-Diels–Alder reaction did not take place, ostensibly because of aromatization of the Diels–Alder adduct.

9.3.7 Smart Materials

A number of polymer networks have been designed that take advantage of the reversible nature of the Diels–Alder reaction to have a thermally controllable character. The change can be an optical property or conductivity or the formation of a new type of polymer network.

Dendrimers, repeatedly branched macromolecules, have been proposed for use as sensors and as drug delivery systems. Reversibly forming dendrimers are desirable for drug delivery because they can be designed to reach a desired target and, upon heating, break apart to release the chemical payload. Although the Diels–Alder chemistry was used previously to prepare dendrimers,[63–65] McElhanon and Wheeler were the first to develop a thermally reversible dendrimer.[17] Up to third-generation dendrimers were reported based on substituted maleimides and disubstituted furans as the dendrons. Dendrimers were observed to break apart upon heating and reform when cooled using ^{1}H NMR. In their work, the formation of Diels–Alder adducts was much slower than its cleavage.

Luo *et al.* prepared smart nonlinear optical (NLO) polymers with Diels–Alder-controlled crosslinks.[41] Organic NLO polymers are desirable for application in electro-optic devices with broad bandwidths and low drive voltages. However, they suffer from low thermal stability and solvent resistance, both problems that could be solved by crosslinking. Polymers chains were functionalized by maleimide and furan, with the maleimide initially protected by a furan. Figure 9.5 shows the reaction scheme for such a system. When crosslinking was desired, the system was heated to 125 °C for 30 min (deprotecting the maleimide) and subsequently cooled (forming the crosslinks between chains). The resulting crosslinked material retained approximately 80% of its electro-optic coefficient, r_{33}, value and was significantly harder and more thermally stable.

Thermoreversible fluorescence has been achieved by functionalizing chromophores with maleimide. Reacting maleimide-bearing chromophores with furan turned on the fluorescence, and the retro-Diels–Alder reaction was used to turn off fluorescence. Additionally,

Figure 9.5 *Schematic of a polymer with thermally controlled crosslinking. When crosslinking is desired, the system is heated and dienophiles are unprotected, allowing them to react with diene-bearing linear polymers. Dienes are shown as notched trapezoids, dienophiles are shown as triangles and Diels–Alder adducts are shown as trapezoids.*

reaction of the chromophore with furfuryl alcohol resulted in a self-organizing structure. At a low (1.0 mg mL^{-1}) concentration, spherical aggregates were formed, and at a high (10 mg mL^{-1}) concentration, dendrimers were recovered. Upon heating to 75–80 °C, the structures disappeared, but would reappear after stirring at room temperature for 6 h.[62]

Kim *et al.* combined these two research areas and developed maleimide-functionalized NLO chromophores that react with polymers bearing pendant anthryl groups.[120] Both crosslinking and linear systems were synthesized, although crosslinked systems are more desirable for reasons mentioned previously. Changing the attaching mode of chromophores on the polymer controls the material's macromolecular architecture. No mention was made regarding reversibility of maleimide–anthryl linkages.

Stimuli-sensitive films are designed to respond to a specific stimulus, such as temperature, autonomously and in a useful way for the given application. Costanzo *et al.* developed a PEG film containing gold nanoparticles that 'bloom' to the surface upon heating.[47] Figure 9.6 is a schematic representation of this system's behavior. Migration of the gold nanoparticles destroyed weak long-range order within the film. Gold nanoparticles were functionalized with a diblock copolymer of maleimide-terminated polystyrene and furan-terminated PEG. Increased temperature cleaved the furan-terminated PEG from the gold nanoparticle and made the gold immiscible in the PEG film, causing gold migration.

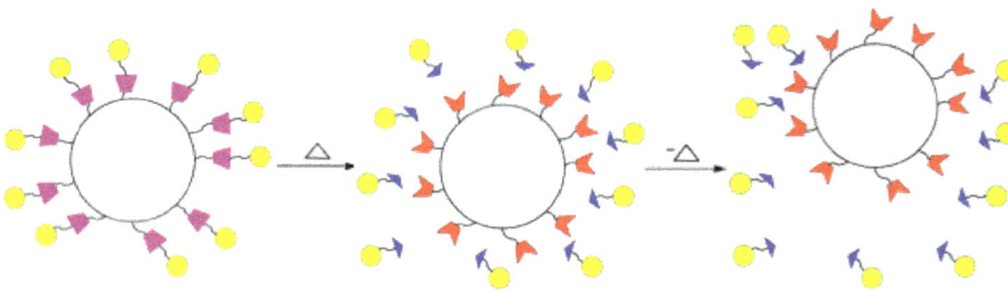

Figure 9.6 *Diels–Alder functionalized particles with controlled dispersion. Upon heating, PEG ligands cleave from the particles, making them immiscible in the system and causing blooming and aggregation. Dienes are shown as notched trapezoids, dienophiles are shown as triangles and Diels–Alder adducts are shown as trapezoids.*

Maleimide-functionalized gold nanoparticles were also prepared by Zhu *et al.* and used in the formation of monolayer protected nanoparticle (MPN) networks.[91] Like Costanzo *et al.*, Zhu *et al.* prepared maleimide-functionalized gold nanoparticles where the maleimide was initially protected with a furan-bearing group. Furan can be removed by heating to 100–110 °C for 12 h and subsequently washing. MPN networks were formed by mixing maleimide- and furan-functionalized gold nanoparticles. These networks, which appeared as large insoluble aggregates, broke apart and became miscible upon heating to 100 °C for 1 h. This reversibility was possible even after 30 cycles.

Costanzo and Beyer reported an optically active PEG film containing silica nanoparticles.[48] These silica nanoparticles were functionalized in a similar fashion to the gold nanoparticles from Costanzo *et al.* and dispersed throughout the PEG film. Upon heating, furan-bearing PEG ligands broke off from maleimide-functionalized silica nanoparticles. Silica nanoparticles then aggregated as a result of their immiscibility in PEG, changing the film from clear to opaque.

9.4 Conclusions

When developing a polymer network based on the Diels–Alder reaction, there are a number of parameters that must be considered, such as operating and reversing temperatures, concentration and location of Diels–Alder bonds, the type of Diels–Alder bonds, and base polymer material properties. These parameters, along with others that have been discussed, determine the kinetic, thermodynamic and thermomechanical behavior of the material. Although the Diels–Alder reaction is considered to be a classic click chemistry, there continue to be new and exciting applications of the Diels–Alder reaction to polymer networks. The range of dienes and dienophiles available as well as the extensive characterization of this reaction in the literature make the Diels–Alder reaction highly desirable for application in novel materials. However, to be able to fully exploit the reversible nature of the reaction in relation to polymer properties and functionality, more work is needed to understand the temperature-dependent equilibrium behavior of these systems, as well as the influence of diffusion limitation imposed by the macromolecular nature of the materials conceived.

References

(1) Hartmuth, C., Kolb, M. G. F., (2001), Click chemistry: diverse chemical function from a few good reactions, *Angewandte Chemie International Edition*, **40** (11), 2004–2021.

(2) Sohár, P., Miklós, F., Csampai, A., Stájer, G., (2001), Preparation of pyrimido[2,1-a]phthalazines and an aminopyrimido[2,1-a]isoindole by retro Diels–Alder reaction, *Journal of the Chemical Society, Perkin Transactions*, **1** (5), 558–564.

(3) Stajer, G., Szabo, A., Sohar, P., Csampai, A., Sillanpaa, R., (2006), A retro Diels–Alder method for the preparation of pyrrolo[1,2-a]pyrimidinediones from diexo-aminooxanorbornenecarboxamide, *Journal of Molecular Structure*, **784** (1–3), 239–243.

(4) Durmaz, H., Colakoglu, B., Tunca, U., Hizal, G., (2006), Preparation of block copolymers via Diels Alder reaction of maleimide- and anthracene-end functionalized polymers. *Journal of Polymer Science Part A: Polymer Chemistry*, **44** (5), 1667–1675.

(5) Durmaz, H., Karatas, F., Tunca, U., Hizal, G., (2006), Preparation of ABC miktoarm star terpolymer containing poly(ethylene glycol), polystyrene, and poly(tert-butylacrylate) arms by

combining Diels–Alder reaction, atom transfer radical, and stable free radical polymerization routes, *Journal of Polymer Science Part A: Polymer Chemistry*, **44** (1), 499–509.

(6) Durmaz, H., Karatas, F., Tunca, U., Hizal, G., (2006), Heteroarm H-shaped terpolymers through the combination of the Diels–Alder reaction and controlled/living radical polymerization techniques, *Journal of Polymer Science Part A: Polymer Chemistry*, **44** (13), 3947–3957.

(7) Nandivada, H., Jiang, X., Lahann, J., (2007), Click chemistry: versatility and control in the hands of materials scientists, *Advances in Materials*, **19** (17), 2197–2208.

(8) Choi, C. K., Tomita, I., Endo, T., (2000), Synthesis of novel pi-conjugated polymer having an enyne unit by palladium-catalyzed three-component coupling polymerization and subsequent retro-Diels–Alder reaction, *Macromolecules*, **33** (5), 1487–1488.

(9) Edelmann, D., Ritter, H., (1993), Synthesis of telechelics with furanyl end-groups by radical polymerisation with azo-initiators and network formation with unsaturated polyesters via Diels–Alder additions, *Makromolekular Chemie*, **194**, 1183.

(10) Wassermann, A., (1965), *Diels–Alder Reactions*. Elsivier Science: Amsterdam.

(11) Francesco Finguelli, A. T. (Ed.), (2002), *The Diels–Alder Reaction: Selected Practical Methods*. John Wiley & Sons Inc.: New York.

(12) Nicolaou, K. C., Snyder, S. A., Montagnon, T., Vassilikogiannakis, G., (2002), The Diels–Alder reaction in total synthesis, *Angewandte Chemie International Edition*, **41** (10), 1668–1698.

(13) Gandini, A., Belgacem, M. N., (1997), Furans in polymer chemistry, *Progress in Polymer Science*, **22**, 1203.

(14) Dewar, M. J. S., Pierini, A. B., (1984), Mechanism of the Diels–Alder reaction. Studies of the addition of maleic anhydride to furan and methylfurans, *Journal of the American Chemical Society*, **106** (1), 203–208.

(15) Laita, H., Boufi, S., Gandini, A., (1997), The application of the Diels–Alder reaction to polymers bearing furan moieties 1. Reactions with maleimides, *European Polymer Journal*, **33** (8), 1203.

(16) Craven, J. M., (1969), Cross-linked thermally reversible polymers produced from condensation polymers with pendant furan groups cross-linked with maleimides. Patent, 3,435,003, 5.

(17) McElhanon, J., Wheeler, D., (2001), Thermally responsive dendrons and dendrimers based on reversible furan-maleimide diels–alder adducts, *Organic Letters*, **3** (17), 2681–2683.

(18) Goussé, C., Gandini, A., Hodge, P., (1998), Application of the Diels–Alder reaction to polymers bearing furan moieties. 2. Diels–Alder and retro-Diels–Alder reactions involving furan rings in some styrene copolymers, *Macromolecules*, **31** (2), 314–321.

(19) Watanabe, M., Yoshie, N., (2006), Synthesis and properties of readily recyclable polymers from bisfuranic terminated poly(ethylene adipate) and multi-maleimide linkers, *Polymer*, **47** (14), 4946–4952.

(20) Liu, Y., Chen, Y., (2007), Thermally reversible cross-linked polyamides with high toughness and self-repairing ability from maleimide- and furan-functionalized aromatic polyamides, *Macromolecular Chemistry and Physics*, **208** (2), 224–232.

(21) Chen, X., Dam, M. A., Ono, K., Mal, A., Shen, H., Nutt, S. R., Sheran, K., Wudl, F., (2002), A Thermally re-mendable cross-linked polymeric material, *Science*, **295** (5560), 1698–1702.

(22) Chen, X., Wudl, F., Mal, A. K., Shen, H., Nutt, S. R., (2003), New thermally remendable highly cross-linked polymeric materials, *Macromolecules*, **36** (6), 1802–1807.

(23) Wudl, F., Chen, X., (2005), Thermally re-mendable cross-linked polymers. Patent, 6,933,361, 21.

(24) Mcelhanon, J., Russick, E., Wheeler, D., Loy, D., Aubert, J., (2002), Removable foams based on an epoxy resin incorporating reversible Diels–Alder adducts, *Journal of Applied Polymer Science*, **85** (7), 1496–1502.

(25) Loy, D. A., Wheeler, D. R., Russick, E., Mcelhanon, J. R., Saunders, R. S., (2002), Methods of making thermally removable epoxies. Patent, 6,337,384, 7.

(26) Loy, D. A., Wheeler, D. R., Mcelhanon, J. R., Saunders, R. S., (2002), Methods of making thermally removable polyurethanes. Patent, 6,403,753, 13.

(27) Kwart, H., King, K., (1968), The reverse Diels–Alder or retrodiene reaction, *Chemistry Review*, **68** (4), 415–447.

(28) Stevens, M. P., (1979), Crosslinking of polystyrene via pendant maleimide groups, *Journal of Polymer Science A Polymer Chemistry*, **17** (11), 3675–3685.
(29) Tesoro, G. C., Sastri, V. R., (1986), Synthesis of siloxane-containing bis(furans) and polymerization with bis(maleimides), *Industrial Engineering Chemistry: Product Research and Development*, **25** (3), 444–448.
(30) Chujo, Y., Sada, K., Saegusa, T., (1990), Reversible gelation of polyoxazoline by means of Diels–Alder reaction, *Macromolecules*, **23** (10), 2636–2641.
(31) Diakoumakos, C. D., Mikroyannidis, J. A., (1992), Polyimides derived from Diels–Alder polymerization of furfuryl-substituted maleamic acids or from the reaction of bismaleamic with bisfurfurylpyromellitamic acids, *Journal of Polymer Science Part A: Polymer Chemistry*, **30** (12), 2559–2567.
(32) Mikroyannidis, J. A., (1992), Synthesis and Diels–Alder polymerization of furfurylidene and furfuryl-substituted maleamic acids, *Journal of Polymer Science Part A: Polymer Chemistry*, **30** (1), 125–132.
(33) Mikroyannidis, J. A., (1992), Furyl-maleimide in situ generated AB-monomers: synthesis, characterization, and Diels–Alder polymerization, *Journal of Polymer Science Part A: Polymer Chemistry*, **30** (9), 2017–2024.
(34) Kuramoto, N., Hayashi, K., Nagai, K., (1994), Thermoreversible reaction of Diels–Alder polymer composed of difurufuryladipate with bismaleimidodiphenylmethane, *Journal of Polymer Science Part A: Polymer Chemistry*, **32** (13), 2501–2504.
(35) Goussé, C., Gandini, A., (1998), Synthesis of 2-furfurylmaleimide and preliminary study of its Diels–Alder polycondensation, *Polymer Bulletin*, **40**, 389–394.
(36) Kamahori, K., Tada, S., Ito, K., Itsuno, S., (1999), Optically active polymer synthesis by diels–alder polymerization with chirally modified Lewis acid catalyst, *Macromolecules*, **32** (3), 541–547.
(37) Imai, Y., Itoh, H., Naka, K., Chujo, Y., (2000), Thermally reversible IPN organic–inorganic polymer hybrids utilizing the Diels–Alder reaction, *Macromolecules*, **33** (12), 4343–4346.
(38) Bibiao, J., Jianjun, H., Wenyun, W., Luxia, J., Xinxian, C., (2001), Synthesis and properties of novel polybismaleimide oligomers, *European Polymer Journal*, **37**, 463.
(39) Small, J. H., Loy, D. A., Wheeler, D. R., Mcelhanon, J. R., Saunders, R. S., (2001), Methods of making thermally removable polymeric encapsulants. Patent, 6,271,335.
(40) Gheneim, R., Perez-Berumen, C., Gandini, A., (2002), Diels–Alder Reactions with novel polymeric dienes and dienophiles: synthesis of reversibly cross-linked elastomers, *Macromolecules*, **35** (19), 7246–7253.
(41) Luo, J., Haller, M., Li, H., Kim, T., Jen, A. K., (2003), Highly efficient and thermally stable electro-optic polymer from a smartly controlled crosslinking process, *Advances in Materials*, **15** (19), 1635–1638.
(42) Goiti, E., Huglin, M., Rego, J., (2003), Thermal breakdown by the Retro Diels–Alder reaction of crosslinking in poly[styrene-co-(furfuryl methacrylate)], *Macromolecules Rapid Communications*, **24** (11), 692–696.
(43) Adachi, K., Achimuthu, A. K., Chujo, Y., (2004), Synthesis of organic–inorganic polymer hybrids controlled by Diels–Alder reaction, *Macromolecules*, **37** (26), 9793–9797.
(44) Goiti, E., Heatley, F., Huglin, M., Rego, J. M., (2004), Kinetic aspects of the Diels–Alder reaction between poly(styrene-co-furfuryl methacrylate) and bismaleimide, *European Polymer Journal*, **40** (7), 1451–1460.
(45) Goiti, E., Huglin, M., Rego, J. M., (2004), Some properties of networks produced by the Diels–Alder reaction between poly(styrene-co-furfuryl methacrylate) and bismaleimide, *European Polymer Journal*, **40** (2), 219–226.
(46) Zhang, X., Li, Z. C., Li, K. B., Du, F. S., Li, F. M., (2004), Multi-maleimides bearing electron-donating chromophores: reversible fluorescence and aggregation behavior, *Journal of the American Chemical Society*, **126** (39), 12200–12201.
(47) Costanzo, P. J., Demaree, J. D., Beyer, F. L., (2006), Controlling dispersion and migration of particulate additives with block copolymers and Diels–Alder chemistry, *Langmuir*, **22**, 10251–10257.

(48) Costanzo, P. J., Beyer, F. L., (2007), Thermoresponsive, optically active films based on Diels–Alder chemistry, *Chemistry Materials*, **19** (25), 6168–6173.
(49) Gacal, B., Durmaz, H., Tasdelen, M. A., Hizal, G., Tunca, U., Yagci, Y., Demirel, A. L., (2006), Anthracene–maleimide-based Diels–Alder 'click chemistry' as a novel route to graft copolymers, *Macromolecules*, **39** (16), 5330–5336.
(50) Gotsmann, B., Duerig, U., Frommer, J., Hawker, C. J., (2006), Exploiting chemical switching in a Diels–Alder polymer for nanoscale probe lithography and data storage, *Advances in Functional Materials*, **16** (11), 1499–1505.
(51) Liu, Y., Hsieh, C., (2006), Crosslinked epoxy materials exhibiting thermal remendablility and removability from multifunctional maleimide and furan compounds, *Journal of Polymer Science A Polymer Chemistry*, **44** (2), 905–913.
(52) Liu, Y., Hsieh, C., Chen, Y., (2006), Thermally reversible cross-linked polyamides and thermo-responsive gels by means of Diels–Alder reaction, *Polymer*, **47** (8), 2581–2586.
(53) Teramoto, N., Arai, Y., Shibata, M., (2006), Thermo-reversible Diels–Alder polymerization of difurfurylidene trehalose and bismaleimides, *Carbohydrate Polymers*, **64** (1), 78–84.
(54) Brock, P., Chaw, M., Dawson, D., Hawker, C., Hedrick, J., Magbitang, T., McKean, D., Miller, R., Palmisano, R., Volksen, W., (2007), Stable encapsulant fluid capable of undergoing reversible Diels–Alder polymerization. Patent, 7,309,754, 16.
(55) Kraiman, E. A., (1959), Maleimide polymers, US Patent, 9 June, 2, 890, 206.
(56) Reeder, J. A., (1967), Polyimides from dimaleimides and bisfulvenes. Patent, 3,334,071, 3.
(57) Alhakimi, G., Klemm, E., Gorls, H., (1995), Synthesis of new polyimides by Diels–Alder reaction of bis(2-pyrone)s with bismaleimides, *Journal of Polymer Science Part A: Polymer Chemistry*, **33** (7), 1133–1142.
(58) Alhakimi, G., Klemm, E., (1995), Synthesis of a tetra(maleimide) as intermediate compound in a linear Diels–Alder Polyaddition of bismaleimides with bis(2-pyrone)s, *Journal of Polymer Science A Polymer Chemistry*, **33**, 4.
(59) Jones, J. R., Liotta, C. L., Collard, D. M., Schiraldi, D. A., (1999), Cross-linking and modification of poly(ethylene terephthalate-co-2,6-anthracenedicarboxylate) by Diels–Alder Reactions with Maleimides, *Macromolecules*, **32** (18), 5786–5792.
(60) Grigoras, M., Colotin, G., (2001), Copolymerization of a bisanthracene compound with bismaleimides by Diels–Alder cycloaddition, *Polymer International*, **50**, 1375–1378.
(61) Vargas, M., Kriegel, R. M., Collard, D. M., Schiraldi, D. A., (2002), Diels–Alder modification of poly(ethylene terephthalate-co-anthracene-2,6-carboxylate) with *N*-substituted maleimides, *Journal of Polymer Science Part A: Polymer Chemistry*, **40** (19), 3256–3263.
(62) Kim, C., Kim, H., Park, K., (2003), Diels–Alder reaction of anthracene and *N*-ethylmaleimide on the carbosilane dendrimer, *Journal of Organometallic Chemistry*, **667**, (1–2), 96–102.
(63) Morgenroth, F., Müllen, K., (1997), Dendritic and hyperbranched polyphenylenes via a simple Diels–Alder route, *Tetrahedron*, **53** (45), 15349.
(64) Morgenroth, F., Kübel, C., Müller, M., Wiesler, U. M., Berresheim, A. J., Wagner, M., Müllen, K., (1998), From three-dimensional polyphenylene dendrimers to large graphite subunits, *Carbon*, **36** (5–6), 833.
(65) Morgenroth, F., Berresheim, A. J., Wagner, M., Müllenö, K., (1998), Spherical polyphenylene dendrimers via Diels–Alder reactions: the first example of an A4B building block in dendrimer chemistry, *Chemical Communications*, (10), 1139.
(66) Yousaf, M. N., Mrksich, M., (1999), Diels–Alder reaction for the selective immobilization of protein to electroactive self-assembled monolayers, *Journal of the American Chemical Society*, **121** (17), 4286–4287.
(67) Yousaf, M. N., Chan, E. W. L., Mrksich, M., (2000), The kinetic order of an interfacial diels–alder reaction depends on the environment of the immobilized dienophile, *Angewandte Chemie International Edition*, **39** (11), 1943–1946.
(68) Kwon, Y., Mrksich, M., (2002), Dependence of the rate of an interfacial Diels–Alder reaction on the steric environment of the immobilized dienophile: an example of enthalpy–entropy compensation, *Journal of American Chemical Society*, **124** (5), 806–812.
(69) Gawalt, E. S., Mrksich, M., (2004), A Substituent Effects Study Reveals the Kinetic Pathway for an Interfacial Reaction, *Journal of American Chemical Society*, **126** (47), 15613–15617.

(70) Upson, R. W., (1955), P-Xylene Bis(5-cyclopentadiene) and Polymer. Patent, 2,726,232, 4.
(71) Herndon, W. C., Grayson, C. R., Manion, J. M., (1967), Retro-Diels–Alder reactions. III. Kinetics of the thermal decompositions of exo- and endo-dicyclopentadiene, *Journal of Organic Chemistry*, **32** (3), 526–529.
(72) Kennedy, J. P., Castner, K. F., (1979), Thermally reversible polymer systems by cyclopentadienylation. II. The synthesis of cyclopentadiene-containing polymers, *Journal of Polymer Science: Polymer Chemistry Edition*, **17** (7), 2055–2070.
(73) Kennedy, J. P., Castner, K. F., (1979), Thermally reversible polymer systems by cyclopentadienylation. I. A model for termination by cyclopentadienylation of olefin polymerization, *Journal of Polymer Science: Polymer Chemistry Edition*, **17** (7), 2039–2054.
(74) Kennedy, J., Carlson, G., (1983), Synthesis, characterization, and Diels–Alder extension of cyclopentadiene telechelic polyisobutylene. IV. alpha,omega-Di(3-cyclopentadienylpropyldimethylsilyl) polyisobutylene, *Journal of Polymer Science: Polymer Chemistry Edition*, **21** (12), 3551–3561.
(75) Murphy, E. B., Bolanos, E., Schaffner-Hamann, C., Wudl, F., Nutt, S. R., Auad, M. L., (2008), Synthesis and characterization of a single-component thermally remendable polymer network: Staudinger and Stille revisited, *Macromolecules*, **41** (9), 3169–3174.
(76) Stille, J. K., Plummer, L., (1961), Polymerization by the Diels–Alder reaction, *Journal of Organic Chemistry*, **26** (10), 4026–4029.
(77) Takeshita, Y., Uoi, M., Hirai, Y., Uchiyama, M., (1974). Thermoplastic plastic composition. Patent, 3,826,760, 9.
(78) Miura, M., Akutsu, F., Ursui, T., Ikebukuro, Y., Nagakubo, K., (1985), Soluble cyclopentadienylated polymers, *Makromolekulare Chemie*, **186**, 473–481.
(79) Salamone, J. C., Chung, Y., Clough, S. B., Watterson, A. C., (1988), Thermally reversible, covalently crosslinked polyphosphazenes, *Journal of Polymer Science Part A: Polymer Chemistry*, **26** (11), 2923–2939.
(80) Canary, S. A., Stevens, M., (1992), Thermally reversible crosslinking of polystyrene via the furan-maleimide Diels–Alder reaction, *Journal of Polymer Science A Polymer Chemistry*, **30** (8), 1755–1760.
(81) Kelen, T., Iván, B., Nagy, T. T., Turcsányi, B., Tüdös, F., Kennedy, J. P., (1978), Reversible crosslinking during thermal degradation of PVC, *Polymer Bulletin*, **1** (2), 79–84.
(82) Iván, B., Nagy, T. T., Kelen, T., Turcsányi, B., Tüdös, F., (1980), Crosslinking and scission in thermooxidative degradation of PVC, *Polymer Bulletin*, **2** (1), 83–88.
(83) Nagy, T. T., Iván, B., Turcsányi, B., Kelen, T., Tüdös, F., (1980), Crosslinking, scission and benzene formation during PVC degradation under various conditions, *Polymer Bulletin*, **3** (11), 613–620.
(84) Neukam, W., Grimme, W., (1978), Anionic [4+2]-cycloreversions leading to the cyanocyclopentadienide ion, *Tetrahedron Letters*, **19** (25), 2201.
(85) Kwart, H., Burchuk, I., (1952), Isomerism and adduct stability in the Diels–Alder Reaction. I. The adducts of furan and maleimide, *Journal of the American Chemical Society*, **74**, 3094–3097.
(86) Papies, O., Grimme, W., (1980), Acceleration of the [4 + 2]-cycloreversion by the alkoxide substituent, *Tetrahedron Letters*, **21**, 2799.
(87) Brand, T., Klapper, M., (1999), Control of viscosity through reversible addition of telechelics via repetitive Diels–Alder reaction in bulk, *Designed Monomers and Polymers*, **2** (4), 287–309.
(88) Edelmann, D., Ritter, H., (1993), Synthesis of telechelics with furanyl end-groups, 2: Radical polymerisation with C–C-bond splitting initiators and network formation with unsaturated polyesters via Diels–Alder additions, *Makromolekulare Chemie*, **194**, 2375–2384.
(89) Ritter, H., Sperber, R., Weisshuhn, C. M., (1993), Reactive comb-like polymers. Kinetic studies of the Diels–Alder reaction of furan-containing comb-like polymers with dimethyl butyndioate by means of 1H NMR spectroscopy, *Makromolekulare Chemie*, **194**, 1721–1731.
(90) Cowie, J. M. G., (1991), *Polymers: Chemistry & Physics of Modern Materials*. CRC Press: Boca Raton, FL, p. 431.

(91) Zhu, J., Kell, A. J., Workentin, M. S., (2006), A retro-Diels–Alder reaction to uncover maleimide-modified surfaces on monolayer-protected nanoparticles for reversible covalent assembly, *Organic Letters*, **8** (22), 4993–4996.
(92) Hao, J., Wang, W., Jiang, B., Cai, X., Jiang, L., (1999), Preparation, solubility and thermal behaviour of new bismaleimides containing silicone linkages, *Polymer International*, **48**, 235–243.
(93) Jenkins, A. D., Kratochvil, P., Stepto, R. F. T., Suter, U. W., (1996), Glossary of basic terms in polymer science, *Pure and Applied Chemistry*, **68** (12), 2287–2311.
(94) Engle, L. P., Wagener, K. B., (1993), A review of thermally controlled covalent bond formation in polymer chemistry, *Journal of Macromolecular Science: Reviews in Macromolecular Chemistry and Physics*, **33** (3), 239–257.
(95) Bergman, S., Wudl, F., (2008), Mendable polymers, *Journal of Materials Chemistry*, **18** (1), 41.
(96) Wool, R. P., (2008), Self-healing materials: a review, *Soft Matter*; DOI: 10.1039/ B711716G.
(97) Leibler, L., Rubinstein, M., Colby, R. H., (1991), Dynamics of reversible networks, *Macromolecules*, **24** (16), 4701–4707.
(98) Flory, P. J., (1941), Molecular size distribution in three dimensional polymers, *I. Gelation.* Journal of American Chemical Society, **63** (11), 3083–3090.
(99) Flory, P. J., (1941), Molecular size distribution in three dimensional polymers. II. Trifunctional branching units, *Journal of American Chemical Society*, **63** (11), 3091–3096.
(100) Stockmayer, W. H., (1943), Theory of molecular size distribution and gel formation in branched-chain polymers, *Journal of Chemistry Phys.*, **11** (2), 45–55.
(101) Stockmayer, W. H., (1944), Theory of molecular size distribution and gel formation in branched polymers II. General cross linking, *Journal of Chemistry Phys.*, **12** (4), 125–131.
(102) Williams, M. L., Landel, R. F., Ferry, J. D., The temperature dependence of relaxation mechanisms in amorphous polymers and other glass-forming liquids. *Journal of the American Chemical Society* **1955**, *77*, 3701–3707.
(103) Chow, S., John M. Whelan, J., (1961), Maleimide polymers. Patent, 2,971,944, 5.
(104) Harris, F. W., Stille, J. K., (1968), Diels–Alder polymers. Polyimides from a biscyclopentadienone and bismaleimides, *Macromolecules*, **1** (5), 463–464.
(105) Wagener, K. B., Engle, L. P., (1993), Thermally reversible polymer linkages. II. Linear addition polymers, *Journal of Polymer Science A Polymer Chemistry*, **31** (4), 865–875.
(106) Kintzel, O., Luger, P., Weber, M., Schlüter, A., (1998), Ring-chain equilibrium between an [18]cyclacene derivative and a ladder oligomer, *European Journal of Organic Chemistry*, (1), 99.
(107) Diakoumakos, C. D., Mikroyannidis, J. A., (1994), Heat-resistant resins derived from cyano-substituted Diels–Alder polymers, *European Polymer Journal*, **30** (4), 465–472.
(108) Blatter, K., Schlueter, A., (1989), Model studies for the synthesis of ribbon-shaped structure by repetitive Diels–Alder reaction, *Chemische Berichte*, **122**, 1351–1356.
(109) Blatter, K., Schlueter, A. D., (1989), Ribbon-shaped structures via repetitive Diels–Alder reaction. A polycatafusene, *Macromolecules*, **22** (8), 3506–3508.
(110) Sedghat-Herati, M. R., (2000), Synthesis and characterization of 5-[methoxypoly(oxyethylene)]-(3*E*)-1,3-pentadiene and its Diels–Alder reactions, *Macromolecules*, **33** (6), 1924–1925.
(111) Hopkins, T. E., Park, M., Chacon, A., Sedaghat-Herati, R., (2005), New poly(oxyethylene) derivatives and their oligo analogues from Diels–Alder reactions of 5-[methoxypoly(oxyethylene)]-(3*E*)-1,3-pentadiene and 5-methoxyethoxy-(3*E*)-1,3-pentadiene, *Journal of Polymer Science Part A: Polymer Chemistry*, **43** (9), 1895–1902.
(112) Goiti, E., Huglin, M., Rego, J. M., (2001), Some observations on the copolymerization of styrene with furfuryl methacrylate, *Polymer*, **42**, 10187–10193.
(113) Peterson, A. M., Jensen, R. E., Palmese, G. R., (2009), Reversibly cross-linked polymer gels as healing agents for epoxy-amine thermosets, *ACS Applied Materials & Interfaces*, **1(5)**, 00–00.
(114) Plaisted, T. A., Nemat-Nasser, S., (2007), Quantitative evaluation of fracture, healing and re-healing of a reversibly cross-linked polymer, *Acta Materialia*, **55** (17), 5684–5696.

(115) White, S. R., Sottos, N. R., Geubelle, P. H., Moore, J. S., Kessler, M., Sriram, S. R., Brown, E. N., Viswanathan, S., (2001), Autonomic healing of polymer composites, *Nature*, **409** (6822), 794–797.

(116) Plaisted, T. A., Amirkhizi, A. V., Arbelaez, D., Nemat-Nasser, S. C., Nemat-Nasser, S., (2003), Self-healing Structural Composites with Electromagnetic Functionality, *Proceedings of SPIE*, pp. 372–381.

(117) Kavitha, A. A., Singha, N. K., (2007), A tailor-made polymethacrylate bearing a reactive diene in reversible Diels–Alder reaction, *Journal of Polymer Science A Polymer Chemistry*, **45** (19), 4441–4449.

(118) Nakano, M., Usuki, A., Okada, A., Kamigaito, O., (1997), Recyclabe polymer, process for producing the same, method for recovering the same, and method for regenerating the same. Patent, 5,643,998, 16.

(119) Reinecke, M., Ritter, H., (1993), Renewable resources, 1: branching and crosslinking of an unsaturated oligoester with furfurylamides and sorbic acid amides via Diels–Alder additions, *Makromolekulare Chemie*, **194**, 2385–2393.

(120) Kim, T., Luo, J., Tian, Y., Ka, J., Tucker, N. M., Haller, M., Kang, J., Jen, A. K., (2006), Diels–Alder 'click chemistry' for highly efficient electrooptic polymers, *Macromolecules*, **39** (5), 1676–1680.

10

Click Chemistry in the Preparation of Biohybrid Materials

Heather J. Kitto, Jan Lauko, Floris P. J. T. Rutjes and Alan E. Rowan

10.1 Introduction

From a relatively small reservoir of molecules, Nature is able to construct an incredibly wide range of materials that have highly specific functions. Scientists have made extensive use of many natural building blocks to construct materials for applications in medicine and nanotechnology. Molecules, such as peptides, proteins, carbohydrates and lipids, can be assembled into large arrays or can be covalently attached to other biomolecules or synthetic scaffolds, such as polymers, viruses or dendrimers, to generate biohybrid materials with a diverse selection of new and improved properties. The formation of biohybrid materials is motivated by numerous structural and functional reasons and involves the exploitation of biological systems at various levels of their natural hierarchical organization, these being monomeric, oligomeric, polymeric or supramolecular. Building blocks, such as nucleobases, oligonucleotides and oligopeptides, are coupled to scaffolds because of their exceptional capability to self-organize – a desirable property for the creation of highly ordered synthetic nanomaterials – whereas those of higher complexity, such as proteins or enzymes, are used more for their biological properties.[1]

The synthesis of biohybrid materials requires methods that are both selective and biocompatible to ensure the primary properties of the building blocks are retained in the final structure. Numerous approaches have been developed with varying degrees of success. The most critical aspect for the synthesis of biohybrid materials is that the reaction is orthogonal in nature, that is, that it only occurs at the desired site with no side reactions. The copper-catalyzed azide–acetylene cycloaddition reaction (CuAAC) has been found to

Click Chemistry for Biotechnology and Materials Science Edited by Joerg Lahann

be ideal for this purpose.[2–4] This reaction is efficient at linking two components, avoiding the construction of large and complicated structures that would require the formation of difficult bonds. The CuAAC reaction is high yielding, highly specific, regioselective and has a high tolerance to other functional groups present within the components. It also allows reactions to proceed in many different solvents, including aqueous media, without the loss of efficiency, and is easy to work up. The versatility and scope of the CuAAC reaction has given enormous potential for the construction of new biohybrid materials as the relative unreactivity of azides and alkynes towards most functional groups ensures that bioconjugation occurs only at the desired positions.[5] This reaction has, therefore, been used extensively for biohybridization in the fields of materials and medicinal sciences in recent years and has been widely reviewed.[1,5–16]

10.2 Polymer-containing Biohybrid Materials

The use of polymers of both natural origin (optimized by billions of years of evolutionary changes) and synthetic design (simpler, but chemically more diverse) is an appealing method for the preparation of biohybrid molecules as they generate materials with interesting and diverse properties. Polymers of various chemical and topological compositions (linear, branched, stars) can be prepared allowing a great deal of freedom to material scientists. The combination of biomacromolecules with synthetic polymers is an attractive method for increasing the *in vivo* and *in vitro* applications of these compounds. The conjugation of polymers with proteins alters the solubility and surface properties of the protein and therefore affects its stability, activity and biocompatibility. The applications for these types of biohybrid compounds include the areas of biosensors,[17,18] artificial enzymes,[19] biometrics,[17,18] photonics[20] and nano-electronic devices.[21,22] The appendage of polymers with an alkyne or an azide moiety allows for their post modification by click chemistry with various materials, such as biomacromolecules (proteins, nucleic acids, polysaccharides, etc.), and opens the door to a vast range of possible biohybrid materials. The decorating of polymers with alkyne and azide functionalities in preparation for clicking can occur by different 'controlled' approaches. End-functionalized polymers can be obtained by the polymerization of a monomer using a functional initiator or by the conversion of an existing functional group of a polymer into an azide or alkyne. Side arm appended clickable polymers can be prepared by the polymerization of azide or (protected) alkyne containing monomers using controlled polymerization techniques to generate a variety of well-defined homo- and block copolymers.[6]

10.2.1 Polymers from Controlled Techniques

One of the first reports, in which controlled polymerization techniques were used in conjunction with click chemistry for the construction of biohybrid materials, was given by Opsteen and Van Hest.[23] Using atom transfer radical polymerization (ATRP), a bromide functionalized polystyrene (PS) chain was obtained, which was then converted into the corresponding azide terminated molecule [Figure 10.1(a)]. The azide reacted under CuAAC conditions with several alkyne-functionalized polymers including poly(ethylene glycol) (PEG), PS and poly(methyl methacrylate) (PMMA) to generate the desired block copolymers in a

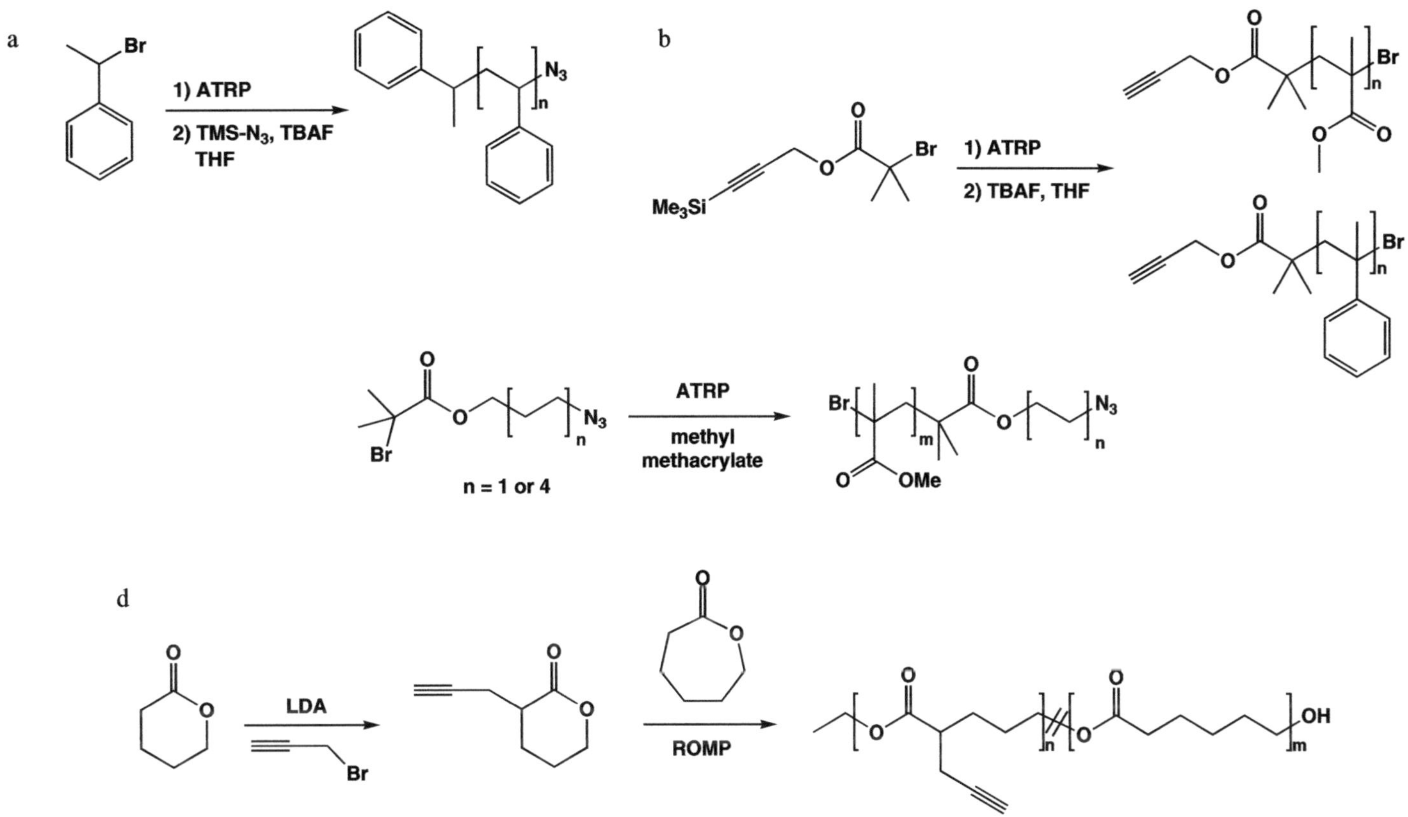

Figure 10.1 *Examples of end- and side-functionalized polymers formed by controlled polymerization techniques for use in the CuAAC reaction from the work of (a, b) Opsteen and van Hest,[23] (c) Haddleton[25] and (d) Emrick.[26]*

modular fashion and in high yields. Alkyne-appended PS and PPMA blocks were also prepared by ATRP, but in this case a functional initiator, a trimethylsilane-protected acetylene, was used [Figure 10.1(b)]. Matyjaszewski and coworkers have, however, reported a case, in which an unprotected acetylene initiator was successfully used.[24] Azide-functionalized initiators are less commonly employed, but have been used by Haddleton and coworkers to prepare azide terminated PMMA [Figure 10.1(c)].[25] After the ATRP was complete, the azide moiety was further functionalized by click chemistry in one-pot setup using the same catalyst for both processes. The group of Emrick, using controlled anionic ring-opening polymerization, prepared side-arm functionalized aliphatic polyesters [Figure 10.1(d)].[26] The homo- and copolymerization of α-propargyl-δ-valerolactone generated polymers that were clicked to PEG and peptide functional groups. The resulting amphiphilic materials were shown to be biocompatible by *in vitro* cytotoxicity evaluation and have the potential for use in biomaterial delivery applications. Later work by the same authors has resulted in the formation of biocompatible zwitterionic polymers by the clicking of the same homopolymer to phosphorylcholine (PC) moieties.[27]

A large number of azide and alkyne-containing end- and side-functionalized polymers, synthesized by controlled polymerization techniques, are presented in Table 10.1. Further details of some specific examples, with their relevant bioconjugation applications, will be given in later sections.

10.2.2 Bio-inspired Polymers via Click Chemistry

Controlled polymerization techniques can, in some cases, limit the choice of monomers that can be employed for this purpose and consequently limit the diversity of the polymer backbone composition. Other techniques can give access to materials with more bio-inspired structures based on polypeptides or DNA. These materials are of growing interest because of their potential applications as drug delivery systems, scaffolds for tissue engineering and repair, and protein mimics.[46–49] Based on short peptides, the polyisocyanides form well-defined, stable β-helical architectures due to the presence of a hydrogen-bonding chain parallel to the covalent polymer backbone. The hydrogen bonding network rigidifies the array resulting in extremely stiff polymers as seen in the AlaAlaOMe polyisocyanide (L,L-PIAA) shown in Figure 10.2(a).[50,51] The polyisocyanides have a 4_1 helical conformation (i.e. four repeat units per helical turn) with an average spacing between the side chain n and $(n + 4)$ of 4.7 Å [Figure 10.2(b)]. The rigidity, in conjunction with the highly organized arrangement of the side arms makes the polyisocyanides favorable materials as scaffolds for the arrangement of many types of chemical motifs, such as biomolecules or fluorescent markers. The rigid polymers can be readily functionalized with azide or acetylene groups, both at end and side arm positions, which allows for post-modification of the rigid polymers with a wide variety of functional moieties. The groups of Rowan and Nolte prepared end-functionalized polyisocyanides for clicking using functionalized nickel initiators.[52] The use of a functional initiator was previously demonstrated in the formation of poly(styrene)–poly(isocyanide) block copolymers; the isocyanide monomer was polymerized by a poly(styrene)-functionalized nickel initiator.[53] The polymerization of the AlaAlaOMe isocyanide monomer with an azide functionalized nickel initiator resulted in azide end-appended polymers, which on reaction with a coumarin dye were found to emit strong fluorescence [Figure 10.2(c)].

Table 10.1 *Azide and alkyne-containing end- and side-functionalized polymers from controlled polymerization techniques*

Entry	Polymer	Polymerization method	References
1	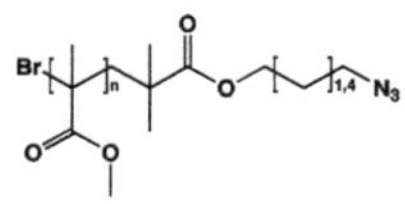	ATRP	25
2	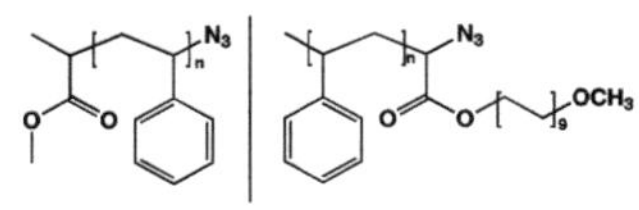	ATRP	28, 29
3	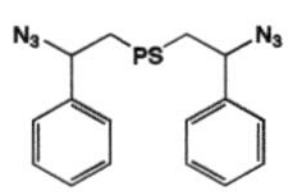	ATRP	30
4	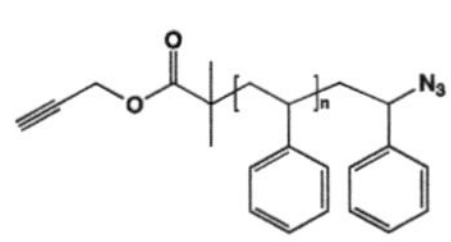	ATRP	24, 31
5	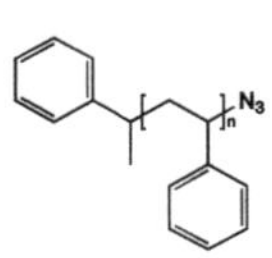	ATRP	32
6	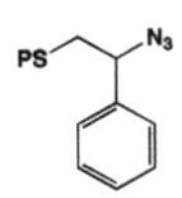	ATRP	33
7	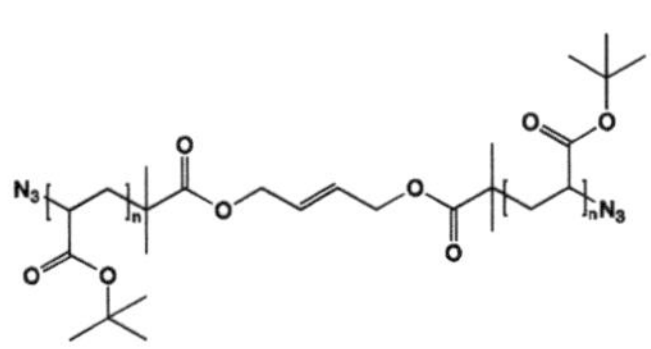	ATRP	34
9	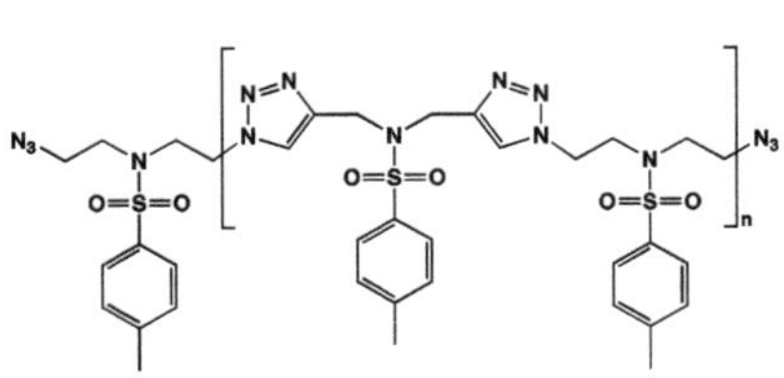	Polyaddition	35
10	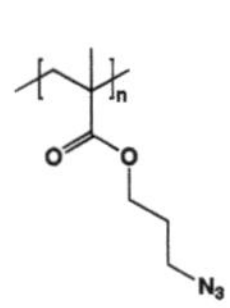	ATRP	36

(continued)

Table 10.1 (*Continued*)

Entry	Polymer	Polymerization method	References
12	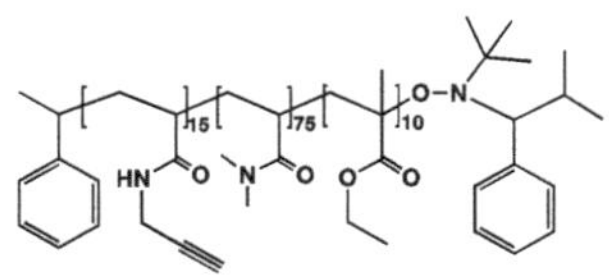	Nitroxide living-radical polymerisation	37, 38
13	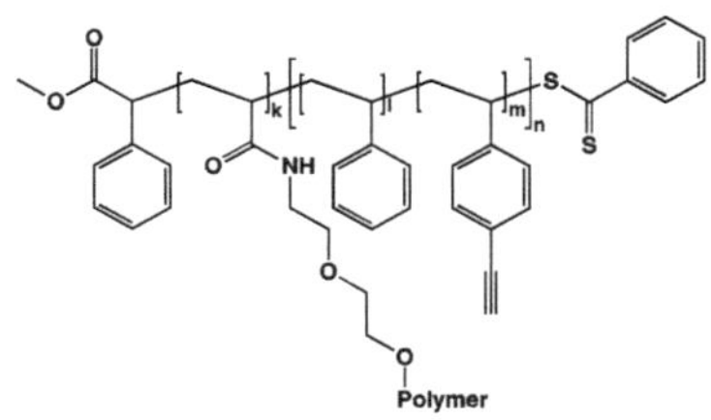	RAFT	38
15	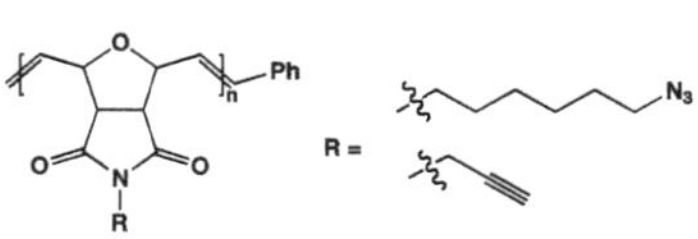	ROMP	39
16	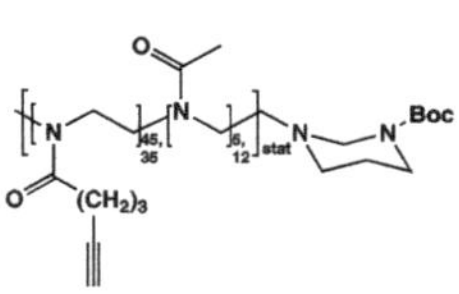	Living cationic ring-opening polymerisation	40, 41
17	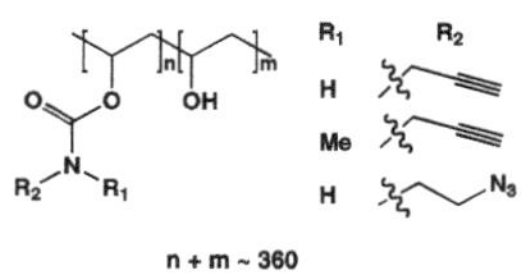	Free-radical polymerisation	42
18	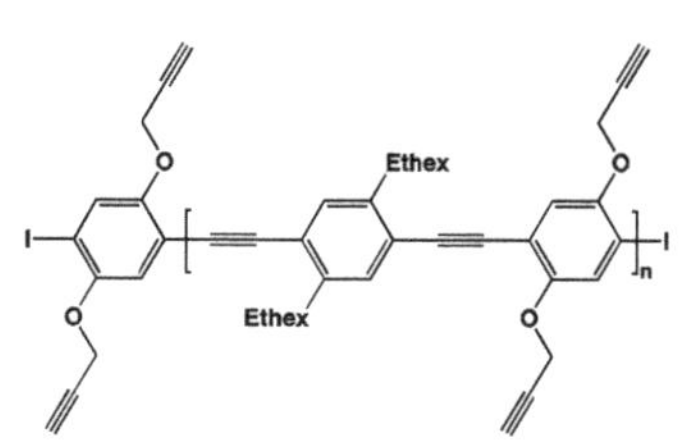	polyaddition	43
19	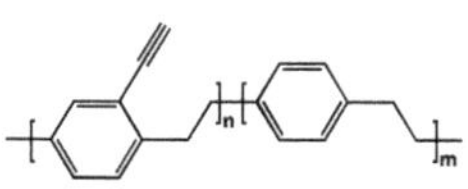	CVD polymerisation	44
20	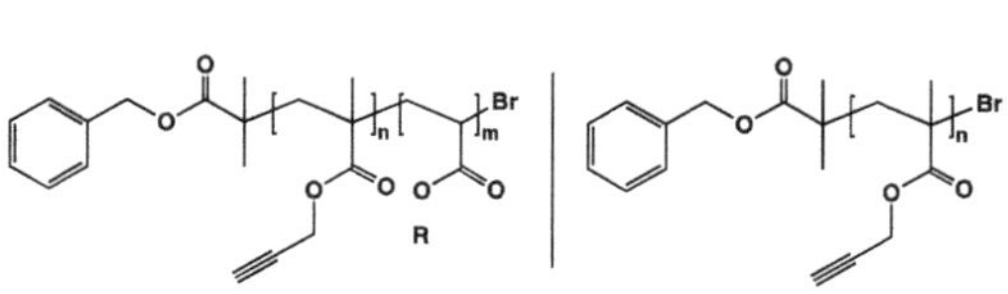	ATRP	45

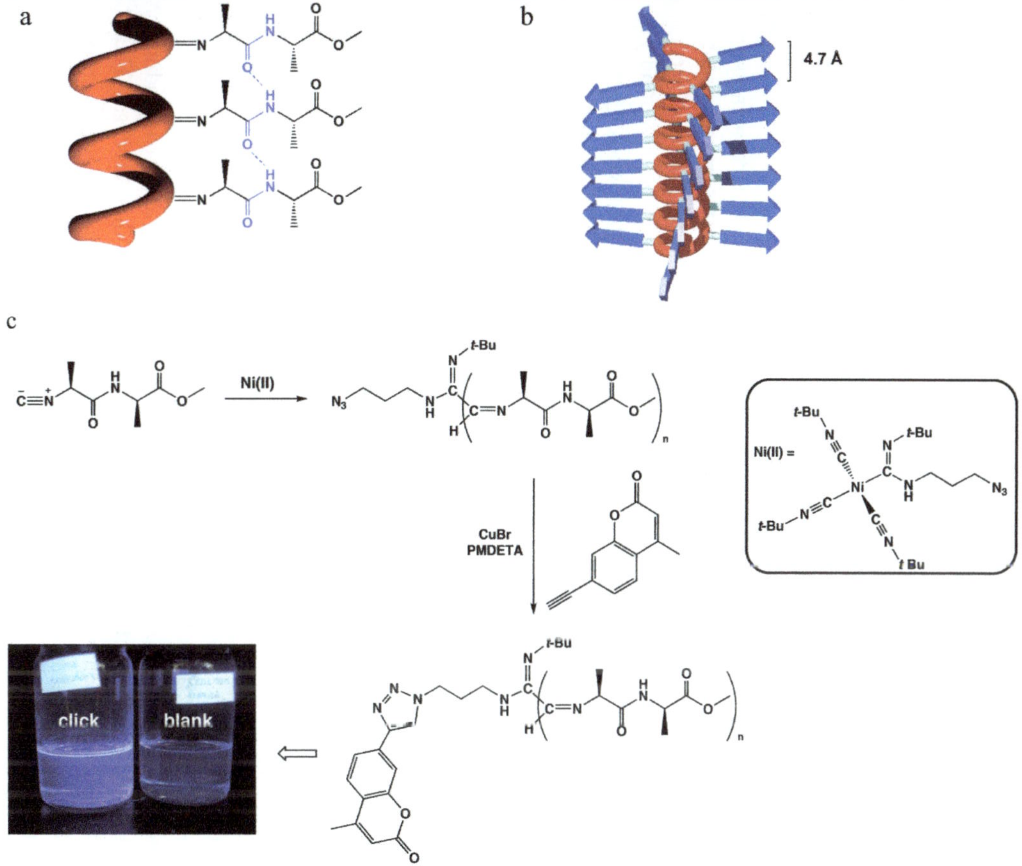

Figure 10.2 *Schematic representations of (a) the hydrogen-bonding network present between the alanine units of the side arms in L,L-PIAA (Reprinted with permission from ref.[50]. Copyright 2001 AAAS) and (b) the 4_1 helical conformation found in polyisocyanides (Reprinted with permission from ref.[51]. Copyright 2002 American Chemical Society). (c) Polymerization of an isocyanide monomer with a functionalized nickel initiator and subsequent clicking with coumarin dye to yield a highly fluorescent polymer.*

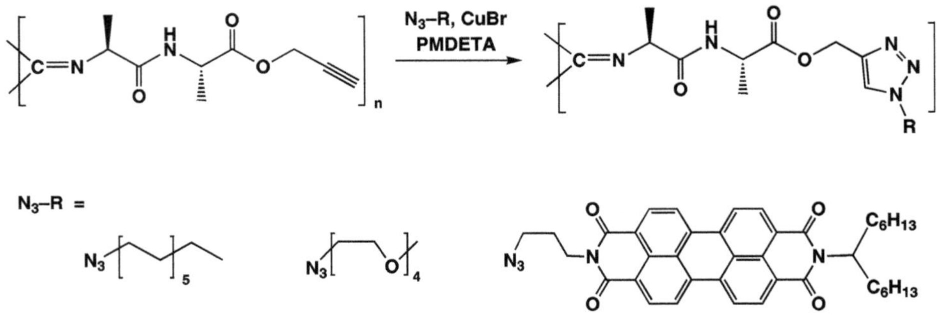

Figure 10.3 The click reaction of L,L-PIAAPE to form highly functionalized side arm appended polyisocyanides. Reprinted with permission from refs.[54] and [55]. Copyright 2007 and 2008 respectively, Royal Society of Chemistry.

Rowan, Nolte and coworkers have also shown that isocyano peptide monomers containing acetylene side arms can be effectively polymerized under nickel conditions generating the appended polymers.[54,55] The polymer, poly(L-isocyanoalanyl-L-alanine prop-2-ynol ester) (L,L-PIAAPE) was easily formed and isolated; however, it was found to be only mildly soluble in chlorinated solvents.[54] The solubility was greatly increased on clicking with azide functionalized aliphatic tails, namely dodecyl azide (Figure 10.3). The clicking of the same scaffold was also employed to prepare polyisocyanides decorate with ethylene glycol chains and fluorescent markers based on perylene and coumarin moieties.[55] These reactions resulted in the formation of water-soluble homopolymers from the ethylene glycol azide and fluorescent water-soluble random copolymers by the co-clicking of ethylene glycol and perylene azides. The random clicking of the polyisocyanide scaffold with a mixture of perylene and coumarin azides resulted in polymers from which both absorption and emission from each chromophore could be observed as evidence by a quenched and blue-shifted emission of the coumarin molecules in close proximity to a perylene molecule. The ability to construct, using these scaffolds, water-soluble and modular systems to which other biomacromolecules can be readily attached offers the possibility of many interesting biologically relevant materials.

The need for reliable DNA sequencing and detection methods is important for the diagnosis of pathogenic and genetic disorders and is therefore an area being extensively researched. A variety of methods are available for the sequencing of specific DNA strands,[56] including the incorporation and detection of fluorescently tagged nucleoside building blocks.[57] The enzymatic replacement of each natural building block with a fluorescently tagged analog can, however, be a challenging exercise. An alternative method involves the incorporation of appropriately modified nucleoside building blocks into DNA strands that can then be post functionalized. Following this idea, Carell and coworkers employed solid-phase DNA synthesis to prepare alkyne-functionalized oligodeoxyribonucleotides (ODNs) for CuAAC post functionalization.[58,59] The modified uridine nucleosides (Figure 10.4) were incorporated into a series of 16-mer ODNs and clicked to azido sugars or fluorescent labels under CuAAC conditions The oligonucleotides prepared with the rigid spacer [Figure 10.4(a)] did not result in complete conversion to the triazole due to steric crowding in cases where there were adjacent alkyne-containing nucleotides, whereas those of containing the flexible

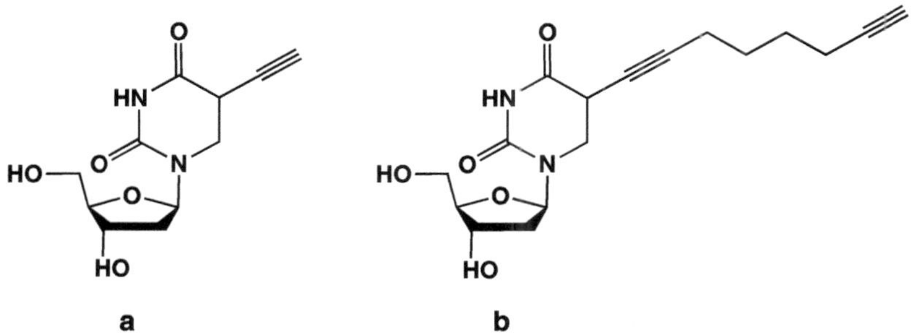

Figure 10.4 Rigid (a) and flexible (b) alkyne-containing modified uridine nucleosides for DNA incorporation. Reprinted with permission from ref.[58]. Copyright 2006 American Chemical Society.

spacer [Figure 10.4(b)] gave full conversion. To determine whether this method was viable for longer DNA fragments, primers containing two flexible click sites were synthesized and used in an enzymatic process (the PCR technique) to generate a range of products from two different plasmids. The click reaction of these strands with fluorescein azide gave only a single product in each case, evidenced by gel electrophoresis with no sign of degradation. This methodology was tested on PCR fragments up to 2000 base pairs.

The same authors made use of sugars, which function as hemi-protected aldehydes, for the sequence-selective metal deposition on DNA.[60,61] Using alkyne-decorated PCR products prepared as above, the clicking of galactose azide gave a modified DNA strand that when silver stained gave yellow/brown spots on a gel; natural DNA was stained under the same conditions but was not evidenced. The ability to deposit silver around the aldehyde-modified DNA is a promising result opening the way to a highly sensitive DNA detection method.

Oligonucleotides exhibit an extraordinary range of bioactivities, but their pharmacological properties, such as their ability to transverse the cell membrane, are often poor. Click chemistry has therefore found a role in the modular construction of biomolecules consisting of these components with the hope of improving their use as therapeutic agents. Kirshenbaum and coworkers used solid-phase synthesis to build up *N*-substituted glycine peptoidic oligomers containing azide and alkyne moieties.[62] Repeated chain elongation, followed by click coupling reactions with a range of azide and alkyne-containing compounds, led to highly functionalized bioconjugate materials (Figure 10.5). A water-soluble estradiol–ferrocene peptoid conjugate prepared by this method demonstrated the potential for application in the modular synthesis of biosensors.

Barthélémy *et al.* have clicked a lipid moiety to oligonucleotides (ON) to increase cellular uptake and allow intracellular delivery.[63] A series of lipid ONs were prepared by the attachement of alkyne-modified lipids derived from cholesterol and octadecanol to an azide-appended nucleotide (Figure 10.6). The click reaction generated the triazole intermediates, which were then converted into the phosphoramidites and further coupled onto the ON chain using solid-phase synthesis. The ON sequence chosen for the study was the 17-mer-2′-*O*-methylribonucleotide antisense (ON17mer) of the hepatitis C virus (HCV) RNA, specifically targeting the subdomain IIId of the internal ribosome entry site

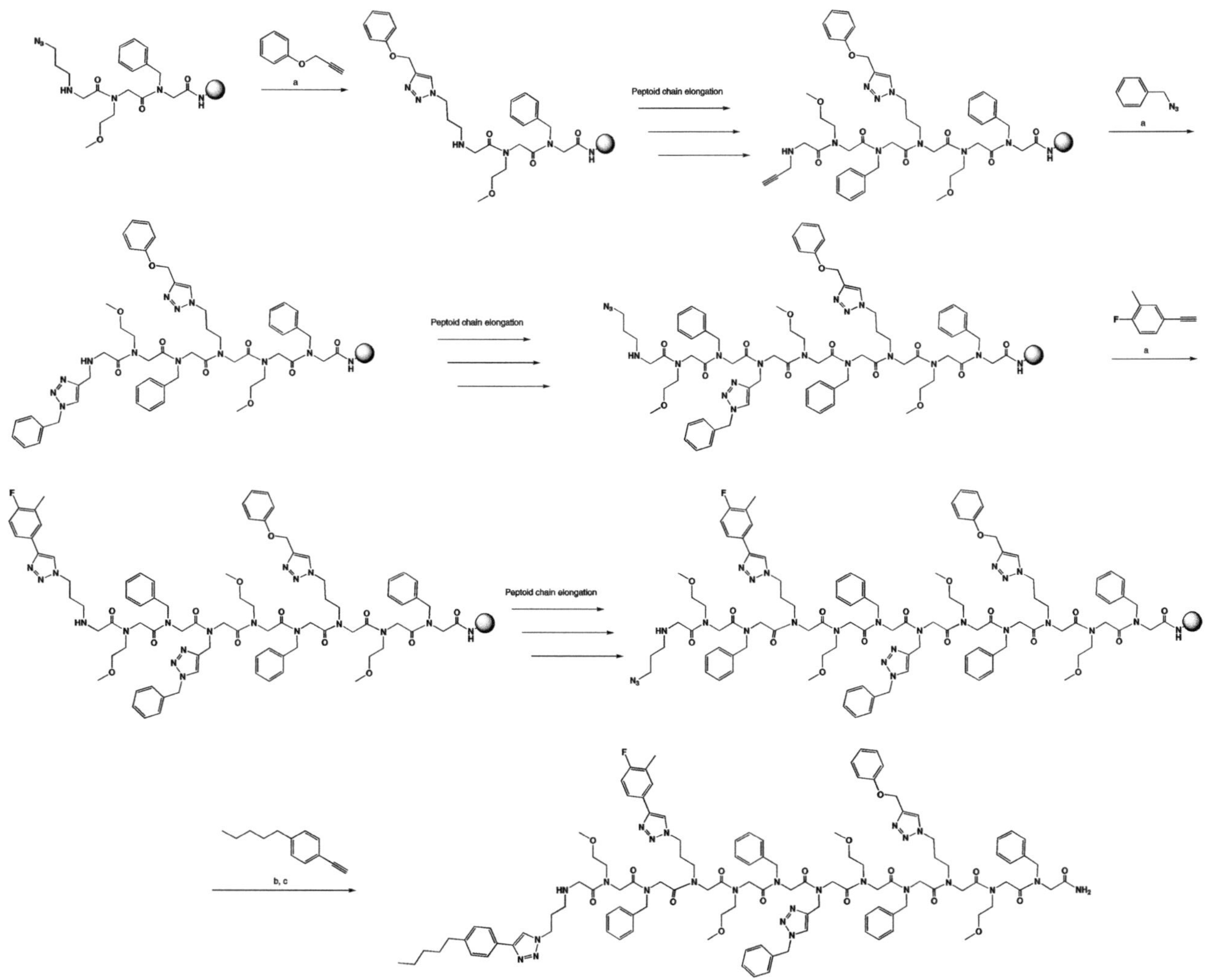

Figure 10.5 N-substituted glycine peptoidic oligomers formed by the repeated chain elongation and subsequent click reactions to generate

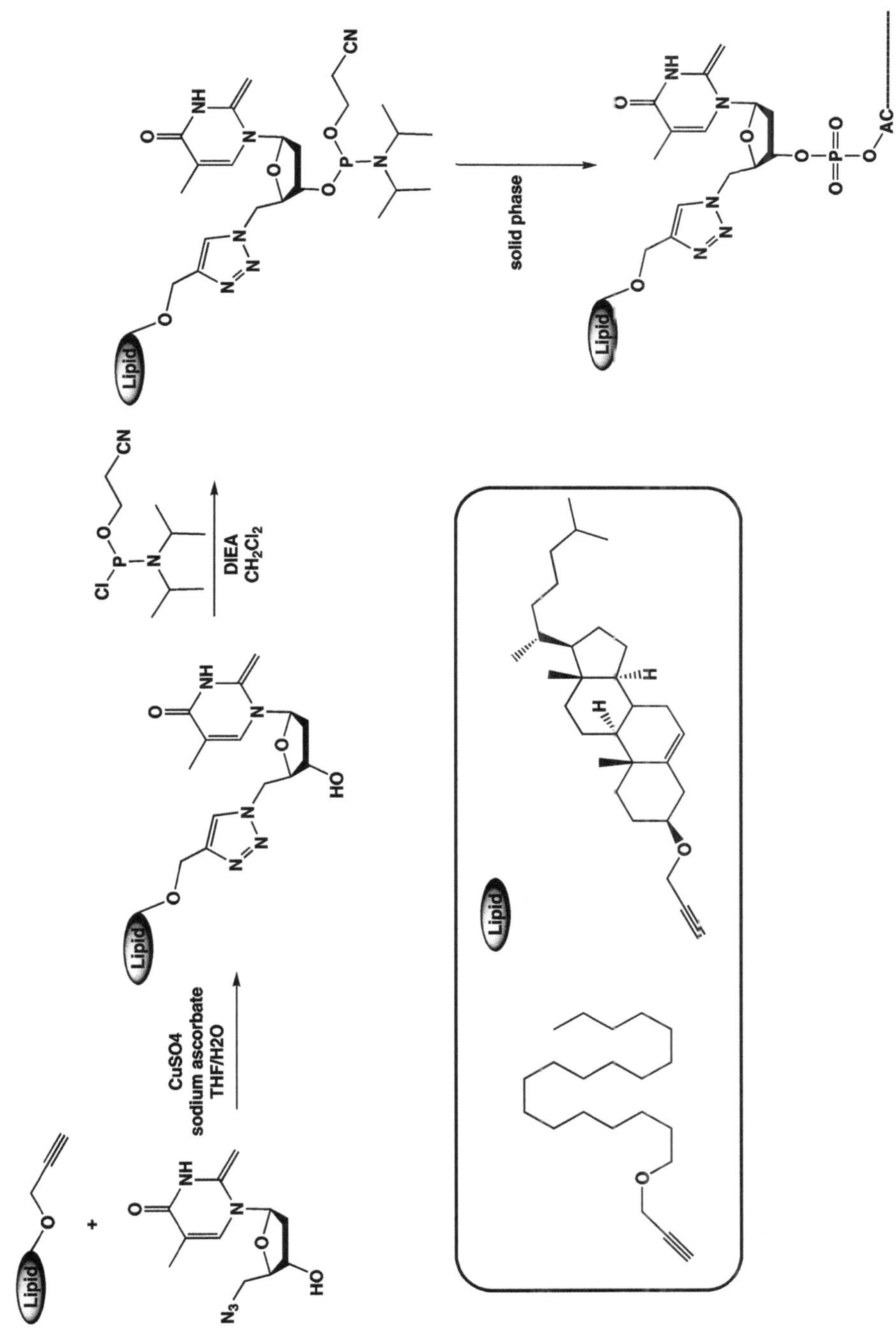

Figure 10.6 *Synthesis of lipid–oligonucleotides. Reprinted with permission from ref.[63]. Copyright 2008 American Chemical Society.*

(IRES). The lipid modification was found to significantly increase the lipophilicity of the oligopeptide, allowing cellular uptake and therefore generating an increase in delivery. Both lipid-containing oligonucleotides induced a dose-dependent reduction of the HCV IRES-dependent translation in the human hepatic cells in which they were tested. The toxicity of the lipid–ON conjugates was found to be negligible.

10.3 Biohybrid Structures based on Protein Conjugates

Structurally, proteins are linear biopolymers biosynthesized through templated processes with unparalleled control of monomer assembly, sequence and molecular weight. Functionally, proteins offer numerous desired activities such as recognition, catalysis and information processing.[11] These factors make proteins attractive building blocks for formation of biologically active materials. They can be used either as the bioactive component or as the macromolecular scaffold for the attachment of other bioactive motifs for applications in biophysical, medicinal and biotechnological disciplines.

The conjugation of a protein to poly(ethylene glycol) (PEG; termed PEGylation) often increases the stability and solubility of biomolecules and has therefore become a frequently employed technique in the field of bioconjugation. Peschke *et al.* have elongated human growth hormone (hGH) at its C-terminus by the addition of a Leu–Ala functionality.[64] The C-terminal amino acid could then be converted into an azide, which was then clicked with various sized PEG groups to give PEGylated hGH derivatives [Figure 10.7(a)]. The PEGylated compounds were prepared in the quest to identify long-acting hGH drugs, which would require less frequent injecting, but have the same relative activity as those currently used. The plasma half-life of hGH is increased upon PEGylation, thereby allowing control over its release. The *in vitro* biological activity of these bioconjugate molecules was determined and was governed by both the size and the shape of the PEG group attached. The cases in which branched PEG groups were attached saw a remarkable drop in activity of the coordinated hGH when compared with wild-type hGH, whereas the linear PEG groups showed only a small decrease in activity.

The use of non-natural amino acids fitted with unique reactive groups is also a powerful technique for the site-specific modification of proteins.[65] The group of Schultz has achieved site-specific PEGylation by the incorporation of non-natural amino acids containing azide functionalities into mutant proteins by genetic engineering.[66] In this way, human superoxide dismutase (SOD) was equipped with an azide group at a specific position in the amino acid sequence. Conjugation with alkyne-terminated PEGs of various lengths resulted in enzymes [Figure 10.7(b)], which showed activity similar to that of the native enzyme; a key enzyme in the processes that prevent the formation of reactive oxygen species in cells.

As seen in the previous example, the use of non-natural amino acids can play an important part in protein engineering.[65,67] The effect of replacing an amino acid with a non-natural analog can lead to increased protein stability[68,69] and large spectral shifts in fluorescent proteins.[70] Tirrell and coworkers utilized this approach for the selective labeling on the cell surface of *E. coli* bacteria by the incorporation of an azido functionalized homoalanine moiety into porin C (OmpC; a protein abundant in the outer membrane of *E. coli*).[71,72] Azidohomoalanine, a methionine surrogate, was metabolically incorporated into OmpC and subsequently clicked with a biotin (Figure 10.8). The biotin-decorated cells could then be stained with avidin allowing discrimination between cells containing the natural and

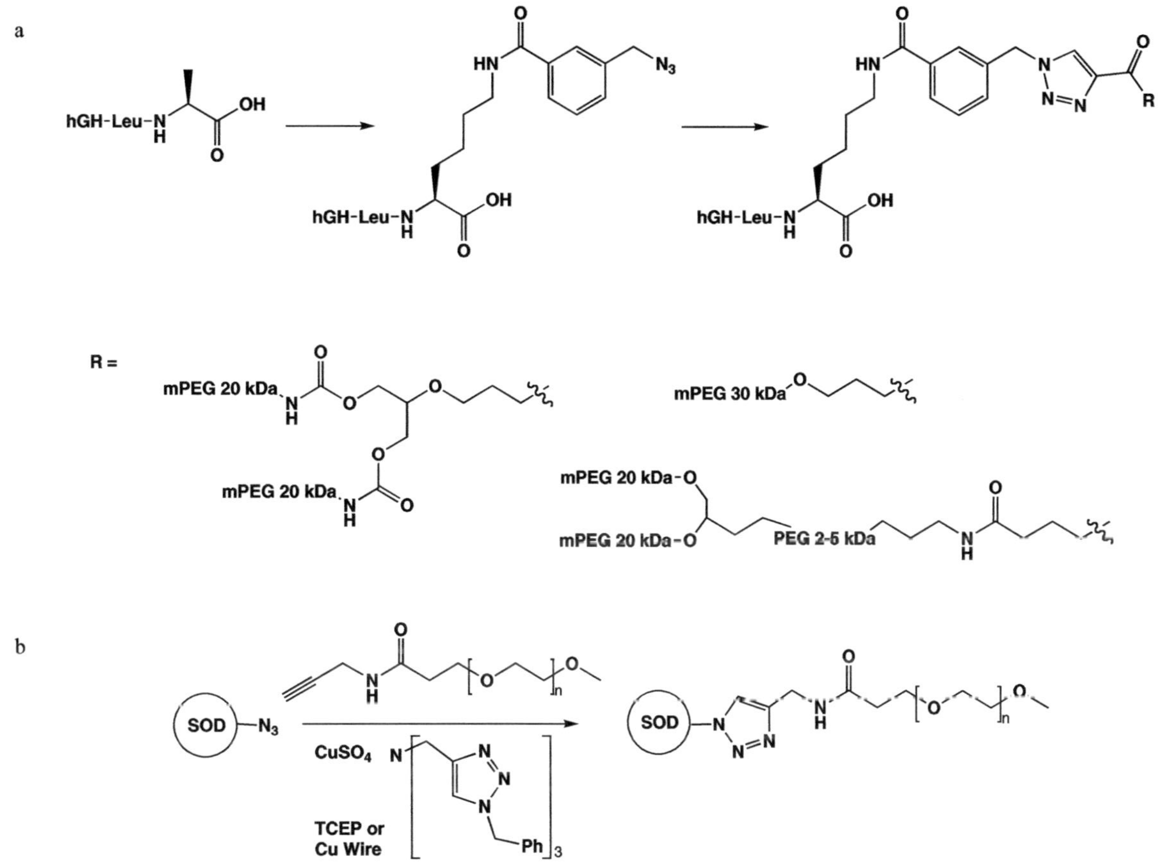

Figure 10.7 *PEGylation of (a) elongated hGH 64 (Reprinted with permission from ref.[64]. Copyright 2007 Elsevier) and (b) SOD with a site-specific incorporated azide functionality (Reprinted with permission from ref.[66]. Copyright 2004 Elsevier) using the CuAAC reaction.*

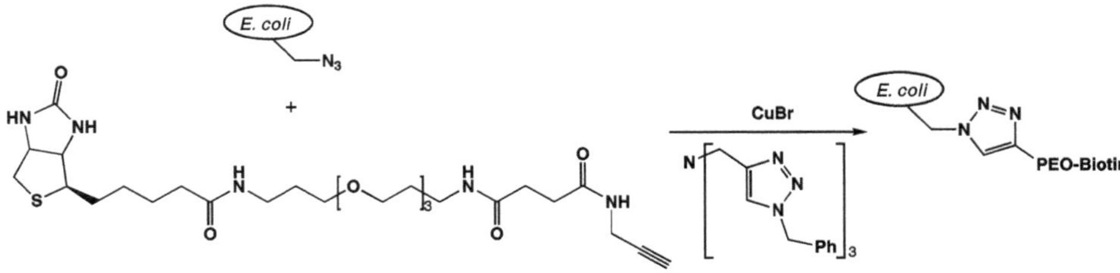

Figure 10.8 Formation of an E. coli–biotin conjugate from a click chemistry reaction. Reprinted with permission from refs.[71] and [72]. Copyright 2003 and 2004 respectively, American Chemical Society.

the unnatural amino acids.[71] More recent experiments were conducted with three different methionine surrogates, azidoalanine, azidonorvaline and azidonorleucine, but using highly pure copper bromide as the catalyst instead of $CuSO_4$/TCEP.[72] The highly active catalyst led to approximately 10-fold more extensive cell labeling than previously observed. This method has been used in practical applications in the discrimination of recent from old proteins in mammalian cells.[73] The incorporation of the azide-bearing amino acid azidohomoalanine is unbiased, nontoxic and was not found to increase protein degradation.

Viruses, self-assembled protein architectures, are often used as macromolecular scaffolds for bioconjugation as they provide robust architectures with multiple functional groups on the exterior.[16] These exterior groups can be used to conjugate biomolecules directly or can be modified by the attachment of an azide or an alkyne for CuAAC derivatization. The group of Finn has used lysine, cysteine and tyrosine residues of the Cowpea Mosaic Virus (CPMV) to introduce azide and alkyne moieties onto the structure.[74–77] These compounds were then successfully functionalized with fluorophores,[74,76,77] peptides,[74] proteins,[74] oligosaccharides[74] and glycopolymers[75] in yields and substrate loadings far superior to those possible with previously established procedures.[77–79] One example[74] involved carbohydrates being attached to the surface of the virus particle with the view of being useful for drug targeting, as well as for the elusive goal of antibody production against carbohydrate epitopes.[80] The azide functionalized tetrasaccharide, which binds the protein galectin-4, an early marker of breast cancer cells,[81,82] was subjected to a CuAAC reaction with a dialkyne fluorescein molecule [Figure 10.9(a)]. The resulting dye–alkyne derivatives were then successfully grafted onto the azide appended CPMV by a second CuAAC reaction [Figure 10.9(b)]. The retention of the activity of the tetrasaccharide was verified by the formation of a gel upon the addition of the conjugate to dimeric galectin-4. In order to determine the *in vivo* effects of CPMV, the virus was derivatized with an alkyne appended gadolinium complex of 1,4,7,10-tetraazacyclododecane-N,N',N'',N'''-tetraacetic acid (DOTA) and the toxicity, biodistribution and pathology were determined in mice.[83] The virus was found to be a safe and nontoxic platform for biomedical applications.

Enzymes are also attractive molecules for the construction of biohybrid materials due to the array of chemical conversions they are capable of catalyzing. This functional activity means that they are of great interest as components in the preparation of biosensors and in the areas of catalytically active materials and surfaces. The catalytic activity of enzymes has been vastly studied in bulk, but only recently have their characteristics been studied at the

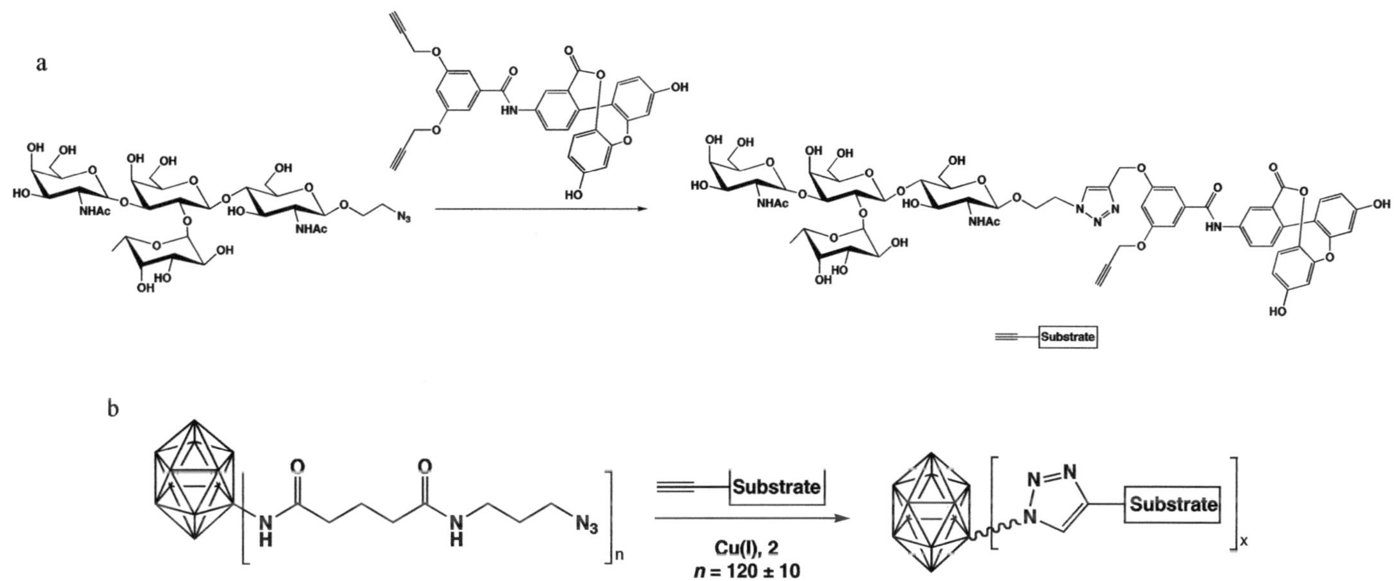

Figure 10.9 *The click reaction of a tetrasaccharide to a fluorescein molecule (a) and the subsequent reaction with CPMV (b) to form a protein conjugate. Reprinted with permission from ref.[74]. Copyright 2005 American Chemical Society.*

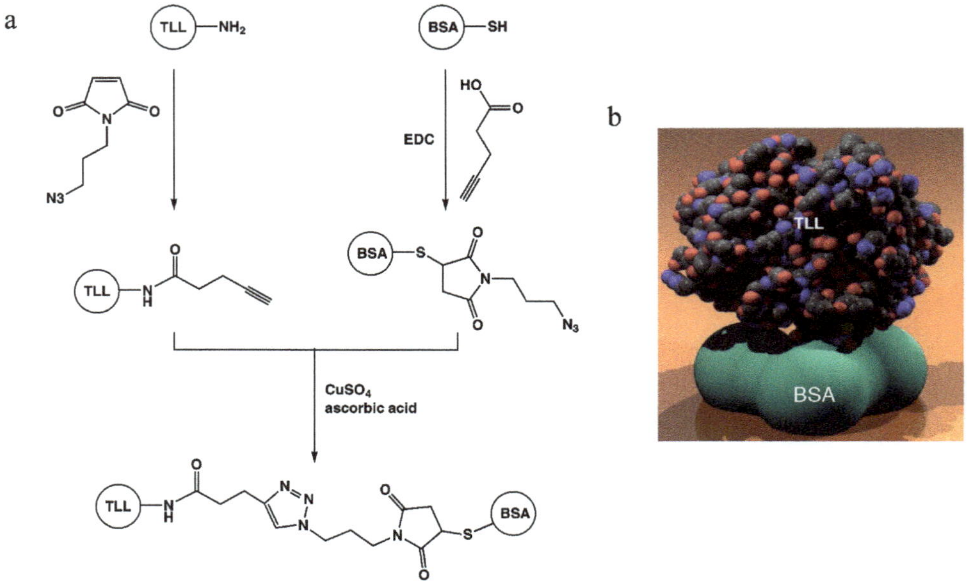

Figure 10.10 *Synthesis of a TLL–BSA heterodimer by means of a CuAAC reaction (a) and schematic representation (b). Reprinted with permission from ref.*[88]. *Copyright 2006 Royal Society of Chemistry.*

single molecule level.[84] In order to conduct the single molecule studies the group of Nolte employed the enzyme lipase B from *Candida Antarctica* (Cal B), which was adsorbed onto a surface and individually monitored by confocal fluorescence spectroscopy while converting the profluorescent BCECF-AM substrate into fluorescent BCECF acid.[85,86] While this method gave a wealth of information, the process of absorbing an enzyme onto the surface was uncontrolled and only a limited number of enzymes were found to remain active. To overcome this, a mutant of thermomyces lanuginosa lipase (TLL) for which the nonspecific adsorption characteristics were too inconsistent for successful single enzyme studies,[87] was used to construct a heterodimer with bovine serum albumin (BSA).[88] The role of the BSA is to act as a 'protein foot' to stick the enzyme onto the surface. The BSA was functionalized with an azide moiety and clicked to a monoalkyne-appended lipase, constructed by derivatization of the single accessible lysine residue (Figure 10.10). Deposition of the protein dimer resulted in all bound enzymes remaining active and exhibiting comparable behavior. The dimer also showed a two-fold increase in catalytic activity in the conversion of the profluorescent substrates 5-(and 6-)-carboxy fluorescein diacetate compared with that of the nonfunctionalized lipase, making this approach an ideal method for the construction of active enzyme surfaces.

10.4 Biohybrid Amphiphiles

The material and self-assembling properties of proteins have been shown to significantly improve on attachment to polymers,[89–92] making the synthesis of polymer–protein conjugates an important field of research for applications in areas such as nanotechnology

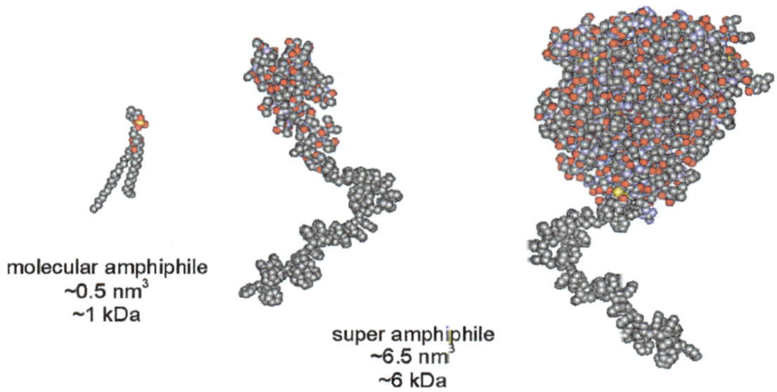

Figure 10.11 *Computer-generated models of molecular, super and giant amphiphiles. Reprinted with permission from ref.[96]. Copyright 2002 American Chemical Society.*

and medicine.[93] The polymers involved are usually water-soluble and in the majority of cases are PEG or PEG analogs. Less well studied is the attachment of a hydrophobic polymer to a biomolecule, a conjugation that results in a polymer biohybrid that is amphiphilic in character. It has been shown in studies of low-molecular weight and super amphiphiles (Figure 10.11) that the shape of the individual molecule determines the structure of the resulting aggregate,[94,95] therefore requiring a highly specific and selective strategy for their preparation. The requirement that, for example, only one tail is attached can be readily controlled using click chemistry making this strategy very attractive.

The Nolte group have formed protein–polystyrene conjugates, 'giant amphiphiles' (Figure 10.11), using the CuAAC reaction[32] and have demonstrated that these compounds exhibit self-assembling properties similar to those of the classical low molecular weight amphiphiles.[96–98] The peptide, H-GlyGlyArg-(7-amino-4-methylcoumarin) (H-GlyGlyArg-AMC) was used as the polar head group as it is easily accessible and contains the AMC fluorophore, a useful tool for characterization. The peptide was functionalized with an alkyne moiety through the reaction of the *N*-terminus with 3-butynyl chloroformate. The apolar PS tail was synthesized by ATRP and subsequent end-group modification. The CuAAC reaction of the two components generated the PS–GlyGlyArg–AMC biohybrid amphiphiles, which were found, by transmission electron microscopy (TEM) and scanning electron microscopy (SEM), to form vesicles in water upon injection from a THF solution [Figure 10.12(a)]. The same PS unit was then coupled to an alkyne-appended BSA molecule generating amphiphiles that self-organized into micellular structures [Figure 10.12(b)].[32] It was shown that the biological function of the protein head groups could be (partially) preserved on conjugation and self-assembly, demonstrating great promise for the construction of biologically active nano-sized assemblies.

Later work from the same authors involved the synthesis of ABC triblock architectures by a cofactor reconstitution approach, as outlined in the schematic in Figure 10.13(a), using well-defined PEG-*b*-PS diblock copolymers and hemeproteins.[99] The PEG-*b*-PS copolymer was chosen for these studies as it is known that this macromolecule, depending on the ratio of the two different blocks, can phase separate into various structures. The diblock copolymer was prepared by the functionalization of monomethoxy PEG with an ATRP initiator and the subsequent polymerization with styrene; the terminal bromine was then converted into

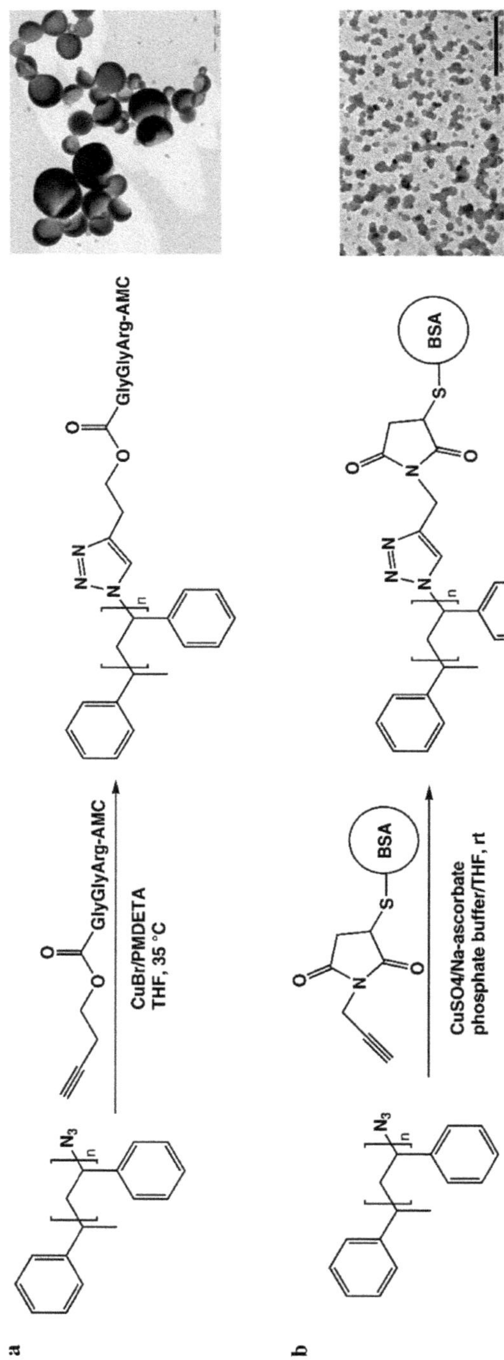

Figure 10.12 *Biohybrid amphiphiles from the click reaction of a PS azide with a tripeptide (a) and a BSA protein (b) and the corresponding TEM images. Reprinted with permission from ref.*[32]*. Copyright 2005 Royal Society of Chemistry.*

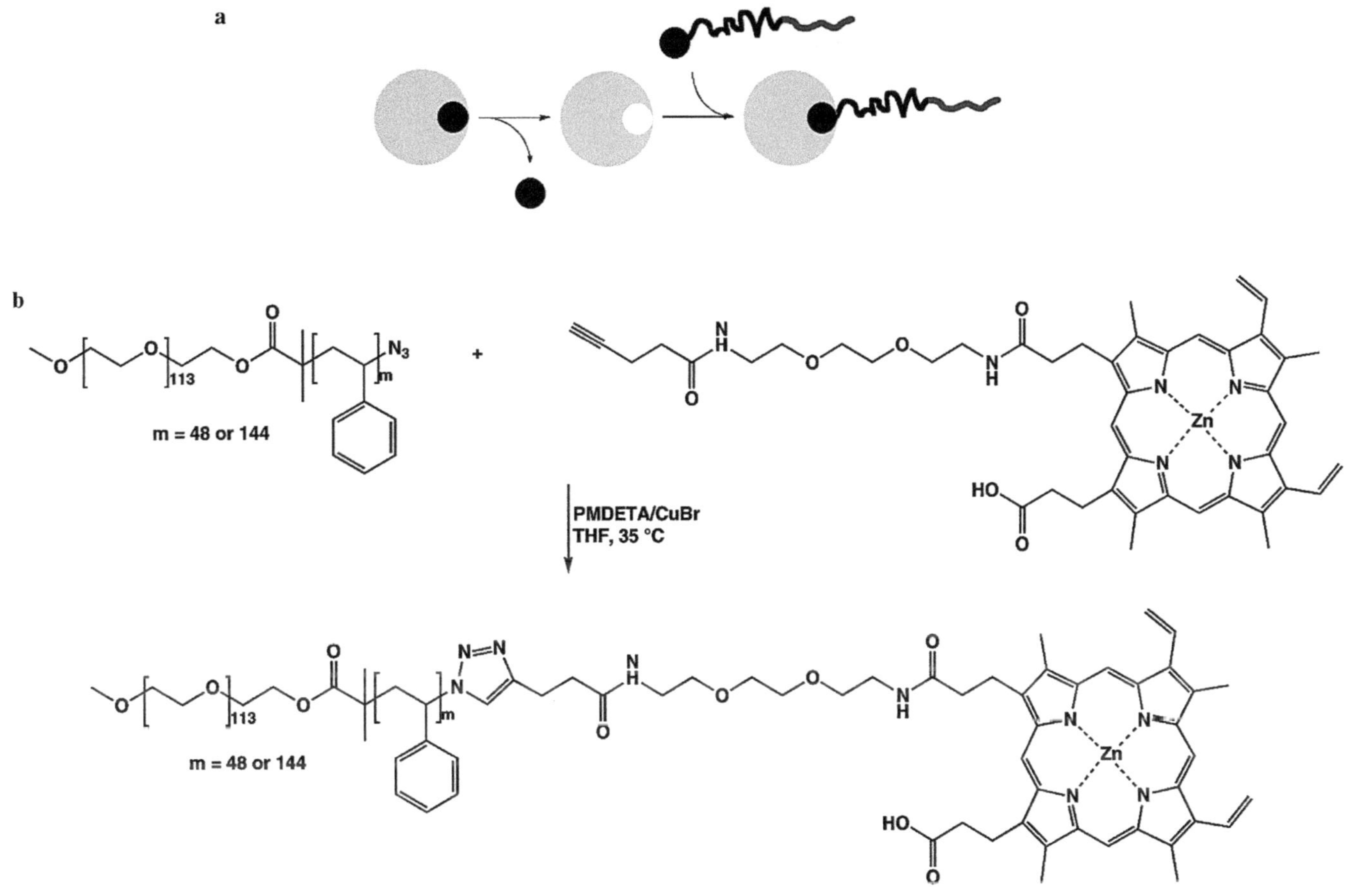

Figure 10.13 *(a) Schematic representation of the reconstitution method employed in the synthesis of ABC triblock polymers and (b) the click reaction of a PEG-b-PS diblock copolymer with heme factor ZnPPIX. Reprinted with permission from ref.[99]. Copyright 2007 American Chemical Society.*

an azide. An acetylene appended heme cofactor, Zinc protoporphyrinIX (ZnPPIX), was coupled to the diblock copolymer under CuAAC conditions [Figure 10.13(b)] and subsequently reconstituted with myoglobin or horseradish peroxidase (HRP). Reconstitution was achieved by the gentle shaking of a mixture of the polymer and the apoenzyme in a plastic tube leading to the gradual obtainment of stable aggregates.

Since the click reaction is performed prior to reconstitution, the latter proceeds elegantly in the absence of the copper catalyst, which has been known to interact with protein structures. Reconstitution with the proteins resulted in a range of more complex aggregate morphologies compared with those observed for PEG-*b*-PS, including micellular rods, vesicles, toroids, figure-eight structures, octopus structures and spheres with a lamellar surface.

The groups of Velonia and Haddleton introduced a post-functionalization approach to giant amphiphiles.[100] This differs from the above method in the sense that a protein is coupled to a hydrophilic polymer generating a water-soluble biohybrid, which can be easily isolated and purified – a difficult feat for some amphiphilic copolymers. The hydrophobicity responsible for the amphiphilic character of the final molecule can then be introduced by means of a click reaction. To demonstrate this, a hydrophilic α-maleimido poly-1-alkyne was prepared using ATRP. This polymer was then coupled to a BSA protein by the Michael addition of the terminal maleimide with a thiol group. The resulting hydrophilic multifunctional bioconjugate was isolated using protein purification and fully characterized before undergoing the CuAAC reaction with hydrophobic azides to generate the amphiphilic species (Figure 10.14). Confocal microscopy and TEM studies showed that these giant amphiphiles exhibit aggregation behavior similar to that reported for the direct coupling of a hydrophobic polymer to a protein. These types of compounds are particularly interesting for potential biomedical and biotechnological applications.

10.5 Glycoconjugates

Carbohydrates, or saccharides, are an essential part of life, whether as an energy source (starch), as structural materials (cellulose and chitin) or as the structural core of nucleic acids. Oligosaccharides, because of their involvement in intracellular and intermolecular communications in the majority of biological and physiological processes, have undergone significant investigation over the last decade in order to define and understand the complexity of multicellular life.[101–105] Because of the presence of polyvalent repeat units, carbohydrates can polymerize in a branched or a linear fashion at a number of linkage positions. This gives rise to many different geometries and a therefore a high degree of complexity. This is evident when it comes to their synthesis, which, in contrast to peptides and nucleic acids, is far from trivial as a result of the large variety of functional groups present and the need for control over chemo- and stereochemical factors. In addition, carbohydrates are often attached to other biomolecules, such as lipids, proteins and nucleic acids, highlighting the need for orthogonal coupling reactions that use mild conditions. Click glycochemistry has proven to be a valuable tool in the construction of glycosylated biohybrid materials and allows the construction of materials that are otherwise unobtainable[8].

10.5.1 Carbohydrate Clusters

A direct logical approach to carbohydrate containing biohybrid materials involves the conjugation of clickable sugar moieties onto functionalized macromolecular materials, such

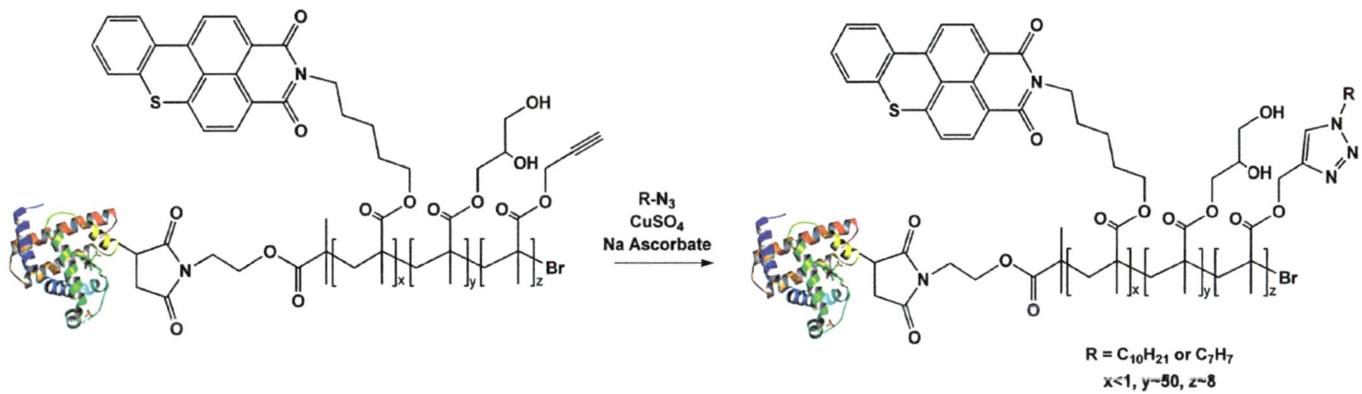

Figure 10.14 *Formation of giant amphiphiles by the post-functionalization approach. Reprinted with permission from ref.*[100]*. Copyright 2007 Royal Society of Chemistry.*

as polymers or dendrimers. Sharpless and coworkers prepared, using the CuAAC reaction, unsymmetrical dendrimers containing both mannose binding units and coumarin fluorescent units.[106] One such biohybrid, based on 2,2-bis(hydroxymethyl)propionic acid, involved the positioning of an alkyne group at the focal point for functionalization with a coumarin chromophore (Figure 10.15). Further alkyne groups were introduced onto the periphery of the dendrimer and coupled with unprotected 2-azido α-D-mannopyranoside molecules. This dendrimer was shown to be a highly efficient dual-purpose recognition/detection agent for the inhibition of pathological conditions, such as hemagglutination conditions, through multivalent interactions. The groups of Liskamp and Pieters also used dendrimers in conjunction with click chemistry. In this case, azidosugars were reacted with a series of alkyne-terminated dendrimers constructed from 3,5-di-(2-aminoethyoxy)-benzoic acid or 3,4,5-tris(3-aminopropoxy) benzoic acid repeat units.[107] Since, in this research, the base dendrimers are easily varied, contain significant rigidity and there is considerable distance between the clickable functional groups, this approach provides a basis for the preparation of a large array of multivalent biomolecular constructs.

In a different approach, Lee and coworkers prepared glycoclusters from an alkyne functionalized carbohydrate core unit.[108] Four individual alkyne groups were introduced onto a methyl β-D-galactopyranoside unit and functionalized, using CuAAC chemistry, with azido linked lactose or *N*-acetyl lactosamine derivatives (Figure 10.16). The resulting sugar-cored glycoclusters were found to be much stronger inhibitors of the RCA_{120} lectin compared with monovalent lactose.

10.5.2 Glycopeptides

The glycoproteins are a class of biomolecules involved in a large number of biological recognitions events.[109] They commonly consist of an oligosaccharide linked, through an *N*- or *O*-atom, to a protein.[80,110,111] This glycosyl–protein bond is intrinsically sensitive towards enzymatic hydrolysis, as is the case with most biological polymers, resulting in limited metabolic stability. In addition, the synthetic assembly of the *O*-glyco-peptides is hindered by the facile elimination of the carbohydrate portion due to β-elimination under basic conditions. The groups of Rutjes[112,113] and Dondoni[114] have independently investigated the incorporation of triazole linkages as stable isosteres for native glycosidic linkages. Although the 1,2,3-triazole functional group does not occur in nature, it is present in diverse biologically active substances exhibiting anti-HIV[115] and anti-bacterial[116] behavior, as well as selective β_3-adrenergic receptor agonism.[117] Interestingly, the triazole moiety has been postulated to act as an amide isosteres in terms of electronic properties and the placement of substituents.[7,118,119] There are, however, noticeable differences between these two functionalities, in particular an increase in the distance between R^1 and R^2 of 1.1 Å for a triazole compared with an amide [Figure 10.17(a)]. The triazole, in addition, possesses a stronger dipole moment (4.83 Debye compared with 3.92 Debye for the amide) – a feature that may enhance peptide bond mimicry by increasing the donor and acceptor properties of hydrogen bonding. The potential of a triazole moiety to act as an imide was demonstrated by Ghadrir and coworkers by the comparison of a triazole-containing octapeptide to a natural peptide.[119] The triazole isostere was found to have similar behavior to the natural peptide with both forming an extended network resulting in solvent-filled nanotubes [Figure 10.17(b)].

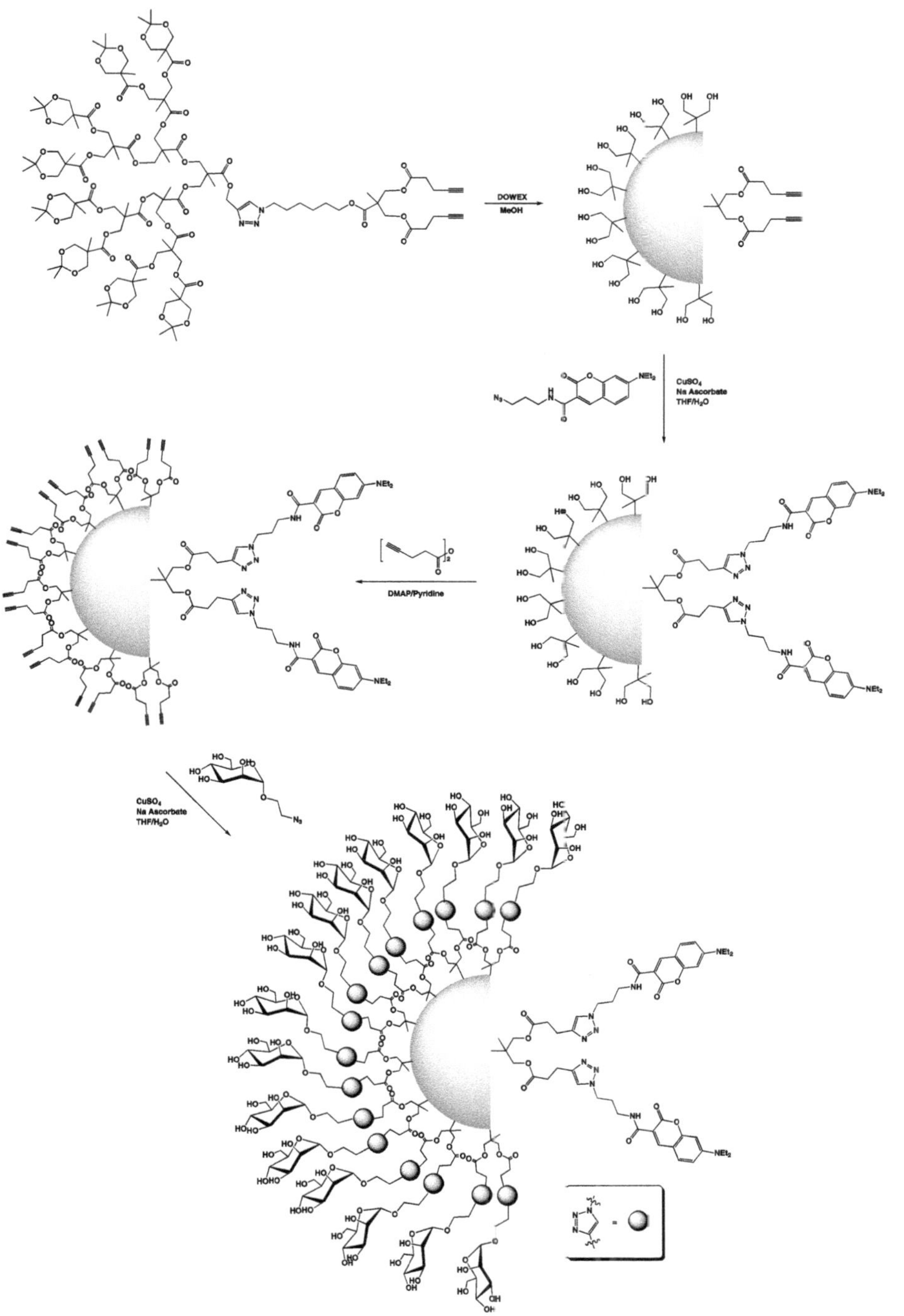

Figure 10.15 *Synthesis of a multivalent, asymmetrical glycodendrimer base on 2,2-bis(hydroxymethyl)propionic acid. Reprinted with permission from ref.*[106]*. Copyright 2005 Royal Society of Chemistry.*

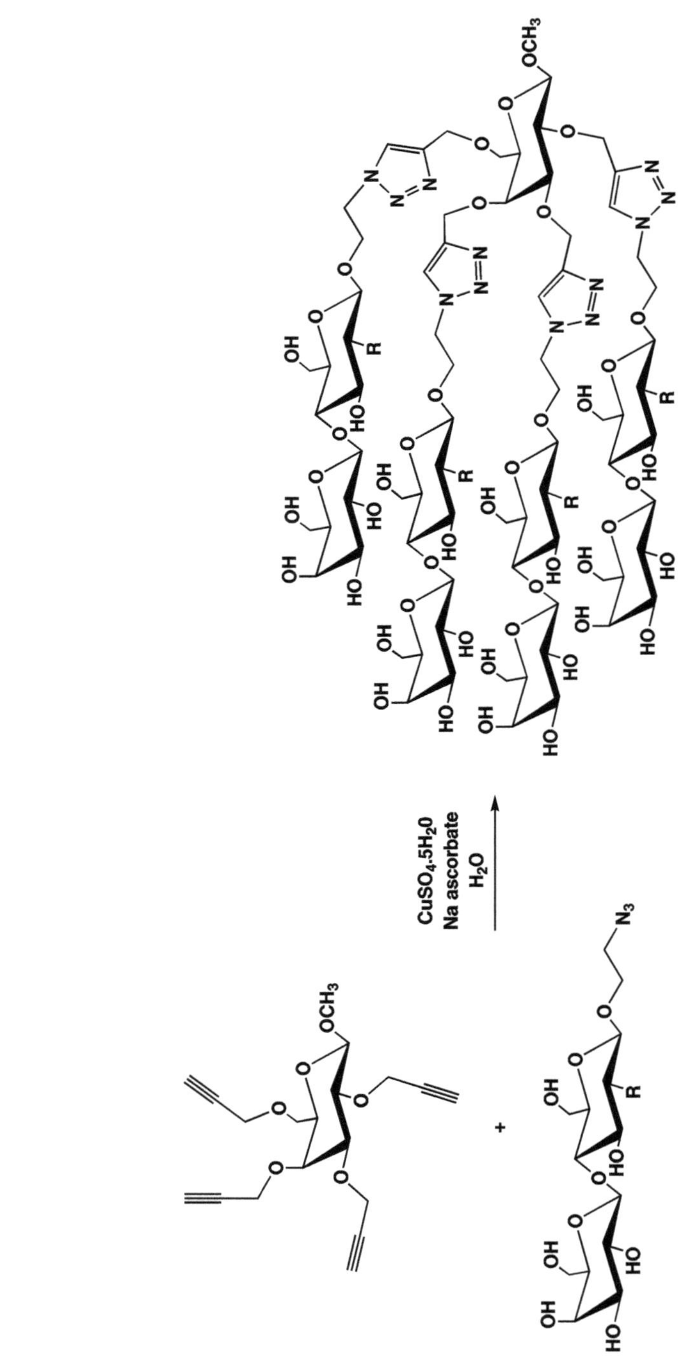

Figure 10.16 *Carbohydrate glycoclusters formed from the click reaction of sugar moieties to a tetraalkyne functionalized carbohydrate core. Reprinted with permission from ref.*[108]*. Copyright 2005 Elsevier.*

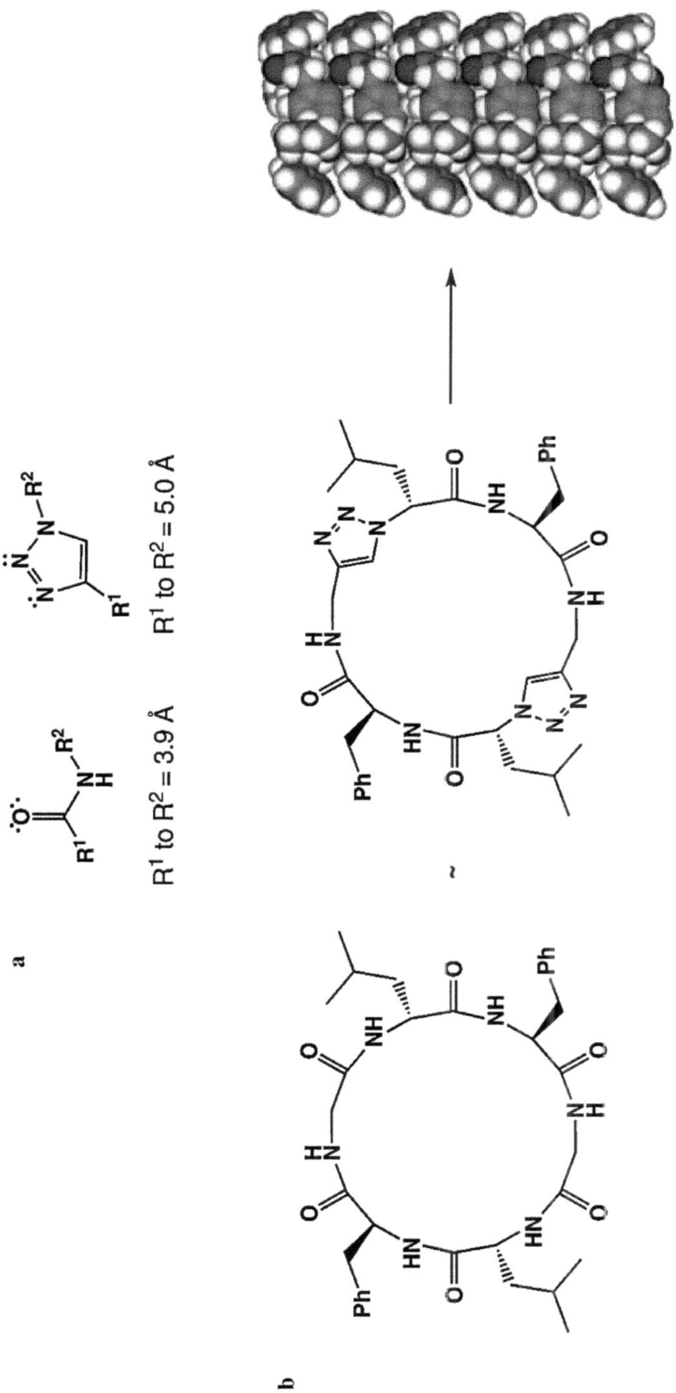

Figure 10.17 (a) Physical properties of a triazole compared with an amide and (b) nanotube formation from the triazole containing octapeptide. Reprinted with permission from ref.[119]. Copyright 2003 American Chemical Society.

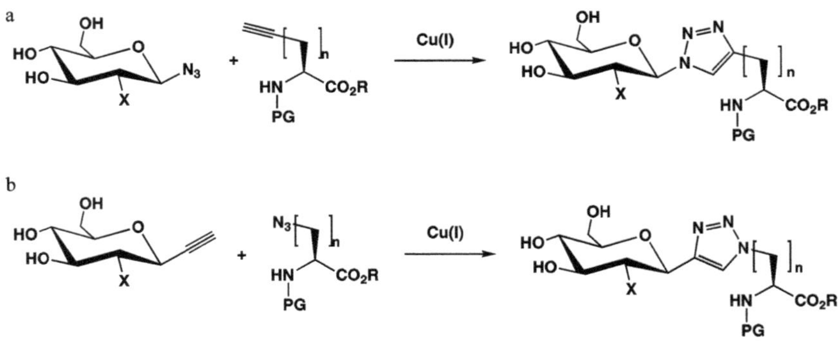

Figure 10.18 *Synthetic strategies for click glycosylation in which the sugar functions as the azide (a) or the alkyne (b). Reprinted with permission from ref.[112]. Copyright 2004 American Chemical Society.*

To determine the synthetic viability of triazolyl glycoamino acids, Rutjes *et al.* performed the CuAAC reaction with a range of anomeric azidoglycosides with *N*-Boc-propargylglycine methyl ester and isolated the triazole-linked products in good yields [Figure 10.18(a)].[112] Variation of the amino acids (either by differing the amino acid or by changing the protecting group of propargylglycine) was then investigated and found to generate the desired products in good yields. In addition, increasing the length of the side chain of the amino acid resulted in no significant change in formation of the triazole linkage. The scope of the click reaction was extended to dipeptides and disaccharides and in both cases the triazole-linked compounds were obtained in satisfactory yields. Finally, the inversion of the alkyne and azide moieties between the carbohydrate and amino acid groups [Figure 10.18(b)] also readily generated the conjugated products in good yields. The described methodology was later successfully combined with enzymatic C-terminal elongation of amino acids, performing click glycosylation either before or after an enzymatic peptide coupling step under the action of alcalase.[113]

To further evaluate the amido isosteric properties of the triazole moiety, Rutjes and coworkers synthesized side chain glycosylated cyclic arginine–glycine–aspartate (*c*RGD) derivatives for biological comparison studies.[120] The RGD peptides are found in proteins of the extracellular matrix, such as vitronectin, fibrinogen and laminin. This motif is specifically recognized by integrins (heterodimeric transmembrane proteins), which link the intracellular cytoskeleton with the extracellular matrix and therefore play an important role in cell signalling, cell–cell adhesion, apoptosis and cell–matrix interactions. To evaluate the binding affinities for $\alpha_v\beta_3$ integrin two glycosylated *c*RGD derivatives, one containing the triazole moiety, and two reference *c*RGD compounds (Figure 10.19) were synthesized by a combination of solid phase and solution phase techniques.

The affinities of *c*(RGDfV), *c*(RGDyV), *c*(RGDy-*N*-1-β-gluco-Asn]) and *c*(RGDy-^{N}TGA) for $\alpha_v\beta_3$ integrin were determined by competitive binding assays using dimeric^{111}In-DOTA-E-[*c*(RGDfK)]$_2$. It was found that the binding of the dimer to $\alpha_v\beta_3$ was inhibited by each compound in a concentration-dependent manner with only relatively small differences between the peptides observed [IC$_{50}$: 65 nM for *c*(RGDfV); 144 nM for *c*(RGDyV); 238 nM for *c*(RGDy-[*N*-1-β-gluco-Asn]); 144 nM for *c*(RGDy-^{N}TGA)],

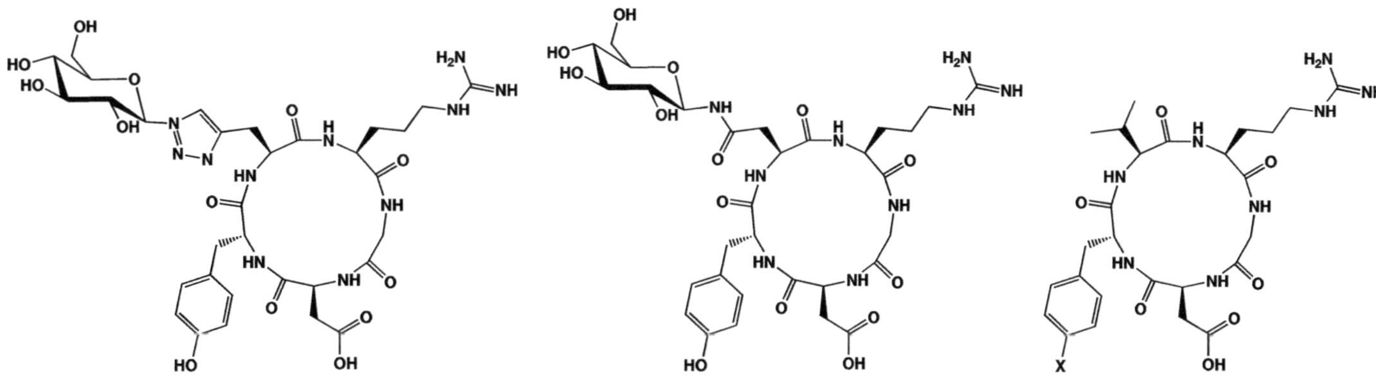

c(RGDy-^{N}TGA)[a]   c(RGDy-[N-1-β-gluco-Asn])   c(RGDfV) X = H
c(RGDyV) X = OH

Figure 10.19 *Cyclic RGD peptides for integrin targeting.* [aN]*TGA = nitrogen-coordinated triazole-linked glycoamino acid. Reprinted with permission from ref.*[120]*. Copyright 2007 American Chemical Society.*

indicating that side-chain modification has only a limited effect on the $\alpha_v\beta_3$ binding affinity of compounds. Additionally, the fact that only small differences are observed between the *c*RGD peptides with carbohydrates attached indicates that the glycoamino acid binding characteristics are nearly unchanged upon substitution of the amide linkage with a triazole. Given that side chain glycosylation of peptides is known to improve the pharmacological properties of hydrophobic/lipophilic peptides,[121] the biodistribution of^{125}I-*c*(RGDyV),^{125}I-*c*(RGDy-[*N*-1-β-gluco-Asn]) and^{125}I-*c*(RGDy-^{N}TGA) in athymic mice with sc $\alpha_v\beta_3$-expressing tumors was determined. All peptides were found to rapidly clear from the blood and 2 h after injection the concentration of the glyco-containing *c*RGDs was higher in the tumor cells than in any of the other tissues examined. In order to determine the nonspecific uptake of the peptides, the biodistribution was also determined in the presence of excess unlabeled DOTA-E-[*c*(RGDfK)]. In this case it was found that, for each of the three compounds being tested, the major part of the uptake in the tumor was $\alpha_v\beta_3$-mediated; the triazole-linked glycopeptide revealed the highest tumor-to-blood ratio, although it showed a lower tumor uptake than that of the amide-linked analog. The carbohydrate-bearing *c*RGD peptides also showed $\alpha_v\beta_3$-mediated uptake in nontargeted organs such as the lung, spleen and intestine. These studies show that tumor uptake is not solely dependent on the binding affinity, but rely also on factors such as blood resident time, molecular weight, structure and charge.

10.5.3 Glycopolymers

The glycopolymers, synthetic macromolecules featuring pendant carbohydrate groups,[12] have been investigated in diverse applications including macromolecular drugs,[122–124] drug delivery systems,[125,126] surface modifiers[127,128] and as models of biological systems.[129] Many of these areas require polymers that have known molecular weight and glycosylation density, as well as the position of glycosylation. Glycopolymers have mostly been synthesized by either the polymerization of a sugar-containing monomer or by the post-functionalization of a pre-formed polymer with sugar moieties.[12] Using the click reaction for the post functionalization of polymers with sugars is an attractive method as libraries of glycopolymers with the same macromolecular features can be obtained by attaching different sugars to the polymer scaffold. This can be of great importance as the effect of the sugar moieties on biological behavior, for example carbohydrate–lectin recognition, can be strongly dependent on the sugar polymer chain length.[130] Haddleton and coworkers have used the post-functionalization of well-defined alkyne containing polymers with sugar azides in their studies of glycopolymers.[45,131,132] A trimethylsilyl methacrylate monomer was polymerized by living radical polymerization (LRP) to form the homopolymer.[45] Sugar azides were coupled to the polymers under CuAAC conditions, with particular focus on sugars able to bind lectins; concavalin A (Con A) was chosen as the model α-mannose-binding lectin as it is involved in a number of biological processes and there are many reports in the literature focusing on its chemical and biological behavior.[133–135] The alkyne functionalized homopolymer was used to form a library of polymers differing only in the amount of Con A-binding mannose ligand and was achieved by the co-clicking of mixtures of mannose- and galactose-based azides (Figure 10.20). The behavior was tested in the presence of Con A and it was observed that the clustering rate and the stoichiometry of the polymer–protein conjugates depended on the epitope density of the polymer,

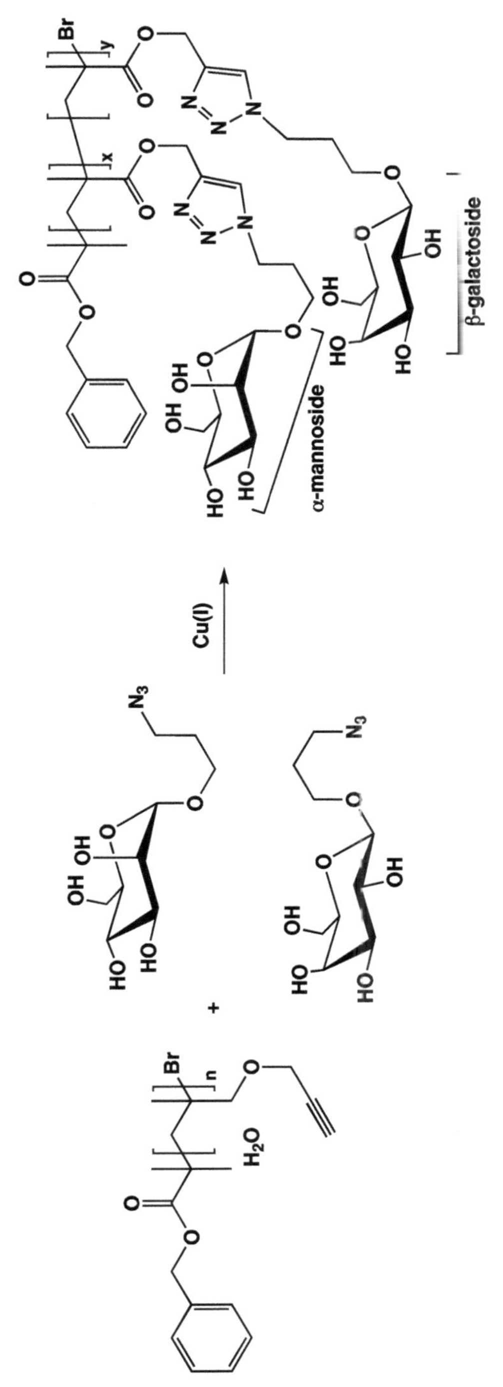

Figure 10.20 *Synthesis of random copolymers employed for Con A binding studies. Reprinted with permission from ref.[45]. Copyright 2006 American Chemical Society.*

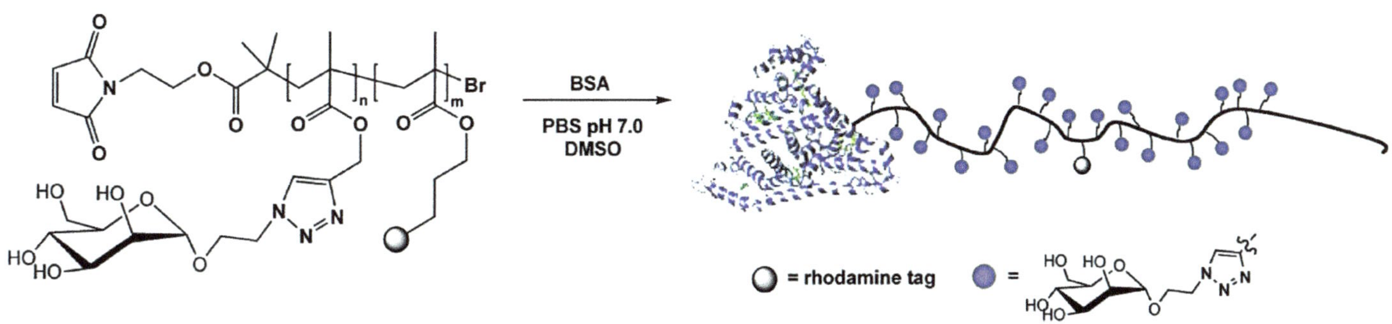

Figure 10.21 *Glycoprotein mimic prepared by the coupling of a clicked glycopolymer to BSA. Reprinted with permission from ref.*[132]*. Copyright 2007 American Chemical Society.*

that is, the number of Con A tetramers bound by each polymer chain was found to increase with an increase in the mannose units attached to that chain. This indicates that the glycopolymers studied here are able to function successfully as multivalent ligands. The interaction of lactose- and galactose-bearing glycopolymers with the lectin *Ricinus Communis* (RCA I) was also studied and the glycopolymers were again found to act as multivalent ligands.[131]

Further work by the same authors resulted in the site-directed conjugation of clicked glycopolymers to form glycoprotein mimics.[132] The maleimide-terminated glycopolymer was prepared by the click reaction of a mannose containing monomer with propargyl methacrylate. This then underwent living radical polymerization with a maleimide-protected initiator in the presence of a fluorescent rhodamine B comonomer (to facilitate characterization) to give the macromolecular intermediate, from which the furan protecting group was removed by a retro-Diels–Alder reaction. The glycopolymer was then conjugated, through the maleimide terminus, to the thiol group of BSA to give the glycoprotein mimic (Figure 10.21). Libraries of glycopolymers were prepared with the co-clicking approach by introducing appropriate mixtures of different sugar azides. The interactions of these glycopolymer–BSA conjugates with recombinant rat mannose-binding lectin (MBL) were then examined by surface plasmon resonance, which revealed clear and dose-dependent MBL binding to the conjugate. The immobilized glycopolymer–BSA conjugates showed a significantly enhanced capacity to activate the complement system (a biochemical cascade that helps clear pathogens from an organism) through the lectin pathway compared with unmodified BSA.

10.6 Conclusions

A variety of biohybrid materials from the copper-catalyzed reaction between azides and alkynes have been described in this chapter. The click reaction has found widespread use for the synthesis of these materials due to the ease with which it is employed and the ability to form products that are otherwise difficult to prepare. Several conjugates formed by the click reaction are already being used in biomedical areas with promising results, and given that this is a reasonably new field of research, there is an exciting future ahead.

References

(1) J.-F. Lutz and H. G. Börner, (2008), Modern trends in polymer bioconjugates design, *Prog. Polym. Sci.*, **33**, 1–39.
(2) H. C. Kolb, M. G. Finn and K. B. Sharpless, (2001), Click chemistry: diverse chemical function from a few good reactions, *Angew. Chem. Int. Edn*, **40**, 2004–2021.
(3) V. V. Rostovtsev, L. G. Green, V. V. Fokin and K. B. Sharpless, (2002), A stepwise Huisgen cycloaddition process: copper(I)-catalyzed regioselective 'ligation' of azides and terminal alkynes, *Angew. Chem. Int. Edn*, **41**, 2596–2599.
(4) C. W. Tornøe, C. Christensen and M. Meldal, (2002), Peptidotriazoles on solid phase: 1,2,3-triazoles by regiospecific copper(I)-catalyzed 1,3-dipolar cycloadditions of terminal alkynes to azides, *J. Org. Chem.*, **67**, 3057–3064.
(5) C. D. Hein, X.-M. Liu and D. Wang, (2008), Click chemistry, A powerful tool for pharmaceutical sciences, *Pharm. Res.*, **25**, 2216–2230.

(6) W. H. Binder and R. Sachsenhofer, (2007), 'Click' chemistry in polymer and materials science, *Macromol. Rapid Commun.*, **28**, 15–54.
(7) V. D. Bock, H. Hiemstra and J. H. van Maarseveen, (2006), CuI-Catalysed alkyne–azide 'click' cycloadditions from a mechanistic and synthetic perspective, *Eur. J. Org. Chem.*, 51–68.
(8) S. Dedola, S. A. Nepogodiev and R. A. Field, (2007), Recent applications of the CuI-catalysed Huisgen azide–alkyne 1,3-dipolar cycloaddition reaction in carbohydrate chemistry, *Org. Biomol. Chem.*, **5**, 1006–1017.
(9) A. J. Dirks, J. J. L. M. Cornelissen, F. L. van Delft, J. C. M. van Hest, R. J. M. Nolte, A. E. Rowan and F. P. J. T. Rutjes, (2007), From (bio)molecules to biohybrid materials with the click chemistry approach, *QSAR Comb. Sci.*, **26**, 1200–1210.
(10) H. C. Kolb and K. B. Sharpless, (2003), The growing impact of click chemistry on drug discovery, *Drug Discov. Today*, **8**, 1128–1137.
(11) B. Le Droumaguet and K. Velonia, (2008), Click chemistry: a powerful tool to create polymer-based macromolecular chimeras, *Macromol. Rapid Commun.*, **29**, 1073–1089.
(12) S. G. Spain, M. I. Gibson and N. R. Cameron, (2007), Recent advances in the synthesis of well-defined glycopolymers, *J. Polym. Sci. Part A: Polym. Chem.*, **45**, 2059–2072.
(13) M. A. Gauthier and H.-A. Klok, (2008), Peptide/protein–polymer conjugates: synthetic strategies and design concepts, *Chem. Commun.*, 2591–2611.
(14) J. A. Johnson, M. G. Finn, J. T. Koberstein and N. J. Turro, (2008), Construction of linear polymers, dendrimers, networks, and other polymeric architectures by copper-catalysed azide-alkyne cycloaddition 'click' chemistry, *Macromol. Rapid Commun.*, **29**, 1052–1072.
(15) J.-F. Lutz, (2007), 1,3-Dipolar cycloadditions of azides and alkynes: a universal ligation tool in polymer and materials science, *Angew. Chem. Int. Edn*, **46**, 1018–1025.
(16) M. Uchida, M. T. Klem, M. Allen, P. Suci, M. Flenniken, E. Gillitzer, Z. Varpness, L. O. Liepold, M. Young and T. Douglas, (2007), Biological containers: protein cages as multifunctional nanoplatforms, *Adv. Mater.*, **19**, 1025–1042.
(17) D. S. Goodsell, (2004), *Bionanotechnology: Lessons from Nature*, Wiley–Liss, Hoboken, NJ.
(18) C. M. Niemeyer and C. A. e. Mirkin, (2004), *Nanobiotechnology: Concepts, Applications and Perspectives*, Wiley–VCH, Weinheim.
(19) J. Bjerre, C. Rousseau, L. Marinescu and M. Bols, (2008), Artificial enzymes, 'Chemzymes': current state and perspectives, *Appl. Microbiol. Biotechnol.*, **81**, 1–11.
(20) I. Ledoux-Rak, A. Dodabalapur and P. Blom, (2002), Novel organic materials and technological advances for photonics, *Synth. Met.*, **127**, 1–2.
(21) J. M. Thijssen and H. S. J. Van Der Zant, (2008), Charge transport and single-electron effects in nanoscale systems, *Nanotechnology*, **3**, 37–64.
(22) A. Zabet-Khosousi and A.-A. Dhirani, (2008), Charge transport in nanoparticle assemblies, *Chem. Rev.*, **108**, 4072–4124.
(23) J. A. Opsteen and J. C. M. van Hest, (2005), Modular synthesis of block copolymers via cycloaddition of terminal azide and alkyne functionalized polymers, *Chem. Commun.*, 57–59.
(24) N. V. Tsarevsky, B. S. Sumerlin and K. Matyjaszewski, (2005), Step-growth 'click' coupling of telechelic polymers prepared by atom transfer radical polymerization, *Macromolecules*, **38**, 3558–3561.
(25) G. Mantovani, V. Ladmiral, L. Tao and D. M. Haddleton, (2005), One-pot tandem living radical polymerisation-Huisgens cycloaddition process ('click') catalysed by *N*-alkyl-2-pyridylmethanimine/Cu(I)Br complexes, *Chem. Commun.*, 2089–2091.
(26) B. Parrish, R. B. Breitenkamp and T. Emrick, (2005), PEG- and peptide-grafted aliphatic polyesters by click chemistry, *J. Am. Chem. Soc.*, **127**, 7404–7410.
(27) B. M. Cooper, D. Chan-Seng, D. Samanta, X. Zhang, S. Parelkar and T. Emrick, (2009), Polyester-graft-phosphorylcholine prepared by ring-opening polymerization and click chemistry, *Chem. Commun.*, 815–817.
(28) J.-F. Lutz, H. G. Börner and K. Weichenhan, (2005), Combining atom transfer radical polymerization and click chemistry: a versatile method for the preparation of end-functional polymers, *Macromol. Rapid Commun.*, **26**, 514–518.

(29) J.-F. Lutz, H. G. Börner and K. Weichenhan, (2006), Combining ATRP and 'click' chemistry: a promising platform toward functional biocompatible polymers and polymer bioconjugates, *Macromolecules*, **39**, 6376–6383.
(30) H. Gao, G. Louche, B. S. Sumerlin, N. Jahed, P. Gloas and K. Matyjaszewski, (2005), Gradient polymer elution chromatographic analysis of α,ω-dihydroxypolystyrene synthesized via ATRP and click chemistry, *Macromolecules*, **38**, 8979–8982.
(31) B. A. Laurent and S. M. Grayson, (2006), An efficient route to well-defined macrocyclic polymers via 'click' cyclization, *J. Am. Chem. Soc.*, **128**, 4238–4239.
(32) A. J. Dirks, S. S. van Berkel, N. S. Hatzakis, J. A. Opsteen, F. L. van Delft, J. J. L. M. Cornelissen, A. E. Rowan, J. C. M. van Hest, F. P. J. T. Rutjes and R. J. M. Nolte, (2005), Preparation of biohybrid amphiphiles via the copper catalysed Huisgen [3+2] dipolar cycloaddition reaction, *Chem. Commun.*, 4172–4174.
(33) H. Gao and K. Matyjaszewski, (2006), Synthesis of star polymers by a combination of ATRP and the 'click' coupling method, *Macromolecules*, **39**, 4960–4965.
(34) J. A. Johnson, D. R. Lewis, D. D. Diaz, M. G. Finn, J. T. Koberstein and N. J. Turro, (2006), Synthesis of degradable model networks via ATRP and click chemistry, *J. Am. Chem. Soc.*, **128**, 6564–6565.
(35) D. D. Diaz, S. Punna, P. Holzer, A. K. McPherson, K. B. Sharpless, V. V. Fokin and M. G. Finn, (2004), Click chemistry in materials synthesis. 1. Adhesive polymers from copper-catalyzed azide-alkyne cycloaddition, *J. Polym. Sci. Part A: Polym. Chem.*, **42**, 4392–4403.
(36) B. S. Sumerlin, N. V. Tsarvsky, G. Lurche, R. Y. Lee and K. Matyjaszewski, (2005), Highly efficient 'click' functionalization of poly(3-azidopropyl methacrylate) prepared by ATRP, *Macromolecules*, **38**, 7540–7545.
(37) M. Malkoch, R. J. Thibault, E. Drockenmuller, M. Messerschmidt, B. Voit, T. P. Russell and C. J. Hawker, (2005), Orthogonal approaches to the simultaneous and cascade functionalization of macromolecules using click chemistry, *J. Am. Chem. Soc.*, **127**, 14942–14949.
(38) R. K. O'Reilly, M. J. Joralemon, C. J. Hawker and K. L. Wooley, (2006), Fluorogenic 1,3-dipolar cycloaddition within the hydrophobic core of a shell cross-linked nanoparticle, *Chem. Eur. J.*, **12**, 6776–6786.
(39) W. H. Binder and C. Kluger, (2004), Combining ring-opening metathesis polymerisation (ROMP) with Sharpless-type 'click' reactions: an easy method for the preparation of side chain functionalised poly(oxynorbornenes), *Macromolecules*, **37**, 9321–9330.
(40) R. Luxenhofer and R. Jordan, (2006), Click chemistry with poly(2-oxazoline)s, *Macromolecules*, **39**, 3509–3516.
(41) D. Thomsen, E. Malmstrom and S. Hvilsted, (2006), Novel polymers with a high carboxylic acid loading, *J. Polym. Sci. Part A: Polym. Chem.*, **44**.
(42) D. A. Ossipov and J. Hilborn, (2006), Poly(vinyl alcohol)-based hydrogels formed by 'click chemistry', *Macromolecules*, **39**, 1709–1718.
(43) B. C. Englert, S. Bakbak and U. H. F. Bunz, (2005), Click chemistry as a powerful tool for the construction of functional poly(p-phenyleneethynylene)s: comparison of pre- and postfunctionalization schemes, *Macromolecules*, **38**, 5868–5877.
(44) H. Nandivada, H.-Y. Chen, L. Bondarenko and J. Lahann, (2006), Reactive polymer coatings that 'click', *Angew. Chem. Int. Edn*, **45**, 3360–3363.
(45) V. Ladmiral, G. Mantovani, G. J. Clarkson, S. Cauet, J. L. Irwin and D. M. Haddleton, (2006), Synthesis of neoglycopolymers by a combination of 'click chemistry' and living radical polymerization, *J. Am. Chem. Soc.*, **128**, 4823–4830.
(46) H. G. Börner and H. Schlaad, (2007), Bioinspired functional block copolymers, *Soft Matter*, **3**, 394–408.
(47) T. J. Demming, (1997), Polypeptide materials: new synthetic methods and applications, *Adv. Mater.*, **9**, 299–311.
(48) H.-A. Klok, (2002), Protein-inspired materials: synthetic concepts and potential applications, *Angew. Chem. Int. Edn*, **41**, 1509–1513.
(49) J. C. M. van Hest and D. A. Tirrell, (2001), Protein-based materials, toward a new level of structural control, *Chem. Commun.*, 1897–1940.

(50) J. J. L. M. Cornelissen, J. J. Donners, R. de Gelder, W. S. Graswinckel, G. A. Metselaar, A. E. Rowan, N. A. J. M. Sommerdijk and R. J. M. Nolte, (2001), β-Helical polymers from isocyanopeptides, *Science*, **293**, 676–680.
(51) P. Samori, C. Ecker, I. Goessl, P. A. J. de Witte, J. J. L. M. Cornelissen, G. A. Metselaar, M. B. J. Otten, A. E. Rowan, R. J. M. Nolte and J. P. Rabe, (2002), High shape persistence in single polymer chains rigidified with lateral hydrogen bonded networks, *Macromolecules*, **35**, 5290–5294.
(52) H. J. Kitto, J. J. L. M. Cornelissen, A. E. Rowan and R. J. M. Nolte, unpublished work.
(53) J. J. L. M. Cornelissen, M. Fischer, N. A. J. M. Sommerdijk and R. J. M. Nolte, (1998), Helical superstructures from charged poly(styrene)–poly(isocyanodipeptide) block copolymers, *Science*, **280**, 1427–1430.
(54) E. Schwartz, H. J. Kitto, R. de Gelder, R. J. M. Nolte, A. E. Rowan and J. J. L. M. Cornelissen, (2007), Synthesis, characterisation and chiroptical properties of 'click'able polyisocyanopeptides, *J. Mater. Chem.*, **17**, 1876–1884.
(55) H. J. Kitto, E. Schwartz, M. Nijemeisland, M. Koepf, J. J. L. M. Cornelissen, A. E. Rowan and R. J. M. Nolte, (2008), Post-modification of helical dipeptido polyisocyanides using the 'click' reaction, *J. Mater. Chem.*, **18**, 5615–5624.
(56) N. L. Rosi and C. A. Mirkin, (2005), Nanostructures in biodiagnostics, *Chem. Rev.*, **105**, 1547–1562.
(57) J. G. Hacia, L. C. Brody, M. S. Chee, S. P. A. Fodor and F. S. Collins, (1996), Detection of heterozygous mutations in BRCA1 using high density oligonucleotide arrays and two-colour fluorescence analysis, *Nat. Genet.*, **14**, 441–447.
(58) J. Gierlich, G. A. Burley, P. M. E. Gramlich, D. M. Hammond and T. Carell, (2006), Click chemistry as a reliable method for the high-density postsynthetic functionalization of alkyne-modified DNA, *Org. Lett.*, **8**, 3639–3642.
(59) P. M. E. Gramlich, C. T. Wirges, A. Manetto and T. Carell, (2008), Postsynthetic DNA modification through the copper-catalyzed azide-alkyne cycloaddition reaction, *Angew. Chem. Int. Edn*, **47**.
(60) G. A. Burley, J. Gierlich, M. R. Mofid, H. Nir, S. Tal, Y. Eichen and T. Carell, (2006), Directed DNA metallization, *J. Am. Chem. Soc.*, **128**, 1398–1399.
(61) C. T. Wirges, P. M. E. Gramlich, K. Gutsmiedl, J. Gierlich, G. A. Burley and T. Carell, (2007), Pronounced effect of DNA hybridization of click reaction efficiency, *QSAR Comb. Sci.*, **26**, 1159–1164.
(62) J. M. Holub, H. Jang and K. Kirshenbaum, (2006), Clickity-click: highly functionalised peptoid oligomers generated by sequential conjugation reactions on solid-phase support, *Org. Biomol. Chem.*, **4**, 1497–1502.
(63) G. Godeau, C. Staedel and P. Barthélémy, (2008), Lipid-conjugated oligonucleotides via 'click chemistry' efficiently inhibit Hepatitis C virus translation, *J. Med. Chem.*, **51**, 4374–4376.
(64) B. Peschke, M. Zundel, S. Bak, T. R. Clausen, N. Blume, A. Pedersen, F. Zaragoza and K. Madsen, (2007), C-terminally PEGylated hGH-derivatives, *Bioorg. Med. Chem.*, **15**, 4382–4395.
(65) R. E. Connor and D. A. Tirrell, (2007), Non-canonical amino acids in protein polymer design, *Polym. Rev.*, **47**, 9–28.
(66) A. Deiters, T. A. Cropp, D. Summerer, M. Mekherji and P. G. Schultz, (2004), Site-specific PEGylation of proteins containing unnatural amino acids, *Bioorg. Med. Chem. Lett.*, **14**, 5743–5746.
(67) T. L. Hendrickson, V. d. Crécy-Lagard and P. Schimmel, (2004), Incorporation of nonnatural amino acids into proteins, *Annu. Rev. Biochem.*, **73**, 147–176.
(68) Y. Tang, G. Ghirlanda, N. Vaidehi, J. Kua, D. T. Mainz, W. A. Goddard III, W. F. DeGrado and D. A. Tirrell, (2001), Stabilization of coiled-coil peptide domains by introduction of trifluoroleucine, *Biochemistry*, **40**, 2790–2796.
(69) Y. Tang and D. A. Tirrell, (2001), Biosynthesis of a highly stable coiled-coil protein containing hexafluoroleucine in an engineered bacterial host, *J. Am. Chem. Soc.*, **123**, 11089–11090.
(70) J. H. Bae, M. Rubini, G. Jung, G. Wiegand, M. H. J. Seifert, M. K. Azim, J.-S. Kim, A. Zumbusch, T. A. Holak, L. Moroder, R. Huber and N. Budisa, (2003), Expansion of the

genetic code enables design of a novel 'gold' class of green fluorescent proteins, *J. Mol. Biol.*, **328**, 1071–1081.
(71) A. J. Link and D. A. Tirrell, (2003), Cell surface labeling of Escherichia coli via the copper(I)-catalyzed [3+2] cycloaddition, *J. Am. Chem. Soc.*, **125**, 11164–11165.
(72) A. J. Link, M. K. S. Vink and D. A. Tirrell, (2004), Presentation and detection of azide functionality in bacterial cell surface proteins, *J. Am. Chem. Soc.*, **126**, 10598–10602.
(73) D. C. Dieterich, A. J. Link, J. Graumann, D. A. Tirrell and E. M. Schuman, (2006), Selective identification of newly synthesized proteins in mammalian cells using bioorthogonal noncanonical amino acid tagging (BONCAT), *Proc. Natl Acad. Sci. USA*, **103**, 9482–9487.
(74) S. S. Gupta, J. Kuzelka, P. Singh, W. G. Lewis, M. Manchester and M. G. Finn, (2005), Accelerated bioorthogonal conjugations: a practical method for the ligations of diverse functional molecules to a polyvalent virus scaffold, *Bioconjugate Chem.*, **16**, 1572–1579.
(75) S. S. Gupta, K. S. Raja, E. Kaltgrad, E. Strable and M. G. Finn, (2005), Virus–glycopolymer conjugates by copper(I) catalysis of atom transfer radical polymerisation and azide–alkyne cycloaddition, *Chem. Commun.*, 4315–4317.
(76) S. Meunier, E. Strable and M. G. Finn, (2004), Crosslinking of and coupling to viral capsid proteins by tyrosine oxidation, *Chem. Biol.*, **11**, 319–326.
(77) Q. Wang, K. S. Raja, K. D. Janda, T. Lin and M. G. Finn, (2003), Blue fluorescent antibodies as reporters of steric accessibility in virus conjugates, *Bioconjugate Chem.*, **14**, 38–43.
(78) Q. Wang, T. Lin, L. Tang, J. E. Johnson and M. G. Finn, (2002), Icosahedral virus particles as addressable nanoscale building blocks, *Angew. Chem. Int. Edn*, **41**, 459–462.
(79) K. S. Raja, Q. Wang, M. J. Gonzalez, M. Manchester, J. E. Johnson and M. G. Finn, (2003), Hybrid virus–polymer materials. 1. Synthesis and properties of PEG-decorated Cowpea Mosaic Virus, *Biomacromolecules*, **4**, 472–476.
(80) O. Seitz, (2000), Glycopeptide synthesis and the effects of glycosylation on protein structure and activity, *Chembiochem*, **1**, 214–246.
(81) M. E. Huflejt and H. Leffler, (2004), Galectin-4 in normal tissues and cancer, *Glycoconjugate J.*, **20**, 247–255.
(82) O. Blixt, S. Head, T. Mondala, C. Scanlan, M. E. Huflejt, R. Alvarez, M. C. Bryan, F. Fazio, D. Calarese, J. Stevens, N. Razi, D. J. Stevens, J. J. Skehel, I. v. Die, D. R. Burton, I. A. Wilson, R. Cummings, N. Bovin, C.-H. Wong and J. C. Paulson, (2004), Printed covalent glycan array for ligand profiling of diverse glycan binding proteins, *Proc. Natl Acad. Sci. USA*, **101**, 17033–17038.
(83) P. Singh, D. Prasuhn, R. M. Yeh, G. Destito, C. S. Rae, K. Osborn, M. G. Finn and M. Manchester, (2007), Bio-distribution, toxicity and pathology of cowpea mosaic virus nanoparticles *in vivo*, *J. Controlled Release*, **120**, 41–50.
(84) H. Engelkamp, N. S. Hatzakis, J. Hofkens, F. C. D. Schryver, R. J. M. Nolte and A. E. Rowan, (2006), Do enzymes sleep and work?, *Chem. Commun.*, 935–940.
(85) K. Velonia, O. Flomenbom, D. Loos, S. Masuo, M. Cotlet, Y. Engelborghs, J. Hofkens, A. E. Rowan, J. Klafter, R. J. M. Nolte and F. C. De Schryver, (2005), Single-enzyme kinetics of CALB-catalyzed hydrolysis, *Angew. Chem. Int. Edn*, **44**, 560–564.
(86) O. Flomenbom, K. Velonia, D. Loos, S. Masuo, M. Cotlet, Y. Engelborghs, J. Hofkens, A. E. Rowan, R. J. M. Nolte, M. V. d. Auweraer, F. C. De Schryver and J. Klafter, (2005), Stretched exponential decay and correlations in the catalytic activity of fluctuating single lipase molecules, *Proc. Natl Acad. Sci. USA*, **102**, 2368–2372.
(87) K. Wannerberger and T. Arnebrant, (1997), Comparison of the adsorption and activity of lipases from *Humicola lanuginosa* and *Candida antarctica* on solid surfaces, *Langmuir*, **13**, 3488–3493.
(88) N. S. Hatzakis, H. Engelkamp, K. Velonia, J. Hofkens, P. C. M. Christianen, A. Svendsen, S. A. Patkar, J. Vind, J. C. Mann, A. E. Rowan and R. J. M. Nolte, (2006), Synthesis and single enzyme activity of a clicked lipase–BSA hetero-dimer, *Chem. Commun.*, 2012–2014.
(89) F. M. Veronese and J. M. Harris, (2002), Introduction and overview of peptide and protein PEGylation, *Adv. Drug Deliv. Rev.*, **54**, 453–456.
(90) P. Caliceti and R. F. Veronese, (2003), Pharmacokinetic and biodistribution properties of poly(ethylene glycol)–protein conjugates, *Adv. Drug Deliv. Rev.*, **55**, 1261–1277.

(91) J. M. Harris and R. B. Chess, (2003), Effect of PEGylation on pharmaceuticals, *Nat. Rev. Drug Discov.*, **2**, 214–221.

(92) A. S. Hoffman and P. S. Stayton, (2004), Bioconjugates of smart polymers and proteins: synthesis and applications, *Macromol. Symp.*, **207**, 139–151.

(93) K. L. Heredia and H. D. Maynard, (2003), Synthesis of protein–polymer conjugates, *Org. Biomol. Chem.*, **5**, 45–53.

(94) M. C. Feiters and R. J. M. Nolte, (2000), *Advances in Supramolecular Chemistry*, Jai Press, Stamford, CT.

(95) H. Shei and A. Eisenberg, (2000), Control of architecture in block-copolymer vesicles, *Angew. Chem. Int. Edn*, **39**, 3310–3312.

(96) K. Velonia, A. E. Rowan and R. J. M. Nolte, (2002), Lipase polystyrene giant amphiphiles, *J. Am. Chem. Soc.*, **124**, 4224–4225.

(97) M. J. Boerakker, J. M. Hannink, P. H. H. Bomans, P. M. Frederik, R. J. M. Nolte, E. M. Meijer and N. A. J. M. Sommerdijk, (2002), Giant amphiphiles by cofactor reconstitution, *Angew. Chem. Int. Edn*, **41**, 4239–4241.

(98) J. M. Hannink, J. J. L. M. Cornelissen, J. A. Farrera, P. Foubert, F. C. D. Schryver, N. A. J. M. Sommerdijk and R. J. M. Nolte, (2001), Protein-polymer hybrid amphiphiles, *Angew. Chem. Int. Edn*, **40**, 4732–4734.

(99) I. C. Reynhout, J. J. L. M. Cornelissen and R. J. M. Nolte, (2007), Self-assembled architectures from biohybrid triblock copolymers, *J. Am. Chem. Soc.*, **129**, 2327–2332.

(100) B. Le Droumaguet, G. Mantovani, D. M. Haddleton and K. Velonia, (2007), Formation of giant amphiphiles by post-functionalization of hydrophilic protein–polymer conjugates, *J. Mater. Chem.*, **17**, 1916–1922.

(101) A. Varki, (1993), Biological roles of oligosaccharides: all of the theories are correct, *Glycobiology*, **3**, 97–130.

(102) R. A. Dwek, (1996), Glycobiology: toward understanding the function of sugars, *Chem. Rev.*, **96**, 683–720.

(103) P. H. Seeberger and W. C. Haase, (2000), Solid-phase oligosaccharide synthesis and combinatorial carbohydrate libraries, *Chem. Rev.*, **100**, 4349–4394.

(104) P. Sears and C.-H. Wong, (2001), Toward automated synthesis of oligosaccharides and glycoproteins, *Science*, **291**, 2344–2350.

(105) L. Joshi, E. Smith and H. Morowitz, (2007), Glycobiology: the sweet language of life, complexity, and morphogenesis: syntax for intermolecular and intercellular communication, *Complexity*, **12**, 9–10.

(106) P. Wu, M. Malkoch, J. N. Hunt, R. Vestberg, E. Kaltgrad, M. G. Finn, V. V. Fokin, K. B. Sharpless and C. J. Hawker, (2005), Multivalent, bifunctional dendrimers prepared by click chemistry, *Chem. Commun.*, 5775–5777.

(107) J. A. F. Joosten, N. T. H. Tholen, F. Ait El Maate, A. J. Brouwer, G. W. van Esse, D. T. S. Rijkers, R. M. J. Liskamp and R. J. Pieters, (2005), High-yielding microwave-assisted synthesis of triazole-linked glycodendrimers by copper-catalysed [3+2] cycloaddition, *Eur. J. Org. Chem.*, 3182–3185.

(108) Y. Gao, A. Eguchi, K. Kakehi and Y. C. Lee, (2005), Synthesis and molecular recognition of carbohydrate-centered multivalent glycoclusters by a plant lectin RCA**120**, *Bioorg. Med. Chem.*, **13**, 6151–6157.

(109) D. G. Large and C. D. Warren (Eds), (1997), *Glycopeptides and Related Compounds: Synthesis, Analysis and Applications*, Marcel Dekker, New York.

(110) H. Kunz, (1987), Synthesis of glycopeptides, partial structures of biological recognition components, *Angew. Chem. Int. Edn*, **26**, 294–308.

(111) R. G. Sprio, (2002), Protein glycosylation: nature, distribution, enzymatic formation, and disease implications of glycopeptide bonds, *Glycobiology*, **12**, 43R–56R.

(112) B. H. M. Kuijpers, S. Groothuys, A. R. Keerweer, P. J. L. M. Quaedflieg, R. H. Blaauw, F. L. van Delft and F. P. J. T. Rutjes, (2004), Expedient synthesis of triazole-linked glycosyl amino acids and peptides, *Org. Lett.*, **6**, 3123–3126.

(113) S. Groothuys, B. H. M. Kuijpers, P. J. L. M. Quaedflieg, H. C. P. F. Roelen, R. W. Wiertz, R. H. Blaauw, F. L. van Delft and F. P. J. T. Rutjes, (2006), Chemoenzymatic synthesis of triazole-linked glycopeptides, *Synthesis*, **18**, 3146–3152.

(114) A. Dondoni, P. P. Giovannini and A. Massi, (2004), Assembling heterocycle-tethered C-glycosyl and α-amino acid residues via 1,3-dipolar cycloaddition reactions, *Org. Lett.*, **6**, 2929–2932.

(115) R. Alvarez, S. Velazquez, A. San-Felix, S. Aquaro, E. D. Clerq, C. F. Perno, A. Karlesson, J. Balzarini and M. J. Camarasa, (1994), 1,2,3-Triazole-2,5-bis-*O*-(tert-butyldimethylsilyl)-β-D-ribofuranosyl-3'-spiro-5''-(4''-amino-1'',2''-oxathiole 2'',2''-dioxide) (TSAO) analogs: synthesis and anti-HIV-1 activity, *J. Med. Chem.*, **37**, 4185–4194.

(116) M. J. Genin, D. A. Allwine, D. J. Anderson, M. R. Barbachyn, D. E. Emmert, S. A. Garmon, D. R. Graber, K. C. Grega, J. B. Hester, D. K. Hutchinson, J. Morris, R. J. Reischer, C. W. Ford, G. E. Zurenko, J. C. Hamel, R. D. Schaadt, D. Stapert and B. H. Yagi, (2000), Substituent effects on the antibacterial activity of nitrogen–carbon-linked (azolylphenyl)oxazolidinones with expanded activity against the fastidious Gram-negative organisms *Haemophilus influenzae* and *Moraxella catarrhalis*, *J. Med. Chem.*, **43**, 953–970.

(117) L. L. Brockunier, E. R. Parmee, H. O. Ok, M. R. Candelore, M. A. Cascieri, L. F. Colwell Jr., L. Deng, W. P. Feeney, M. J. Forrest, G. J. Hom, D. E. MacIntyre, L. Tota, M. J. Wyvratt, M. H. Fisher and A. E. Weber, (2000), Human β3-adrenergic receptor agonists containing 1,2,3-triazole-substituted benzenesulfonamides, *Bioorg. Med. Chem. Lett.*, **10**, 2111–2114.

(118) V. D. Bock, D. Speijer, H. Hiemstra and J. H. van Maarseveen, (2007), 1,2,3-Triazoles as peptide bond isosteres: synthesis and biological evaluation of cyclotetrapeptide mimics, *Org. Biomol. Chem.*, **5**, 971–975.

(119) W. S. Horne, C. D. Stout and M. R. Ghadiri, (2003), A heterocyclic peptide nanotube, *J. Am. Chem. Soc.*, **125**, 9372–9376.

(120) B. H. M. Kuijpers, S. Groothuys, A. C. Soede, O. Laverman, O. C. Boerman, F. L. van Delft and F. P. J. T. Rutjes, (2007), Preparation and evaluation of glycosylated arginine–glycine–asparate (RGD) derivatives for integrin targeting, *Bioconjugate Chem.*, **18**, 1847–1854.

(121) A. M. Sinclair and S. Elliott, (2005), Glycoengineering: the effect of glycosylation on the properties of therapeutic proteins, *J. Pharm. Sci.*, **94**, 1626–1635.

(122) E. J. Gordon, L. E. Strong and L. L. Kiessling, (1998), Glycoprotein-inspired materials promote the proteolytic release of cell surface image-L-Selectin, *Bioorg. Med. Chem.*, **6**, 1293–1299.

(123) T. Yoshida, T. Akasaka, T. Choi, K. Hattori, B. Yu, T. Mimura, Y. Kaneko, H. Nakashima, E. Aragaki, M. Premanathan, N. Yamamoto and T. J. Uryu, (1999), Synthesis of polymethacrylate derivatives having sulfated maltoheptaose side chains with anti-HIV activities, *J. Polym. Sci. Part A: Polym. Chem.*, **37**, 789–800.

(124) S.-K. Choy, M. Mammen and G. M. Whitesides, (1997), Generation and *in situ* evaluation of libraries of poly(acrylic acid) presenting sialosides as side chains as polyvalent inhibitors of influenza-mediated hemagglutination, *J. Am. Chem. Soc.*, **119**, 4103–4111.

(125) Y. Yun, D. J. Goetz, P. Yellen and W. Chen, (2004), Hyaluronan microspheres for sustained gene delivery and site-specific targeting, *Biomaterials*, **25**, 127–157.

(126) J. Li, S. Zacharek, X. Chen, J. Wang, W. Zhang, A. Janczuk and P. G. Wang, (1999), Bacteria targeted by human natural antibodies using α-gal conjugated receptor-specific glycopolymers, *Bioorg. Med. Chem.*, **7**, 1549–1558.

(127) G. Wulff, H. Schmidt and L. Zhu, (1999), Generating hydrophilic surfaces on standard polymers after copolymerization with low amounts of protected vinyl sugars, *Macromol. Chem. Phys.*, **200**, 744–782.

(128) G. Wulff, H. Schmidt and L. Zhu, (1997), Investigations on surface-modified bulk polymers. 1. Copolymers of styrene with a styrene moiety containing a sugar monomer, *Macromolecules*, **30**, 4533–4539.

(129) M. Ambrosi, N. R. Cameron, B. G. Davis and S. Stolnik, (2005), Investigation of the interaction between peanut agglutinin and synthetic glycopolymeric multivalent ligands, *Org. Biomol. Chem.*, **3**, 1476–1480.

(130) M. Kanai, K. H. Mortell and L. L. Kiessling, (1997), Varying the size of multivalent ligands: the dependence of concanavalin A binding on neoglycopolymer length, *J. Am. Chem. Soc.*, **119**, 9931–9932.

(131) J. Geng, J. Lindqvist, G. Mantovani, G. Chen, C. T. Sayers, G. J. Clarkson and D. M. Haddleton, (2007), Well-defined poly(*N*-glycosyl 1,2,3-triazole) multivalent ligands: design, synthesis and lectin binding studies, *QSAR Comb. Sci.*, **26**, 1220–1228.

(132) J. Geng, G. Mantovani, L. Tao, J. Nicolas, G. Chen, R. Wallis, D. A. Mitchell, B. R. G. Johnson, S. D. Evans and D. M. Haddleton, (2007), Site-directed conjugation of 'click' glycopolymers to from glycoprotein mimics: binding to mammalian lectin and induction of immunological function, *J. Am. Chem. Soc.*, **129**, 15156–15163.
(133) G. P. Phondke, K. B. Sainis and N. N. Joshi, (1983), Electrokinetic studies on concanavalin A as a tool to probe the surface characteristics of differentiated lymphoid cells, *J. Biosci.*, **5** (suppl. 1), 137–148.
(134) S. L. Mironov, (1992), Concanavalin A: a tool to change intracellular pH, *Trends Neurosci.*, **15**, 13.
(135) S. S. Lin and I. B. Levitan, (1991), Concanavalin a: a tool to investigate neuronal plasticity, *Trends Neurosci.*, **14**, 273–277.

11

Functional Nanomaterials using the Cu-catalyzed Huisgen Cycloaddition Reaction

Sander S. van Berkel, Arnold W.G. Nijhuis, Dennis W.P.M. Löwik and Jan C.M. van Hest

11.1 Introduction

Nanomaterials represent various materials with the resembling property that their size is 'nano' in at least one dimension. The definition of 'nano' is somewhat arbitrary, but in the majority its magnitude is considered to vary from approximately 10 to 100 nm. Nanomaterials are utilized in many applications. Colloidal gold (gold nanoparticles in suspension), for instance, can be used to stain glass with an intense red color as was done in ancient times. Recent examples are zinc nanoparticles in sun block creams and cancer medicine delivered by liposomes. For the future, countless applications of nanoparticles are anticipated. For these applications nanomaterials need to be connected to the macroscopic world via their exterior. Hence, the modification of nanomaterials to acquire the desired exterior functionality is of major importance to utilize nanomaterials effectively in practical applications.[1] Modification of nanomaterials to obtain specific functionality is often only feasible when highly selective and efficient reactions are used. Sharpless classified reactions that meet such challenges as click reactions.[2] In respect of click chemistry the Cu-catalyzed Huisgen 1,3-dipolar cycloaddition of azides and terminal alkynes has received a lot of attention.[3,4] This cycloaddition unites azides and alkynes regio-specifically to yield 1,4-disubstituted 1,2,3-triazoles. In this review the functionalization of nanosized materials via this conjugation method is described going from inorganics such as silicon, CdSe, magnetic and gold nanoparticles, to carbon-based materials such as fullerenes and carbon nanotubes. Moreover

Click Chemistry for Biotechnology and Materials Science Edited by Joerg Lahann

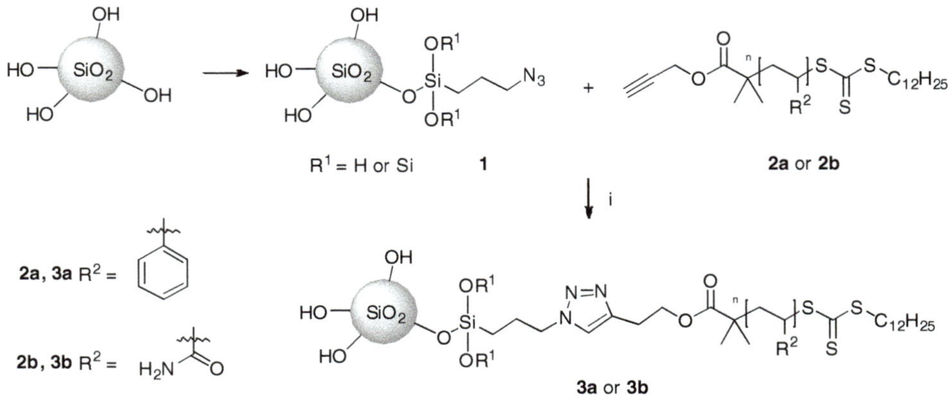

Scheme 11.1 Grafting of acetylene polymers prepared via RAFT to nanoparticle **1**. Reaction conditions: (i) $CuSO_4$–sodium ascorbate; DMSO; 50 °C.[8]

organic and bio-organic, self-assembled systems will be discussed, viz. polymeric micelles, polymersomes, liposomes and virus particles that have also been employed as a scaffold in the Cu-catalyzed azide alkyne cycloaddition.

11.2 Inorganic Nanoparticles

When inorganic materials are downsized to nanometer dimensions new properties, found neither in bulk nor in molecular systems, arise. For example the electronic, optical and magnetic properties of metal and semiconducting nanoparticles strongly depend on their size and shape.[5] Moreover, surface functionality is an important aspect of such nanoparticles due to the relatively large surface-to-volume ratio. In this section click chemistry employed to functionalize silicon, CdSe, magnetic and gold nanoparticle surfaces is surveyed.

11.2.1 Silicon-based Nanoparticles

Modified organic colloidal silica is of particular interest due to the ease of synthesis of well-defined particles and finds applications in stationary chromatography phases, coatings and sensors.[6,7] To render silica nanoparticles compatible with polymer systems Brittain and coworkers have modified silica nanoparticles with azides and attached a variety of acetylene functionalized macromolecules.[8] 3-Bromopropyltrichlorosilane was deposited on colloidal silica, with dimensions of 75–100 nm. Under Schlenk conditions, the bromides were exchanged for azides to obtain azide–silica-particle **1** as shown in Scheme 11.1. Acetylene functional polystyrene (PS) **2a** and polyacrylamide (PAAm) **2b** were synthesized via a reversible addition fragmentation technique (RAFT) with acetylene functional initiators. Employing $CuSO_4$–ascorbic acid-catalyzed click reactions, these polymers were grafted to yield silica nanoparticles **3a** and **3b**.

Complementary to this 'grafting-to' approach, polymers were also 'grafted from' nanoparticle **1**.[9] A RAFT initiator was clicked onto nanoparticle **1** and subsequently polymerization was initiated. 'Grafting-from' has the advantage that low molecular weight monomers diffuse to the surface, while 'grafting to' concerns the mobility of high molecular weight polymers. For that reason the grafting density is generally higher when the 'grafting from' approach is applied. With elemental analysis grafting densities of 0.29 groups nm^{-2} were found for the 'grafting to' approach whereas 0.68 groups nm^{-2} were found for 'grafting from'. Interestingly, when the cycloaddition reaction and polymerization were performed simultaneously in a tandem process an intermediate grafting density was obtained.[10] CuBr–N,N,N',N'',N''-pentamethyldiethylenetriamine (PMDETA) was used as catalyst in this tandem process because it proved to be compatible with the RAFT polymerization. Using 0.01 mol% catalyst a grafting density of 0.51 groups nm^{-2} according to elemental analysis was found. When 0.1 mol% copper catalyst was used a grafting density of 0.70 groups nm^{-2} was observed. With this large quantity of catalyst available the click reaction was faster and thus shorter RAFT polymers were clicked to nanoparticle **1** from where the polymerization continued. Using more catalyst in the tandem process thus made it resemble more the 'grafting-from' approach.

11.2.2 Cadmium Selenide-based Nanoparticles

Semiconducting nanoparticles such as CdSe are also known as quantum dots and have attracted widespread research interest because of their tuneable optical properties and potential applications in the field of biomedical imaging, photovoltaic cells and nanoelectronics.[1,11] Alkane phosphinoxide-ligands used to stabilize CdSe nanoparticles can be replaced with azide-modified phosphinoxide-ligands via ligand-desorption using pyridine and subsequent ligand exchange (Scheme 11.2).[12] In the same way alkane phosphinoxide ligands were replaced with acetylene phosphinoxide ligands.[13] Employing either CuBr–tris-(benzyltriazolylmethyl)amine (TBTA) catalysis or purely thermal conditions (95 °C) for the Huisgen cycloaddition reactions a variety of molecules, depicted in Scheme 11.2, bearing acetylene or azide groups were attached to nanoparticles **5** and **6** respectively.[12,13] Applying heat led to an inseparable mixture of 1,4- and 1,5-regioisomers as expected while the Cu catalyzed reactions provided exclusively 1,4-regioisomer in quantitative yields. To check if substrates were really covalently attached to the nanoparticles, nanoparticle **7c** was used. Nanoparticle **7c** contains barbituric acid which can bind to the Hamilton bisaminopyridine receptor as shown in Figure 11.1(c). Self-assembled monolayers on gold wafers containing the Hamilton receptor were treated with the barbituric acid nanoparticles **7c**. As a control experiment nonfunctional nanoparticles **4** were used. AFM measurements showed a dense layer of nanoparticles covering the gold wafer in the case where the nanoparticles **7c** were used [Figure 11.1(a)]. Relatively little nonspecifically bound particles were present in the control experiment [Figure 11.1(b)].

11.2.3 Ferric Oxide-based Nanoparticles

Magnetic iron oxide nanoparticles with the appropriate surface modifications can be used in a variety of applications. Examples of the use of magnetic nanoparticles (MNP) include magnetic resonance imaging contrast agents, biosensing, drug-delivery and hyperthermia.[14] Oleic acid ligands, coating magnetic nanoparticles, can be replaced with

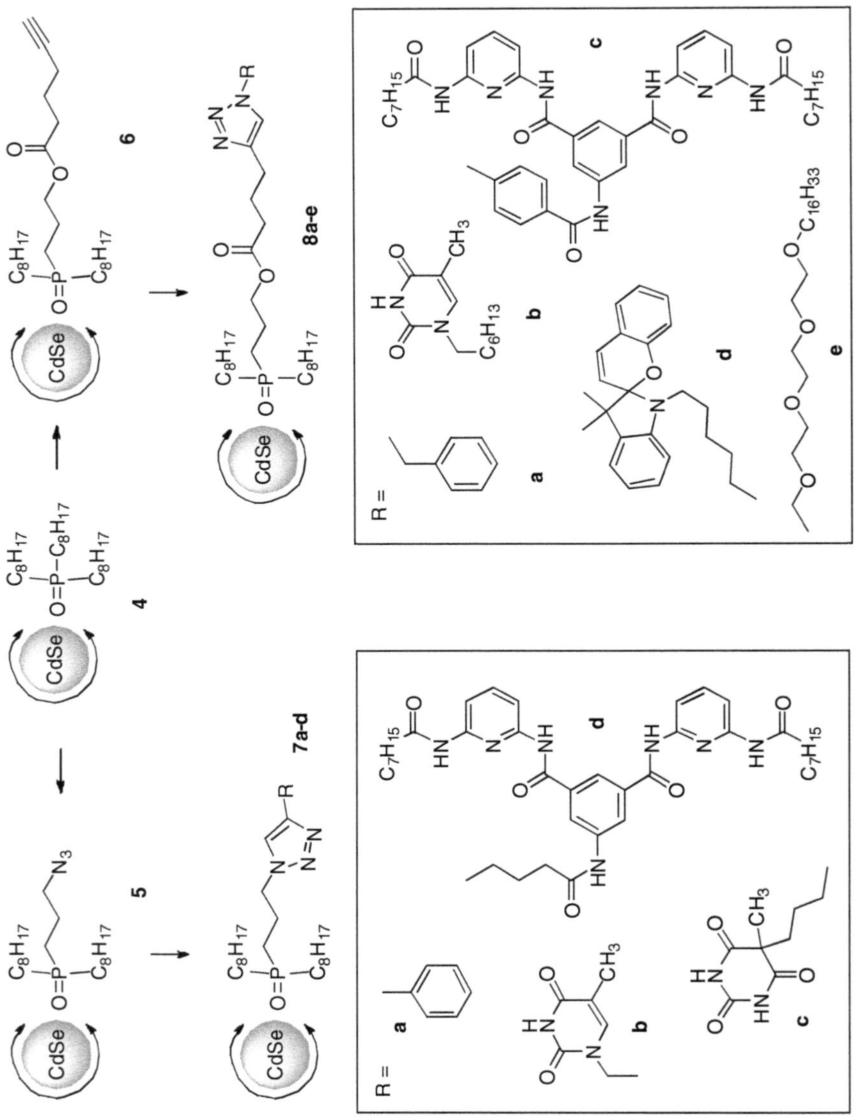

Scheme 11.2 *Replacement of phosphinoxide-ligands with azide or acetylene derivatives, and click reactions of functional molecules with nanoparticles* ***5*** *and* ***6***.[12,13]

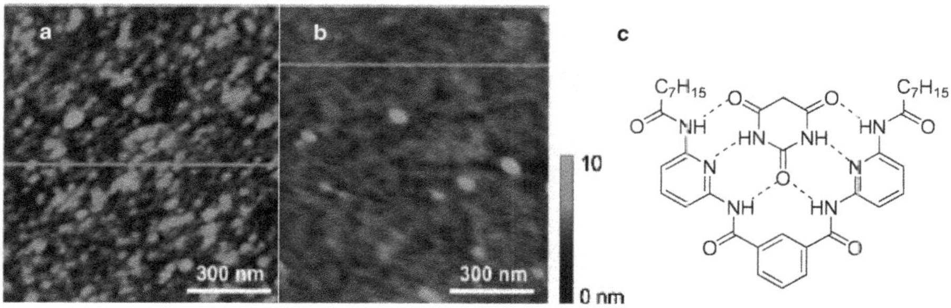

Figure 11.1 Gold surface containing the Hamilton receptor treated with nanoparticles **7d** (a) and **4** (b). Interaction between barbituric acid and the Hamilton receptor (c). Reprinted with permission from ref.[13]. Copyright 2007 Royal Society of Chemistry.

acetylene (**9**) or azide (**10**) containing ligands, providing functionality for further modification (Scheme 11.3).[15]

MNP **9** was attached to benzyl azide and MNP **10** to chloropentyne via the click reaction, using $CuSO_4$–sodium ascorbate as catalyst. After recovering the particles via extraction with organic solvents FTIR showed loss of azide respectively acetylene and appearance of triazole absorbance bands, indicating that the click reaction had proceeded. α-Acetylene-poly(*tert*-butyl acrylate) acid was clicked with MNP **10** under similar conditions to demonstrate that complex functionalities could be introduced. Again the FTIR spectrum showed disappearance of the azide peak.

Another application of these MNPs was explored by Gao and coworkers who investigated the possibility of recycling Pd catalysts.[16] They coated MNPs with a layer of silica and subsequently attached an azide functionalized silane to obtain MNP **11** (Scheme 11.4). Attachment of acetylene dipyridyl to MNP **11** was performed via a CuI-catalyzed click reaction. The corresponding Pd–dipyridyl complex was formed by refluxing the dipyridyl MNP together with $PdCl_2(MeCN)_2$ in toluene. MNP **12** was then used to catalyze Suzuki coupling reactions between aryl halides and aryl boronic acid. MNP **12** was easily removed

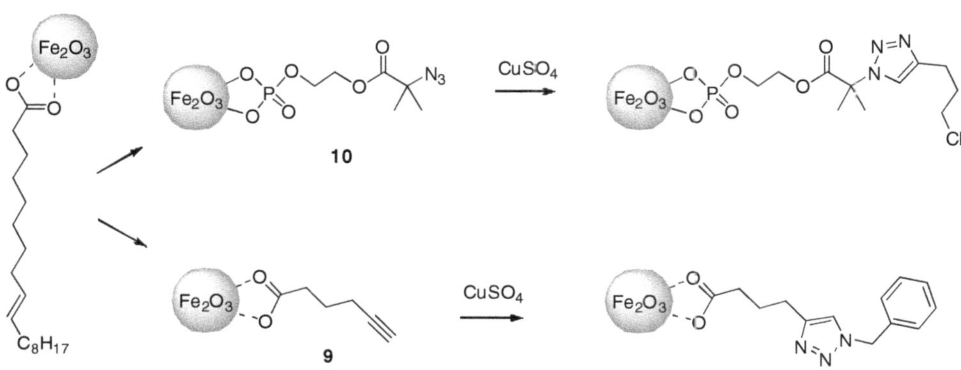

Scheme 11.3 Replacement of oleic acid from Fe_2O_3 with azide or acetylene functionalized ligands.[15]

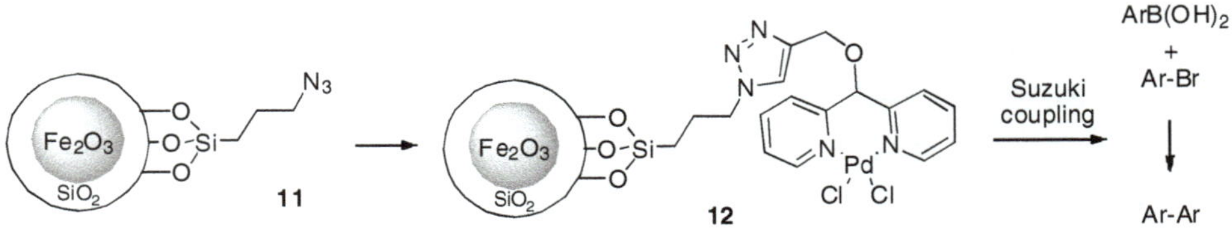

Scheme 11.4 *Functionalization of MNP with Pd catalyst. Reaction conditions for the click reaction: CuI; DIPEA; DMF–THF (1:1).*[16]

afterwards by applying an external magnetic field and it was shown that after several consecutive runs the complex retained activity.

Von Maltzahn *et al.* attached the so-called Lyp-1 targeting peptide to MNPs for precisely targeting biological locations.[17] LyP-1 binds to p32 mitochondrial proteins that are overexpressed in human cancers. The peptide sequence CGNKRTRGC was modified with an acetylene and a fluorophore as shown in Figure 11.2(a). Azido functional MNPs were coated with peptide **13** via copper-catalyzed Huisgen cycloaddition. The obtained MNP-**13** conjugates [Figure 11.2(b)] were stable *in vivo* for more than 5 h and were shown to accumulate in MDA-MB-435 cancer cells, see Figure 11.2(c), whereas control azido-bearing particles did not.

Several acetylene functional proteins, i.e. enhanced green fluorescent protein (EGFP),[18] maltose binding protein (MBP)[18] and human serum albumin (HSA),[19] were clicked to azide functional magnetic nanoparticles. Acetylene–EGFP and MBP were prepared via the expressed protein ligation method.[20] HSA was modified with an alkyne at the only

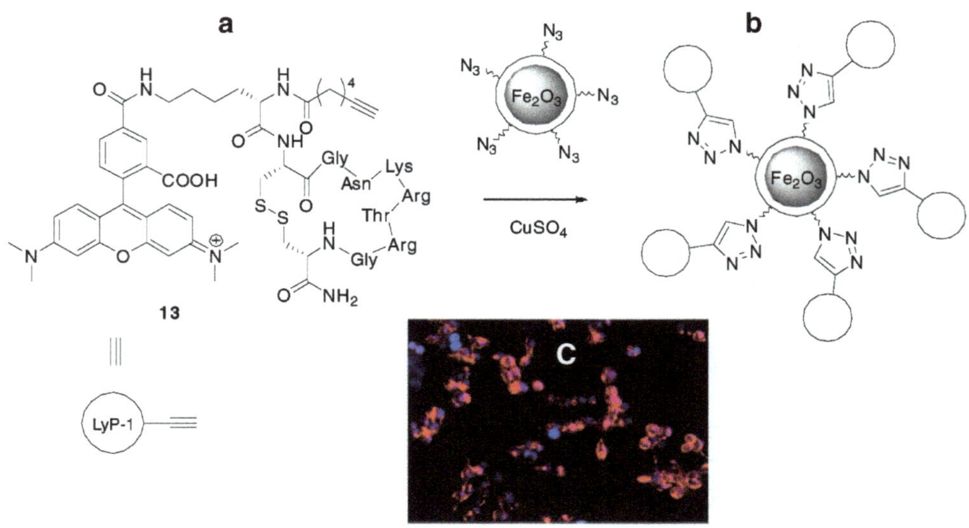

Figure 11.2 *(a) Modified LyP-1 targeting peptide. (b) MNP–**13** conjugate. (c) Fluorescence imaging of MDA-MB-435 cancer cells incubated with MNP–**13** conjugate (red), nuclear stain cells (blue). Reprinted with permission from ref.*[17]*. Copyright 2008 American Chemical Society.*

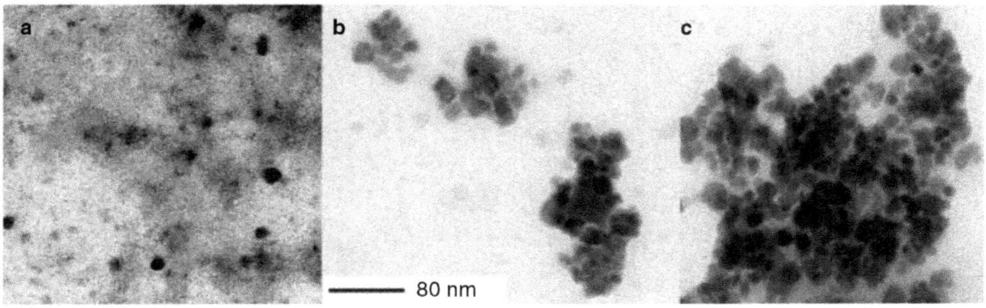

Figure 11.3 *TEM images of different stages of HSA-MNP agglomeration in response to addition of anti-HSA: (a) 0.02 mg ml^{-1} HSA-MNP in 0.1 M phosphate buffer (pH 7.0); (b) same as (a) with addition of 0.01 mg ml^{-1} anti-HSA; (c) same as A with addition of 0.10 mg ml^{-1} anti-HSA. Reprinted with permission from ref.*[19]*. Copyright 2008 Royal Society of Chemistry.*

cysteine residue present, via Michael addition of an alkyne-bearing acrylamide. EGFP with an acetylene attached to its C terminus was clicked with azido MNPs similar to particles **11** employing the Huisgen cycloaddition. Covalently bonded to the MNP, EGFP was shown to exhibit the expected fluorescence signal at 530 nm. MBP was attached to MNP in two different ways, site-specifically at its C terminus via the click reaction or randomly via amide bond formation with lysine and arginine residues. Both MBP-MNPs were incubated with biotinylated maltose and the bound maltose was visualized with fluorescently labeled streptavidin-Cy3. The fluorescence intensity of the site specifically attached MBP-MNP was approximately twice as high as the intensity of the MBP-MNP attached via random amide bond formation. The decreased fluorescence was explained by the fact that the random attachment can cause blocking of the binding site. Finally HSA-MNP conjugates were prepared via the click reaction. Solutions of HSA-MNP were incubated with different amounts of anti-HSA polyclonal antibody (which can bind two HSA units). The level of agglomeration of the HSA-MNP corresponded nicely to the amount of antibody added, as is shown in Figure 11.3.

11.2.4 Gold-based Nanoparticles

Gold nanoparticles possess chemically and physically interesting properties that potentially makes them employable in the field of medical diagnostics, catalysis and imaging.[21] To prevent gold nanoparticles from aggregating they are usually stabilized with alkanethiol ligands. To add functionality to gold nanoparticles the alkanethiol ligands can be replaced with functional ligands in an exchange process, yielding particles with diverse properties such as high water solubility and fluorescence. A drawback is the need to synthesize the individual thiol ligands separately. Therefore post-exchange modification is often performed instead, using a variety of reactions such as nucleophilic substitution, amide bond formation and olefin metathesis. Cycloaddition click chemistry is utilized in this respect as a complementary method to introduce new functionalities to gold nanoparticles under mild and orthogonal conditions. To modify gold nanoparticles via the CuAAC reaction, azides or alkynes have to be present at the nanoparticle surface. In Scheme 11.5 a variety of alkyne

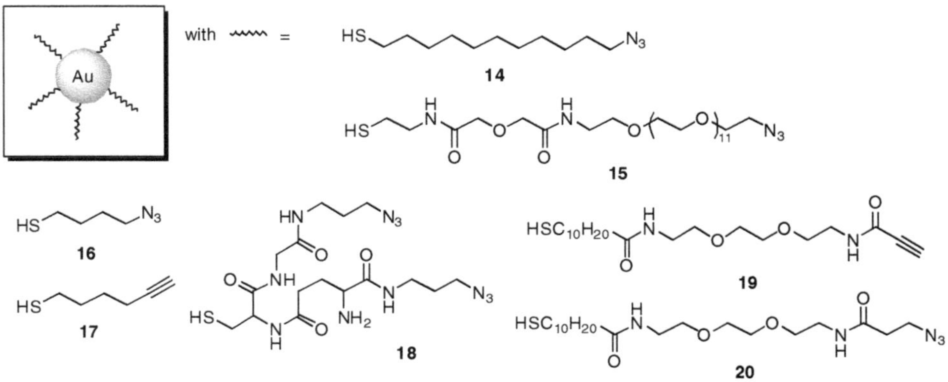

Scheme 11.5 *Examples of clickable gold nanoparticles (**14**,[22–25] **15**,[26] **16**,[27] **17**,[27] **18**,[28] **19**,[29] **20**[29]).*

and azide functional ligands are depicted, which have been used to replace alkane thiol ligands and hence introduce the desired functionality.

Williams and coworkers partially exchanged decanethiol ligands on gold nanoparticles with azide-functionalized ligands to obtain azide-functionalized gold nanoparticles **14** soluble in organic solvents.[23] These nanoparticles precipitate from solutions containing polar solvents, i.e. THF–H_2O, DCM–EtOH and dioxane. Because of the hydrophobicity of the particles, only click catalysts soluble in organic solvents were explored, including CuI, (PPh3)3-CuBr and CuBr–2,6-lutidine. Nevertheless, all catalysts tested were found to cause extensive particle aggregation. Others who used identical gold nanoparticles also observed this aggregation phenomenon.[25] Therefore, cycloaddition reactions were performed in the absence of a catalyst. To provide a more electron-withdrawing environment, which is known to enhance the rate of triazole formation, a series of electron poor acetylene containing molecules was used (Table 11.1).

Williams and coworkers reported conversions of azides to triazoles typically ranging from 5 to 15% after 60 h reaction time. Yields improved when dioxane–hexane 1:1 instead of neat dioxane was used as solvent. From varying the solvent system they concluded that the reactivity primarily depended on the solubility of the gold particles and alkynyl compounds. In a later report Williams *et al.* claimed that the conversion rate is more dependent on relative surface coverage with azide terminated alkanethiols, determined by quantitative FTIR spectroscopy, than on solvent and ligand length.[24] When the relative amount of azide-terminated ligands was lower, higher overall yields were observed. This phenomenon was attributed to steric interactions, leading to a reduced accessibility of azides as the click reaction progresses.

Sommer *et al.* showed that a catalyzed click reaction could be performed on the same type of gold nanoparticles under microwave conditions, which allowed the reaction times to be kept under 10 min.[22] A mixture of dioxane–*t*-BuOH–H_2O (1:1:0.5) was used as solvent and $CuSO_4$–sodium ascorbate as the catalytic system. With these short reaction times particle aggregation caused by the catalyst appeared to be negligible. However, when

Table 11.1 *Acetylene containing (small) molecules attached to gold nanoparticles via cycloaddition reactions*

Without catalyst, yields 5–54%[23]	Without catalyst[25]	$CuSO_4$–Na ascorbate catalyst, MW, 10 min, yields 78–100%[22]
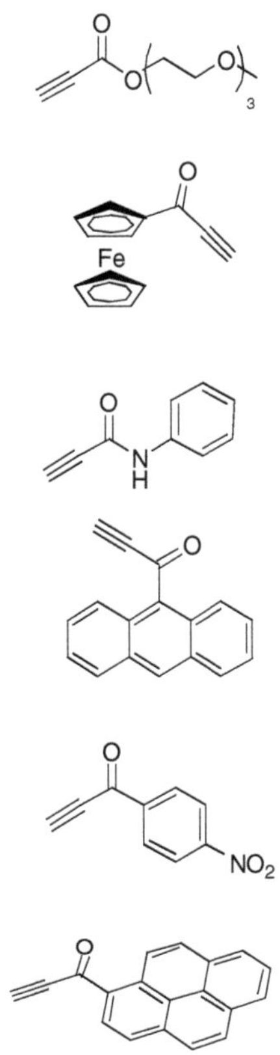	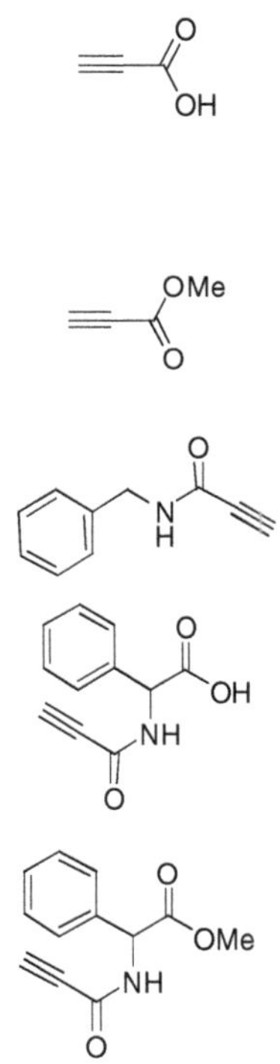	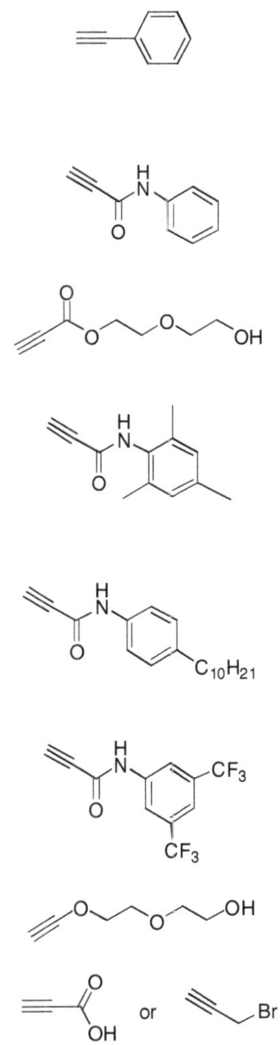

the particles were treated for more than 15 min particle aggregation was also observed. A variety of alkynes (Table 11.1) was clicked to the gold nanoparticles in a yield of 78–100%. In addition, Sommer and coworkers attached an *N*-heterocyclic carbene Pd complex to the gold nanoparticles and tested its catalytic activity in a Suzuki-type reaction. Using this

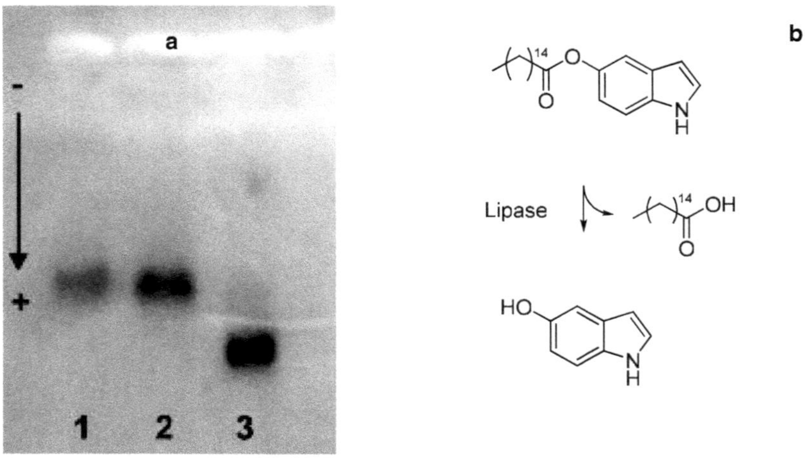

Figure 11.4 *(a) 1.5% agarose gel electrophoresis of AzNPs after click reaction. Lane 1: azide nanoparticles. Lane 2: azide nanoparticle-lipase reaction without copper. Lane 3: azide nanoparticle–lipase reaction with copper. (b) Hydrolysis of nonfluorescent ester, 5-O-palmitoylindole, to highly fluorescent 5-OH-indole. Reprinted with permission from ref.*[26]*. Copyright 2006 American Chemical Society.*

supported Pd catalyst a series of aryl chlorides was coupled with phenyl boronic acid in yields ranging from 85 to 99%.

Click reactions, however, can also be performed in water, thus gold nanoparticles coated with water-soluble ligand **15** were synthesized by Brust and coworkers.[26] They attached acetylene functionalized *thermomyces lanuginosus* lipase to azide-functionalized water-soluble gold nanoparticles. The lipase was genetically altered to express only one solvent-accessible lysine which was modified with carbodiimide chemistry to provide the required acetylene functionality. Subsequently, the click reaction was performed using $CuSO_4$–ascorbic acid. The gold nanoparticle–lipase conjugate was compared with two control samples by agarose gel electrophoresis to demonstrate the covalent connection of the lipase with the gold nanoparticles [Figure 11.4(a)]. The activity of the bound lipase was measured as function of the cleavage of nonfluorescent ester, 5-*O*-palmitoylindole, to highly fluorescent 5-OH-indole [Figure 11.4(b)]. Covalently attached to the gold nanoparticles the lipase retained its enzymatic activity. However, because the exact concentration of bound lipase per nanoparticle was unknown, the activity could not be compared with that of the free lipase. Reasoning the other way around, assuming that the activity is neither increased nor decreased upon binding, the lipase loading to the gold nanoparticle could be calculated. Based on the measured activity, the attachment of seven lipase molecules per nanoparticle was calculated.

In addition to spherical shaped gold nanoparticles, azide-functionalized gold nanorods of length to diameter ratio ~5 were also used to attach enzymes, viz. trypsin.[30] Nanorods coated with poly(4-styrenesulfonic acid-*co*-maleic acid) sodium salt (PSS-*co*-PMA) were synthesized and the carboxylic acid groups were subsequently used to attach NH_2-PEG-N_3 with the use of EDC as coupling reagent. Using $CuSO_4$–ascorbic acid as catalyst the nanorods were decorated with trypsin. The activity of this bioconjugate was determined by incubation with casein. The amount of tyrosine released is a measure for the activity of trypsin. Attached to the gold nanorod trypsin retained 57% of its biological activity compared with free trypsin. When trypsin was attached to the nanorods electrostatically or via direct EDC-coupling the activity was less then 20%. It should be noted, however, that in the latter two cases the poly(ethylene glycol) (PEG) linker was lacking, which provides more flexibility and thus increases the accessibility of trypsin.

In the above-mentioned studies, gold nanoparticles were successfully linked with enzymes and small molecules. A converse approach is to label larger structures with gold nanoparticles. Fischler *et al.* clicked azide functionalized gold nanoparticles with alkyne modified DNA strands.[28] To this end gold nanoparticles of approximately 3 nm coated with glutathione bisazide ligand **18** were synthesized. Acetylene functional DNA strands were obtained by replacing all thymine bases of a 294 base pair long fragment with alkyne-modified derivatives in a polymerase chain reaction (PCR) experiment. TBTA complexes of Cu(I) were used as catalyst in the click reaction to couple the gold nanoparticles to the DNA strand. In order to prevent cross-linking of the DNA strands as a result of the attachment of multiple DNA strands to one gold particle a 1000 fold excess of azide–gold nanoparticles was used. TEM images showed the nanoparticles to be equidistantly (2.8 ± 0.5 nm) assembled on the DNA (Figure 11.5). The shell of the gold particles and the DNA strand lack contrast in TEM and are therefore not visible. The spacing of the particles corresponded nicely with the expected value of twice the thickness of the shell, which is approximately 1.4 nm.

Vogge *et al.* synthesized gold nanorods with either acetylene (**17**) or azide functionalized ligands (**16**).[27] Applying a 1:1 concentration of the acetylene and azide rods in a Cu-catalyzed 1,3-dipolar cycloaddition reaction, chainlike assemblies were formed [Figure 11.6(a)]. This preferred end-to-end assembly of gold nanorods was previously observed by Caswell *et al.* and attributed to preferential ligand displacement at the (111) faces at the end of the rods.[31] If one of the functional rods was in excess compared to the other one, more complex assemblies were obtained. Zhou *et al.* used a mixture of azide functionalized gold nanoparticles and acetylene functionalized gold nanoparticles to visually detect Cu(II) ions.[29] They used gold nanoparticles **19** and **20** in a 1:1 ratio and added $CuSO_4$ to the

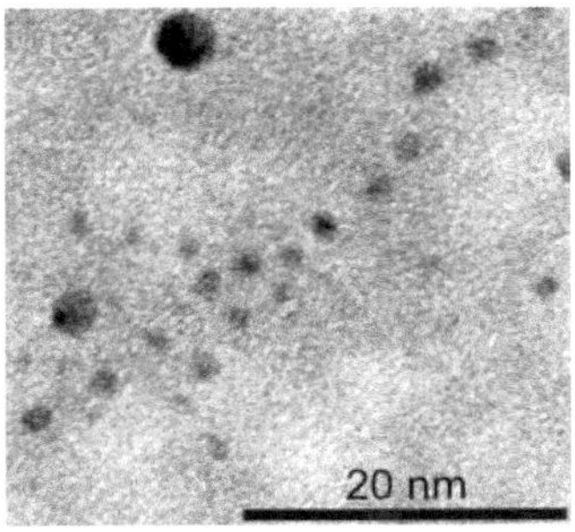

Figure 11.5 *TEM micrograph of one-dimensionally and equidistantly assembled gold nanoparticles on DNA immobilized on a TEM foil. Reprinted with permission from ref.[28]. Copyright 2008 Royal Society of Chemistry.*

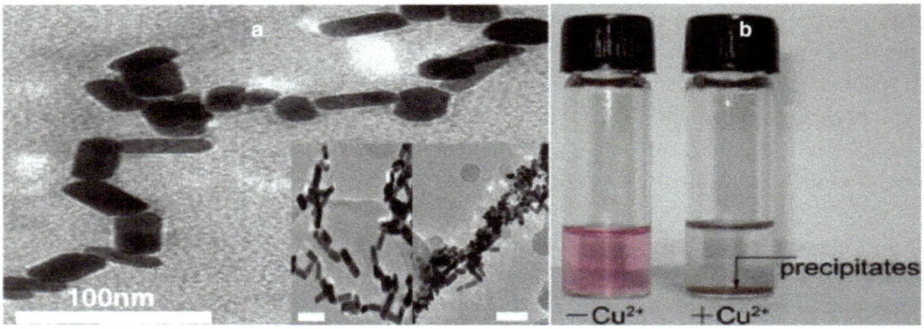

Figure 11.6 *(a) TEM pictures of chain like assemblies of gold nanorods obtained after the click reaction with 1:1 concentrations of nanoparticles* ***17*** *and* ***18****. Insert: the more complex assemblies if one of the compounds is present in excess. Reprinted with permission from ref.[27]. Copyright 2007 Elsevier. (b) Solution containing gold nanoparticles* ***19*** *and* ***20*** *before and after the addition of Cu(II) and sodium ascorbate. Reprinted with permission from ref.[29]. Copyright 2008 Wiley-VCH.*

red solution. As a result of the cross-linking of the particles the red color of the solution disappeared [Figure 11.6(b)].

With this assay Cu(II) ions could be detected with the naked eye down to a concentration of 50 μM. It was shown that the assay was insensitive to a variety of other metal ions and did not work without sodium ascorbate. Finally, 2.5 and 12 nm sized gold nanoparticles with acetylenes on their surface were conjugated with azide functionalized single walled nano tubes (SWNTs), as was demonstrated with electron microscopy (Figure 11.7).[32] The larger structures shown in Figure 11.7 were proposed to be bundles of SWNTs, since single SWNTs have a diameter of approximately 1 nm. The attached gold particles caused the semiconducting nanotubes to behave as conductors.

11.3 Carbon-based Nanomaterials

In the previous section was described how acetylene-labeled gold nanoparticles were conjugated with azide functionalized SWNTs. SWNTs are graphite-like allotropes of carbon

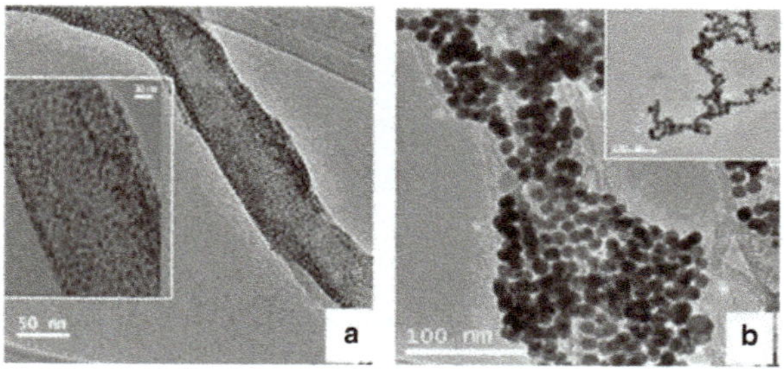

Figure 11.7 *Gold nanoparticles decorated with SWNTs via CuAAC: (a) 2.5 nm; (b) 12 nm. Reprinted with permission from ref.[32]. Copyright 2008 Institute of Physics.*

arranged in a cylindrical form; comparable allotropes arranged in a spherical form are referred to as fullerenes. Because of the versatile properties of SWNTs and fullerenes, they have often been utilized as molecular building blocks.

11.3.1 Fullerenes

Fullerenes show a wide range of interesting chemical and physical properties which make them attractive objects of investigation. However, the range of suitable reactions that can be applied to functionalize fullerenes is relatively limited, because of their high reactivity toward nucleophiles and their appropriate 2π character for the application in cycloaddition reactions.[33] To prevent side reactions with fullerene conjugates, often the fullerenes are introduced in the final step of a process. This turned out to be challenging for both acetylene and azide functional fullerenes since purification of a mixture of fullerenes with a different number of functional groups is usually impossible.

Zhang *et al.* were able to functionalize a C_{60} fullerene with an acetylene via a Diels–Alder reaction with 2-trimethylsiloxy-1,3-butadiene, followed by reduction of the resulting cyclohexanone and subsequent esterification of the corresponding cyclohexanol (Scheme 11.6).[34] Using CuBr–PMDETA as catalyst acetylene-C_{60} was reacted with monofunctionalized polystyrene-azide (PS-N_3) of three different molecular weights (2, 6 and 10 kg mol^{-1}). These conjugates showed an unusual SEC behavior as depicted in Figure 11.8. Compared with unmodified 2 kg mol^{-1} PS-N_3 the retention volume of the fullerene conjugate was larger, as a result of which the molecular weight distribution appeared lower. For the 6 kg mol^{-1} polymer this observation was less obvious while the 10 kg mol^{-1} polymer did not show any such behavior. Apparently the C_{60} has a negative impact on the hydrodynamic volume of the polymer, especially in the case of the 2 kg mol^{-1} polymer. It was speculated that the lack of solubility of C_{60} in the solvent (THF) causes the polymer to shield the fullerene from the environment, which consequentially leads to a decreased hydrodynamic volume. In the case of the longer chains this effect is less significant.

To demonstrate that click reactions could be performed with fullerenes, Niergarten and coworkers prepared mono-acetylene, di-acetylene and di-azide functional C_{60} fullerenes.[35] Click reactions with either benzyl azide or 3-phenyl-1-propyne were utilized to functionalize these fullerenes. When the fullerenes did not dissolve well, side reactions (most probably cycloadditions of the azides to the fullerene core) were observed. These side reactions were

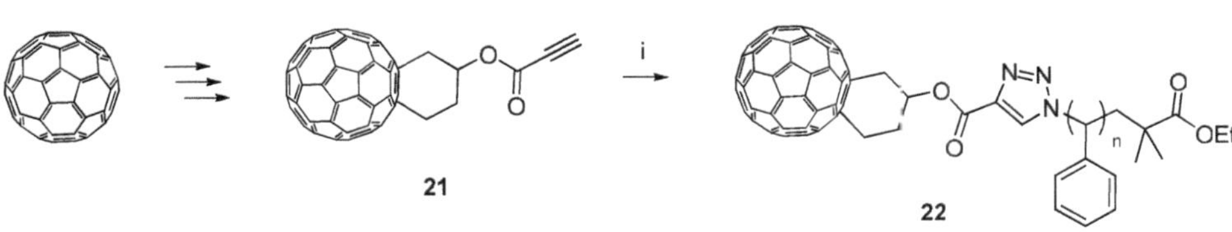

*Scheme 11.6 Modification of C_{60} with an acetylene group to acetylene–C_{60} **21** and subsequent click reaction with azide terminated PS to PS–C_{60} **22**. Reaction conditions: (i) CuBr, PMDETA, toluene, room temperature (r.t.).*[34]

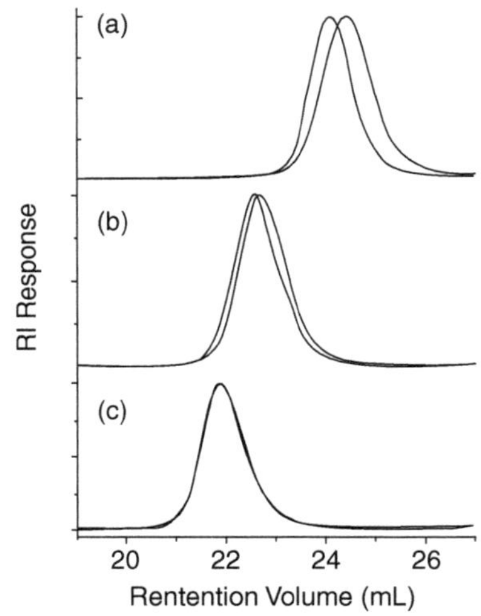

Figure 11.8 *Overlay of SEC traces of PS-N_3 and PS-C_{60}. (a) $M_n = 2000$; (b) $M_n = 6100$; (c) $M_n = 10\,000$ for PS. Reprinted with permission from ref.[34]. Copyright 2008 American Chemical Society.*

reduced in the case of the fullerene bis-adduct because of its lower chemical reactivity. Niergarten also linked azides with acetylene fullerenes; mono-acetylene fullerene was attached to a di-azide fullerene and the expected triple fullerene conjugate was obtained according to mass spectrometry. Besides mono- and di-functional fullerenes a symmetrical addition pattern of C_{60} is also possible (the octahedral addition pattern with six adducts). Most previously reported C_{60} hexakis adducts were prepared via a one-pot synthesis using cycloaddition reactions with malonates. For small molecules this is an effective procedure, but for more complex structures yields are often found to be low. Therefore, fullerene hexakis-adducts with an octahedral addition pattern were prepared (Scheme 11.7), containing two azides per adduct, hence 12 in total.[36] With azides present on the fullerene six different acetylene-containing molecules, including a Zn(II)-porphyrin, were grafted to the fullerene in 12-fold with yields ranging from 56 to 81%.

In biological molecular recognition events the multivalent display of saccharides is an important principle. Such a multivalent display of functional groups can be achieved in a straightforward manner starting from fullerenes. Shiga-like toxins for example were shown to be neutralized efficiently by a pentavalent, water-soluble carbohydrate.[37] The AB_5 shiga-like toxins are produced by *Escherichia coli* serotype O157:H7 and cause dysentery in humans. Shiga-like toxin is built up from one A subunit responsible for its toxicity and five B subunits arranged in a C_5-symmetrical way for recognition of specific cell types. Isobe *et al.* have reported a C_5-symmetrical fullerene functionalized with five monosaccharides via sulfide linkages.[38] When they wanted to apply their methodology to larger saccharides the reaction was sluggish and many by-products were formed. To overcome this problem they synthesized a fullerene-bearing five acetylene moieties.[39] This C_5-symmetric acetylene C_{60} was attached to a variety of molecules including a trisaccharide using the CuAAC reaction. Under the mild click conditions yields of 86% were achieved.

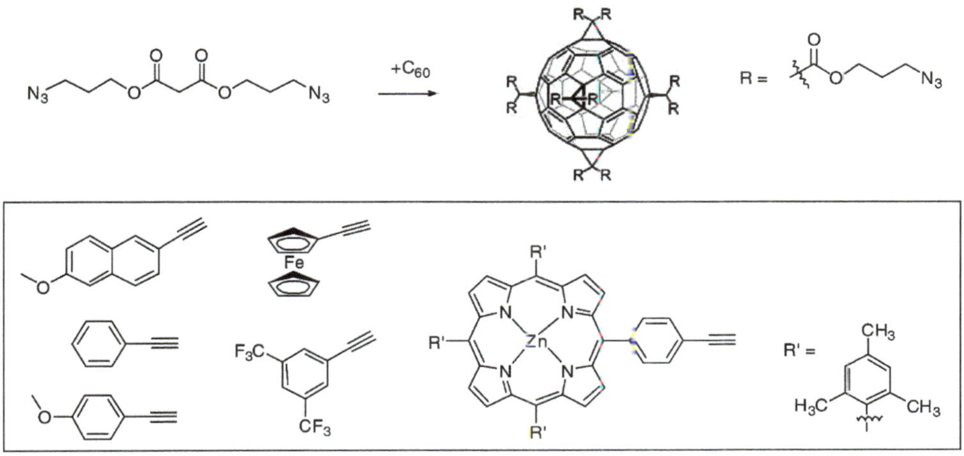

Scheme 11.7 Functionalization of C_{60} with 12 azides in an octahedral addition pattern and the acetylene molecules applied in the click reaction.[36]

The obtained penta-trisaccharide is expected to be an effective receptor for shiga-like toxin (Figure 11.9).

11.3.2 Carbon Nanotubes

Single walled carbon nanotubes are cylindrical sheets of graphite with a diameter of usually several nanometers and a possible length of up to several millimeters.[40] SWNTs are among the stiffest and strongest fibers known and possess remarkable electronic properties along

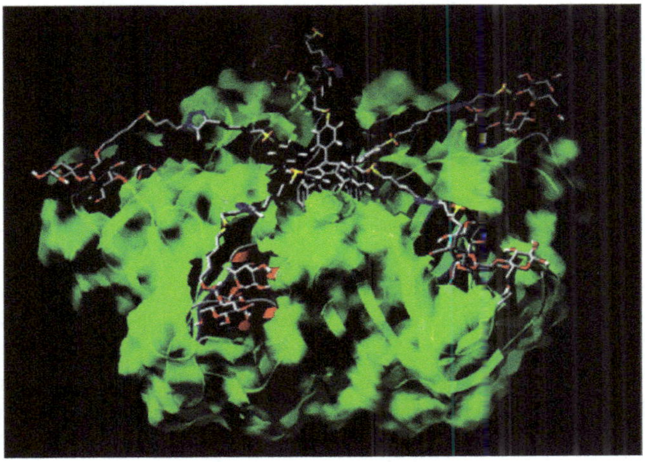

Figure 11.9 Molecular model of shiga-like toxin (the molecular surface shown in green) complexed with a C_{60} trisaccharide conjugate (stick model).[39]

with other unique characteristics, for example SWNTs can be semi-conducting or metallic depending on the rearrangement of their hexagon rings. In the section on metal nanoparticles it has already been described how SWNTs were coupled to gold nanoparticles via the Huisgen cycloaddition reaction.

Because of their graphite-like structure, SWNTs are insoluble in most organic and aqueous solvents. Much research effort has been put into trying to overcome this problem. Adronov and coworkers have used the click reaction to graft azide–PS on acetylene-SWNTs (Scheme 11.8) in order to increase solubility in the solvents in which PS is soluble.[41] Reactions were catalyzed with $(PPh_3)_3CuBr$ or CuI–1,8-diazabicyclo[5.4.0-undec-7-een] (DBU) in DMF; the system with CuI–DBU gave the best results. After treating this PS–SWNT conjugate with acetyl sulfate the material dissolved in water.[42] To demonstrate their increased solubility SWNT-PS and SWNT-sulfonated-PS (SWNT-PSS) were dissolved in an H_2O–DCM mixture; Figure 11.10 shows that SWNT-PS dissolves in DCM and SWNT-sulfonated-PS in H_2O.

Moreover, cyclodextrins (CDs) were successfully applied by Zheng *et al.* to functionalize SWNTs. The CD-SWNTs were prepared using CuI–DBU in DMF as solvent (Scheme 11.8).[43] Cyclodextrins are known to bind various organic and biological guests within their hydrophobic cavity in aqueous solution and CD-SWNTs hybrids are therefore anticipated as interesting supramolecular building blocks. To test the binding ability of the covalently linked CDs, the association of quinine with the CD-SWNTs was examined by fluorescence spectroscopy. As can be seen in Figure 11.11, the fluorescence of quinine was quenched upon the addition of the CD-SWNTs, indicating binding of the CD-SWNTs to quinine.

Phthalocyanines are planar electron-rich aromatic macrocycles that possess remarkably high extinction coefficients in the red/near-infrared region. A combination of such an electron donor molecule with the electron accepting carbon framework of SWNTs might lead to new photoactive materials. Such materials are potentially applicable in the development of, for example, efficient photovoltaic cells. Campidelli *et al.* applied $CuSO_4$–sodium ascorbate to react azide zinc–phthalocyanine (ZnPc) with acetylene–SWNT to obtain a single SWNT–ZnPc conjugate **23** (Scheme 11.8).[44] To test the photovoltaic properties of the complex, SWNT–ZnPc conjugate **23** was deposited on indium tin oxide (ITO)-coated glass slides. These glass slides were built into a photoelectrochemical cell as depicted in Figure 11.12. Reproducible photocurrents with monochromatic incident photon-to-current conversion efficiency (IPCE) values of 17.3% were measured for the SWNT–ZnPc conjugate. The monochromatic IPCE value for SWNTs coated with trimethylsilyl-protected 4-ethynylbenzene (a precursor for the SWNT-ZnPc conjugate) was more than 50% less.

Another class of graphite-like structures was prepared by Mynar *et al.* They made a graphitic diblock nanotube with an azide-covered surface via self-assembly of compound **24** (Figure 11.13).[45]

The diblock nanotubes were attached to dendron **25** via the CuAAC reaction as a suspension in THF–MeOH–hexane. AFM images showed that the diameter of dendron conjugated assemblies increased 4 nm compared to the nonfunctionalized nanotubes. The size of the dendron is roughly 2 nm and therefore this observation indicates that the exterior of the graphitic diblock nanotube is fully occupied with dendron **25**.

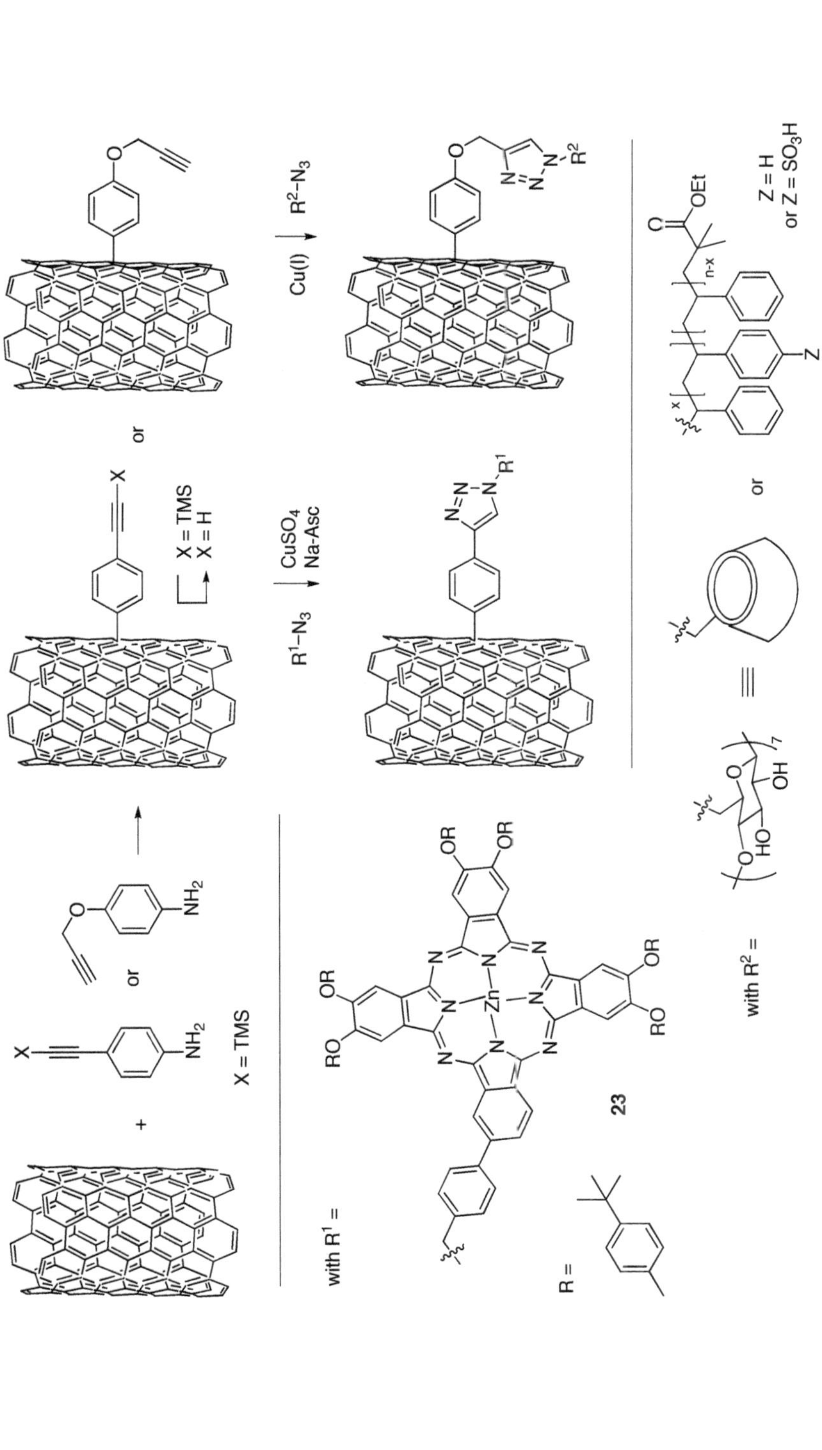

Scheme 11.8 *Functionalization of SWNTs with alkyne moieties and subsequent click reaction with azido-zinc-phthalocyanine,[44] azido-PS, azido-PSS[42] and azido-β-cyclodextrin.[43]*

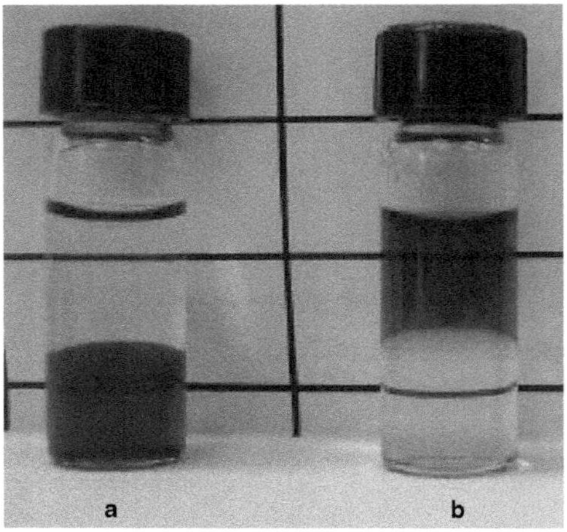

Figure 11.10 *Photograph of two separate SWNTs samples in CH_2Cl_2–H_2O. (a) Polystyrene functionalized SWNTs; (b) sulfonated polystyrene functionalized SWNTs. Reprinted with permission from ref.[42]. Copyright 2007 Elsevier.*

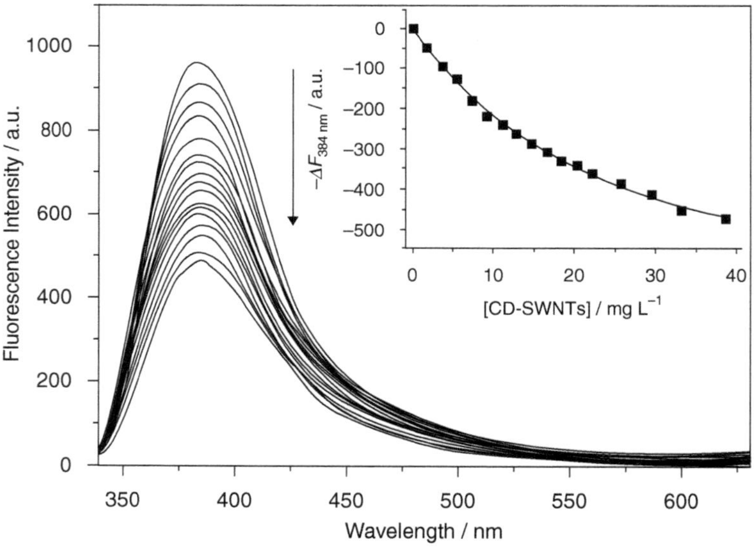

Figure 11.11 *Fluorescence spectral changes of quinine (2.34×10^{-5}mol l^{-1}) upon addition of CD-SWNTs (0–39 mg l^{-1} from top to bottom) in aqueous buffer solution (pH = 7.4, 25 °C). Insert: differential fluorescence intensity (ΔF) vs the concentration of CD-SWNTs relationship. The excitation wavelength is 330 nm. Reprinted with permission from ref.[43]. Copyright 2008 Springer Science and Business Media.*

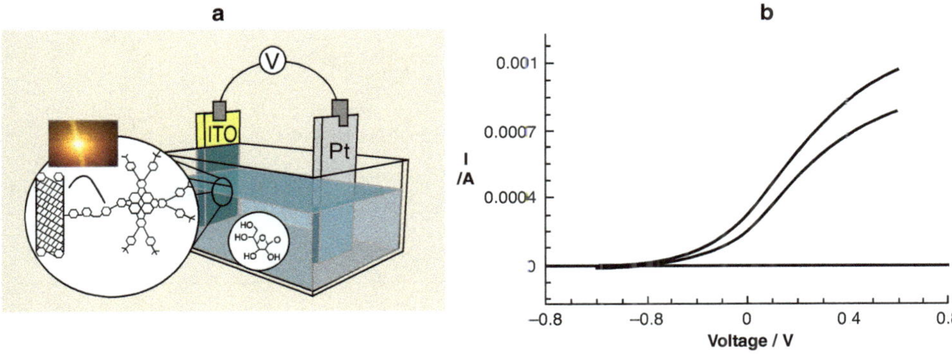

Figure 11.12 *(a) Schematic representation of the photoelectrochemical cell used for the IPCE measurements. (b) I–V characteristics of the SWNT–ZnPc conjugate under white light illumination, gray line, and in the dark, black line. Three-electrode setup, 0.1 M Na_3PO_4, 1 mM sodium ascorbate, N_2 purged. Voltages measured vs an Ag–AgCl reference electrode (0.1 M KCl). Reprinted with permission from ref.[44]. Copyright 2008 American Chemical Society.*

11.4 Self-assembled Organic Structures

In addition to the self-assembled nanotubes described in the previous section, here the functionalization of different types of self-assembled organic structures is surveyed. Huisgen cycloaddition reactions performed to functionalize liposomes, polymersomes and polymer

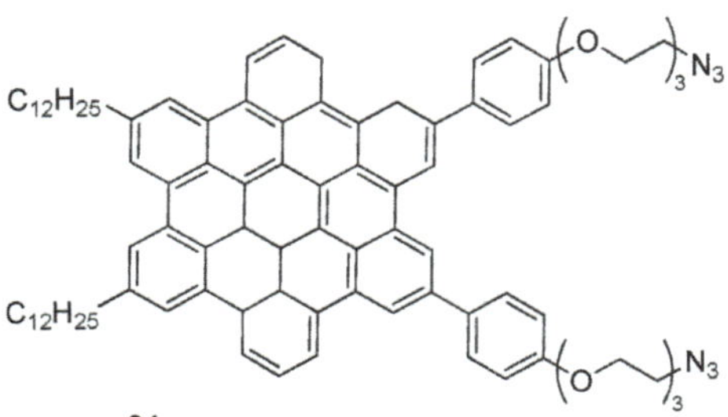

24

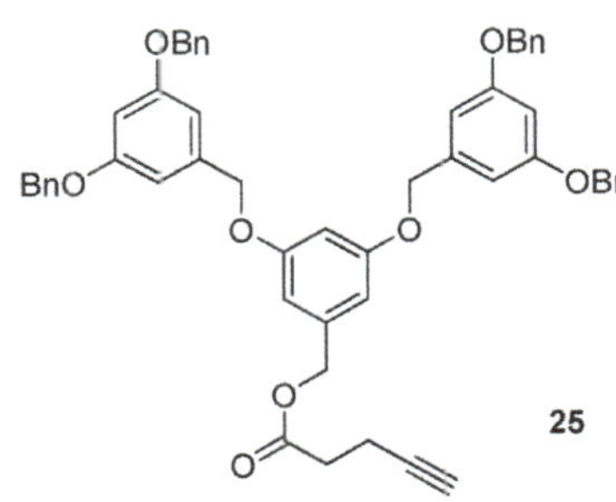

25

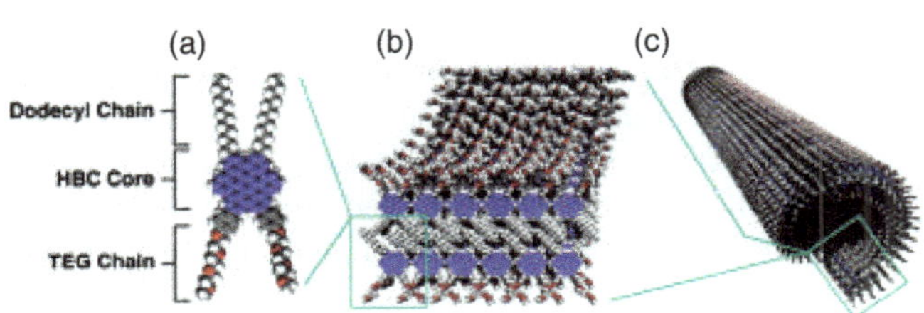

Figure 11.13 *Schematic structures of (a)* ***24****, (b) self-assembled bilayer tape, (c) graphitic nanotube and structures of Hexa-peri-hexabenzocoronene (HBC)* ***24*** *and dendron* ***25****. Reprinted with permission from ref.[45]. Copyright 2008 American Chemical Society.*

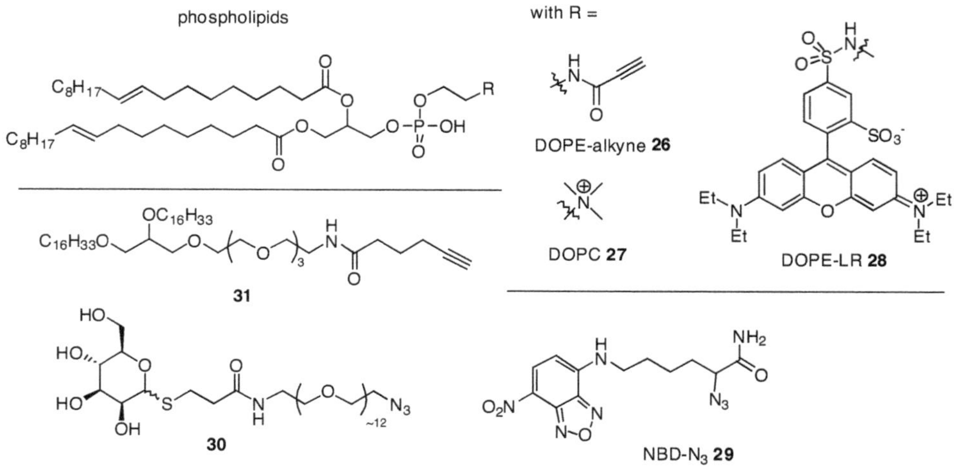

Scheme 11.9 *Structures of phospholipids **26, 27, 28** and **31**, NBD derivative **29** and mannose derivative **30**.*[47,48]

micelles will be discussed. Liposomes are structures similar to cells composed of an (aqueous) solution enclosed by a membrane. They are formed through self-assembly of naturally occurring phospholipids into a bilayer structure. Another example of amphiphilic molecules that can self-assemble are block-copolymers composed of a hydrophilic and a hydrophobic segment. They can assemble into, for example, polymersomes, the polymeric equivalent of liposomes, and polymeric micelles.

11.4.1 Liposomes

Surface modification of liposomes with for example peptides or proteins is used to obtain liposomes that can for instance be used for targeting to specific cells.[46] To perform surface modification of liposomes with the click reaction, Kros and coworkers prepared liposomes bearing alkyne groups at their surface.[47] This surface modification was done by preparing liposomes from a mixture of three different phospholipids, i.e. DOPE-acetylene (**26**), DOPC (**27**) and DOPE-LR (lissamine rhodamine) (**28**) (Scheme 11.9). The different phospholipids were mixed in a ratio of 50:49:1 DOPE-acetylene–DOPC–DOPE-LR. Unilamelllar liposomes (**32**) with an average diameter in the range of 110–120 nm were formed. Liposomes obtained with this procedure were attached to nitro-benzoxadiazole (NBD) derivative **29** via the Huisgen cycloaddition reaction using CuBr as catalyst. After attachment of **29** a Förster resonance energy transfer (FRET) effect between NBD and LR was observed, indicative that the modification occurred at the liposome surface. (Scheme 11.10). In the control sample without a copper catalyst only a minor FRET effect was observed, probably due to random presence of the NBD.

Hassane *et al.* prepared acetylene containing liposomes **33** (Scheme 11.10) from DPPC, DPPG, cholesterol and 5–10 mol% acetylene modified lipid **31** (Scheme 11.9).[48] The obtained liposomes were clicked with azido-mannose **30** using $CuSO_4$, sodium ascorbate and bathophenanthroline disulfonic acid disodium salt (BPhT) as catalyst system.

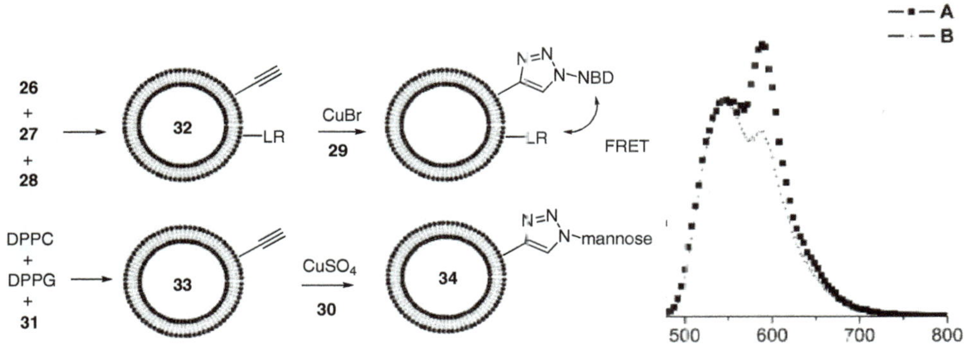

Scheme 11.10 *Top left: CuAAC reaction of LR and acetylene functional liposomes with NBD-derivative **29**. Right: FRET (excitation λ = 470 nm) spectra (**A** with CuBr and **B** for negative control without CuBr).[47] Bottom left: Attachment of azido-mannose **30** to liposome **33**.[48]*

To determine the mannose accessibility when conjugated to the liposome surface, concanavalin A (ConA), a sugar binding protein, was added. After addition of ConA to liposomes **34** in HEPES-buffered saline (HBS) (pH 6.5), an increase in turbidity was observed which was assessed by measuring the increase in optical density at 360 nm. Addition of unbound mannose resulted in immediate decrease in turbidity. This phenomenon was not observed for control liposomes lacking mannose.

11.4.2 Polymersomes

Polymersomes are architectures formed by self-assembly of amphiphilic block copolymers into bilayer structures that bear a resemblance to liposomes. Polymersomes, however, show a much greater stability compared with their low molecular weight counterparts.[49]

Van Hest and coworkers synthesized a diblock copolymer of poly(acrylic acid) (PAA) and polystyrene (PS) terminated with an azide group at the PAA side (Scheme 11.11).[50] After self-assembly of amphiphilic block copolymer **35**, polymersome **36** was formed with azide groups present at its periphery.

Azide containing polymersome **36** was further functionalized with various acetylene-containing substrates. Using TBTA or BPhT as ligands for the Cu(I) catalyzed cycloaddition, dansyl-acetylene was conjugated to the azide polymersomes. To measure the degree

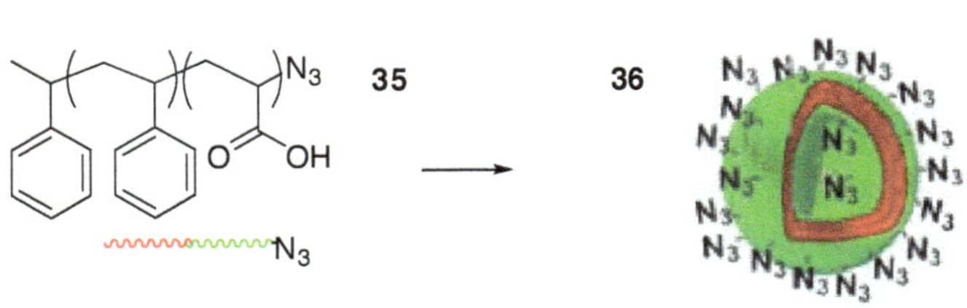

Scheme 11.11 *Structures of copolymer **35** (PS-b-PAA) and azide containing polymersome **36**.[50]*

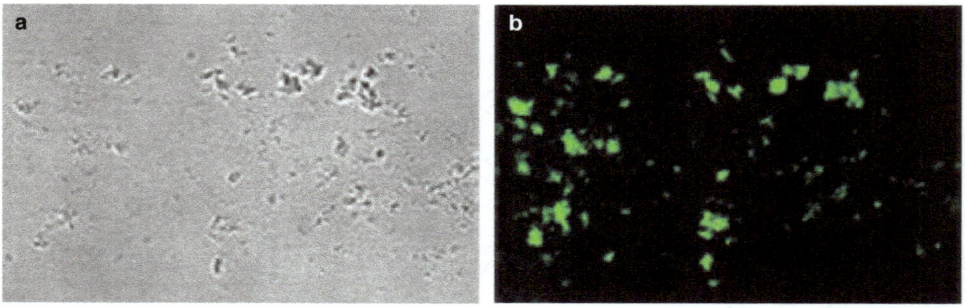

Figure 11.14 Confocal laser-scanning microscopy images [transmission (A) and fluorescence excited at 488 nm (B)].[50]

of functionalization, a reference compound (dansyl-PAA-*b*-PS) was synthesized. By comparison of the fluorescence of re-dissolved polymersomes and the reference compound it was calculated that about 23% of all azides were converted. Since the interior azides are unavailable for reaction, approximately 40–50% of the theoretical available azide-groups had reacted. To prove that the dansyl was really covalently attached, SEC traces of the re-dissolved block copolymers were recorded. After conjugation with dansyl the block copolymers were detected at 345 nm whereas unreacted block copolymer **35** did not give a signal at this wavelength.

Under the same conditions, enhanced green fluorescent protein (EGFP) was clicked to azido-polymersomes **36**. EGFP was equipped with acetylenes by reaction of pentynoic acid *N*-succinimidyl ester with one or more of the 20 available lysine residues. The EGFP-polymersomes showed fluorescence, which was detected by confocal laser scanning microscopy (Figure 11.14). When copper was omitted, the polymersomes did not show any fluorescence after dialysis.

Additionally, van Hest and coworkers designed an acetylene-PEG-*b*-PS block copolymer for the co-aggregation into PS-*b*-PIAT {polystyrene-poly[L-isocyanoalanine(2-thiophen-3-yl-ethyl)amide]}-based polymersomes.[51] Polymersomes prepared from PS-*b*-PEG or PS-*b*-PAA have the advantage that they are easily post modified. However, polymersomes based on these polymer constructs are impermeable for most organic substrates. PS-*b*-PIAT-based polymersomes, in contrast, are permeable for organic substrates. To obtain acetylene functionalized PS-*b*-PIAT polymersomes, acetylene-PEG-*b*-PS was co-aggregated into its bilayer membrane. Based on a calibration curve measured with SEC, the incorporation efficiency of acetylene-PEG-*b*-PS into PS-*b*-PIAT polymersomes was calculated to be 85%. Finally, an azido-functional lipase (azido-CalB) was clicked to acetylene polymersomes in the presence of $CuSO_4$, sodium ascorbate and bathophenanthroline. Hydrolysis of 6,8-difluoro-7-hydroxy-4-methylcoumarin octanoate (DiFMU octanoate) by *Candida antarctica* lipase B (CalB), yielding fluorescent difluoro-coumarin, was used to check the protein's activity.[52] CalB polymersomes showed the expected enzymatic activity whereas control experiments did not (Figure 11.15).

Another type of polymersome was constructed by Li *et al.*, by mixing azide-terminated polybutadiene-*b*-poly(ethylene oxide) (PBD-*b*-PEO-N_3) with hydroxyl terminated PBD-*b*-PEO-OH to obtain azide functional polymersomes.[53] These polymersomes were

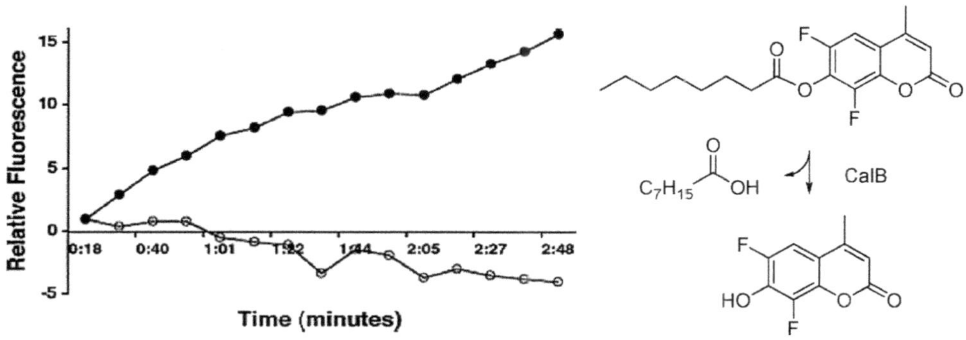

Figure 11.15 *Activity measurement of azido-functional CalB conjugated to a polymersome surface (•) compared with a blank experiment where acetylene-PEG-b-PS was not embedded in the polymersomes (○), via monitoring the hydrolysis of DiFMU octanoate to yield fluorescent difluoro-coumarin. Reprinted with permission from ref.[52]. Copyright 2008 Wiley-VCH.*

subsequently attached to an acetylene functional polyester dendron via the CuAAC reaction (Figure 11.16). To prove that the dendron was covalently attached, it was modified with a rhodamine derivative before performing the click reaction. This modification was done in such a way that approximately one rhodamine moiety was introduced per dendron. The yield of the conjugation was measured as a function of the PBD-*b*-PEO-N_3 concentration relative to PBD-*b*-PEO-OH. Again 50% of PBD-*b*-PEO-N_3 was expected to be inaccessible at the polymersome interior.

At low concentrations the conjugation yields were higher, as anticipated based upon the expected accessibility and crowdedness. The authors suggested that some of the interior azides had moved to the surface during the 24 h course of the reaction. Typically, increasing the amount of azide block copolymer in the mixture gave a decreased conjugation yield,

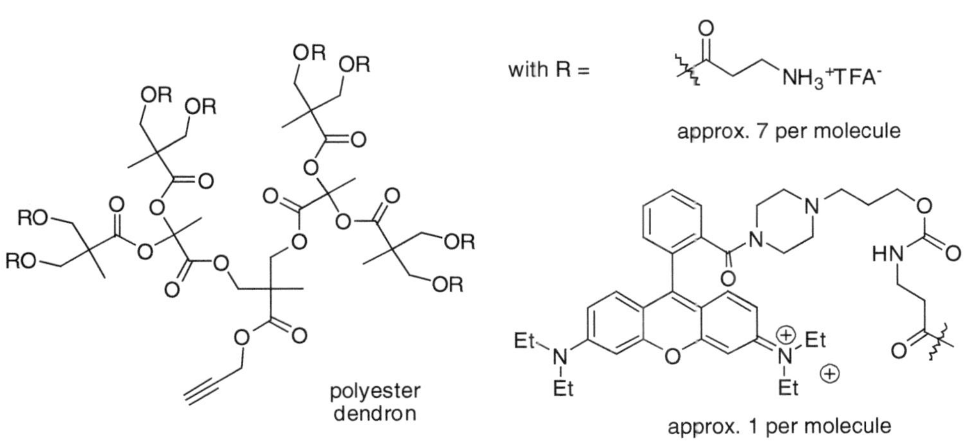

Figure 11.16 *Polyester dendron labeled with approximately one rhodamine moiety per molecule.*[53]

probably due to steric hindrance between the dendrons. Moreover, a deformation could be observed for polymersomes containing more than 40% of the azide amphiphile.

11.4.3 Micelles and Cross-linked Nanoparticles

As mentioned in the introduction of this chapter, besides polymersomes, amphiphilic block copolymers can also assemble into polymeric micelles. When polymeric micelles are cross-linked, well-defined and stable nanoparticles emerge. The dimensions and other properties of these cross-linked nanoparticles can easily be altered by changing the amphiphilic block copolymer properties. Polymeric micelle systems have attracted a lot of interest, e.g. as drug delivery vehicles or nanoreactors.[54]

Wooley and coworkers prepared micelles **38a–b** from PAA-*b*-PS **37** and used the PAA block for cross-linking and the introduction of azides or acetylenes (Scheme 11.12).[55] Block copolymer **37** was synthesized via consecutive nitroxide-mediated polymerization of *tert*-butyl acrylate and styrene. Afterwards the *tert*-butyl esters were removed with trifluoro acetic acid in dichloromethane. Micelles were formed by addition of water to a solution of the block copolymer in THF. After the formation of micelles, either azides (**38a**) or acetylenes (**38b**) were introduced to the micelle shell by EDC-coupling of respectively 3−azidopropylamine or propargylamine to the carboxylic acid functionalities of the PAA. The nanoparticles were subsequently shell cross-linked through amidation with 2,2′−(ethylenedioxy)bis(ethylamine). The authors showed that azide or acetylene functional fluorescein could be covalently linked to the cross-linked shell with the appropriate functionality through the Cu-catalyzed Huisgen cycloaddition reaction.

To prove that the fluorescein was linked covalently, a purified sample was centrifuged at 5000 rpm and emission spectra were recorded at different positions across the cell. The expected absorption maximum was only visible at the cell bottom and not at the middle or top. For the authors this was indicative that there was no free dye in solution and thus the dye was bound covalently to the cross-linked micelles.

To introduce azides or acetylenes to the core, the PS block was modified.[55,56] The incorporation of azides to the core was achieved via polymer **39** and after assembly into micelles the chlorides were replaced with azides. Acetylene functional core micelles were prepared via polymer **40**, which was deprotected to polymer **41** and subsequently self-assembled. Using the CuAAC reaction acetylene dansyl was conjugated with micelle **42a** and azide coumarin was attached to **42b**; in both cases IR and NMR spectroscopy confirmed a covalent attachment to the coronas.

Acetylene functional micelles were also shell cross-linked[57] and core cross-linked[58] via the 1,3-dipolar cycloaddition reaction with azide dendrimers by Wooley and coworkers. Micelles **38b** and **42b** were prepared and were subsequently subjected to click reactions with the azide dendrimers shown in Scheme 11.13. After the addition and subsequent purification by dialysis, dynamic light scattering (DLS) showed that the size of cross-linked particles was independent of concentration and temperature. Furthermore, it was shown the shell was only effectively cross-linked with the first-order dendrimer whereas the core was cross-linked for all dendrimer generations. It was suggested that the hydrophobic nature of the dendrimers prevented efficient cross-linking of the hydrophilic shell.

Functionalities introduced at the shell or in the core of polymer micelles are likely to be sterically and electronically hindered. To mimic a more solution-like environment

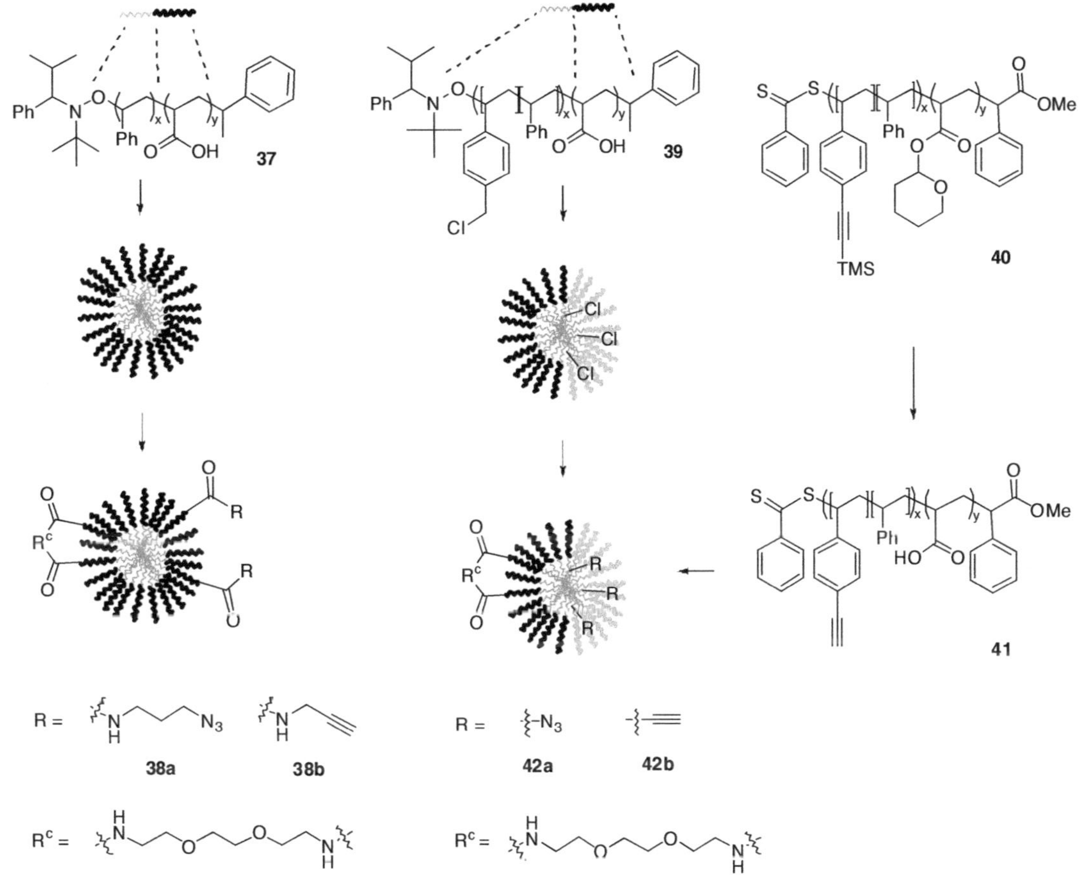

Scheme 11.12 *Preparation of cross-linked micelles functionalized in their shell with azides **38a** or acetylenes **38b** and functionalized in their core with azides **42a** and acetylenes **42b**.*[55–57]

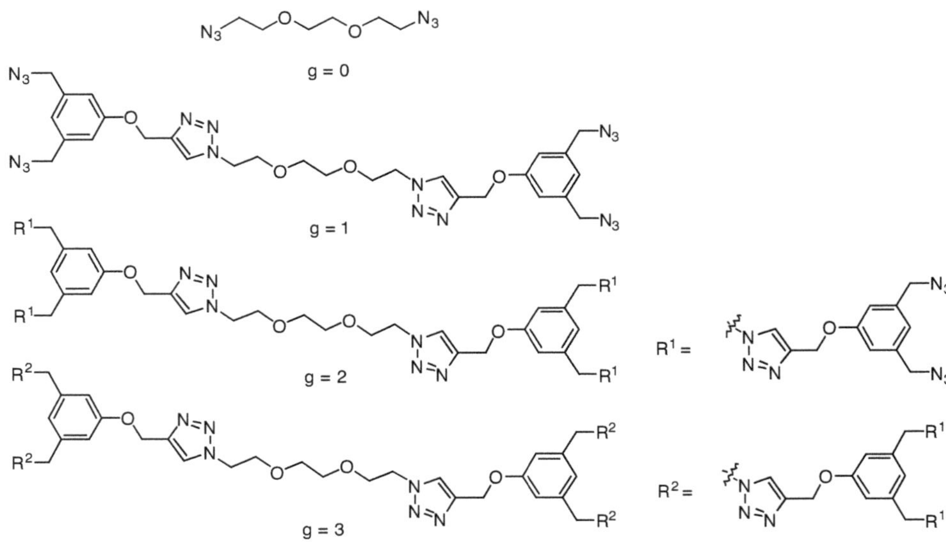

Scheme 11.13 *Dendrimers of generation (g) 0–3 used to cross-link polymeric micelles* ***38b*** *and* ***42b***.[57]

attachment should only occur at the hydrophilic termini of the block copolymers. Therefore, PAA-*b*-PS block copolymers terminated with azide or acetylene functionalities at the PAA chain ends were synthesized (not shown).[59] Azide-terminated copolymers were prepared via a nitroxide-mediated polymerization, using a chloride-functional initiator, of which the chloride was subsequently replaced by an azide. Acetylene copolymers were obtained with a controlled RAFT polymerization starting with an acetylene functional initiator. Micelles were prepared from mixtures of the copolymers with and without functional termini, which were subsequently cross-linked through amidation and decorated with fluorescein dye molecules using the CuAAC reaction. The fluorescence intensity depending on pH was measured for both the free alkyne-functional fluorescein and fluorescein attached to azide micelles from which the authors concluded that fluorescein conjugated to the termini behaves as if free in solution.

Jiang *et al.* described the cross-linking via the click reaction of poly(*N*,*N*-dimethylacrylamide)-*b*-poly(*N*-isopropylacrylamide-*co*-3-azidopropylacrylamide) [PDMA*b*-(NIPAM-*co*-AzPAM)] **43** (Scheme 11.14).[60] The NIPAM-*co*-AzPAM block is a thermo-responsive polymer with a lower critical solution temperature (LCST). Above the LCST, the NIPAM-co-AzPAM segment of block copolymer **43** starts to aggregate and micelles with a NIPAM-*co*-AzPAM core form.

Cross-linking of the core was achieved via a click reaction of dipropargyl ether with the azides present in the micelle interior. To prove cross-linking, the optical density as a function of temperature was measured. Dissolved block copolymer showed a transmittance of about 100%, which decreased to approximately 80% above the aggregation temperature. The transmittance of the cross-linked micelles did not change over the measured interval, indicating their stability in this temperature range.

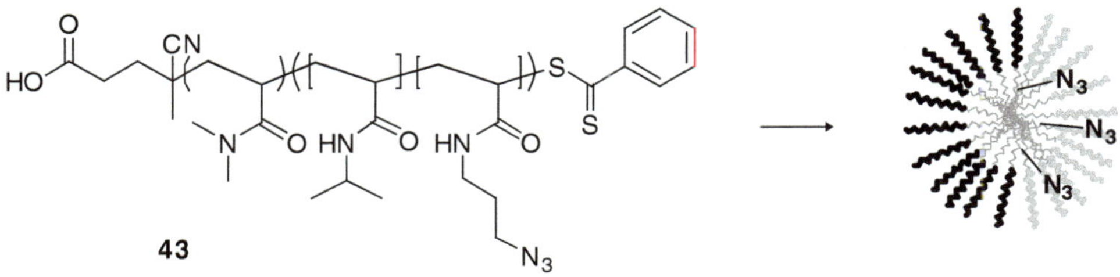

Scheme 11.14 *Thermo-responsive block copolymer* **43**.[60]

11.5 Virus Particles

The block copolymers from which polymersomes and polymer micelles are formed always show a distribution of molecular weight. Even if building blocks are monodisperse, like phospholipids, the self-assembled structures still display a distribution in size. The number of building blocks in liposomes, polymersomes or polymer micelles always varies, even under complete equilibrium conditions. Monodisperse building blocks are, however, encountered in biological systems in the form of proteins. Well-defined protein building blocks can assemble into highly organized 3D systems as is the case with virus particles. Virus particles contain a protein coat formed by self-assembly of proteins and exhibit an unparalleled monodispersity at this size scale. Virus particles are usually 50–300 nm in size and the plant viruses in particular are extremely suitable for modification and employable as well-defined biological nanoparticles.

An example of a well defined and characterized, easily available virus is the Cowpea Mosaic Virus (CPMV). The CPMV particles are 30 nm in diameter and are formed by the assembly of 60 subunits around single-stranded viral genomic RNA. Each subunit is a complex of a large and a small protein of 42 and 24 kDa. Finn and coworkers decorated the CPMV capsid with azides or alkynes at either reactive lysine or cysteine residues (Scheme 11.15).[61]

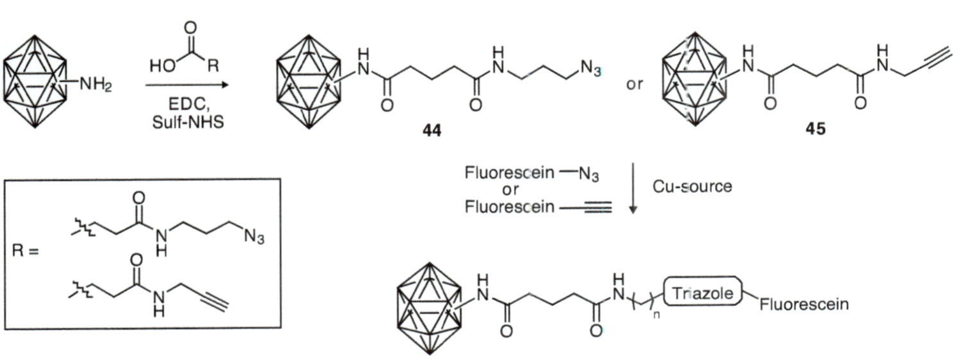

Scheme 11.15 *Preparation of azide- or acetylene-functionalized virus capsids.*[61]

Fluorescein derivatives with complementary functionality to the particles were used in a CuAAC reaction with $CuSO_4$–tris(triazole)amine, tris(carboxyethyl)phosphine (TCEP) and Cu wire as catalysts. Here TCEP was used as water-soluble reducing agent because addition of sodium ascorbate or *p*-hydroquinone as reductants resulted in disassembly of the virus capsids (later other conditions without TCEP were preferred[62–64]). Varying these conditions, a 100% conversion could be achieved for attachment of acetylene fluorescein to azide–CPMV and 80% was realized for the opposite combination. Independently, Wang and coworkers studied the attachment of various hemicyanine dyes using a similar approach.[65]

The Finn group reported optimized conditions for the click reaction to virus particles.[62] Using $Cu(MeCN)_4OTf$, $Cu(MeCN)_4PF_6$ or CuBr and sulfonated bathophenanthroline as catalyst, carbohydrates, peptides, poly(ethylene oxide) and transferrin were attached in high yields and substrate loadings. The carbohydrates were accessible when attached to CPMV, as was verified by the formation of a gel upon the addition of galectin-4. Peptides (KIRGDTFAGF and GLPLKDNYKK) were modified with an acetylene moiety for click reactions, and fluorescein to make them easy to detect. Sodium dodecyl sulfate–poly(acrylamide) gel electrophoresis (SDS–PAGE) analysis by UV irradiation showed modification of the particles with the selected peptides. This method for attachment of a fluorescent marker was also used to show the conjugation of PEG chains to the CPMV particles. Finally transferrin, an 80 kDa blood plasma protein for iron delivery, was attached to the virus particles via the cycloaddition reaction. TEM images showed an increase in particle size; the CPMV particles remained still intact after conjugation with transferrin (Figure 11.17).

Attachment of glycopolymers to, for example, a protein carrier could be a valuable tool in the study of carbohydrate-based cellular processes, which quite often depend on

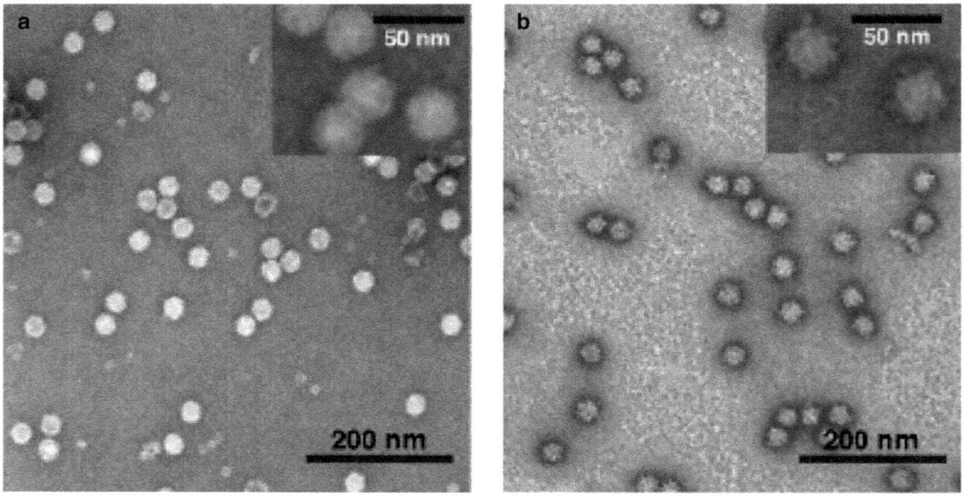

Figure 11.17 *(a) Negative-stained TEM of wild-type CPMV. (b) Negative-stained TEM of CPMV conjugated with transferrin. Automated measurement of the particles showed the average diameter to be 30 ± 1 nm for wild-type and 46 ± 5 nm for CPMV–transferrin. Reprinted with permission from ref.[62]. Copyright 2005 American Chemical Society.*

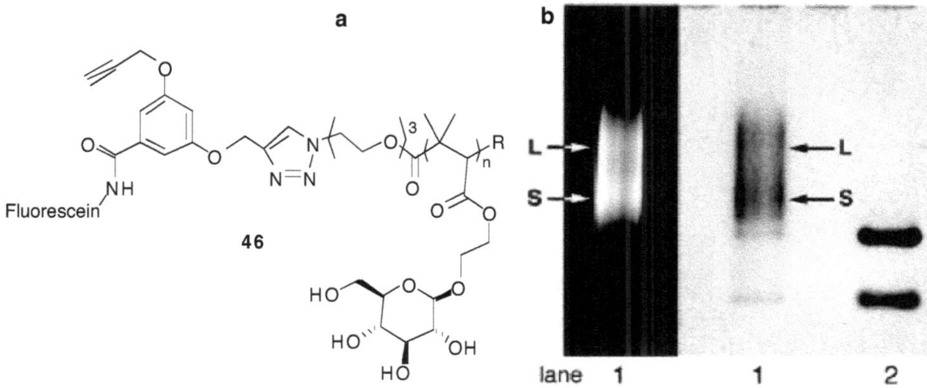

Figure 11.18 (a) Glycopolymer prepared via ATRP. (b) SDS–PAGE of the glycopolymer conjugate (lane 1) and WT-CPMV (lane 2). The arrows mark the center of the bands derived from the small (S) and large (L) subunits; their broad nature derives from the polydispersity of the polymer and the possibility for more than one attachment of polymer per protein subunit. Reprinted with permission from ref.[63]. Copyright 2005 Royal Society of Chemistry.

multivalency. Finn and coworkers synthesized glycopolymer **46** connected to a fluorescein moiety via atom transfer radical polymerization (ATRP) and conjugated it to CPMV particles via the CuAAC reaction using CuOTf–sulfonated bathophenanthroline as catalyst.[63] Figure 11.18 shows an SDS–PAGE gel of the glycopolymer–CPMV conjugates, demonstrating that both the large and small coat protein were functionalized with the glycopolymers (the broad bands arise from the polydispersity of the polymer). Besides glycopolymers, tetra- and hexasaccharides (e.g. globo-H) were also arrayed on the exterior surface of CPMV.[66] Particles coated with saccharides were injected into chickens to evaluate their IgY immune response. IgY was isolated from eggs and glycan-binding was determined on printed micro arrays containing 200 or 264 glycans.

Normal expression of the folate receptor (FR) in cells is low, but is upregulated in a variety of tumors. To target CPMV to a tumor cell with high FR expression poly(ethylene oxide)–folic acid was attached to CPMV particles **45**.[64] A PEG-spacer was incorporated to accomplish efficient cellular recognition of the folic acid (FA) units. CuOTf–sulfonated bathophenanthroline was used as catalyst for the click reaction. KB tumor cells (a human nasopharyngeal carcinoma cell line with high expression of FR) were incubated with different modified CPMV particles. Using flow cytometry an increased binding of CPMV-PEG-FA was established [Figure 11.19(a)].

Finally, CPMV particles conjugated with a gadolinium-tetraazacyclododecanetetraacetic acid Gd(DOTA) analog were prepared by Finn and coworkers.[67] Gd(DOTA) is a paramagnetic Gd complex used in magnetic resonance imaging as contrast agent. Acetylene functional Gd(DOTA) analog **47** [Figure 11.19(b)] was clicked with azide–CPMV **44** using Cu–bathophenanthroline as catalyst. With inductive coupled plasma optical emission spectrometry (ICP-OES), it was determined that 223 ± 20 Gd complexes were attached per CPMV particle.

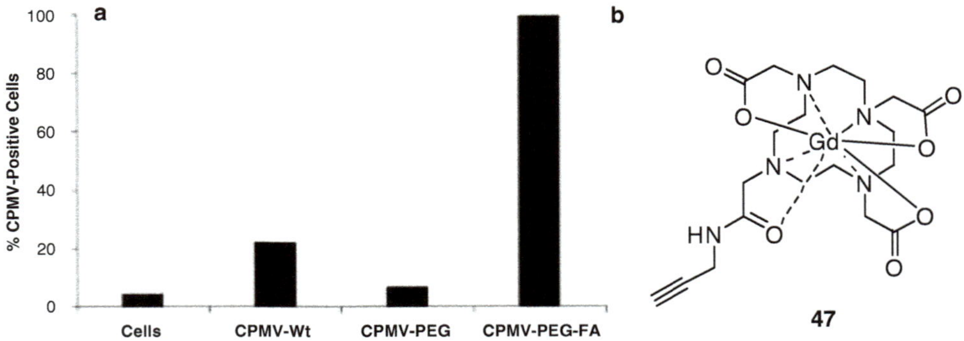

Figure 11.19 *(a) Percentage of increased binding of CPMV-PEG-FA to KB cells over controls. (b) Structure of acetylene functional Gd(DOTA). Reprinted with permission from ref.*[64]. *Copyright 2007 Elsevier.*

Another type of virus particle is the tobacco mosaic virus (TMV). TMV is a rod-like virus consisting of 2130 protein subunits arranged helically around genomic single RNA strands to form a 300 nm sized assembly. Francis and coworkers reported the exterior modification of TMV with diazonium coupling reactions.[68] Because this coupling procedure is difficult to apply using complex molecules, Bruckman *et al.* modified the exterior tyrosine molecules of TMV with the diazonium salt generated from 3-ethynylaniline *in situ* to obtain acetylene functional TMV.[69] Azide functional TMV was prepared by conjugating 1,4-bis-azidomethylbenzene to acetylene-TMV. A library of acetylene or azide functional (bio)molecules was successfully attached via the copper-catalyzed Huisgen cycloaddition reaction, see Scheme 11.16.

11.6 Conclusions

In this chapter the functionalization of a variety of nanoparticles with the copper-catalyzed Huisgen cycloaddition reaction between azides and alkynes has been described. The popularity of click chemistry in this field of science is based on its ease of execution, efficiency and orthogonality. Quite some examples can be found in which this method outperforms more traditional conjugation or functionalization procedures. It is also clear that there is still no general catalytic system, which requires an optimization of click conditions for each novel conjugation strategy. However, there are many opportunities ahead for functional particles prepared by the Huisgen cycloaddition reaction, also spurred on by new developments in the field. One such a recent development is the introduction of Cu-free click methods, which will further increase the application potential to areas such as microelectronics and the biomedical field, in which Cu is regarded as an undesirable impurity. Since most scientific achievements are of very recent date, it is clear that we have only witnessed the start of the effects click chemistry will have on the preparation of well-defined nanoparticles of both inorganic and organic origin.

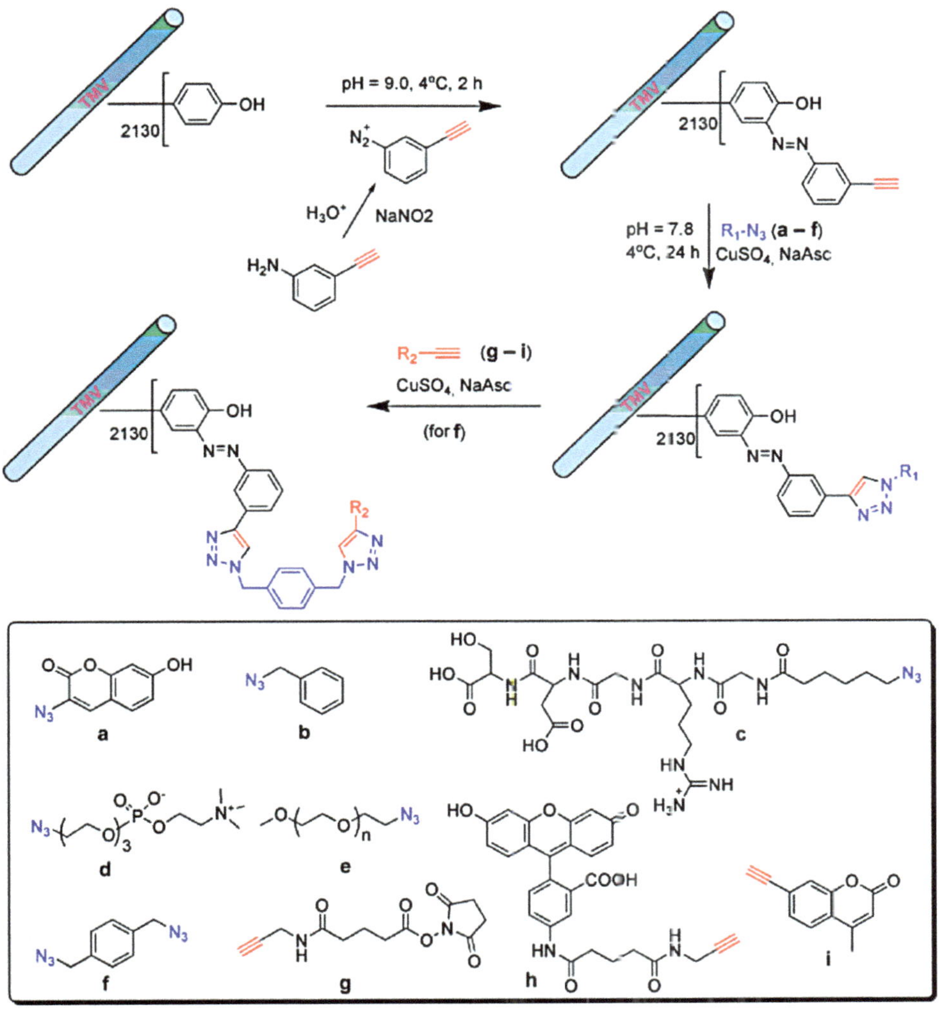

Scheme 11.16 *Bioconjugation of TMV by means of CuAAC reactions. Reproduced with permission from M. A. Bruckman, (2008), Surface modification of tobacco mosaic virus with 'click' chemistry,* ChemBioChem, ***9****, 519–523.*[69]

References

(1) C. Burda, X. Chen, R. Narayanan, and M. A. El-Sayed, (2005), Chemistry and properties of nanocrystals of different shapes, *Chem. Rev.*, **105**, 1025–1102.

(2) H. C. Kolb, M. G. Finn, and K. B. Sharpless, (2001), Click chemistry: diverse chemical function from a few good reactions, *Angew. Chem., Int. Edn*, **40**, 2004–2021.

(3) V. V. Rostovtsev, L. G. Green, V. V. Fokin, and K. B. Sharpless, (2002), A stepwise Huisgen cycloaddition process: copper(I)-catalyzed regioselective 'ligation' of azides and terminal alkynes, *Angew. Chem., Int. Edn*, **41**, 2596–2599.

(4) C. W. Tornøe, C. Christensen, and M. Meldal, (2002), Peptidotriazoles on solid phase: [1,2,3-triazoles by regiospecific copper(I)-catalyzed 1,3-dipolar cycloadditions of terminal alkynes to azides, *J. Org. Chem.*, **67**, 3057–3064.
(5) C. B. Murray, C. R. Kagan, and M. G. Bawendi, (2000), Synthesis and characterization of monodisperse nanocrystals and close-packed nanocrystal assemblies, *Annu. Rev. Mater. Sci.*, **30**, 545–610.
(6) H. Zou, S. Wu, and J. Shen, (2008), Polymer/silica nanocomposites: preparation, characterization, properties, and applications, *Chem. Rev.*, **108**, 3893–3957.
(7) B. Radhakrishnan, R. Ranjan, and W. J. Brittain, (2006), Surface initiated polymerizations from silica nanoparticles, *Soft Matter*, **2**, 386–396.
(8) R. Ranjan, and W. J. Brittain, (2007), Combination of living radical polymerization and click chemistry for surface modification, *Macromolecules*, **40**, 6217–6223.
(9) R. Ranjan, and W. J. Brittain, (2008), Synthesis of high density polymer brushes on nanoparticles by combined RAFT polymerization and click chemistry, *Macromol. Rapid Commun.*, **29**, 1104–1110.
(10) R. Ranjan, and W. J. Brittain, (2007), Tandem RAFT polymerization and click chemistry: an efficient approach to surface modification, *Macromol. Rapid Commun.*, **28**, 2084–2089.
(11) A. P. Alivisatos, W. Gu, and C. Larabell, (2005), Quantum dots as cellular probes, *Annu. Rev. Biomed. Eng*, **7**, 55–76.
(12) W. H. Binder, L. Petraru, R. Sachenshofer, and R. Zirbs, (2006), Synthesis of surface-modified nanoparticles via cycloaddition-reactions, *Monatsh. Chem.*, **137**, 835–841.
(13) W. H. Binder, R. Sachsenhofer, C. J. Straif, and R. Zirbs, (2007), Surface-modified nanoparticles via thermal and Cu(I)-mediated 'click' chemistry: Generation of luminescent CdSe nanoparticles with polar ligands guiding supramolecular recognition, *J. Mater. Chem.*, **17**, 2125–2132.
(14) S. Laurent, D. Forge, M. Port, A. Roch, C. Robic, L. Vander Elst, and R. N. Muller, (2008), Magnetic iron oxide nanoparticles: synthesis, stabilization, vectorization, physicochemical characterizations, and biological applications, *Chem. Rev.*, **108**, 2064–2110.
(15) M. A. White, J. A. Johnson, J. T. Koberstein, and N. J. Turro, (2006), Toward the syntheses of universal ligands for metal oxide surfaces: controlling surface functionality through click chemistry, *J. Am. Chem. Soc.*, **128**, 11356–11357.
(16) G. Lv, W. Mai, R. Jin, and L. Gao, (2008), Immobilization of dipyridyl complex to magnetic nanoparticle via click chemistry as a recyclable catalyst for Suzuki cross-coupling reactions, *Synlett*, 1418–1422.
(17) G. von Maltzahn, Y. Ren, J.-H. Park, D.-H. Min, V. R. Kotamraju, J. Jayakumar, V. Fogal, M. J. Sailor, E. Ruoslahti, and S. N. Bhatia, (2008), *In vivo* tumor cell targeting with 'click' nanoparticles, *Bioconjugate Chem.*, **19**, 1570–1578.
(18) P.-C. Lin, S.-H. Ueng, S.-C. Yu, M.-D. Jan, A. K. Adak, C.-C. Yu, and C.-C. Lin, (2007), Surface modification of magnetic nanoparticle via Cu(I)-catalyzed alkyne-azide [2 + 3] cycloaddition, *Org. Lett.*, **9**, 2131–2134.
(19) L. Polito, D. Monti, E. Caneva, E. Delnevo, G. Russo, and D. Prosperi, (2008), One-step bioengineering of magnetic nanoparticles via a surface diazo transfer/azide–alkyne click reaction sequence, *Chem. Commun. (Cambridge)*, 621–623.
(20) P.-C. Lin, S.-H. Ueng, M.-C. Tseng, J.-L. Ko, K.-T. Huang, S.-C. Yu, A. Kumar Adak, Y.-J. Chen, and C.-C. Lin, (2006), Site-specific protein modification through Cu(I)-catalyzed 1,2,3-triazole formation and its implementation in protein microarray fabrication, *Angew. Chem., Int. Edn*, **45**, 4286–4290.
(21) M.-C. Daniel, and D. Astruc, (2004), Gold nanoparticles: assembly, supramolecular chemistry, quantum-size-related properties, and applications toward biology, catalysis, and nanotechnology, *Chem. Rev.*, **104**, 293–346.
(22) W. J. Sommer, and M. Weck, (2007), Facile functionalization of gold nanoparticles via microwave-assisted 1,3 dipolar cycloaddition, *Langmuir*, **23**, 11991–11995.
(23) D. A. Fleming, C. J. Thode, and M. E. Williams, (2006), Triazole cycloaddition as a general route for functionalization of Au nanoparticles, *Chem. Mater.*, **18**, 2327–2334.
(24) C. J. Thode, and M. E. Williams, (2008), Kinetics of 1,3-dipolar cycloaddition on the surfaces of Au nanoparticles, *J. Colloid Interface Sci.*, **320**, 346–352.

(25) W. Limapichat, and A. Basu, (2008), Reagentless functionalization of gold nanoparticles via a 3 + 2 Huisgen cycloaddition, *J. Colloid Interface Sci.*, **318**, 140–144.
(26) J. L. Brennan, N. S. Hatzakis, T. R. Tshikhudo, N. Dirvianskyte, V. Razumas, S. Patkar, J. Vind, A. Svendsen, R. J. M. Nolte, A. E. Rowan, and M. Brust, (2006), Bionanoconjugation via click chemistry: the creation of functional hybrids of lipases and gold nanoparticles, *Bioconjugate Chem.*, **17**, 1373–1375.
(27) R. Voggu, P. Suguna, S. Chandrasekaran, and C. N. R. Rao, (2007), Assembling covalently linked nanocrystals and nanotubes through click chemistry, *Chem. Phys. Lett.*, **443**, 118–121.
(28) M. Fischler, A. Sologubenko, J. Mayer, G. Clever, G. Burley, J. Gierlich, T. Carell, and U. Simon, (2008), Chain-like assembly of gold nanoparticles on artificial DNA templates via 'click chemistry', *Chem. Commun. (Cambridge)*, 169–171.
(29) Y. Zhou, S. Wang, K. Zhang, and X. Jiang, (2008), Visual detection of copper(II) by azide- and alkyne-functionalized gold nanoparticles using click chemistry, *Angew. Chem., Int. Edn*, **47**, 7454–7456.
(30) A. Gole, and C. J. Murphy, (2008), Azide-derivatized gold nanorods: functional materials for 'click' chemistry, *Langmuir*, **24**, 266–272.
(31) K. K. Caswell, J. N. Wilson, U. H. F. Bunz, and C. J. Murphy, (2003), Preferential end-to-end assembly of gold nanorods by biotin–streptavidin connectors, *J. Am. Chem. Soc.*, **125**, 13914–13915.
(32) R. Voggu, S. Pal, S. K. Pati, and C. N. R. Rao, (2008), Semiconductor to metal transition in SWNTs caused by interaction with gold and platinum nanoparticles, *J. Phys.: Condens. Matter*, **20**, 215211.
(33) A. Hirsch, and M. Brettreich, (2005), *Fullerenes – Chemistry and Reactions*, Wiley-VCH, Weinheim.
(34) W.-B. Zhang, Y. Tu, R. Ranjan, R. M. Van Horn, S. Leng, J. Wang, M. J. Polce, C. Wesdemiotis, R. P. Quirk, G. R. Newkome, and S. Z. D. Cheng, (2008), 'Clicking' fullerene with polymers: synthesis of [60]fullerene end-capped polystyrene, *Macromolecules*, **41**, 515–517.
(35) J. Iehl, R. Pereira de Freitas, and J.-F. Nierengarten, (2008), Click chemistry with fullerene derivatives, *Tetrahedron Lett.*, **49**, 4063–4066.
(36) J. Iehl, R. Pereira de Freitas, B. Delavaux-Nicot, and J.-F. Nierengarten, (2008), Click chemistry for the efficient preparation of functionalized [60]fullerene hexakis-adducts, *Chem. Commun. (Cambridge)*, 2450–2452.
(37) P. I. Kitov, J. M. Sadowska, G. Mulvey, G. D. Armstrong, H. Ling, N. S. Pannu, R. J. Read, and D. R. Bundle, (2000), Shiga-like toxins are neutralized by tailored multivalent carbohydrate ligands, *Nature*, **403**, 669–672.
(38) H. Isobe, H. Mashima, H. Yorimitsu, and E. Nakamura, (2003), Synthesis of fullerene glycoconjugates through sulfide connection in aqueous media, *Org. Lett.*, **5**, 4461–4463.
(39) H. Isobe, K. Cho, N. Solin, D. B. Werz, P. H. Seeberger, and E. Nakamura, (2007), Synthesis of fullerene glycoconjugates via a copper-catalyzed huisgen cycloaddition reaction, *Org. Lett.*, **9**, 4611–4614.
(40) D. Tasis, N. Tagmatarchis, A. Bianco, and M. Prato, (2006), Chemistry of carbon nanotubes, *Chem. Rev.*, **106**, 1105–1136.
(41) H. Li, F. Cheng, A. M. Duft, and A. Adronov, (2005), Functionalization of single-walled carbon nanotubes with well-defined polystyrene by 'click' coupling, *J. Am. Chem. Soc.*, **127**, 14518–14524.
(42) H. Li, and A. Adronov, (2007), Water-soluble SWCNTs from sulfonation of nanotube-bound polystyrene, *Carbon*, **45**, 984–990.
(43) Z. Guo, L. Liang, J.-J. Liang, Y.-F. Ma, X.-Y. Yang, D.-M. Ren, Y.-S. Chen, and J.-Y. Zheng, (2008), Covalently β-cyclodextrin modified single-walled carbon nanotubes: a novel artificial receptor synthesized by 'click' chemistry, *J. Nanopart. Res.*, **10**, 1077–1083.
(44) S. Campidelli, B. Ballesteros, A. Filoramo, D. Díaz Díaz, G. de la Torre, T. Torres, G. M. A. Rahman, C. Ehli, D. Kiessling, F. Werner, V. Sgobba, D. M. Guldi, C. Cioffi, M. Prato, and J.-P. Bourgoin, (2008), Facile decoration of functionalized single-wall carbon nanotubes with phthalocyanines via 'click chemistry', *J. Am. Chem. Soc.*, **130**, 11503–11509.

(45) J. L. Mynar, T. Yamamoto, A. Kosaka, T. Fukushima, N. Ishii, and T. Aida, (2008), Radially diblock nanotube: site-selective functionalization of a tubularly assembled hexabenzocoronene, *J. Am. Chem. Soc.*, **130**, 1530–1531.
(46) V. P. Torchilin, (2005), Recent advances with liposomes as pharmaceutical carriers, *Nat. Rev. Drug Discov.*, **4**, 145–160.
(47) S. Cavalli, A. R. Tipton, M. Overhand, and A. Kros, (2006), The chemical modification of liposome surfaces via a copper-mediated [3 + 2] azide–alkyne cycloaddition monitored by a colorimetric assay, *Chem. Commun. (Cambridge)*, 3193–3195.
(48) F. Said Hassane, B. Frisch, and F. Schuber, (2006), Targeted liposomes: convenient coupling of ligands to preformed vesicles using 'click chemistry', *Bioconjugate Chem.*, **17**, 849–854.
(49) D. E. Discher, and F. Ahmed, Polymersomes, (2006), *Annu. Rev. Biomed. Eng*, **8**, 323–341.
(50) J. A. Opsteen, R. P. Brinkhuis, R. L. M. Teeuwen, D. W. P. M. Löwik, and J. C. M. v. Hest, (2007), 'Clickable' polymersomes, *Chem. Commun. (Cambridge)*, 3136–3138.
(51) S. F. M. van Dongen, M. Nallani, S. Schoffelen, J. J. L. M. Cornelissen, R. J. M. Nolte, and J. C. M. van Hest, (2008), A block copolymer for functionalisation of polymersome surfaces, *Macromol. Rapid Commun.*, **29**, 321–325.
(52) D. M. Vriezema, P. M. L. Garcia, N. Sancho Oltra, N. S. Hatzakis, S. M. Kuiper, R. J. M. Nolte, A. E. Rowan, and J. C. M. van Hest, (2007), Positional assembly of enzymes in polymersome nanoreactors for cascade reactions, *Angew. Chem., Int. Edn*, **46**, 7378–7382.
(53) B. Li, A. L. Martin, and E. R. Gillies, (2007), Multivalent polymer vesicles via surface functionalization, *Chem. Commun. (Cambridge)*, 5217–5219.
(54) D. M. Vriezema, M. Comellas Aragones, J. A. A. W. Elemans, J. J. L. M. Cornelissen, A. E. Rowan, and R. J. M. Nolte, (2005), Self-assembled nanoreactors, *Chem. Rev.*, **105**, 1445–1490.
(55) R. K. O'Reilly, M. J. Joralemon, K. L. Wooley, and C. J. Hawker, (2005), Functionalization of micelles and shell cross-linked nanoparticles using click chemistry, *Chem. Mater.*, **17**, 5976–5988.
(56) R. K. O'Reilly, M. J. Joralemon, C. J. Hawker, and K. L. Wooley, (2006), Fluorogenic 1,3-dipolar cycloaddition within the hydrophobic core of a shell cross-linked nanoparticle, *Chem. Eur. J.*, **12**, 6776–6786.
(57) M. J. Joralemon, R. K. O'Reilly, C. J. Hawker, and K. L. Wooley, (2005), Shell click-cross-linked (SCC) nanoparticles: a new methodology for synthesis and orthogonal functionalization, *J. Am. Chem. Soc.*, **127**, 16892–16899.
(58) R. K. O'Reilly, M. J. Joralemon, C. J. Hawker, and K. L. Wooley, (2007), Preparation of orthogonally-functionalized core click cross-linked nanoparticles, *New J. Chem.*, **31**, 718–724.
(59) R. K. O'Reilly, M. J. Joralemon, C. J. Hawker, and K. L. Wooley, (2006), Facile syntheses of surface-functionalized micelles and shell cross-linked nanoparticles, *J. Polym. Sci., Part A: Polym. Chem.*, **44**, 5203–5217.
(60) X. Jiang, J. Zhang, Y. Zhou, J. Xu, and S. Liu, (2008), Facile preparation of core-cross-linked micelles from azide-containing thermoresponsive double hydrophilic diblock copolymer via click chemistry, *J. Polym. Sci., Part A: Polym. Chem.*, **46**, 860–871.
(61) Q. Wang, T. R. Chan, R. Hilgraf, V. V. Fokin, K. B. Sharpless, and M. G. Finn, (2003), Bioconjugation by copper(I)-catalyzed azide-alkyne [3 + 2] cycloaddition, *J. Am. Chem. Soc.*, **125**, 3192–3193.
(62) S. Sen Gupta, J. Kuzelka, P. Singh, W. G. Lewis, M. Manchester, and M. G. Finn, (2005), Accelerated bioorthogonal conjugation: a practical method for the ligation of diverse functional molecules to a polyvalent virus scaffold, *Bioconjugate Chem.*, **16**, 1572–1579.
(63) S. Sen Gupta, K. S. Raja, E. Kaltgrad, E. Strable, and M. G. Finn, (2005), Virus–glycopolymer conjugates by copper(I) catalysis of atom transfer radical polymerization and azide–alkyne cycloaddition, *Chem. Commun. (Cambridge)*, 4315–4317.
(64) G. Destito, R. Yeh, C. S. Rae, M. G. Finn, and M. Manchester, (2007), Folic acid-mediated targeting of cowpea mosaic virus particles to tumor cells, *Chem. Biol.*, **14**, 1152–1162.
(65) W.-H. Zhan, H. N. Barnhill, K. Sivakumar, H. Tian, and Q. Wang, (2005), Synthesis of hemicyanine dyes for 'click' bioconjugation, *Tetrahedron Lett.*, **46**, 1691–1695.

(66) E. Kaltgrad, S. Sen Gupta, S. Punna, C. Y. Huang, A. Chang, C.-H. Wong, M. G. Finn, and O. Blixt, (2007), Anti-carbohydrate antibodies elicited by polyvalent display on a viral scaffold, *ChemBioChem*, **8**, 1455–1462.
(67) D. E. Prasuhn, R. M. Yeh, A. Obenaus, M. Manchester, and M. G. Finn, (2007), Viral MRI contrast agents: coordination of Gd by native virions and attachment of Gd complexes by azide–alkyne cycloaddition, *Chem. Commun. (Cambridge)*, 1269–1271.
(68) T. L. Schlick, Z. Ding, E. W. Kovacs, and M. B. Francis, (2005), Dual-surface modification of the tobacco mosaic virus, *J. Am. Chem. Soc.*, **127**, 3718–3723.
(69) M. A. Bruckman, (2008), Surface modification of tobacco mosaic virus with 'click' chemistry, *ChemBioChem*, **9**, 519–523.

12

Copper-catalyzed 'Click' Chemistry for Surface Engineering

Himabindu Nandivada and Joerg Lahann

12.1 Introduction

In the past few decades, there has been rapid development in the field of surface engineering, relating to the control of structure and properties of surfaces, which is of utmost importance for applications such as cell biology, tissue engineering, microfluidics, optics and electronics.[1,2] Surface attributes (such as wettability, charge and surface reactivity) depend on the chemical and physical details of the molecular structure at the interface. At the same time, the ability to modify inorganic surfaces with organic molecules or biological ligands is also a common requirement for a host of applications.

Surface modification or functionalization via covalent coupling is one of the major strategies being explored by surface scientists. Surface functionalization reaction involves a solid surface interacting with the reactant in liquid or vapor phase and may involve complicated steric and kinetic effects. At the same time, most of the coupling reactions available for surface chemistry are limited by incomplete conversions, nonspecificity, harsh reaction conditions and side reactions. In this respect, copper-catalyzed Huisgen's 1,3-dipolar cycloaddition between terminal alkyne and azide groups has proven to be an excellent choice due to its superior properties such as mild reaction conditions, high conversions, selectivity and reproducibility.[3–8] In general, this so-called click reaction is a reaction between terminal alkyne and azide groups to yield 1,4-disubstituted 1,2,3-triazoles. It is compatible with a wide range of functional groups except for groups which disrupt the catalytic activity of copper.[9] Alkyne–azide click reaction demonstrates high reactivity in heterogeneous

Click Chemistry for Biotechnology and Materials Science Edited by Joerg Lahann

reaction systems, so it is useful for surface reactions. This also implies that the solvent and catalyst system utilized during click reaction is quite important.[10]

This chapter focuses on strategies being employed to fabricate alkyne and azide functionalized surfaces and also highlights the applicability of click reaction for surface reactions with a specific focus on conjugation of biological ligands such as saccharides, oligonucleotides, proteins and peptide sequences.

12.2 Synthesis of Alkyne or Azide-functionalized Surfaces

The first step is to create surfaces amenable to copper-catalyzed Huisgen's 1,3-dipolar cycloaddition, i.e., surfaces containing alkyne or azide groups. This has been achieved by utilizing several methods such as self-assembled monolayers (SAMs), spin-coating, layer-by-layer assembly (LbL), lithography and polymeric methods.

12.2.1 Self-assembled Monolayers of Alkanethiolates

Self-assembled monolayers are monomolecular films, which are formed by spontaneous organization of active surfactant molecules onto specific solid substrates.[11] The most commonly used SAM system is the gold–alkanethiolate system due its ease of formation and the ability to use gold as an electrode. Typically the self-assembling molecules are modified and then assembled; however this procedure is often cumbersome and may not form the desired monolayer. Hence various surface coupling reactions have been used to modify SAMs after assembly. Modification of SAMs is limited by the thermal stability of the gold–thiol bond, because thiols desorb at higher temperatures, typically above 100 °C. These monolayers are also sensitive to pH conditions and solvents. Therefore Huisgen's 1,3-dipolar cycloaddition is an attractive option for surface reactions on monolayers. Copper-catalyzed click reaction was first applied to mixed monolayers of azidoundecanethiol and decanethiol on gold by Collman *et al.*[12] These monolayers were subsequently reacted with alkyne-functionalized ferrocene compounds (Figure 12.1). Collman *et al.* also assembled several other azide-terminated monolayers and the reactive azide surface coverage and rate of click reaction were determined via electrochemical and spectroscopic techniques.[13] Other researchers have also used gold-alkanethiolate SAMs in conjunction with alkyne–azide click chemistry for the introduction of several ligands to the monolayer surface under mild conditions (discussed later).[14–20]

12.2.2 Self-assembled Monolayers of Silanes and Siloxanes

Another self-assembling system that has been studied extensively is alkylsiloxane [CH_3-$(CH_2)_n$-SiO_x] monolayers on silica. The high reactivity of the surface anchor group makes it difficult to functionalize the opposite end of the monolayer. Lummerstorfer and Hoffmann created azide-containing siloxane monolayers on silica by nucleophillic substitution of bromine-terminated monolayers and subsequently reacted the monolayers with substituted acetylenes.[21] Several attempts have also been made to functionalize the silane monolayers using click chemistry.[22–24] Organic silanes are of special interest since they form stable monolayer films and can be attached to substrates with hydroxyl or oxide groups such as glass and silica. Ostaci *et al.* synthesized alkyne-containing silane monolayers on silicon

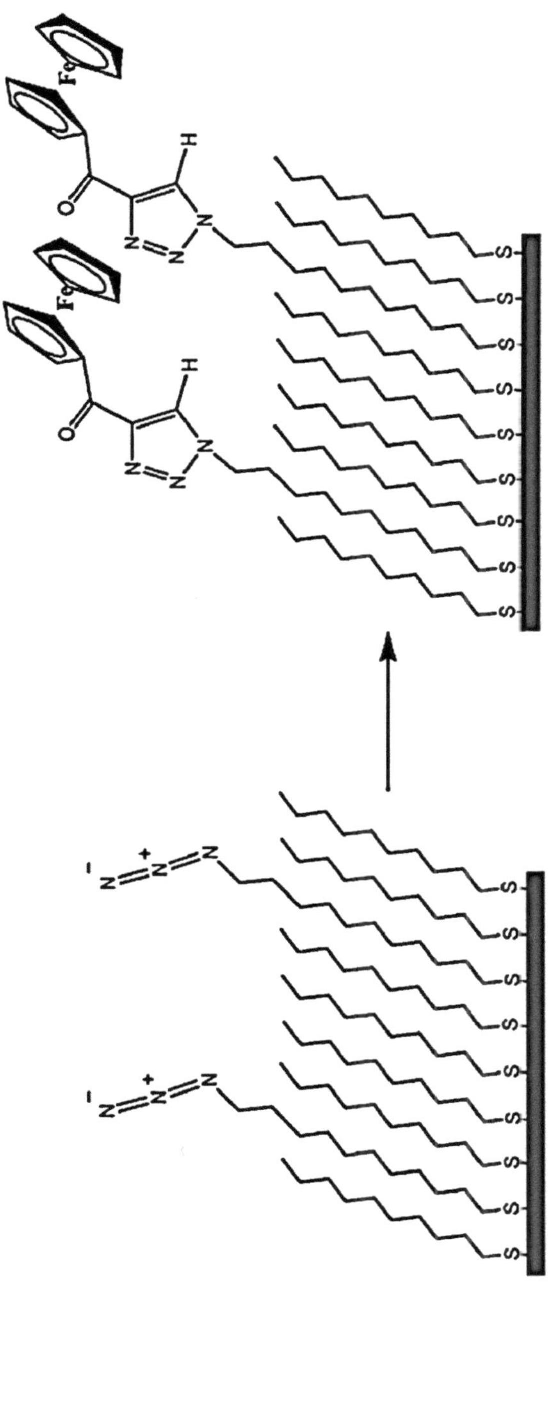

Figure 12.1 Schematic showing the immobilization of ferrocenepropynone onto mixed monolayers of azidoundecanethiol and decanethiol via click reaction. Reprinted from ref.[12]. Copyright 2004 American Chemical Society.

substrates and grafted ω-azido-modified linear polymer brushes [such as poly(ethylene glycol), polystyrene and poly(methylmethacrylate)] onto the monolayer-coated substrates (Figure 12.2).[23] It was further observed that the silane monolayer also acted as a passivation layer preventing nonspecific adsorption of the polymers. A combination of silanization and click chemistry was also used to modify glass microfluidic channels with preformed polymers by Prakash *et al.*[25] They covalently attached alkyne-terminated linear and dendritic polymers to glass substrates and conducted electroosmotic flow measurements in these channels. Covalent attachment of the polymers was seen to alter the surface charge in the channel (zeta potential) and in turn affect the fluid transport properties.

Haensch *et al.* used click chemistry to immobilize alkyne functionalized supramolecular terpyridine units on to azide-terminated silane monolayers.[26] Terpyridine moieties typically form complexes with a range of metal ions, which influences the physiochemical properties of the surface. In this case, the terpyridine ligand, 4′-(4-ethynylphenyl)-2,2′:6′,2″-terpyridine was synthesized and conjugated with Fe(II), which prevented complexation between the copper catalyst and terpyridine. The Fe(II) protected supramolecular ligand was immobilized onto 11-bromoundecyltrichlorosilane monolayers and deprotected under suitable conditions to obtain free terpyridine ligands on the substrate.

To address the important issue of passivation of native silicon surfaces, Rohde *et al.* acetylated silicon surfaces in a two-step process and coupled azide-functionalized benzoquinone using click chemistry.[27] Benzoquinone was further electrochemically activated to yield an amine-terminus, which was used for subsequent functionalization with ligands such as ferrocene carboxylic acid and biotin. During this procedure, minimal oxidation of silicon was observed, so it can be used in sensor applications, which are highly sensitive to the presence of a SiO_2 layer. Interestingly, Evrard *et al.* functionalized carbon electrodes using the electrochemical reduction of diazonium salts of phenylazide or phenylacetylene, which resulted in a grafted layer containing azide or acetylene groups respectively.[28] This method provided higher control over the density of the functional group and the surface retained its reactivity.

Click chemistry has been utilized to PEGylate porous silicon (pSi) substrates, which are of significance in biomedical applications as drug delivery vehicles and for biosensing.[29] Britchet and coworkers functionalized porous Si surfaces with acetylene groups via hydrosilation with 1,6-heptadiyne and then reacted with PEG–azide. The PEGylated pSi surface demonstrated a dramatic improvement in wettability. On the other hand, Schlossbauer and coworkers incorporated azide groups on mesoporous silica by simply treating it with sodium azide and then immobilized alkyne-containing trypsin.[30] They further demonstrated that the surface immobilized enzyme trypsin retained its activity.

12.2.3 Block Copolymers

Another common technique for the functionalization of surfaces, such as glass and silicon oxide is spin-coating.[31–33] The spin-coating technique is preferred for a variety of applications, because the thickness of the film can be precisely controlled. Rengifo *et al.* reported the use of spin-coating for the self-assembly of alkyne-functionalized diblock copolymers on various substrates, which was probed by subsequent click reaction with fluorescently-labeled azide ligands.[31] The diblock copolymer sequence, α-alkyne-ω-Br-poly(*tert*-butylacrylate-*b*-methylmethacrylate) [poly(^{t}BA-MMA)] was selected such that

Figure 12.2 *Grafting of linear polymer chains onto alkyne-functionalized silane monolayers.*[23]

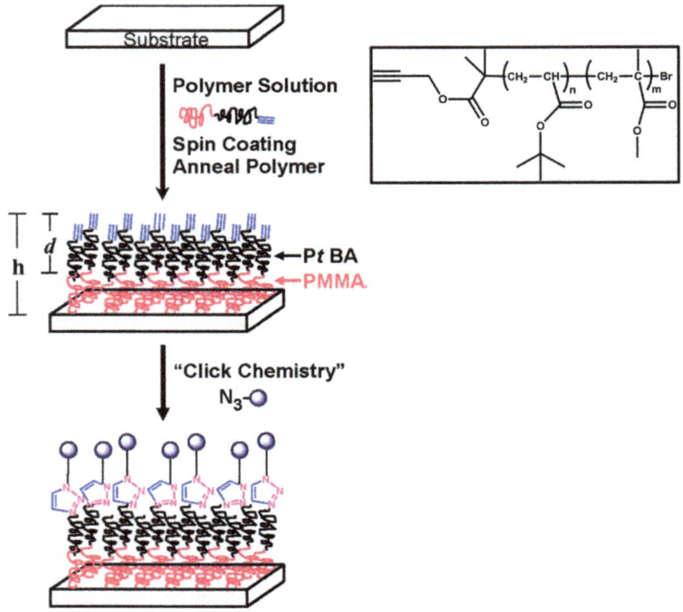

Figure 12.3 *Click reaction on films of spin-coated diblock copolymer, α-alkyne-ω-Br-poly(*tert-*butylacrylate*-b-*methylmethacrylate) [poly(^{t}BA-MMA)]. Inset: chemical structure of the copolymer. Reprinted with permission from ref.*[31]*. Copyright 2008 American Chemical Society.*

the first block (MMA) physisorbed strongly onto the substrate and the second sequence (^{t}BA) possessed a low surface tension, such that it remained on the surface, thus imparting a layered structure (Figure 12.3). The alkyne groups were attached to the second block and thus were presented at the surface for further reaction. The density of reactive groups on the film surface was controlled by the thickness of the film, which in turn was dependent on the molecular weight of the copolymer, concentration of the spin-coating solution and the spinning speed. This unique copolymer system was applied for the quantitative immobilization of azide-terminated 20-mer DNA molecules.[32]

Fleischmann *et al.* synthesized a terpolymer containing styrene (base of the polymer film), 4-(ethynyl)styrene (part containing the alkyne groups) and glycidyl methacrylate (anchor to the silicon substrate) using nitroxide mediated radical polymerization (NMRP).[33] This terpolymer was spin-coated on silicon substrates and the reactivity was probed with fluorescently-labeled azides.

12.2.4 Layer-by-layer Films

Layer-by-layer (LbL) assembly of composite films is generally based on charge interactions or hydrogen bonding interactions between sequential polymer layers. Crosslinking of the films using covalent reactions between layers can increase the stability of the films. The other advantage of incorporating covalent reactions between layers is that it enables the assembly of polymers with similar electrostatic properties (noncharged or nonhydrogen

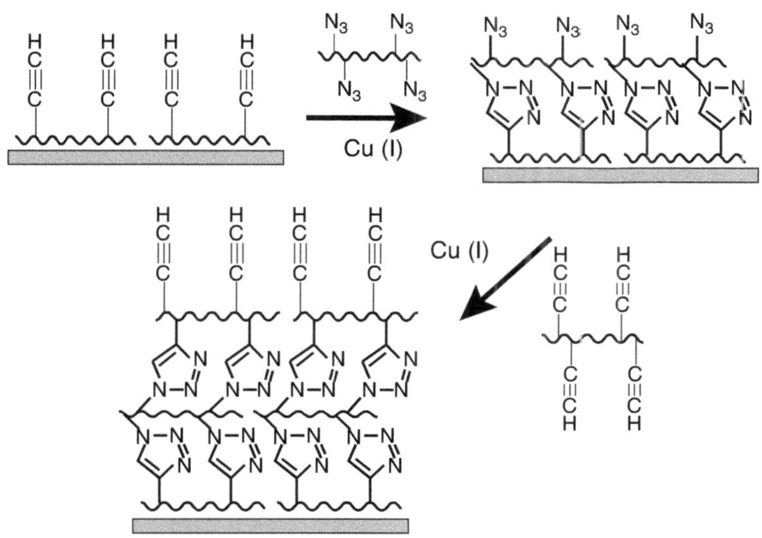

Figure 12.4 *Utilization of click reaction for the layer-by-layer assembly of thin polymer films with alkyne and azide functional groups. Reprinted with permission from ref.*[34]*. Copyright 2006 American Chemical Society.*

bonding), albeit with different reactive groups, thus creating single-component films. Taking advantage of these aspects, Such *et al.* synthesized multilayer films via LbL assembly of alkyne and azide functionalized versions of the same polyelectrolyte (polyacrylic acid), using copper-catalyzed click reaction as the crosslinking reaction (Figure 12.4).[34] Similarly, LbL assembly of alkyne- and azide-functionalized poly(n-isopropylacrylamide) was achieved by click chemistry.[35] The entire assembled film was subsequently covalently attached to an alkyne-functionalized polyethylene surface.

12.2.5 Chemical Vapor Deposition Polymerization

Surface reaction on thin polymer films requires robust attachment of the films to the flat substrate. In this respect chemical vapor deposition (CVD) has been used to fabricate functionalized coatings with good adhesion towards a wide variety of substrates.[36] This vapor-based process provides a solvent-free environment, good film adhesion and generates conformal coatings. More recently CVD has been extended to create alkyne-derivatized polymer coatings.[37,38] Nandivada *et al.* synthesized a reactive coating containing alkyne functional groups, poly(4-ethynyl-*p*-xylylene-co-*p*-xylylene), via CVD polymerization of 4-ethynyl-[2.2]paracyclophane.[37] Azide-bearing biotinylated ligands were conjugated to this reactive coating using microcontact printing and the reaction was probed with fluorescently-labeled streptavidin (Figure 12.5). In a different study, Im *et al.* synthesized an alkyne-functionalized polymer coating using an initiated-CVD (iCVD) process.[38] Using a single-step approach, a commercially available monomer, propargyl methacrylate, was polymerized to form poly(propargyl methacrylate). This polymer was also patterned using e-beam lithography to form nanometer patterns and the reactivity of the polymer was demonstrated by click reaction with azide-functionalized biotin.

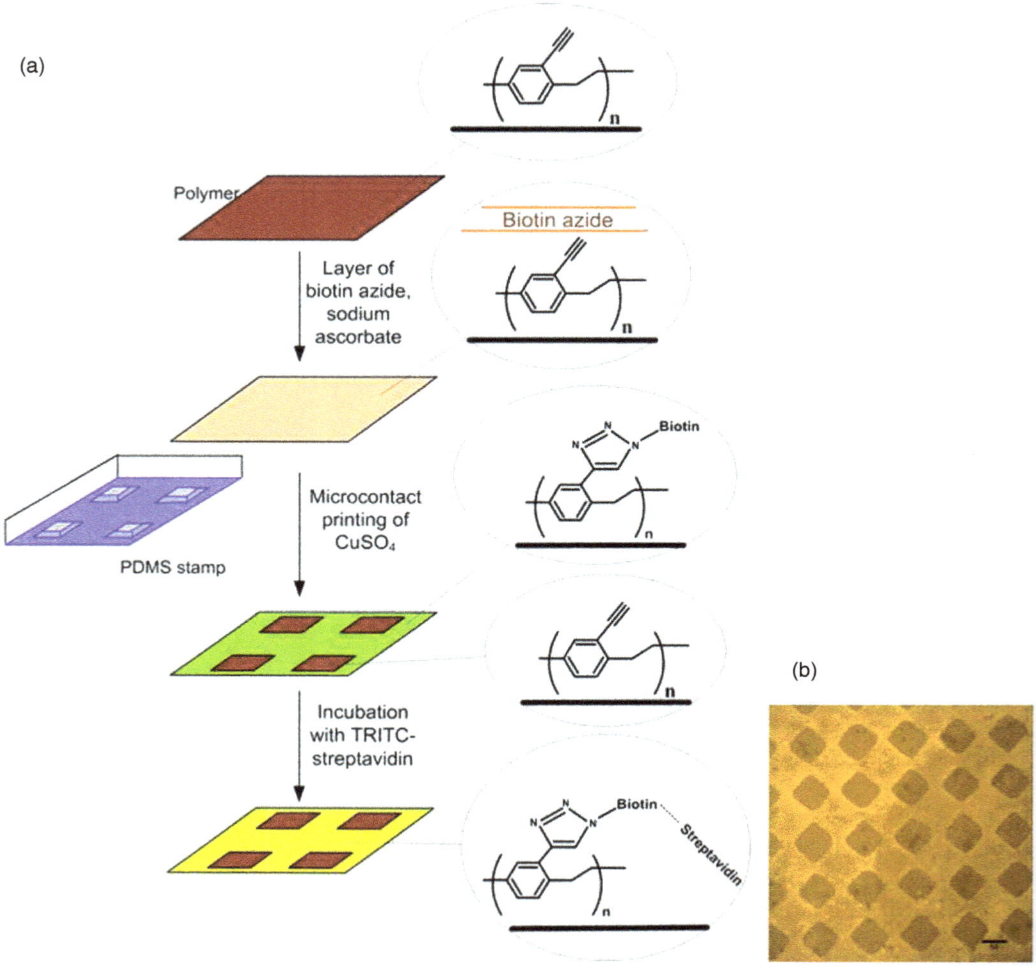

Figure 12.5 *(a) Schematic showing alkyne-functionalized polymer created using CVD and patterned via microcontact printing of biotin azide. (b) A fluorescence micrograph demonstrating the binding of TRITC-streptavidin on biotin azide patterns. Reprinted with permission from ref.[37]. Copyright 2006 Wiley-VCH.*

12.2.6 Fiber Networks

Apart from planar or flat substrates, click chemistry has also been employed to functionalize fiber networks. Shi *et al.* demonstrated the immobilization of a testis-specific protease (TSP50) on biodegradable polymer fibers.[39] Ultrathin biodegradable lactide fibers containing propargyl groups were created using electrospinning and azide-functionalized TSP50 molecules were conjugated to this fiber via click chemistry. Another example of fiber modification was demonstrated by Bhaskar *et al.*, where biphasic fibers containing alkyne groups in only one phase were fabricated using electrohydrodynamic co-jetting of alkyne-modified poly(lactide-co-glycolide) (PLGA) with unmodified PLGA.[40] The

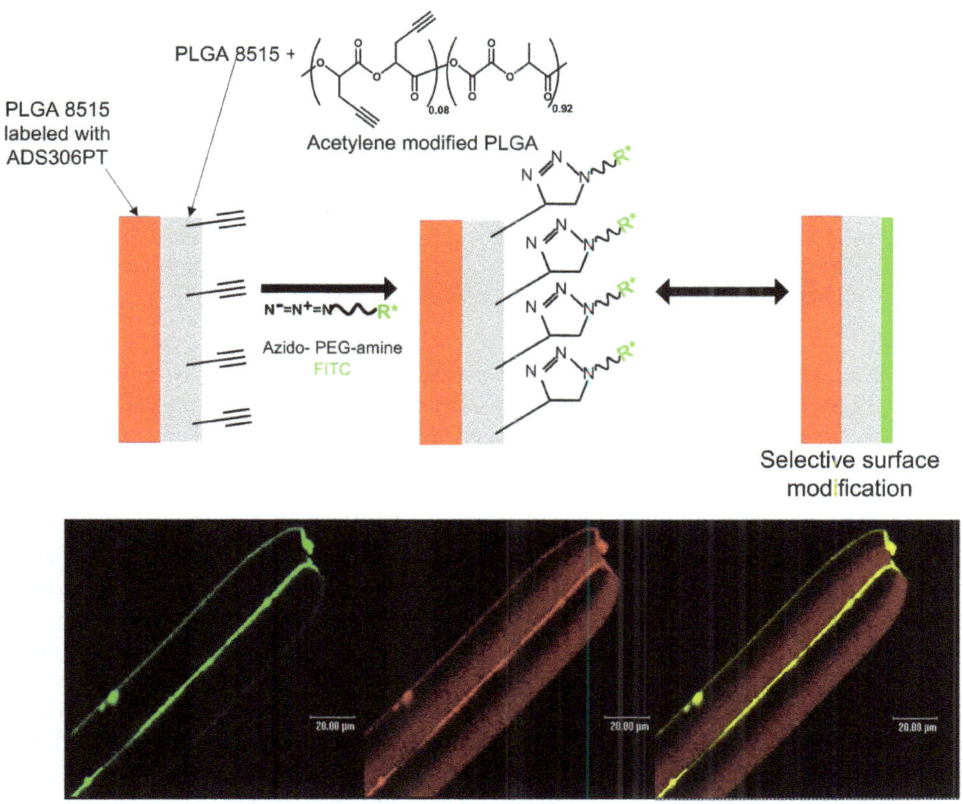

Figure 12.6 *Schematic and fluorescence micrographs showing the selective click modification of the biphasic fibers containing alkyne-functionalized PLGA in only one phase. Reprinted with permission from ref.[40]. Copyright 2008 Wiley-VCH.*

fibers were further selectively modified with an azide-functionalized fluorescent dye (Figure 12.6).

The use of copper-catalyzed 1,3-dipolar cycloaddition reaction has not been limited to biodegradable fiber surfaces. Krouit *et al.* employed click reaction to graft polycaprolactone macromolecular chains to the surface of cellulose fibers.[41] Cotton fibers have also been modified by Chen *et al.* to incorporate alkyne groups and further reacted with azide-functionalized methyl methacrylate polymers.[42]

12.3 Spatially Controlled Click Chemistry

The ability to spatially pattern a surface is quite important for electronics as well as biotechnological applications. The versatile click reaction is compatible with microcontact printing, which is a soft-lithographic process frequently employed to create micro or nanoscale patterns by depositing molecules on surfaces using a patterned stamp.[37,43–45] For instance, Nandivada *et al.* microcontact printed the copper catalyst onto a thin layer of

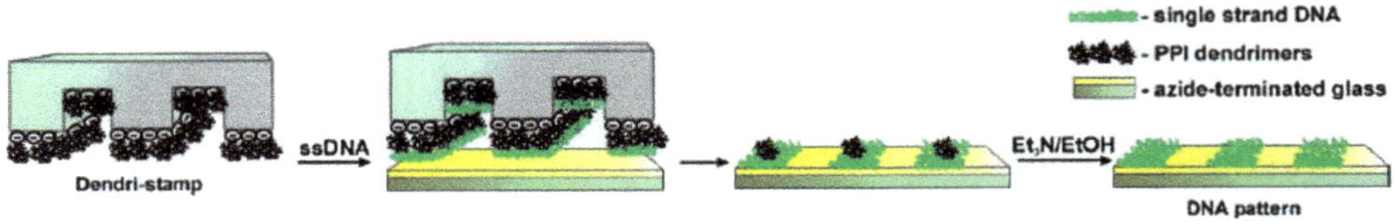

Figure 12.7 *Schematic description of catalyst-free click reaction via microcontact printing of alkyne-modified ss-DNA onto azide-functionalized substrate. The oxidized PDMS stamp is inked with dendrimers, incubated with ss-DNA and stamped onto azide functionalized surfaces. Reprinted with permission from ref.[46]. Copyright 2007 Wiley-VCH.*

biotinylated azide ligands adsorbed on vapor-based alkyne-functionalized polymer coatings.[37] By decoupling the catalyst and the reactants, the reaction was catalyzed only in the regions where the copper catalyst was deposited, thus creating patterns of covalently bound biotin. On the other hand, Rozkiewicz *et al.* performed the reaction without the copper catalyst, solely relying on the high concentration of reactants during microcontact printing as a driving force for the reaction.[43] Briefly, bromo-terminated SAMs were treated with sodium azide to create azide-terminated monolayers and alkyne-functionalized ligands were directly microcontact printed onto this monolayer. In the past, proximity of the ligands on the surface has been shown to lead to an enhancement of the reaction rate thus improving the efficiency of surface functionalization reactions.[9] This catalyst-free microcontact printing technique was further extended to pattern alkyne-containing oligonucleotides onto azide-modified glass slides.[46] To create surface patterns, a layer of positively-charged dendrimers was first inked onto the PDMS stamp, which promoted the binding of single-stranded DNA (ss-DNA) 'ink' to the stamp surface. This stamp was then brought into contact with the azido-substrate without the presence of the catalyst and click reaction was initiated (Figure 12.7). Furthermore, the covalently immobilized oligonucleotides were hybridized with their complementary strands.

The scope of copper-catalyzed click reactions has been broadened to create nanoscale patterns on solid surfaces via dip-pen nanolithography.[47] AFM tips were used to deliver Cu(I) catalyst and azide-functionalized dendrons to alkyne-functionalized surfaces. This technique takes advantage of the fact that shorter distances between the reactants may lead to an enhancement of the reaction rate. The main advantage of this approach is that multiple azides can be delivered sequentially or using different AFM tips to the same alkynated surface.

Furthermore, fluorescent molecules were patterned using a technique called scanning electrochemical microscopy (SECM), where a gold microelectrode was used to generate Cu(I) ions locally to catalyze the click reaction between alkyne-functionalized fluorophores and azide groups on a nonconductive glass substrate.[48]

12.4 Copper-catalyzed Click Chemistry for Bioimmobilization

The immobilization of biomolecules on surfaces is of tremendous interest for a wide variety of applications such as biosensors, microarrays, bioactive implant surfaces and tissue engineering. Preservation of the biomolecular activity after the reaction is a key attribute for a successful bioconjugation reaction. This typically requires mild reaction conditions and absence of cross-reactivity between the functional groups present. By definition, click chemistry represents a collection of reactions with mild operating conditions, high yields and nonreactivity towards other functional groups. Therefore, in this respect, alkyne–azide click reaction is attracting a lot of attention from material and surface scientists for bioconjugation, specifically due to the inactivity of alkyne and azide groups towards other functional groups present in biomolecules. Conjugation of molecules like carbohydrates, oligonucleotide probes, proteins and peptide sequences has been successfully demonstrated using click reaction on surfaces and will be discussed in further detail.

Several reports have described the immobilization of biotin on flat surfaces via click chemistry and have taken advantage of the highly specific albeit noncovalent biotin–streptavidin binding.[37,49,50] For example, Lee *et al.* used click chemistry to functionalize polymeric nanobrushes with azide end-groups.[50] Ethylene glycol-based polymer films were synthesized using surface-initiated ATRP onto initiator-containing SAMs. Subsequently the bromide-presenting polymer was reacted with sodium azide to introduce azide groups on the surface, which were then reacted with alkyne-containing biotin. This ethylene glycol-based polymer film demonstrated nonbiofouling characteristics combined with specific reactivity towards alkyne-functionalized ligands.

Glycan arrays provide an opportunity to study the complex protein–sugar interactions and enhance our understanding of role of glycans present on the cell surface.[51] Microarrays also allow the screening of multiple ligands simultaneously with minimal use of material. Copper-catalyzed click chemistry presents a robust strategy to covalently link saccharide molecules to a flat surface, thus mimicking the cell surface expressing these glycans.[52] Sugar-modified SAMs or glyco-SAMs have been used as glycan arrays because they provide better control over the density and orientation of the saccharide molecules and can be characterized after immobilization via surface analysis techniques. For example, Zhang *et al.* employed alkyne–azide click reaction to immobilize azide-functionalized sugars (mannose, lactose, α-galactose) on to alkyne-containing SAMs.[20] This method is much simpler than the direct assembly of pre-synthesized thiol-terminated sugar molecules, which require complex synthesis procedures. The unique platform displaying sugar-functionalized SAMs was further used to study binding interactions between sugars and lectins by employing electrochemical characterization and surface plasmon resonance spectroscopy. This approach represents a label-free technique to elucidate real-time structure–activity information. Similarly, Kleinart *et al.* extended this study by synthesizing a series of functionalized thiol molecules and comparing the assembly of pre-formed glyco molecules with the previously described 'click on SAM' approach.[19] Miura *et al.* studied the interactions between pathogenic protein Alzheimer amyloid-β (Aβ) and monosaccharide displaying silane-based monolayers which were created using Huisgen's 1,3-cycloaddition.[53] This study enabled the estimation of the core saccharide interacting structure of the Aβ protein.

Click conjugation has also been used to reversibly capture azide-modified saccharides on an alkyne-functionalized microtiter plate.[54,55] A disulfide bridge was included in the linker to enable the cleavage and release of the captured oligosaccharide molecule for further characterization utilizing a reductive treatment with a thiol (dithiothreitol, DTT) (Figure 12.8). Using this technique, a breast cancer antigen, Globo-H, was captured and analyzed after cleavage, thus demonstrating the utility of this method for biosensor applications.[55]

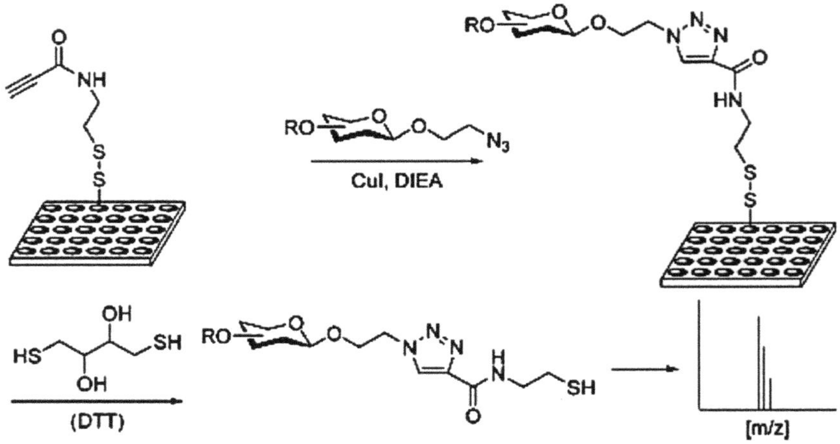

Figure 12.8 *Alkyne-functionalized microtiter plates which captured azide-modified saccharides. Saccharides were cleaved by DTT for further MS analyses. Reprinted with permission from ref.[55]. Copyright 2004 American Chemical Society.*

Copper-catalyzed Huisgen's 1,3-dipolar cycloaddition has also been used in conjunction with another reaction from the click family, namely Diels–Alder reaction, for the immobilization of carbohydrates on solid surfaces.[49] Sun *et al.* synthesized a short heterobifunctional PEG linker with alkyne and cyclodiene terminal groups on either side. This linker was conjugated to maleimidocaproyl-functionalized substrates via Diels–Alder reaction leaving the alkyne-terminal end for subsequent alkyne–azide click reaction with azide-functionalized ligands. This 'dual-click' approach was used for the successful immobilization of biomolecules such as biotin, lactose and *r*-thrombomodulin. Success of the immobilization step was further confirmed using antibody-binding via surface plasmon resonance (SPR) spectroscopy.

Immobilization of saccharide molecules has also been achieved via microcontact printing of alkyne-functionalized carbohydrates onto azido SAMs.[44] Michel and Ravoo microcontact printed carbohydrate (mannose, glucose, galactose and maltose) conjugates with alkyne functionality and used the corresponding lectins to probe the sugars. These arrays provide important information regarding structure–function relationships, which may ultimately lead to better understanding of the immune responses.

On the other hand, DNA arrays are generally created using nucleophilic–electrophilic reactions, which are hampered by the lack of efficiency, chemoselectivity and reproducibility. Immobilization of DNA further requires an aqueous reaction environment. An elegant strategy utilizing copper-catalyzed click chemistry for the fabrication of oligonucleotide arrays was demonstrated by Devaraj *et al.*[18] Alkyne-functionalized oligodeoxyribonucleotides were synthesized and conjugated with azide-terminated monolayers in the presence of a triazolylamine copper ligand, tris(bezyltriazolylmethyl)amine [Cu(I)TBTA] as the catalyst. Cu(I)TBTA accelerates the cycloaddition reaction without damaging the structure of DNA as opposed to free Cu(I), which disrupts the oligonucleotide structure in the presence of reactive oxygen species. On the other hand, Seo *et al.* created a photocleavable DNA array using click chemistry where azido-labeled DNA was attached to alkyne-modified glass surfaces (Figure 12.9).[56] Furthermore, DNA–polymerase-extension

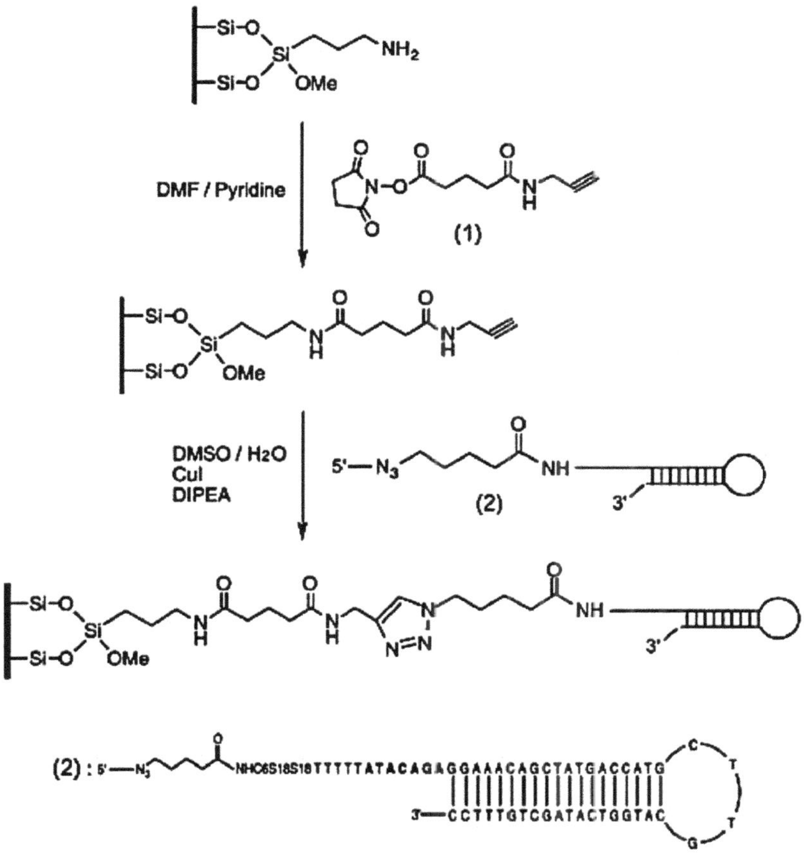

Figure 12.9 *Schematic demonstrating the construction of a DNA chip using copper catalyzed click reaction. Alkyne-functionalized glass slides form a covalent bond with 5′-azide-modified DNA. Reprinted with permission from ref.[56]. Copyright 2004 National Academy of Sciences, USA.*

reaction was accurately performed on this array to incorporate fluorescent nucleoside analogs.

Protein microarrays can be used to study protein–protein and protein–ligand interactions. Immobilized proteins are more robust than noncovalently bound proteins and also an enhanced sensitivity can be achieved.[57] However it is challenging to maintain the activity and conformation of the proteins during immobilization reactions. Some research groups have demonstrated the immobilization of proteins using copper-catalyzed click chemistry.[58,59] Azide- or alkyne-modified proteins were covalently bound to alkynated or azidated glass slides via click chemistry.[59] Interestingly, it was observed that immobilization of the alkyne-functionalized protein onto an azide-presenting surface was more efficient than the other way around. This may indicate that alkyne groups were coupling with the copper ions during the reaction, thereby reducing the catalytic effects.

A fascinating approach for the fabrication of a density gradient of cell-adhesion peptides by 'clicking' RGD azide-peptides onto an alkyne-gradient substrate has been reported by

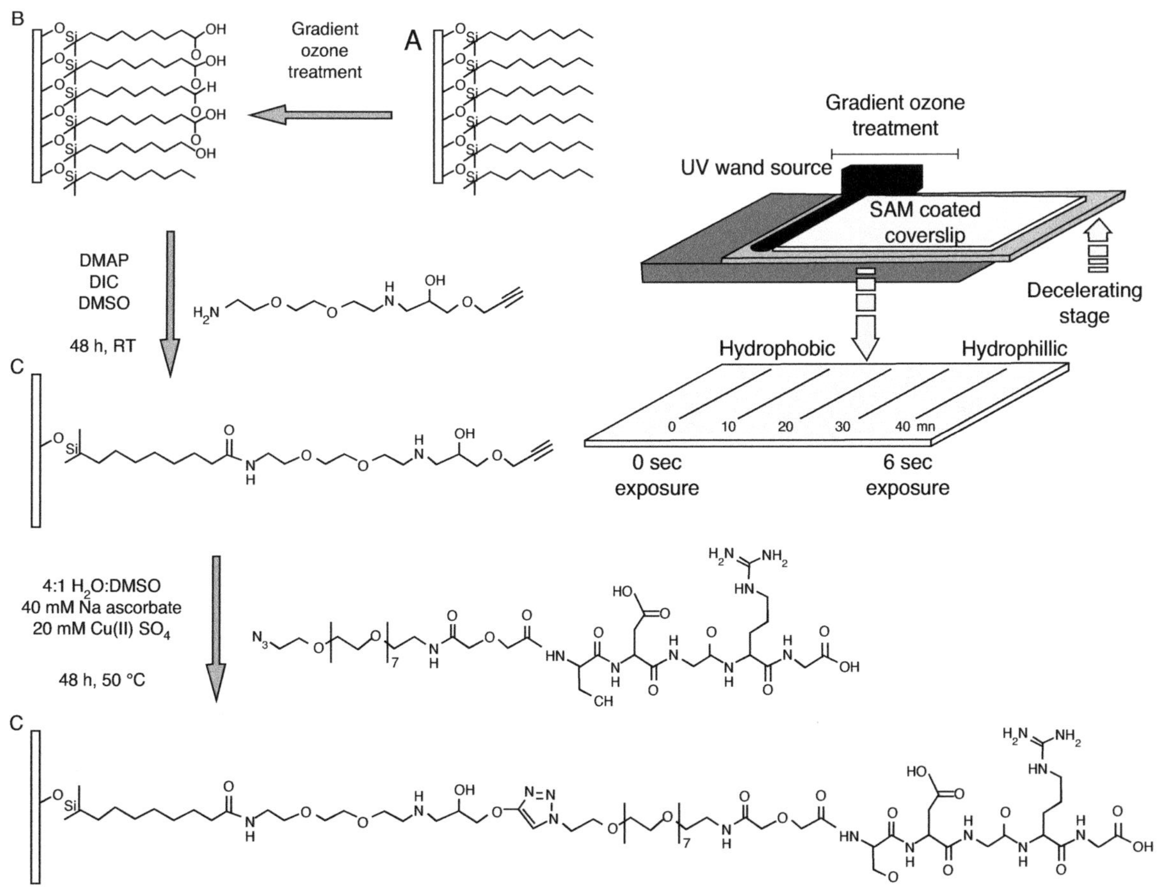

Figure 12.10 *Creation of a gradient substrate for RGD peptide immobilization by click chemistry. (A) Variable UV oxidation generated (B) a carboxyl gradient which was converted to (C) an alkyne gradient and (D) RGD azide–peptide was covalently immobilized by click cycloaddition reaction. Reprinted with permission from ref.[60]. Copyright 2007 Wiley-VCH.*

Gallant *et al.*[60] Briefly, variable UV oxidation followed by a bifunctional linker was used to synthesize an alkyne-functionalized gradient (Figure 12.10). Subsequent reaction with an RGD azide-modified peptide resulted in a gradient of peptides, which was able to modulate smooth muscle cell attachment. The density of the immobilized moiety depends on the density of the functional group and the efficiency of the coupling chemistry.

12.5 Summary

In this chapter, the use of copper-catalyzed click chemistry has been discussed in the context of surface engineering. Alkyne–azide click reactions were successfully used for coupling of different organic compounds and biomolecules to a wide variety of substrates. These generally robust reactions have gained immense popularity among surface scientists owing to their functional group tolerance, mild reaction conditions and high reactivity. Numerous strategies are being employed to incorporate the alkyne and azide functionalities onto the surface.

References

(1) Langer, R. and Tirrell, D.A., (2004), Designing materials for biology and medicine, *Nature*, **428**, 487–492.
(2) Hubbell, J.A., (1999), Bioactive biomaterials, *Curr. Opin. Biotechnol.*, **10**, 123–129.
(3) Kolb, H.C., Finn, M.G. and Sharpless, K.B., (2001), Click chemistry: diverse chemical function from a few good reactions, *Angew. Chem. Int. Edn*, **40**, 2004–2021.
(4) Meldal, M., (2008), Polymer 'Clicking' by CuAAC reactions, *Macromol. Rapid Commun.*, **29**, 1016–1051.
(5) Meldal, M. and Tornoe, C.W., (2008), Cu-catalyzed azide–alkyne cycloaddition, *Chem. Rev.*, **108**, 2952–3015.
(6) Rostovtsev, V.V., Green, L.G., Fokin, V.V. and Sharpless, K.B., (2002), A stepwise Huisgen cycloaddition process: copper(I)-catalyzed regioselective 'ligation' of azides and terminal alkynes, *Angew. Chem. Int. Edn*, **41**, 2596–2599.
(7) Nandivada, H., Jiang, X.W. and Lahann, J., (2007), Click chemistry: versatility and control in the hands of materials scientists, *Adv. Mater.*, **19**, 2197–2208.
(8) Devaraj, N.K. and Collman, J.P., (2007), Copper catalyzed azide–alkyne cycloadditions on solid surfaces: applications and future directions, *QSAR Comb. Sci.*, **26**, 1253–1260.
(9) Binder, W.H. and Sachsenhofer, R., (2008), Click chemistry in polymer and material science: an update, *Macromol. Rapid Commun.*, **29**, 952–981.
(10) Binder, W.H. and Sachsenhofer, R., (2007), 'Click' chemistry in polymer and materials science, *Macromol. Rapid Commun.*, **28**, 15–54.
(11) Ulman, A., (1996), Formation and structure of self-assembled monolayers, *Chem. Rev.*, **96**, 1533–1554.
(12) Collman, J.P., Devaraj, N.K. and Chidsey, C.E.D., (2004), 'Clicking' functionality onto electrode surfaces, *Langmuir*, **20**, 1051–1053.
(13) Collman, J.P., Devaraj, N.K., Eberspacher, T.P.A. and Chidsey, C.E.D., (2006), Mixed azide-terminated monolayers: a platform for modifying electrode surfaces, *Langmuir*, **22**, 2457–2464.
(14) Lee, J.K., Chi, Y.S. and Choi, I.S., (2004), Reactivity of acetylenyl-terminated self-assembled monolayers on gold: triazole formation, *Langmuir*, **20**, 3844–3847.
(15) Haensch, C., Ott, C., Hoeppener, S. and Schubert, U.S., (2008), Combination of different chemical surface reactions for the fabrication of chemically versatile building blocks onto silicon surfaces, *Langmuir*, **24**, 10222–10227.

(16) Haensch, C., Hoeppener, S. and Schubert, U.S., (2008), Chemical surface reactions by click chemistry: coumarin dye modification of 11-bromoundecyltrichlorosilane monolayers, *Nanotechnology*, **19**, 7.

(17) Wang, L. *et al.*, (2009), Covalent grafting nitrophenyl group on Au surface via click reaction: Assembling process and electrochemical behaviors, *Electrochem. Commun.*, **11**, 339–342.

(18) Devaraj, N.K. *et al.*, (2005), Chemoselective covalent coupling of oligonucleotide probes to self-assembled monolayers, *J. Am. Chem. Soc.*, **127**, 8600–8601.

(19) Kleinert, M., Winkler, T., Terfort, A. and Lindhorst, T.K., (2008), A modular approach for the construction and modification of glyco-SAMs utilizing 1,3-dipolar cycloaddition, *Org. Biomol. Chem.*, **6**, 2118–2132.

(20) Zhang, Y. *et al.*, (2006), Carbohydrate–protein interactions by 'clicked' carbohydrate self-assembled monolayers, *Anal. Chem.*, **78**, 2001–2008.

(21) Lummerstorfer, T. and Hoffmann, H., (2004), Click chemistry on surfaces: 1,3-dipolar cycloaddition reactions of azide-terminated monolayers on silica, *J. Phys. Chem. B*, **108**, 3963–3966.

(22) Ciampi, S. *et al.*, (2007), Functionalization of acetylene-terminated monolayers on Si(100) surfaces: a click chemistry approach, *Langmuir*, **23**, 9320–9329.

(23) Ostaci, R.V. *et al.*, (2008), Polymer brushes grafted to 'passivated' silicon substrates using click chemistry, *Langmuir*, **24**, 2732–2738.

(24) Ciampi, S., Le Saux, G., Harper, J.B. and Gooding, J.J., (2008), Optimization of click chemistry of ferrocene derivatives on acetylene-functionalized silicon(100) surfaces, *Electroanalysis*, **20**, 1513–1519.

(25) Prakash, S., Long, T.M., Selby, J.C., Moore, J.S. and Shannon, M.A., (2007), 'Click' modification of silica surfaces and glass microfluidic channels, *Anal. Chem.*, **79**, 1661–1667.

(26) Haensch, C. *et al.*, (2008), Reversible supramolecular functionalization of surfaces: terpyridine ligands as versatile building blocks for noncovalent architectures, *Langmuir*, **24**, 12981–12985.

(27) Rohde, R.D., Agnew, H.D., Yeo, W.S., Bailey, R.C. and Heath, J.R., (2006), A non-oxidative approach toward chemically and electrochemically functionalizing Si(111), *J. Am. Chem. Soc.*, **128**, 9518–9525.

(28) Evrard, D., Lambert, F., Policar, C., Balland, V. and Limoges, B., (2008), Electrochemical functionalization of carbon surfaces by aromatic azide or alkyne molecules: a versatile platform for click chemistry, *Chem. Eur. J.*, **14**, 9286–9291.

(29) Britcher, L., Barnes, T.J., Griesser, H.J. and Prestidge, C.A., (2008), PEGylation of porous silicon using click chemistry, *Langmuir*, **24**, 7625–7627.

(30) Schlossbauer, A., Schaffert, D., Kecht, J., Wagner, E. and Bein, T., (2008), Click Chemistry for high-density biofunctionalization of mesoporous silica, *J. Am. Chem. Soc.*, **130**, 12558–12559.

(31) Rengifo, H.R., Chen, L., Grigoras, C., Ju, J.Y. and Koberstein, J.T., (2008), 'Click-functional' block copolymers provide precise surface functionality via spin coating, *Langmuir*, **24**, 7450–7456.

(32) Chen, L. *et al.*, (2008), Spin-on end-functional diblock copolymers for quantitative DNA immobilization, *Biomacromolecules*, **9**, 2345–2352.

(33) Fleischmann, S. *et al.*, (2008), Modification of polymer surfaces by click chemistry, *Macromol. Rapid Commun.*, **29**, 1177–1185.

(34) Such, G.K., Quinn, J.F., Quinn, A., Tjipto, E. and Caruso, F., (2006), Assembly of ultrathin polymer multilayer films by click chemistry, *J. Am. Chem. Soc.*, **128**, 9318–9319.

(35) Bergbreiter, D.E. and Chance, B.S., (2007), 'Click'-based covalent layer-by-layer assembly on polyethylene using water-soluble polymeric reagents, *Macromolecules*, **40**, 5337–5343.

(36) Yoshida, M., Langer, R., Lendlein, A. and Lahann, J., (2006), From advanced biomedical coatings to multi-functionalized biomaterials, *Polym. Rev.*, **46**, 347–375.

(37) Nandivada, H., Chen, H.-Y., Bondarenko, L. and Lahann, J. Reactive polymer coatings that 'click', (2006), *Angew. Chem. Int. Edn*, **45**, 3360–3363.

(38) Im, S.G. *et al.*, (2008), A Directly patternable, click-active polymer film via initiated chemical vapor deposition, *Macromol. Rapid Commun.*, **29**, 1648–1654.

(39) Shi, Q., Chen, X.S., Lu, T.C. and Jing, X.B., (2008), The immobilization of proteins on biodegradable polymer fibers via click chemistry, *Biomaterials*, **29**, 1118–1126.

(40) Bhaskar, S., Roh, K.H., Jiang, X.W., Baker, G.L. and Lahann, J., (2008), Spatioselective modification of bicompartmental polymer particles and fibers via Huisgen 1,3-dipolar cycloaddition, *Macromol. Rapid Commun.*, **29**, 1655–1660.

(41) Krouit, M., Bras, J. and Belgacem, M.N., (2008), Cellulose surface grafting with polycaprolactone by heterogeneous click-chemistry, *Eur. Polym. J.*, **44**, 4074–4081.

(42) Chen, G.J. *et al.*, (2007), Synthesis of azide/alkyne-terminal polymers and application for surface functionalisation through a [2 + 3] Huisgen cycloaddition process, 'click chemistry'', *Soft Matter*, **3**, 732–739.

(43) Rozkiewicz, D.I., Janczewski, D., Verboom, W., Ravoo, B.J. and Reinhoudt, D.N., (2006), 'Click' chemistry by microcontact printing, *Angew. Chem. Int. Edn*, **45**, 5292–5296.

(44) Michel, O. and Ravoo, B.J., (2008), Carbohydrate microarrays by microcontact 'click' chemistry, *Langmuir*, **24**, 12116–12118.

(45) Spruell, J.M. *et al.*, (2008), Heterogeneous catalysis through microcontact printing, *Angew. Chem. Int. Edn*, **47**, 9927–9932.

(46) Rozkiewicz, D.I. *et al.*, (2007), Transfer printing of DNA by 'Click' chemistry, *Chembiochem*, **8**, 1997–2002.

(47) Long, D.A., Unal, K., Pratt, R.C., Malkoch, M. and Frommer, J., (2007), Localized 'click' chemistry through dip-pen nanolithography, *Adv. Mater.*, **19**, 4471–4473.

(48) Ku, S.-Y., Wong, K.-T. and Bard, A.J., (2008), Surface patterning with fluorescent molecules using click chemistry directed by scanning electrochemical microscopy, *J. Am. Chem. Soc.*, **130**, 2392–2393.

(49) Sun, X.L., Stabler, C.L., Cazalis, C.S. and Chaikof, E.L., (2006), Carbohydrate and protein immobilization onto solid surfaces by sequential Diels–Alder and azide–alkyne cycloadditions, *Bioconjugate Chem.*, **17**, 52–57.

(50) Lee, B.S. *et al.*, (2007), Surface-initiated, atom transfer radical polymerization of oligo(ethylene glycol) methyl ether methacrylate and subsequent click chemistry for bioconjugation, *Biomacromolecules*, **8**, 744–749.

(51) Hanson, S.R., Greenberg, W.A. and Wong, C.H., (2007), Probing glycans with the copper(I)-catalyzed [3 + 2] azide–alkyne cycloaddition, *QSAR Comb. Sci.*, **26**, 1243–1252.

(52) Dedola, S., Nepogodiev, S.A. and Field, R.A., (2007), Recent applications of the Cu(I)-catalysed Huisgen azide–alkyne 1,3-dipolar cycloaddition reaction in carbohydrate chemistry, *Org. Biomol. Chem.*, **5**, 1006–1017.

(53) Miura, Y., Yamauchi, T., Sato, H. and Fukuda, T., (2008), The self-assembled monolayer of saccharide via click chemistry: formation and protein recognition, *Thin Solid Films*, **516**, 2443–2449.

(54) Fazio, F., Bryan, M.C., Blixt, O., Paulson, J.C. and Wong, C.-H., (2002), Synthesis of sugar arrays in microtiter plate, *J. Am. Chem. Soc.*, **124**, 14397–14402.

(55) Bryan, M.C. *et al.*, (2004), Covalent display of oligosaccharide arrays in microtiter plates, *J. Am. Chem. Soc.*, **126**, 8640–8641.

(56) Seo, T.S. *et al.*, (2004), Photocleavable fluorescent nucleotides for DNA sequencing on a chip constructed by site-specific coupling chemistry, *Proc. Natl Acad. Sci. USA*, **101**, 5488–5493.

(57) Camarero, J.A., (2008), Recent developments in the site-specific immobilization of proteins onto solid supports, *Biopolymers*, **90**, 450–458.

(58) Gauchet, C., Labadie, G.R. and Poulter, C.D., (2006), Regio- and chemoselective covalent immobilization of proteins through unnatural amino acids, *J. Am. Chem. Soc.*, **128**, 9274–9275.

(59) Lin, P.C. *et al.*, (2006), Site-specific protein modification through Cu-I-catalyzed 1,2,3-triazole formation and its implementation in protein microarray fabrication, *Angew. Chem. Int. Edn*, **45**, 4286–4290.

(60) Gallant, N.D., Lavery, K.A., Amis, E.J. and Becker, M.L., (2007), Universal gradient substrates for 'click' biofunctionalization, *Adv. Mater.*, **19**, 965–969.

13

Click Chemistry in Protein Engineering, Design, Detection and Profiling

Daniela C. Dieterich and A. James Link

13.1 Introduction

The definition of click chemistry as first posited by Kolb, Finn, and Sharpless[1] is a set of high-yielding, energetically favorable, highly modular organic transformations that can occur in benign solvents, particularly water. Though originally conceived as a novel philosophy for synthetic organic chemistry (specifically the construction of small molecule libraries), many of the characteristics of a click reaction are desirable in other branches of chemistry, as evidenced by the wide array of topics covered in this book. Specifically, high-yielding reactions that can be carried out in water, but also at physiological temperatures, are of much interest in biology. The Huisgen [3 + 2] cycloaddition between azides and alkynes[2] is mentioned by Kolb *et al.* as being 'as good as a reaction can get', but it typically requires temperatures well above what can be tolerated by cells and proteins, limiting its use in biological applications. All of this changed once it was discovered that Cu(I) could catalyze the azide–alkyne cycloaddition.[3,4] The copper-catalyzed version of the reaction works well at ambient temperatures (and at even lower temperatures), can be run in water and in common buffer systems for biological applications and can tolerate atmospheric oxygen. Taken together, these aspects render the copper-catalyzed azide–alkyne cycloaddition (CuAAC) nearly ideal for bioconjugation applications. CuAAC has subsequently been used for the functionalization of many classes of biomolecules including sugars,[5] nucleic acids[6] and proteins, the topic of this chapter.

Click Chemistry for Biotechnology and Materials Science Edited by Joerg Lahann

The array of molecules that can be conjugated to proteins using CuAAC is nearly limitless in large part because of the high selectivity and synthetic simplicity of adding azides or alkynes to a molecule of interest. A survey of the literature indicates that CuAAC has been used to decorate proteins with molecules including affinity handles such as biotin,[7] sugars,[8] DNA,[9] fluorophores,[10] polymers,[11] as well as more exotic molecules like gadolinium complexes useful in magnetic resonance imaging.[12] The other side of creating bioconjugates using CuAAC is the necessity to introduce the azide or alkyne moiety into the protein of interest. Neither the azide nor the alkyne is present in the canonical set of proteinogenic amino acids, which is simultaneously a blessing and a curse. On one hand, azides or alkynes must be grafted onto proteins in some fashion before bioconjugation via CuAAC can occur. On the other hand, if the azides or alkynes can be added in a controlled or site-specific fashion, there is the potential to create protein conjugates with a high degree of specificity. The specificity of CuAAC and its 'bioorthogonality' are what sets CuAAC apart from canonical bioconjugation techniques such as reacting lysines with activated esters or reacting thiols with maleimides.

In the first part of the chapter, the posttranslational addition of azides and alkynes to proteins will be discussed (Figure 13.1). Both chemical and enzymatic methods that add these functional groups to proteins will be detailed as well as the applications of the protein conjugates produced by these methods. The second segment of this chapter will focus on the cotranslational addition of azides and alkynes into proteins via the *in vivo* incorporation of unnatural amino acids (Figure 13.1). Finally, the last section of this chapter will detail the BONCAT technology, which uses CuAAC to label and analyze the newly synthesized pool of proteins of the cell in a time-resolved fashion.

13.2 Posttranslational Functionalization of Proteins with Azides and Alkynes

The first use of CuAAC for bioconjugation was carried out by Wang *et al.* to produce uniformly labeled cowpea mosaic virus (CPMV) particles.[13] The protein capsid of CPMV consists of 60 identical two-protein units forming a rigid, icosahedral, 30 nm particle that the Finn group has thoroughly explored as a functional nanomaterial.[14] Wild-type CPMV contains a single surface-exposed lysine residue per asymmetrical unit with exceptional reactivity toward electrophiles.[15] Alternatively, a genetically engineered version of CPMV with a single exposed cysteine residue (per asymmetric unit) has been designed and produced.[16] These particles present 60 reactive handles in the form of either amine or a thiol groups, the workhorses of traditional protein conjugation. Reactive linkers were added to the amines or thiols in a quantitative fashion resulting in particles labeled with 60 azides or alkynes. Finally, these functionalized particles were further reacted under CuAAC conditions with fluorescein derivatives functionalized with the appropriate reaction partner. The fluorophore-labeled particles were analyzed using quantitative chromatography. Complete labeling of all 60 reactive sites was achieved with azide-labeled particles after 16 h at 4 °C. A reaction with the opposite polarity (alkyne-labeled particles reacting with azide-labeled dye) was less efficient, but could still be driven to near completion by increasing the concentration of the copper catalyst.

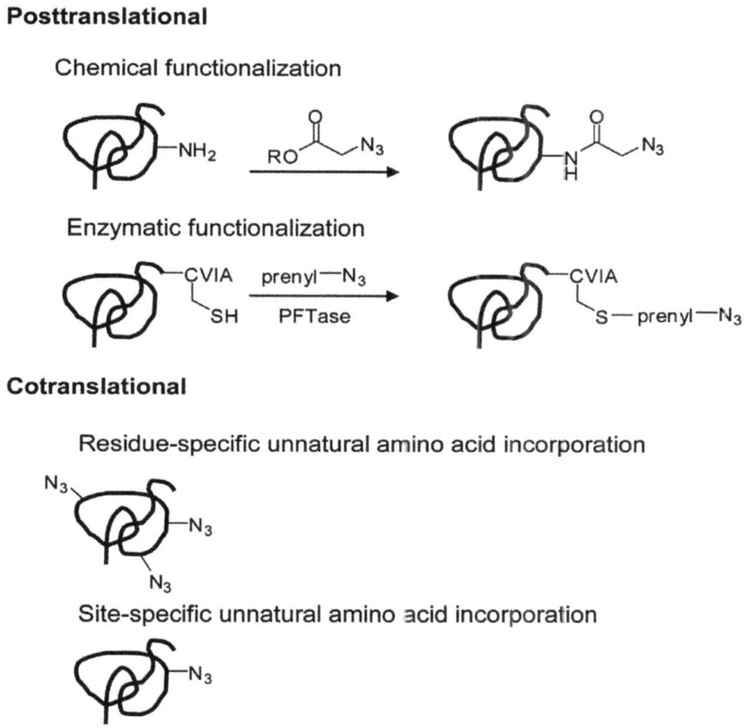

Figure 13.1 *A sampling of methods for the covalent introduction of azides and alkynes into proteins. Methods can be classified as either posttranslational or cotranslational. Examples are given for azide incorporation into proteins; the same methods can be used for introduction of alkynes. From top: reaction of azide-bearing activated ester with lysine residue, functionalization of the C-terminus of a protein via protein farnesyltransferase (PFTase), multiple site incorporation of azide-bearing unnatural amino acids and site-specific (single site) incorporation of azide-bearing unnatural amino acids.*

Although the output of these seminal experiments (fluorophore-labeled virus particles) could have been obtained more easily using traditional protein bioconjugation, this first example of protein bioconjugation using CuAAC is significant for several reasons. The authors claim that the kinetics of CuAAC bioconjugation are comparable to other bioconjugation reactions, such as the conjugation of thiols to maleimides, thus establishing CuAAC as a viable bioconjugation method. This report also demonstrated that the conditions required for CuAAC are gentle enough for the maintenance of tertiary (and even quaternary) protein structure. Lastly, this report introduced the *tris*-triazolylamine family of ligands. These ligands, particularly the benzyl derivative TBTA,[17] serve the dual purpose of both activating the Cu(I) ion for catalysis and protecting it from disproportionation. The TBTA ligand thus further improves the kinetics of CuAAC bioconjugations and also allows for long reaction times required in some applications. Its use has proved critical for the success of many subsequent applications of CuAAC for protein labeling.

Another seminal use of CuAAC in protein conjugation was carried out by Speers and Cravatt.[18] CuAAC was used to advance the field of activity-based protein profiling (ABPP). In ABPP, reactive substrate analogs are incubated with complex proteomic mixtures in order to covalently label enzymes at their active site.[19] The promise of ABPP is to rapidly assign functions to proteins, one of the main challenges in the postgenomic era. Traditional ABPP used probes in which the substrate analog was directly conjugated to a fluorescent dye or affinity handle such as biotin. The large size of the dye or biotin, however, was viewed as an impediment to ABPP since these molecules may sterically interfere with the insertion of the probe into the active site. To circumvent this limitation, the authors synthesized a phenyl sulfonate probe derivatized with an azide group, which consists of only three atoms and is much smaller than the dyes typically used in ABPP probes. The phenyl sulfonate probe is known to specifically label the active site of several enzymes including a glutathione-*S*-transferase (GSTO 1-1), aldehyde dehydrogenases (ALDH-1) and enoyl CoA hydratases (ECH-1). The probe was incubated with several complex proteomic mixtures including mammalian cell lysates and mouse tissue homogenates. Following this incubation, the entire protein mixture was subjected to CuAAC conditions with an alkyne-tagged rhodamine dye. Following separation of the proteins by electrophoresis, the labeled proteins were readily detected by fluorescence scanning of the gel. In addition to these *in vitro* demonstrations of CuAAC-enabled ABPP, the authors also injected the azide-labeled phenyl sulfonate probe into live mice to test whether CuAAC-enabled ABPP could function *in vivo*. Remarkably, following treatment of homogenates of the heart tissue from these mice with CuAAC conditions and the rhodamine alkyne dye, labeled ECH-1 protein was readily detected via gel electrophoresis and fluorescence scanning. These results underscore the 'silence' of the azide group toward biomolecules since the probe was able to survive the metabolism of a live animal and find its protein target. Moreover, the probe demonstrated no apparent toxicity for the animal. This work also demonstrates the incredible tolerance of the CuAAC reaction. The reactions were successful despite being carried out in a cellular milieu composed of thousands of different molecules. In a subsequent report,[20] the same authors present several important refinements to CuAAC-enabled ABPP including the interesting observation that switching the polarity of the reaction such that the probe carries the alkyne group and the dye carries the azide leads to lower levels of background labeling in proteomic samples. In this report, the importance of being able to perform ABPP *in vivo* is also underscored as the authors report several proteins that are labeled by the probe in intact cells but not in cell lysates.

Both of the examples discussed so far have relied on chemical derivatization of the protein before carrying out bioconjugation via CuAAC. More recently, there have been several reports of enzymatic or semisynthetic transformations of proteins in order to introduce the requisite azide or alkyne moiety. Kalia and Raines carried out a study[21] in which a variation on the expressed protein ligation[22] protocol was used to install the azide functionality specifically to the C-terminus of RNAse A (Figure 13.2). Expressed protein ligation relies on the formation of an electrophilic thioester during the *N–S* acyl shift of the intein protein splicing reaction. The authors evaluated a selection of nucleophiles for their ability to attack a model thioester, and selected the hydrazine moiety for further study. An RNAse A-intein-chitin binding domain tripartite fusion was expressed in *E. coli* and immobilized on chitin beads. Addition of a bifunctional small molecule containing both the azide and hydrazine moieties to the beads effected the cleavage of RNAse A and the addition of

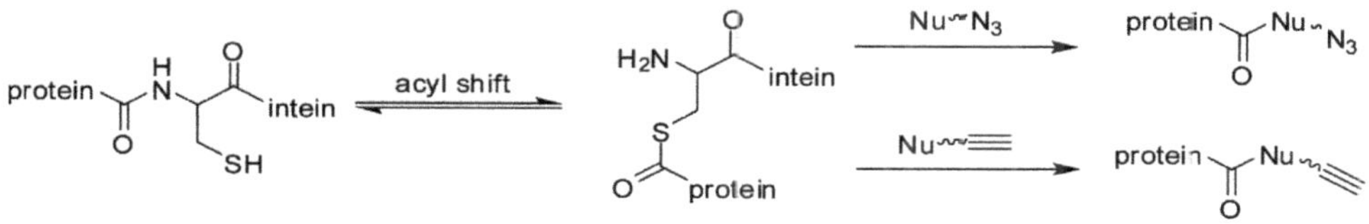

Figure 13.2 *Introduction of azide or alkyne functionality to the C-terminus of a protein via intein-mediated thioester formation. An electrophilic thioester is formed by an N–S acyl shift. The thioester can be attacked by an azide- or alkyne-modified nucleophile, resulting in a single, amide-linked modification on the C-terminus of the protein.*

the azide group to its C-terminus in a single step. The authors report a yield of about 1 mg/l of bacterial culture of the azide-labeled protein, which is exceptional considering the simplicity of the protocol. Lastly, the authors used CuAAC to covalently label azido-RNAse A with alkyne-modified fluorescein. Gel electrophoresis and fluorescence scanning as well as mass spectrometry confirmed that the protein was specifically labeled.

A similar expressed protein ligation strategy was used by Lin *et al.* to label *E. coli* maltose-binding protein (MBP) and green fluorescent protein (GFP) at their C-termini.[23] In contrast to Kalia and Raines, the authors used a nucleophile consisting of either the azide or alkyne linked to the carboxyl group of the amino acid cysteine via an amide bond. Several different azide-labeled small molecules, including fluorescein, biotin, glucosamine, a glycopeptide and a diazide linker, were conjugated to the alkyne-modified MBP using CuAAC. Notably, the authors claim that CuAAC carried out at 4 °C without the TBTA ligand led to poor product yields in these reactions. However, the addition of TBTA to the reaction and raising the temperature to 25 °C led to nearly quantitative modification of the protein in only 6 h. Lin *et al.* also used CuAAC to attach azide- and alkyne-modified proteins to glass surfaces in order to test the potential of CuAAC in the fabrication of protein microarrays. GFP labeled with either an azide or an alkyne was readily attached to surfaces decorated with the appropriate reaction partner, although significantly more protein was attached to the surface when alkynyl-GFP was reacted with an azido surface. Alkynyl-MBP was also covalently linked to the azido glass surface, and binding experiments with biotinyl-maltose indicated that the surface-bound MBP retained activity. The importance of being able to include a single, C-terminal anchoring moiety was underscored in experiments comparing the CuAAC-based protocol to traditional surface anchoring in which proteins are randomly linked to the substrate via surface accessible lysines or arginines. The CuAAC-based surface attachment of MBP led to significantly more active protein (as measured by binding of biotinyl-maltose) on the surface when compared with the conventional random-amide formation method.

Two groups have described an alternative way to specifically introduce an azide or alkyne group near the C-terminus of a protein via posttranslational prenylation catalyzed by the protein farnesyl transferase (PFTase).[24,25] The natural function of PFTase is to add the farnesyl moiety to cysteine residues within a four amino acid motif, CaaX (a = aliphatic amino acid, X = Ala, Ser, Met, Asn), found at the C-terminus of a protein. The yeast PFTase accepts azide and alkyne analogs of farnesyl diphosphate allowing for the facile modification of recombinant proteins with the CaaX motif at the C-terminus. Gauchet *et al.* immobilized alkyne-farnesylated GFP and glutathione reductase (GST) on an azide-modified glass surface using CuAAC.[25] Duckworth *et al.* immobilized GFP modified with an azidofarnesyl analog on agarose beads that had been modified with alkyne groups. Notably, these authors point out that the endogenous proteome (the entity of all proteins in a cell) of *E. coli* lacks any proteins with the C-terminal CaaX motif, allowing the farnesylation reaction to be carried out in a specific fashion on total cell lysates. The orthogonality of the farnesylation step toward the proteome of *E. coli* coupled with the orthogonality and robustness of the CuAAC chemistry leads to a potent combination that should enable advances in the construction of protein microarrays.

The examples presented in this section demonstrate the diversity of methods to introduce the azide and alkyne moieties into proteins as well as the plethora of applications possible with these modified proteins. All of these approaches, however, rely upon some posttranslational modification of the protein by either chemical or enzymatic means. This 'extra step' in the conjugation of proteins via CuAAC is critical in certain applications, such as ABPP, but its elimination could speed up bioconjugation protocols and place the azide and alkyne on equal footing with amines and thiols, the workhorses of traditional bioconjugation. Research in the past two decades into the incorporation of unnatural amino acids into proteins has facilitated the direct, cotranslational incorporation of azides and alkynes (as well as a host of other functional groups). These technologies and their applications will be discussed in the next section of this chapter.

13.3 Cotranslational Functionalization of Proteins with Azides and Alkynes

The language of proteins is the set of 20 canonical amino acids, and a veritable encyclopedia of protein structures and functions have been assembled by Nature using this language. However, curious researchers have not been satisfied with Nature's language for proteins, and have sought to add new words to it in the form of unnatural amino acids. Efforts to add new amino acids to proteins can be classified into two distinct classes. In residue-specific unnatural amino acid incorporation, the codon(s) for one of the 20 natural amino acids is reassigned to an unnatural amino acid.[26] Residue-specific incorporation is also referred to as selective pressure incorporation by Budisa and colleagues.[27] The second class of unnatural amino acid incorporation experiments is referred to as site-specific incorporation in which a twenty-first amino acid is added to the set of proteinogenic amino acids. Site-specific unnatural amino acid incorporation is also a form of codon reassignment, but instead of reassigning sense codons to the unnatural amino acid, one of the three nonsense codons is reclaimed and used to code for a twenty-first amino acid.[28,29]

The residue-specific incorporation technique has been used for more than 50 years[30] and relies on the use of auxotrophic strains of *E. coli* along with a chemically defined growth medium. Site-specific incorporation was first demonstrated as an *in vitro* protein expression technique in which the unnatural amino acid was chemically ligated to a tRNA that could decode the amber nonsense codon.[28] Site-specific incorporation of an unnatural amino acid *in vivo* was first demonstrated by Furter in 1998.[29] The key discovery in this work was that an orthogonal aminoacyl-tRNA synthetase (aaRS)–tRNA pair is required for the incorporation of the twenty-first amino acid. Furter introduced a yeast phenylalanine

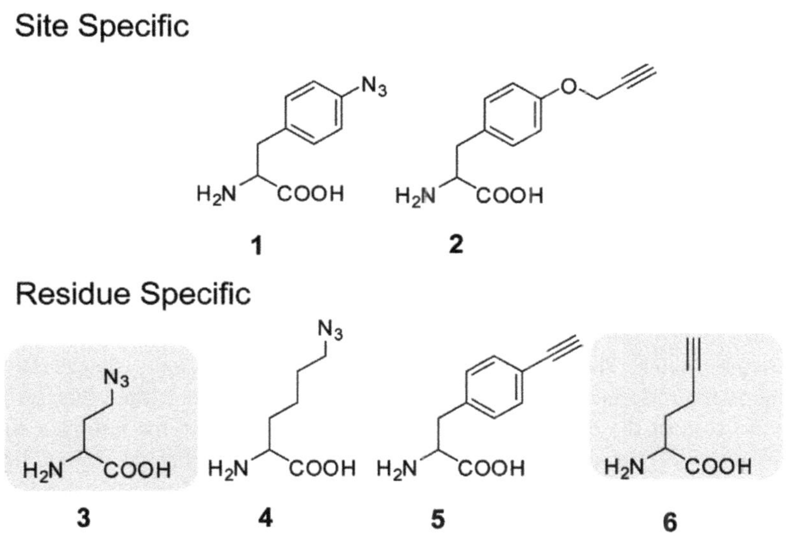

*Figure 13.3 Azido and alkynyl unnatural amino acids that can be incorporated into proteins in vivo. **1**, p-azidophenylalanine; **2**, O-propargyltyrosine; **3**, azidohomoalanine; **4**, homopropargylglycine; **5**, p-ethynylphenylalanine; **6**, azidonorleucine. **1** and **2** have been incorporated in a site-specific fashion while **3–6** have been incorporated in a residue-specific fashion; **3** and **6** (boxed) are substrates for the wild-type aminoacyl-tRNA synthetase activity of the cell while **1**, **2**, **4** and **5** require engineered aminoacyl-tRNA synthetase activities.*

tRNA ($tRNA^{Phe}$) with an altered anticodon and the yeast phenylalanyl-tRNA synthetase (PheRS) into *E. coli* cells. The addition of these two components to the cells led to efficient incorporation of fluorophenylalanine in response to a single amber nonsense codon within the test protein, dihydrofolate reductase. Subsequently, Schultz and colleagues have greatly expanded upon this methodology by generating high-throughput screening algorithms for the directed evolution of both the orthogonal tRNA and aaRS.[31] These advances have made possible the site-specific incorporation of dozens of different unnatural amino acids into proteins expressed in *E. coli* and in other organisms.[32] Both residue- and site-specific incorporation techniques have been used to deliver the azide and alkyne functionalities to proteins (Figure 13.3), enabling a host of different applications.

As mentioned above, one of the advantages of directly incorporating an azido- or alkynyl-amino acid into a protein is that the protein emerges from the cell ready for bioconjugation. No extra posttranslational modification steps are necessary. Deiters *et al.* used site-specific incorporation to introduce the amino acids *p*-azidophenylalanine (**1**, Figure 13.3) and *O*-propargyltyrosine (**2**, Figure 13.3) into human superoxide dismutase (SOD) expressed in the yeast *S. cerevisiae*.[10] These modified proteins were reacted with azide- or alkyne-modified dansyl or fluorescein dyes and imaged by fluorescence scanning after gel electrophoresis. The same authors demonstrated site-specific modification of human SOD in which *p*-azidophenylalanine had been incorporated with an alkyne-poly(ethylene glycol)

(PEG) reagent.[33] It is noteworthy that *p*-azidophenylalanine can also be incorporated into proteins in a residue-specific fashion using an active-site variant of the *E. coli* phenylalanyl-tRNA synthetase.[34]

Schoffelen *et al.* used residue-specific incorporation of the amino acid azidohomoalanine (AHA, **3**, Figure 13.3) in place of methionine to produce an azide-modified lipase from *Candida antarctica*.[11] AHA has been demonstrated to be an excellent surrogate for methionine in protein synthesis, and near-quantitative replacement of the five methionine residues of the lipase with AHA was readily achieved. Subsequent modification of this enzyme via CuAAC with either an alkyne-dansyl dye or an alkyne-PEG led to protein with only a single modification. Mass spectrometric analysis confirmed that this modification occurred only at the *N*-terminus of the protein despite the fact that there are five methionine residues in the protein. The rationale for this selective modification is that the *N*-terminal methionine is the only methionine residue exposed to the solvent; the other residues are buried in the core of the protein. Furthermore, the modified enzyme retains a significant fraction of its wild-type activity. It is important to point out that the TBTA ligand described earlier was not used in these studies. A sulfonated bathophenanthroline ligand with significantly improved water solubility over TBTA[35] was used instead. Though both site-specific and residue-specific unnatural amino acid incorporation techniques can be used to produce proteins for bioconjugation via CuAAC, protein production via residue-specific incorporation is simpler and does not suffer from the protein yield limitations sometimes observed with site-specific incorporation.[31] The relatively low abundance of methionine in the proteome coupled with its hydrophobicity that often confines it to the core of proteins makes residue-specific incorporation of AHA in place of methionine a promising avenue for ‘pseudo-site-specific’ modification of proteins via CuAAC.

The ability of CuAAC to tolerate complex biological milieu has been discussed above. In another example of the tolerance of CuAAC to biological conditions, Link and Tirrell set out to use CuAAC to address azide-labeled proteins in their native cellular context.[36,37] As a model system, AHA was incorporated into a variant of the *E. coli* outer membrane OmpC. The modified protein was properly targeted to the outer membrane, and subsequent reaction of the whole cells with a biotin–alkyne reagent under gentle CuAAC conditions led to extensive and specific functionalization of the cell surface with biotin as determined by western blotting experiments on outer membrane fraction of the cells. The biotinylated cells were also readily differentiable from unlabeled cells when stained with fluorescent streptavidin and subjected to flow cytometry. These experiments again demonstrate the versatility of the CuAAC chemistry; the CuAAC reaction is unhindered by the complex environment of the bacterial outer membrane, which contains lipids, proteins, polysaccharides and other molecules. These experiments formed the basis for further investigations by Link *et al.* in which the cell surface display of unnatural amino acids was exploited in a high-throughput screening method for the identification of novel methionyl-tRNA synthetase (MetRS) activity.[38] A saturation mutagenesis library of MetRS was screened for the ability to incorporate the long-chain amino acid azidonorleucine (**4**, Figure 13.3) using a flow cytometric screen, and a variant of MetRS that can activate azidonorleucine efficiently was discovered. In this screening protocol, the copper-catalyzed reaction was eschewed in favor of a cyclooctyne reagent developed in the Bertozzi laboratory[39] because of the observation of toxicity of the copper catalyst toward *E. coli*. The increased sensitivity of the cells to copper in this case may be attributed to perturbations of the outer membrane of the cells via overexpression of the OmpC protein since wild-type *E. coli* can typically tolerate up to 10 mM concentrations of copper ions. Beatty *et al.* also demonstrated the tolerance of the CuAAC chemistry to a complex cellular environment in experiments in which the unnatural amino acid *p*-ethynylphenylalanine (**5**, Figure 13.3) was incorporated into the recombinant cytosolic protein barstar.[40] Cells expressing the substituted barstar were treated with a fluorogenic azidocoumarin dye[41] and were rendered fluorescent after treatment with CuAAC reagents. Similar experiments were carried out in fixed mammalian cells in which the methionine surrogate homopropargylglycine (HPG, **6**, Figure 13.3) was incorporated throughout the proteome.[42]

The Finn group has also used the residue-specific incorporation of the methionine analogs AHA and HPG in their studies of viral capsids as functional nanomaterials. Prasuhn *et al.* incorporated AHA into genetically engineered mutants of the coat protein of bacteriophage Qb such that the azide groups from AHA are only displayed on the interior of the assembled particle.[12] CuAAC was used to covalently ligate a gadolinium complex to the assembled particles, which were used in a study of the effect of surface charge on the plasma clearance of the particles from mice. The same group also incorporated AHA into a surface-exposed position of the Qb particle and could address 90% (over 300 moieties per particle) of the displayed azide moieties with a fluorescein–alkyne reagent via CuAAC.[43] Despite this heavy functionalization, the recovery of intact particles was efficient, again underscoring the mild nature of the CuAAC reaction conditions.

A particularly elegant use of residue-specific incorporation of AHA followed by bioconjugation via CuAAC was described in a recent paper by van Kasteren *et al.*[8] The authors were interested in being able to mimic multiple posttranslational modifications in a site-specific fashion within a protein. To achieve this goal, they generated a mutant version of the SSβG protein, an enzyme with LacZ-type galactosidase activity, in which all of the methionine residues except one were replaced with the nearly isosteric isoleucine. Additionally, one of the cysteine residues found naturally in the protein was mutated to serine creating a protein with a single methionine and a single cysteine. Despite all of the mutations, the mutant SSβG retained its enzymatic activity. The protein was expressed under conditions that led to near-quantitative replacement of methionine with AHA, resulting in a protein with two orthogonal site-specific handles for chemical modification. As a first test, the authors installed glucose at the cysteine thiol via a disulfide-forming reaction and galactose at the azide sidechain of AHA via CuAAC. Both modifications proceeded nearly quantitatively and under gentle conditions. Notably, optimized CuAAC conditions were employed which eliminated the reducing agents used in many applications of CuAAC. These reducing agents, such as ascorbic acid or tris(carboxyethyl) phosphine (TCEP), are incompatible with the thiol chemistry. The authors also generated SSβG with a mimic of tyrosine sulfonation at the cysteine moiety and either a specific trisaccharide or tetrasaccharide at the AHA moiety. The doubly modified protein was designed to be a mimic of the human protein P-selectin-glycoprotein-ligand-1 (PSGL-1). The chemically modified protein was competent in binding to human selectin, and it was demonstrated that both modifications were necessary for optimal binding. Most impressively, the doubly modified SSβG could be used *in vivo* as a sensor of either acute or chronic inflammation in rat cortex. In a similar application, the protein could be used to detect a malarial infection in a mouse model. These studies demonstrate the power of CuAAC-based bioconjugation when used in conjunction with conventional cysteine-based bioconjugation. The techniques described should be generally useful for biologists interested in mimicking multiple posttranslational modifications to proteins. This is an application for which traditional bioconjugation to lysines and cysteines simply would not work because of the abundance (and functional importance) of lysine residues in proteins. The relatively low abundance of methionine in the proteome along with some clever genetic engineering allowed once again for the 'pseudo-site-specific' incorporation of the azide group into a protein.

13.4 BONCAT: Identification of Newly Synthesized Proteins via Noncanonical Amino Acid Tagging

In the last section of this chapter we want to discuss the application of CuAAC to proteomic profiling of cultured cells and tissues. On the molecular level proteins drive all major cellular functions. Cells, tissues, and living organisms all are dynamic entities responding to perturbations in their environment by changing the set of proteins the proteome they express either through posttranslational modifications of existing proteins or via adjustments in protein synthesis and degradation. For example, the composition of protein complexes and networks, such as the NMDA glutamate receptor complex or signaling cascades, can be regulated by the addition of *de novo* synthesized proteins or removal of existing proteins through changes in the activity pattern. Disruptions or imbalances in cellular protein homeostasis cause severe disorders of which the various types of cancers or fragile X mental retardation are just a few prominent examples. Therefore, a tremendous challenge for researchers and health professionals alike is the comparison of two or more proteomes, for example the cancerous vs noncancerous state, to eventually pinpoint and isolate the exact cellular malfunction.

Currently, various forms of mass spectrometry (MS) based proteomic profiling tools are the prime candidate approaches to characterize expression and functional modification profiles of proteins. Modern proteomics, however, faces several major challenges: first, proteins display an undisputed heterogeneity and cannot be amplified like their genomic counterparts, aggravating their identification in complex mixtures. Second, in-depth identification of a cell's entire proteome, let alone the comparison to another proteome, is unarguably a difficult feat with an estimated number of approximately 10 000 different proteins in a single mammalian cell.[44] Third, copy numbers of proteins from different mammalian cells and tissues vary with a predicted dynamic range of up to six orders of magnitude, and this number is even several orders of magnitude larger in plasma samples. Compared with subfemtomolar sensitivity in the analysis of a single purified protein, the effective identification of low-abundance proteins in complex mixtures is several orders of magnitude lower due to limited dynamic range and sequencing speed of current mass spectrometry instruments.[45] Hence, no single proteome of a mammalian cell or lower eukaryotic microorganism, such as yeast, has been completely characterized so far. A recent proteomic profiling study identified 5111 proteins in murine embryonic stem cells.[46]

How can these obstacles be mastered? How can one achieve in-depth identification and capture temporal and spatial proteome dynamics associated with changes in a cell's activation pattern, or its developmental stage if not all proteins – high or low in abundance – can be identified with equal chances and accuracy?

Biochemical and analytical approaches to reduce proteome complexity and to increase the dynamic range of protein identification use fractionation and affinity-purification tools on protein and peptide levels prior to MS analysis. In particular, fractionation methods of whole organelles (mitochondria[47] and nucleolus[48]) and compartments, such as the postsynaptic density of neurons[49,50] as well as affinity-purified protein complexes,[49,51,52] have been successfully used to enhance in-depth proteomic analysis.

However, not only qualitative knowledge but also quantitative knowledge of a proteome is important to understand a cell's activation state and overall phenotype. Therefore, extensive efforts have been dedicated to the development of differential proteomic profiling approaches to compare proteomes with one another and to obtain relative quantification of individual proteins among samples. These methods include differential 2D gel electrophoresis[53,54] (DIGE), isotope-coded affinity tags[55] (ICAT) or isobaric tags for relative and absolute quantification[56] (iTRAQ), quantitative proteomic analysis using samples from cells grown in ^{14}N- or ^{15}N-media[57] and stable isotope labeling by amino acids in cell culture[48,58] (SILAC).

Furthermore, the combination of MS with affinity purification for different posttranslational modifications[59–63] decreases sample complexity by enrichment of a specific subpopulation of the proteome. While posttranslational modifications such as phosphorylation or ubiquitination readily provide a suitable handle for enrichment of the 'phosphoproteome' or for proteins destined for degradation, reducing sample complexity by selectively enriching for newly synthesized proteins is troublesome, since all proteins – old and new – share the same pool of 20 amino acids. Nonetheless, the specific enrichment and identification of recently synthesized proteins would complement the range of differential proteomic profiling methods already available, and add another level of separation and simplification of complex protein mixtures, deepening our insights into the spatial and temporal dynamics of proteomes.

To provide the proteomics community with this added feature of selecting for newly synthesized proteins, the BONCAT (bio-orthogonal noncanonical amino acid tagging) technology was developed. The core of the BONCAT technique capitalizes on the manifold potential of small bioorthogonal chemo-selective groups (for a review see Prescher and Bertozzi[64]). In the first step of BONCAT, newly synthesized proteins are labeled using the azide-bearing unnatural amino acid azidohomoalanine (AHA, **3**, Figure 13.3), endowing them with novel azide functionality, which distinguishes them from the pool of pre-existing proteins (Figure 13.4). Employing CuAAC, the reactive azide group of AHA is covalently coupled to an alkyne-bearing affinity-tag in the second step. This tag enables the subsequent detection, affinity purification and MS identification of AHA-labeled proteins. The enrichment for newly synthesized proteins decreases the complexity of the sample, fostering the identification of proteins expressed at low levels.

Although Dieterich *et al.* used an alkyne-biotin-FLAG tag in the original application of BONCAT,[65,66] researchers may wish to substitute the biotin and the FLAG epitope for other affinity moieties. While the biotin moiety is used for avidin-based affinity purification, the FLAG epitope provides sites for trypsin cleavage, allowing immediate proteolysis of proteins on the affinity resin, bypassing the need for a separate elution step. Furthermore, the FLAG epitope can be used as an alternative purification module if native biotinylation of proteins is a concern. In this case affinity-purified AHA-tagged proteins can be eluted using high-salt conditions or competition with the FLAG-peptide. The increasing number

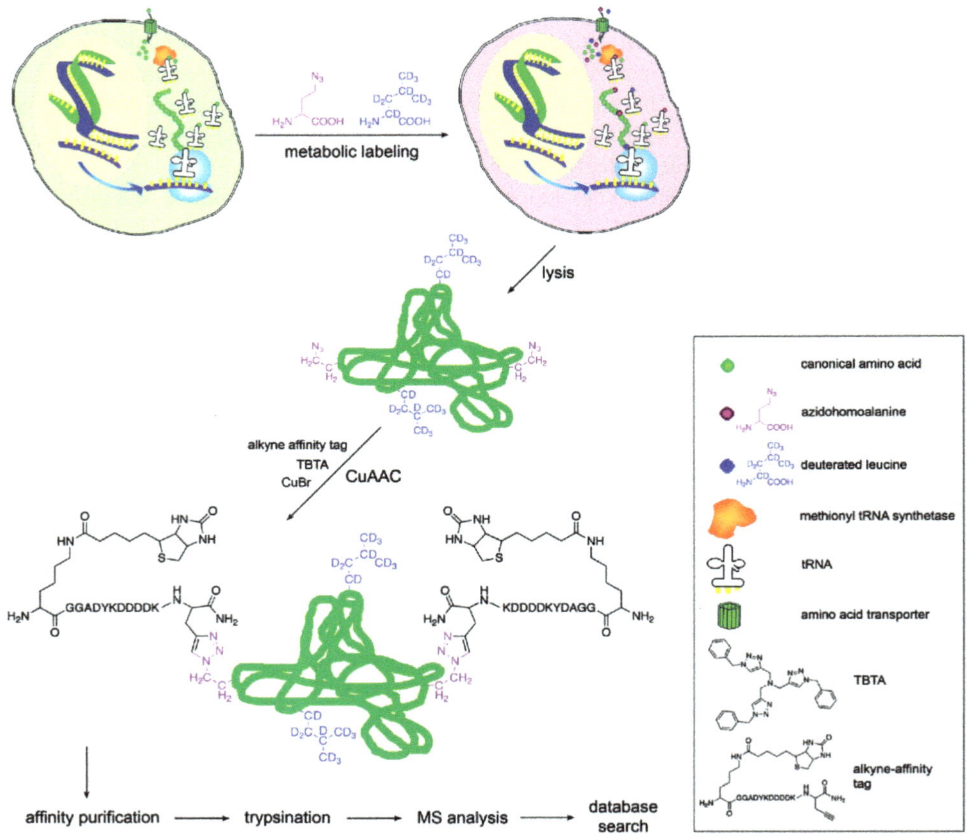

Figure 13.4 *A schematic depiction of the BONCAT approach. Metabolic cotranslational labeling with azidohomoalanine and an isotopically heavy amino acid (shown here is deuterated leucine) confers bioorthogonal functionalization to newly synthesized proteins. After incubation, cells are directly lysed or, alternatively, a subcellular fractionation for biochemical enrichment of specific cellular compartments is performed prior to lysis. Lysates are coupled to an alkyne-bearing affinity tag, followed by affinity purification. Purified proteins are digested with a protease, most commonly trypsin, and the resulting peptides are analyzed by mass spectrometry to obtain experimental spectra. Different search algorithms are used to match the acquired spectra to protein sequences.*

of new functional linkers, such as photocleavable or acid-labile groups, promises to add to the versatility and specificity of affinity tags for BONCAT applications.

After tryptic digestion of avidin-bound or eluted proteins, peptides bearing the tryptic remains of tagged AHA can serve as an immediate validation of candidate proteins. In the event of failed tagging, i.e. unligated AHA, the mass loss of AHA over methionine marks this peptide as derived from a true newly synthesized candidate protein. To increase the chances of detecting metabolically modified peptides co-labeling cells with deuterated

L-leucine ($d_{10}L$) or any other isotopically heavy amino acid allows, in conjunction with the modification derived from the introduction of AHA, the validation of candidate proteins.

As described in the previous section of this chapter, AHA is an effective surrogate for methionine, and does not require any cellular manipulations to be accepted as a substrate by the methionyl-tRNA synthetase.[7,67] Methionine is an essential amino acid in mammals, and, therefore, the prerequisite for effective depletion and, thus, increased incorporation rates for AHA into newly synthesized proteins are favorable. However, identification of newly synthesized proteins with BONCAT is limited to proteins that possess at least one methionine residue, excluding the 1.02% of all entries in a human protein database, which do not contain a single methionine. Given that 5.08% of the human proteome possess only a single, *N*-terminal methionine and that this residue may be subject to removal by posttranslational processing, at least 94% of the mammalian proteome are candidates for identification by BONCAT.[65] Interestingly, we have found no bias toward methionine-rich proteins in the proteomes characterized so far.

In general, labeling with AHA is very similar to the traditional metabolic labeling with radioactive amino acids (^{35}S-labeled methionine or cysteine) and can be performed in any biology or chemistry laboratory. BONCAT has been tested by us in a variety of cell lines, primary neuronal cells as well as organotypic brain slice cultures (unpublished observations). We found in all systems tested that the presence and incorporation of AHA is nontoxic and does not affect global rates of protein synthesis or degradation. CuAAC can be performed on denatured proteins in the presence of detergents, such as SDS, promoting the identification of diverse classes of proteins, i.e. membranous and soluble, acidic and basic as well as high and low molecular weight proteins. Indeed, the identified proteins display a broad range of functional and biochemical diversity[65] in terms of size or isoelectric point. Since the coupling reaction withstands harsh buffer conditions, even membrane proteins can be tagged and identified. Finally, AHA-tagged newly synthesized proteins show undisturbed subcellular distribution, for example tagged histone proteins could be identified from and found in the nuclear fraction.

Lastly, the BONCAT principle is not restricted to MS-based proteomic profiling, but the spatial fate of newly synthesized proteins can directly be visualized by using fluorescent CuAAC tags as Beatty and coworkers have demonstrated.[42] It also promises the possibility to work in combination with other proteomic approaches to directly compare different proteomes in a single MS experiment or to facilitate the identification of even more specific subpopulations of the proteome. Moreover, subcellular fractionation and immunopurification of protein complexes can be followed by BONCAT to assess the temporal and spatial dynamics of certain subcellular compartments, organelles and protein–protein interaction networks.

13.5 Conclusions and Future Prospects

As evidenced by the breadth of the examples provided in this chapter, the preeminent click chemistry reaction, CuAAC, has done much to advance the fields of protein conjugation, protein engineering and even proteomics. Protein engineers now have a viable alternative to traditional cysteine- and lysine-based bioconjugation strategies that can function in even the most complex biological environments. The relative synthetic ease of introducing the azide and alkyne moieties into small molecules should ensure the continued growth of CuAAC as a bioconjugation method by researchers across many fields of biology.

The use of a copper catalyst is a potential limitation of CuAAC in some biological applications. If CuAAC is being performed on live cells, the concentration of copper needed to effect CuAAC may be toxic to the cells. Similarly, if a protein bioconjugate is being produced for therapeutic use, the removal of all traces of copper is critical. To address these needs, Bertozzi and others[39,68,69] have developed copper-free versions of the azide–alkyne ligation in which the alkyne is activated by ring strain. These conjugation strategies, which are discussed in detail in Chapter 3, promise to enable even more applications in bioconjugation and proteomics.

What is the future of click chemistry in protein engineering? One can imagine that new click reactions beyond CuAAC will come to the forefront and may be used either as a replacement for or in conjunction with CuAAC. In the more immediate future, protein engineers should take advantage of the fact that CuAAC transcends any single discipline, percolating freely through biological chemistry, organic chemistry and materials science. Many of the future uses of CuAAC in protein science may come in interfacing proteins with materials, both soft and hard. Such work is already underway, as evidenced by the use of CuAAC to construct a protein microarray described earlier in this chapter.[23] One could also envision covalently interfacing redox active proteins with electronic materials via CuAAC for sensing applications. On the soft materials front, CuAAC is an ideal chemistry for the attachment of targeting proteins to liposomes or polymeric nanoparticles currently being explored for drug delivery applications. The simplicity and reliability of CuAAC has led to a plethora of uses in protein science described herein in a period of just over five years. These same aspects ensure that CuAAC will continue to be a versatile, well-used tool for protein engineers (and biologists in general) over the next five years and beyond.

References

(1) H. C. Kolb, M. G. Finn, and K. B. Sharpless, (2001), Click chemistry: diverse chemical function from a few good reactions, *Angew. Chem. Int. Edn*, **40**, 2004–2021.

(2) R. Huisgen, G. Szeimies, and L. Mobius, (1967), 1.3-Dipolare Cycloadditionen 32: Kinetik Der Additionen Organischer Azide an CC-Mehrfachbindungen, *Chem. Ber.-Recueil*, **100**, 2494–2507.

(3) V. V. Rostovtsev, L. G. Green, V. V. Fokin, and K. B. Sharpless, (2002), A stepwise Huisgen cycloaddition process: copper(I)-catalyzed regioselective 'ligation' of azides and terminal alkynes, *Angew. Chem. Int. Edn*, **41**, 2596–2599.

(4) C. W. Tornoe, C. Christensen, and M. Meldal, (2002), Peptidotriazoles on solid phase: [1,2,3]-triazoles by regiospecific copper(I)-catalyzed 1,3-dipolar cycloadditions of terminal alkynes to azides, *J. Org. Chem.*, **67**, 3057–3064.

(5) N. J. Agard, J. M. Baskin, J. A. Prescher, A. Lo, and C. R. Bertozzi, (2006), A comparative study of bioorthogonal reactions with azides, *ACS Chem. Biol.*, **1**, 644–648.

(6) T. S. Seo, X. P. Bai, H. Ruparel, Z. M. Li, N. J. Turro, and J. Y. Ju, (2004), Photocleavable fluorescent nucleotides for DNA sequencing on a chip constructed by site-specific coupling chemistry, *Proc. Natl Acad. Sci. USA*, **101**, 5488–5493.

(7) A. J. Link, and D. A. Tirrell, (2003), Cell surface labeling of *Escherichia coli* via copper(I)-catalyzed [3 + 2] cycloaddition, *J. Am. Chem. Soc.*, **125**, 11164–11165.

(8) S. I. van Kasteren, H. B. Kramer, H. H. Jensen, S. J. Campbell, J. Kirkpatrick, N. J. Oldham, D. C. Anthony, and B. G. Davis, (2007), Expanding the diversity of chemical protein modification allows post-translational mimicry, *Nature*, **446**, 1105–1109.

(9) B. P. Duckworth, Y. Chen, J. W. Wollack, Y. Sham, J. D. Mueller, T. A. Taton, and M. D. Distefano, (2007), A universal method for the preparation of covalent protein–DNA conjugates for use in creating protein nanostructures, *Angew. Chem. Int. Edn*, **46**, 8819–8822.
(10) A. Deiters, T. A. Cropp, M. Mukherji, J. W. Chin, J. C. Anderson, and P. G. Schultz, (2003), Adding amino acids with novel reactivity to the genetic code of Saccharomyces cerevisiae, *J. Am. Chem. Soc.*, **125**, 11782–11783.
(11) S. Schoffelen, M. H. L. Lambermon, M. B. van Eldijk, and J. C. M. van Hest, (2008), Site-specific modification of *Candida antarctica* lipase B via residue-specific incorporation of a non-canonical amino acid, *Bioconjugate Chem.*, **19**, 1127–1131.
(12) D. E. Prasuhn, P. Singh, E. Strable, S. Brown, M. Manchester, and M. G. Finn, (2008), Plasma clearance of bacteriophage Q beta particles as a function of surface charge, *J. Am. Chem. Soc.*, **130**, 1328–1334.
(13) Q. Wang, T. R. Chan, R. Hilgraf, V. V. Fokin, K. B. Sharpless, and M. G. Finn, (2003), Bioconjugation by copper(I)-catalyzed azide–alkyne [3 + 2] cycloaddition, *J. Am. Chem. Soc.*, **125**, 3192–3193.
(14) Q. Wang, T. W. Lin, L. Tang, J. E. Johnson, and M. G. Finn, (2002), Icosahedral virus particles as addressable nanoscale building blocks, *Angew. Chem. Int. Edn*, **41**, 459–462.
(15) Q. Wang, E. Kaltgrad, T. W. Lin, J. E. Johnson, and M. G. Finn, (2002), Natural supramolecular building blocks: wild-type cowpea mosaic virus, *Chem. Biol.*, **9**, 805–811.
(16) Q. Wang, T. W. Lin, J. E. Johnson, and M. G. Finn, (2002), Natural supramolecular building blocks: cysteine-added mutants of cowpea mosaic virus, *Chem. Biol.*, **9**, 813–819.
(17) T. R. Chan, R. Hilgraf, K. B. Sharpless, and V. V. Fokin, (2004), Polytriazoles as copper(I)-stabilizing ligands in catalysis, *Org. Lett.*, **6**, 2853–2855.
(18) A. E. Speers, G. C. Adam, and B. F. Cravatt, (2003), Activity-based protein profiling *in vivo* using a copper(I)- catalyzed azide–alkyne [3 + 2] cycloaddition, *J. Am. Chem. Soc.*, **125**, 4686–4687.
(19) A. E. Speers, and B. F. Cravatt, (2004), Chemical strategies for activity-based proteomics, *Chembiochem*, **5**, 41–47.
(20) A. E. Speers, and B. F. Cravatt, (2004), Profiling enzyme activities *in vivo* using click chemistry methods, *Chem. Biol.*, **11**, 535–546.
(21) J. Kalia, and R. T. Raines, (2006), Reactivity of intein thioesters: appending a functional group to a protein, *Chembiochem*, **7**, 1375–1383.
(22) T. W. Muir, (2003), Semisynthesis of proteins by expressed protein ligation, *Annu. Rev. Biochem.*, **72**, 249–289.
(23) P. C. Lin, S. H. Ueng, M. C. Tseng, J. L. Ko, K. T. Huang, S. C. Yu, A. K. Adak, Y. J. Chen, and C. C. Lin, (2006), Site-specific protein modification through Cu-I-catalyzed 1,2,3-triazole formation and its implementation in protein microarray fabrication, *Angew. Chem. Int. Edn*, **45**, 4286–4290.
(24) B. P. Duckworth, J. H. Xu, T. A. Taton, A. Guo, and M. D. Distefano, (2006), Site-specific, covalent attachment of proteins to a solid surface, *Bioconjugate Chem.*, **17**, 967–974.
(25) C. Gauchet, G. R. Labadie, and C. D. Poulter, (2006), Regio- and chemoselective covalent immobilization of proteins through unnatural amino acids, *J. Am. Chem. Soc.*, **128**, 9274–9275.
(26) A. J. Link, and D. A. Tirrell, (2005), Reassignment of sense codons *in vivo*, *Methods*, **36**, 291–298.
(27) N. Budisa, (2004), Prolegomena to future experimental efforts on genetic code engineering by expanding its amino acid repertoire, *Angew. Chem. Int. Edn*, **43**, 6426–6463.
(28) C. J. Noren, S. J. Anthony-Cahill, M. C. Griffith, and P. G. Schultz, (1989), A general-method for site-specific incorporation of unnatural amino-acids into proteins, *Science*, **244**, 182–188.
(29) R. Furter, (1998), Expansion of the genetic code: site-directed *p*-fluoro-phenylalanine incorporation in *Escherichia coli*, *Protein Sci.*, **7**, 419–426.
(30) D. B. Cowie, and G. N. Cohen, (1957), Biosynthesis by *Escherichia-coli* of active altered proteins containing selenium instead of sulfur, *Biochim. Biophys. Acta*, **26**, 252–261.
(31) L. Wang, A. Brock, B. Herberich, and P. G. Schultz, (2001), Expanding the genetic code of *Escherichia coli*, *Science*, **292**, 498–500.

(32) L. Wang, and P. G. Schultz, (2005), Expanding the genetic code, *Angew. Chem. Int. Edn*, **44**, 34–66.
(33) A. Deiters, T. A. Cropp, D. Summerer, M. Mukherji, and P. G. Schultz, (2004), Site-specific PEGylation of proteins containing unnatural amino acids, *Bioorg. Med. Chem. Lett.*, **14**, 5743–5745.
(34) K. Kirshenbaum, I. S. Carrico, and D. A. Tirrell, (2002), Biosynthesis of proteins incorporating a versatile set of phenylalanine analogues, *Chembiochem*, **3**, 235–237.
(35) W. G. Lewis, F. G. Magallon, V. V. Fokin, and M. G. Finn, (2004), Discovery and characterization of catalysts for azide–alkyne cycloaddition by fluorescence quenching, *J. Am. Chem. Soc.*, **126**, 9152–9153.
(36) A. J. Link, M. L. Mock, and D. A. Tirrell, (2003), Non-canonical amino acids in protein engineering, *Curr. Opin. Biotechnol.*, **14**, 603–609.
(37) A. J. Link, M. K. S. Vink, and D. A. Tirrell, (2004), Presentation and detection of azide functionality in bacterial cell surface proteins, *J. Am. Chem. Soc.*, **126**, 10598–10602.
(38) A. J. Link, M. K. S. Vink, N. J. Agard, J. A. Prescher, C. R. Bertozzi, and D. A. Tirrell, (2006), Discovery of aminoacyl-tRNA synthetase activity through cell-surface display of noncanonical amino acids, *Proc. Natl Acad. Sci. USA*, **103**, 10180–10185.
(39) N. J. Agard, J. A. Prescher, and C. R. Bertozzi, (2004), A strain-promoted [3 + 2] azide-alkyne cycloaddition for covalent modification of biomolecules in living systems, *J. Am. Chem. Soc.*, **126**, 15046–15047.
(40) K. E. Beatty, F. Xie, Q. Wang, and D. A. Tirrell, (2005), Selective dye-labeling of newly synthesized proteins in bacterial cells, *J. Am. Chem. Soc.*, **127**, 14150–14151.
(41) K. Sivakumar, F. Xie, B. M. Cash, S. Long, H. N. Barnhill, and Q. Wang, (2004), A fluorogenic 1,3-dipolar cycloaddition reaction of 3-azidocoumarins and acetylenes, *Org. Lett.*, **6**, 4603–4606.
(42) K. E. Beatty, J. C. Liu, F. Xie, D. C. Dieterich, E. M. Schuman, Q. Wang, and D. A. Tirrell, (2006), Fluorescence visualization of newly synthesized proteins in mammalian cells, *Angew. Chem. Int. Edn*, **45**, 7364–7367.
(43) E. Strable, D. E. Prasuhn, A. K. Udit, S. Brown, A. J. Link, J. T. Ngo, G. Lander, J. Quispe, C. S. Potter, B. Carragher, *et al.*, (2008), Unnatural amino acid incorporation into virus-like particles, *Bioconjugate Chem.*, **19**, 866–875.
(44) A. Pandey, and M. Mann, (2000), Proteomics to study genes and genomes, *Nature*, **405**, 837–846.
(45) L. de Godoy, J. Olsen, G. de Souza, G. Li, P. Mortensen, and M. Mann, (2006), Status of complete proteome analysis by mass spectrometry: SILAC labeled yeast as a model system, *Genome Biol.*, **7**, R50.
(46) J. Graumann, N. C. Hubner, J. B. Kim, K. Ko, M. Moser, C. Kumar, J. Cox, H. Scholer, and M. Mann, (2008), Stable isotope labeling by amino acids in cell culture (SILAC) and proteome quantitation of mouse embryonic stem cells to a depth of 5,111 proteins, *Mol. Cell. Proteomics*, **7**, 672–683.
(47) V. K. Mootha, J. Bunkenborg, J. V. Olsen, M. Hjerrild, J. R. Wisniewski, E. Stahl, M. S. Bolouri, H. N. Ray, S. Sihag, M. Kamal, *et al.*, (2003), Integrated analysis of protein composition, tissue diversity, and gene regulation in mouse mitochondria, *Cell*, **115**, 629–640.
(48) J. S. Andersen, Y. W. Lam, A. K. Leung, S. E. Ong, C. E. Lyon, A. I. Lamond, and M. Mann, (2005), Nucleolar proteome dynamics, *Nature*, **433**, 77–83.
(49) M. O. Collins, H. Husi, L. Yu, J. M. Brandon, C. N. Anderson, W. P. Blackstock, J. S. Choudhary, and S. G. Grant, (2006), Molecular characterization and comparison of the components and multiprotein complexes in the postsynaptic proteome, *J. Neurochem.*, **97** (suppl. 1), 16–23.
(50) S. P. Schrimpf, V. Meskenaite, E. Brunner, D. Rutishauser, P. Walther, J. Eng, R. Aebersold, and P. Sonderegger, (2005), Proteomic analysis of synaptosomes using isotope-coded affinity tags and mass spectrometry, *Proteomics*, **5**, 2531–2541.
(51) A. C. Gavin, M. Bosche, R. Krause, P. Grandi, M. Marzioch, A. Bauer, J. Schultz, J. M. Rick, A. M. Michon, C. M. Cruciat, *et al.*, (2002), Functional organization of the yeast proteome by systematic analysis of protein complexes, *Nature*, **415**, 141–147.

(52) Y. Ho, A. Gruhler, A. Heilbut, G. D. Bader, L. Moore, S. L. Adams, A. Millar, P. Taylor, K. Bennett, K. Boutilier, *et al.*, (2002), Systematic identification of protein complexes in *Saccharomyces cerevisiae* by mass spectrometry, *Nature*, **415**, 180–183.

(53) W. M. Freeman, and S. E. Hemby, (2004), Proteomics for protein expression profiling in neuroscience, *Neurochem. Res.*, **29**, 1065–1081.

(54) K. S. Lilley, and D. B. Friedman, (2004), All about DIGE: quantification technology for differential-display 2D-gel proteomics, *Expert Rev. Proteomics*, 1, 401–409.

(55) S. P. Gygi, B. Rist, S. A. Gerber, F. Turecek, M. H. Gelb, and R. Aebersold, (1999), Quantitative analysis of complex protein mixtures using isotope-coded affinity tags, *Nat. Biotechnol.*, **17**, 994–999.

(56) P. L. Ross, Y. N. Huang, J. N. Marchese, B. Williamson, K. Parker, S. Hattan, N. Khainovski, S. Pillai, S. Dey, S. Daniels, *et al.*, (2004), Multiplexed protein quantitation in Saccharomyces cerevisiae using amine-reactive isobaric tagging reagents, *Mol. Cell. Proteomics*, **3**, 1154–1169.

(57) M. P. Washburn, R. Ulaszek, C. Deciu, D. M. Schieltz, and J. R. Yates, 3rd, (2002), Analysis of quantitative proteomic data generated via multidimensional protein identification technology, *Anal. Chem.*, **74**, 1650–1657.

(58) S. E. Ong, B. Blagoev, I. Kratchmarova, D. B. Kristensen, H. Steen, A. Pandey, and M. Mann, (2002), Stable isotope labeling by amino acids in cell culture, SILAC, as a simple and accurate approach to expression proteomics, *Mol. Cell. Proteomics*, **1**, 376–386.

(59) E. J. Chang, V. Archambault, D. T. McLachlin, A. N. Krutchinsky, and B. T. Chait, (2004), Analysis of protein phosphorylation by hypothesis-driven multiple-stage mass spectrometry, *Anal. Chem.*, **76**, 4472–4483.

(60) B. A. Garcia, J. Shabanowitz, and D. F. Hunt, (2005), Analysis of protein phosphorylation by mass spectrometry, *Methods*, **35**, 256–264.

(61) E. C. Peters, A. Brock, and S. B. Ficarro, (2004), Exploring the phosphoproteome with mass spectrometry, *Mini Rev. Med. Chem.*, **4**, 313–324.

(62) M. Mann, and O. N. Jensen, (2003), Proteomic analysis of post-translational modifications, *Nat. Biotechnol.*, **21**, 255–261.

(63) P. Xu, and J. Peng, (2006), Dissecting the ubiquitin pathway by mass spectrometry, *Biochim. Biophys. Acta*, **1764**, 1940–1947.

(64) J. A. Prescher, and C. R. Bertozzi, (2005), Chemistry in living systems, *Nat. Chem. Biol.*, **1**, 13–21.

(65) D. C. Dieterich, A. J. Link, J. Graumann, D. A. Tirrell, and E. M. Schuman, (2006), Selective identification of newly synthesized proteins in mammalian cells using bioorthogonal noncanonical amino acid tagging (BONCAT), *Proc. Natl Acad. Sci. USA*, **103**, 9482–9487.

(66) D. C. Dieterich, J. J. Lee, A. J. Link, J. Graumann, D. A. Tirrell, and E. M. Schuman, (2007), Labeling, detection and identification of newly synthesized proteomes with bioorthogonal noncanonical amino-acid tagging, *Nat. Protocols*, **2**, 532–540.

(67) K. L. Kiick, E. Saxon, D. A. Tirrell, and C. R. Bertozzi, (2002), Incorporation of azides into recombinant proteins for chemoselective modification by the Staudinger ligation, *Proc. Natl Acad. Sci. USA*, **99**, 19–24.

(68) E. M. Sletten, and C. R. Bertozzi, (2008), A hydrophilic azacyclooctyne for Cu-free click chemistry, *Org. Lett.*, **10**, 3097–3099.

(69) X. H. Ning, J. Guo, M. A. Wolfert, and G. J. Boons, (2008), Visualizing metabolically labeled glycoconjugates of living cells by copper-free and fast huisgen cycloadditions, *Angew. Chem. Int. Edn*, **47**, 2253–2255.

14

Fluorogenic Copper(I)-catalyzed Azide–Alkyne Cycloaddition Reactions and their Applications in Bioconjugation

Céline Le Droumaguet and Qian Wang

14.1 Click Reaction for Bioconjugation Applications

Bioconjugation has recently emerged as a fast growing technology that affects almost every discipline of life sciences. It aims at the ligation of two or more molecules (or supramolecules) to form a new complex with the combined properties of its individual components.[1] Taking advantage of the outstanding reaction profile, 'click chemistry', in particular the Cu(I)-catalyzed or copper-free alkyne-azide cycloaddition reactions,[2–4] was embraced by bioconjugation chemistry in its early developmental stages.

The first demonstration in bioconjugation application of the Cu(I)-catalyzed alkyne–azide cycloaddition (CuAAC) reaction was reported by Finn, Sharpless, Fokin and coworkers.[5] Cowpea mosaic virus (CPMV), a nonenveloped icosahedral plant virus, was chosen as a protein prototype and successfully labeled at all 60 identical protein asymmetric units of the capsid. Tris(carboxylethyl)phosphine (TCEP), a water-soluble reducing agent, was used to reduce Cu(II) to Cu(I) at 4 °C. Addition of tris[(1-benzyl-1*H*-1,2,3-triazol-4-yl)methyl]amine ligand (TBTA, **1**)[6] drastically enhanced the reaction rate and an almost quantitative amount of modified proteins was able to be recovered after the reaction (Figure 14.1). The azide and alkyne moieties could be attached to lysine, cysteine or tyrosine residues, and the reaction is quite inert to the structures to be conjugated.[5,7–13]

Click Chemistry for Biotechnology and Materials Science Edited by Joerg Lahann

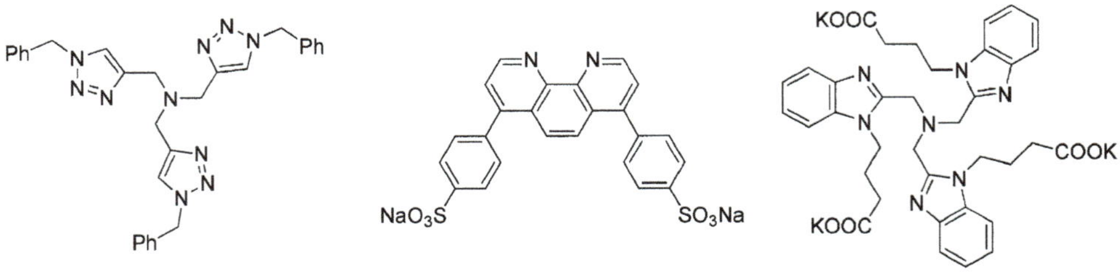

1 2 3

Figure 14.1 Structures of the most well-known ligands used for CuAAC-based bioconjugation reactions.

Many ligands and catalytic systems have been developed to make CuAAC reactions suitable for a bioconjugation process, where mild reaction conditions and high reaction efficiency are necessary. Besides ligand **1**, pyridine-containing compounds and benzimidazole-related ligands were found to be effective accelerating ligands to the CuAAC reaction.[14,15] The water-soluble bathophenanthrolinedisulfonic acid **2** and benzimidazole tricarboxylate **3** were confirmed to be remarkably reliable to catalyze a rapid and high-yielding synthesis of functionalized triazoles with an extremely low quantity of copper ion at room temperature or 4 °C (Figure 14.1). Being suitable for most of bio-platforms, CuAAC reactions have been followed by other groups for different bioconjugation applications. For example, Cravatt *et al.* used CuAAC reaction for activity-based protein profiling, where the proteomes (enzymes) of human breast cancer cells were labeled *in vivo*.[16,17] In this study, enzymes were functionalized with azides, homogenized and reacted with tetramethylrhodamine-alkynes. The labeled enzymes were detected and quantified, giving similar results with standard activity-based protein labeling. Tirrell *et al.* incorporated nonnatural azido-amino acids into the *E. coli* cell membrane protein OmpC, which were successfully modified with biotin–alkyne via a CuAAC reaction.[18,19] Schultz and coworkers introduced azido-amino acids or alkyne-amino acids into proteins in yeast[20,21] and the pIII protein of M13 filamentous phage.[22] These were sequentially reacted with fluorescent dyes or polyethylene glycols via click reaction. Ju and coworkers have applied the click reaction to fluorescently label DNA.[23,24] In addition, the CuAAC reaction has been applied to immobilize oligonucleotides on glass substrates in well-defined micropatterns.[25,26]

14.2 Significance of Fluorogenic Reactions in Bioconjugation

One important application of bioconjugation is to modify cellular components selectively with signaling probes for the research of *in vivo* imaging, proteomics, cell biology and functional genomics.[27–29] A multistep procedure is commonly employed: the cellular entity is first attached with a detectable tag, such as fluorescent dyes and biotin followed with purification of the ligated product and then detection of the conjugated tag with the target protein. However, excess prelabeled reagents (i.e. fluorescent dyes and biotin) are generally difficult to be removed from the intracellular environment or from tissues of living organisms, which prohibits the application of a multistep labeling procedure in many biological

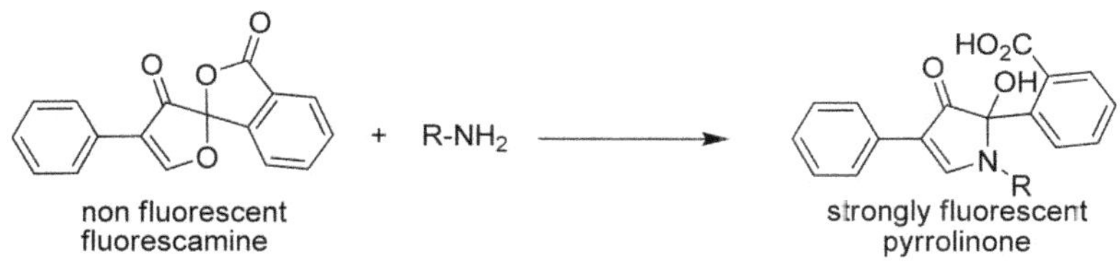

Scheme 14.1 *Fluorogenic reaction between fluorescamine and a primary amine.*

applications. An ideal alternative is a chemoselective process that is orthogonal to biological components, and the ligated product will afford strong detectable signal while the unbound reagent does not contribute to any background. Therefore, the fluorogenic reaction, a process where non- or weakly fluorescent reagents meet each other to give rise to visible fluorescence, would be invaluable for many bioimaging applications.

To date a wide array of fluorescent sensors and switches have been synthesized to recognize important events of chemistry, biology and materials. For example, selective ligands and ionophores for cations and ions are well established in optical sensing and *in vivo* probing.[30] Additionally, many fluorogenic dyes have been developed to detect neutral analytes based on noncovalent interactions.[31–33] In comparison, fewer reagents are available for covalent modification of biomolecules with high specificity and fluorogenic properties. Fluorescamine is one of the best-known reagents; it is intrinsically nonfluorescent but reacts rapidly with primary aliphatic amines to yield a blue-green-fluorescent pyrrolinone (Scheme 14.1). It has been broadly used in protein labeling, protein sequencing, determination of protein concentration and detection of low-molecule-weight amines in chromatography.[34–38]

Table 14.1 lists a few other commercially available fluorogenic probes which are able to tag biomolecules containing functional groups like primary amines, thiols or carbonyls, as well as DNA or RNA.[30] However, since amine, thiol and carbonyl groups are the most abundant functional units in biosystems, all these reagents can hardly afford selective modification of a targeted biospecies under the complicated intracellular conditions. In order to distinguish the target protein among the surrounding components, genetically encoded tags such as green fluorescent protein (GFP) and its variants are routinely applied.[39] Although the development of the GFP technology in the past two decades has enabled the use of GFP (or its variants) to signal physiological activation and indicate its chemical environment, GFP is still potentially perturbative because its size (238 amino acids) is often larger than the protein of interest. Therefore, Tsien and coworkers designed a short peptide domain composed of six to 20 amino acids containing the sequence Cys–Cys–Xaa–Xaa–Cys–Cys (where Xaa is a non-cysteine amino acid), and this domain can be genetically incorporated into the protein of interest.[40,41] Two fluorogenic dyes, the bisarsenical derivatives of fluorescein (**FLAsH-EDT$_2$**) and resorufin (**ReAsH-EDT$_2$**), were synthesized (Scheme 14.2). The membrane-permeating **FLAsH-EDT$_2$** and **ReAsH-EDT$_2$** are nonfluorescent but become brightly fluorescent upon binding to the tetracysteine motif.[41] This fluorogenic reaction, designed by Tsien *et al.*, is a powerful strategy to label proteins; however there is still a growing interest in visualizing biomolecules that are not amenable to such genetic modification. Furthermore, it is still not applicable for certain applications due to the potential cytotoxicity of bisarsenical compounds at high concentration.

Table 14.1 *Structure of representative commercially available fluorogenic reagents*

Compound[a]	Structure	Reaction partner
MDPF	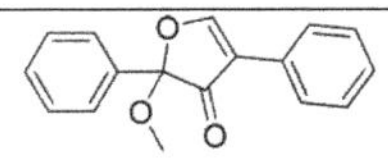	Primary amine
NPA	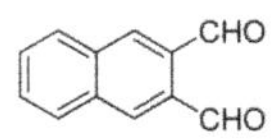	
NBD-Cl	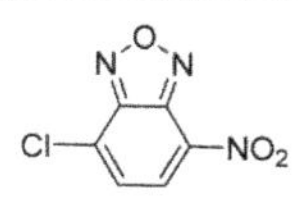	Primary amine, thiol
ABD-F	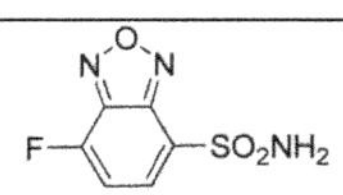	Thiol
D-346[b]	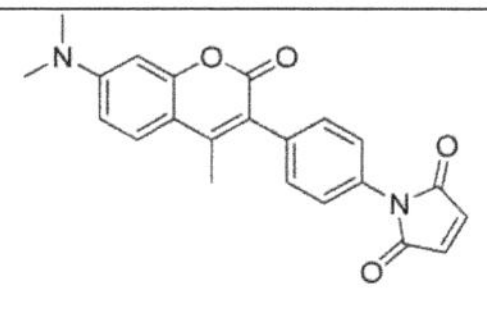	
D-100[b]	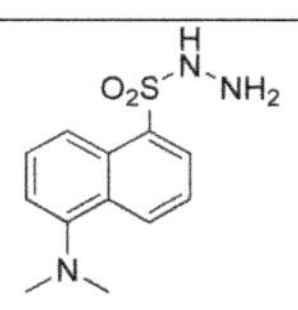	Carbonyl
DBD-H	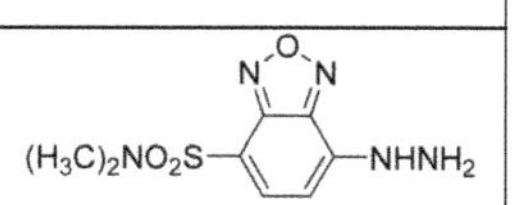	
Ethidium bromide	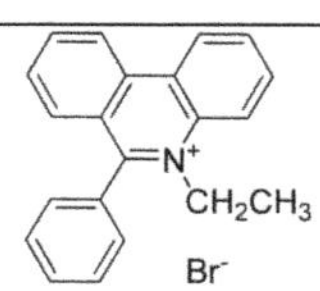	DNA

[a] MDPF, 2-methoxy-2,4-diphenyl-3(2H)-furanone; NPA, naphthalene-2,3-carboxaldehyde; NBD-Cl, 4-chloro-7-nitro-2,1,3-benzoxadiazole; ABD-F, 4-fluoro-7-aminosulfonylbenzofuran; D-346, 7-diethylamino-3(4′-maleimidylphenyl)-4-methylcoumarin; D-100, 5-dimethylamino-naphthalene-1-sulfonyl hydrazine; DBD-H, 4-(*N,N*-dimethylam- inosulfonyl)-7-hydrazino-1,3-benzoxadiazole.

[b] Compound references in Molecular Probes catalog.

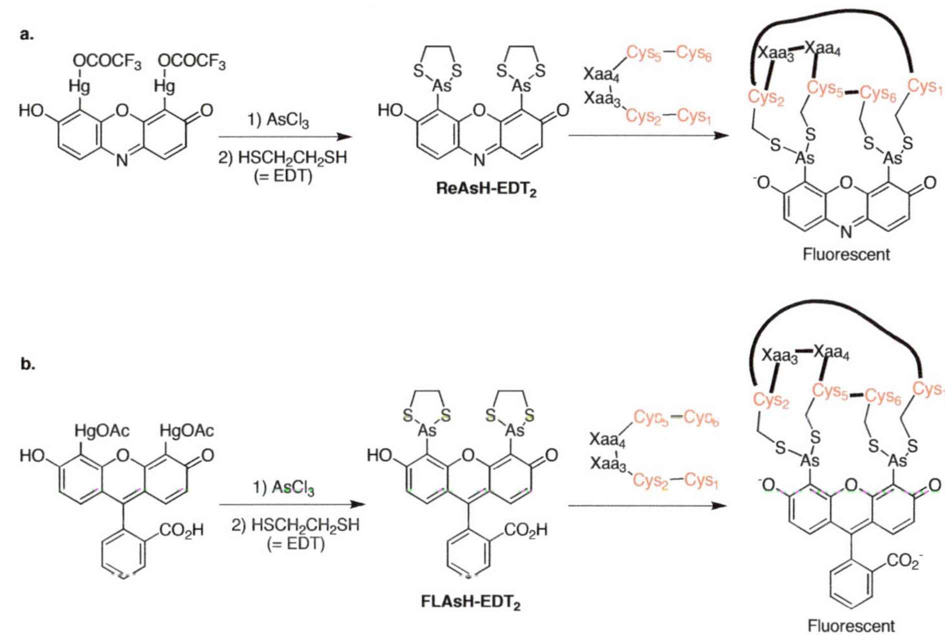

Scheme 14.2 *Non-fluorescent bisarsenical dyes* ***ReAsH-EDT$_2$*** *(a) and* ***FLAsH-EDT$_2$*** *(b) and proposed structures of complexes with an a-helical tetracysteine-containing peptide or protein domain. Reprinted with permission from B. A. Griffin* et al.*, (1998),* Science*,* ***281****, 269–271. Copyright 1998 AAAS.*

Being a highly energetic functional group, the organic azide is stable and unreactive with most biomolecules under physiological conditions as demonstrated by Bertozzi and coworkers with a modified version of the Staudinger reaction (see Chapter 3 for further details).[42] In brief, the product of the classical Staudinger reaction between a phosphine and an azide is an aza-ylide. Hydrolysis of the aza-ylide produces an amine and a phosphine oxide. The Bertozzi group elegantly placed an electrophilic trap adjacent to phosphine, which could react with the aza-ylide to form a stable amide adduct via an intramolecular electrophilic addition reaction. This reaction is highly efficient and specific even in the presence of water and a variety of other functional groups. They further designed a fluorogenic Staudinger reaction using a coumarin core.[43] The 3-position of the coumarin core is known to strongly influence its fluorescent properties. In compound **4**, the lone pair on phosphine quenches the fluorescence of coumarin [Scheme 14.3(a)]. After the formation of phosphorous oxide through the modified Staudinger reaction, the electron-donating phosphorous is switched into an electron-withdrawing functionality, and the fluorescence was activated.

Another fluorogenic Staudinger reaction was also reported for live-cell imaging.[44] In this study, phosphine compound **6** was synthesized with fluorescein-based fluorophore that is quenched intramolecularly by an ester-linked fluorescence resonance energy transfer (FRET) quencher. The reaction performed between **6** and benzyl azide in aqueous solution gave a compound whose fluorescence was enhanced 170-fold due to the free of the quencher [Scheme 14.3(b)]. With that in mind, HeLa cells were incubated with peracetylated *N*-azidoacetylmannosamine for 40 h in order to introduce *N*-azidoacetyl sialic acid into their surface cells, and the cells incubated with **6** for 8 h at 37 °C. After the incubation, fluorescence microscopy images showed highly localized fluorescence on the cell surfaces with little to no background with the cells that were not labeled with azidoacetylmannosamine.[44] The work by Bertozzi and coworkers highlighted that potential of using an azido group as the anchorage for bioconjugation and the merits of the fluorogenic reactions in real-time imaging of cellular components.

14.3 CuAAC-based Fluorogenic Reaction

As a prototype of 'click chemistry',[45–48] the recent advance of CuAAC reaction affords superior regioselectivity and almost quantitative transformation under extremely mild conditions.[2,49] Alkyne and azide groups are very small in size, are highly energetic, and have a particularly narrow distribution of reactivity. They can be conveniently introduced to organic compounds, and are quite insensitive to solvent and pH. Therefore, the CuAAC reaction becomes an ideal candidate to develop new fluorogenic reactions for the bioconjugation purpose. Figure 14.2 shows a schematic illustration of fluorogenic 1,3-cycloadditions between azides and alkynes, which have been reported to covalently link two biomolecules or supramolecular complexes for imaging or as reporters to monitor the ligation efficiency. In most situations, some prefluorophores are designed as the starting materials and the fluorescent signals can be triggered by the formation of triazole rings.

The coumarin was simultaneously chosen to develop fluorogenic CuAAC reactions by Fahrni *et al.* and Wang *et al.*[50,51] Coumarins are easy to synthesize and biocompatible and their photophysical properties are well known: substitution by electron withdrawing group at the 3-position and substitution by electron donating group at the 7-position strongly

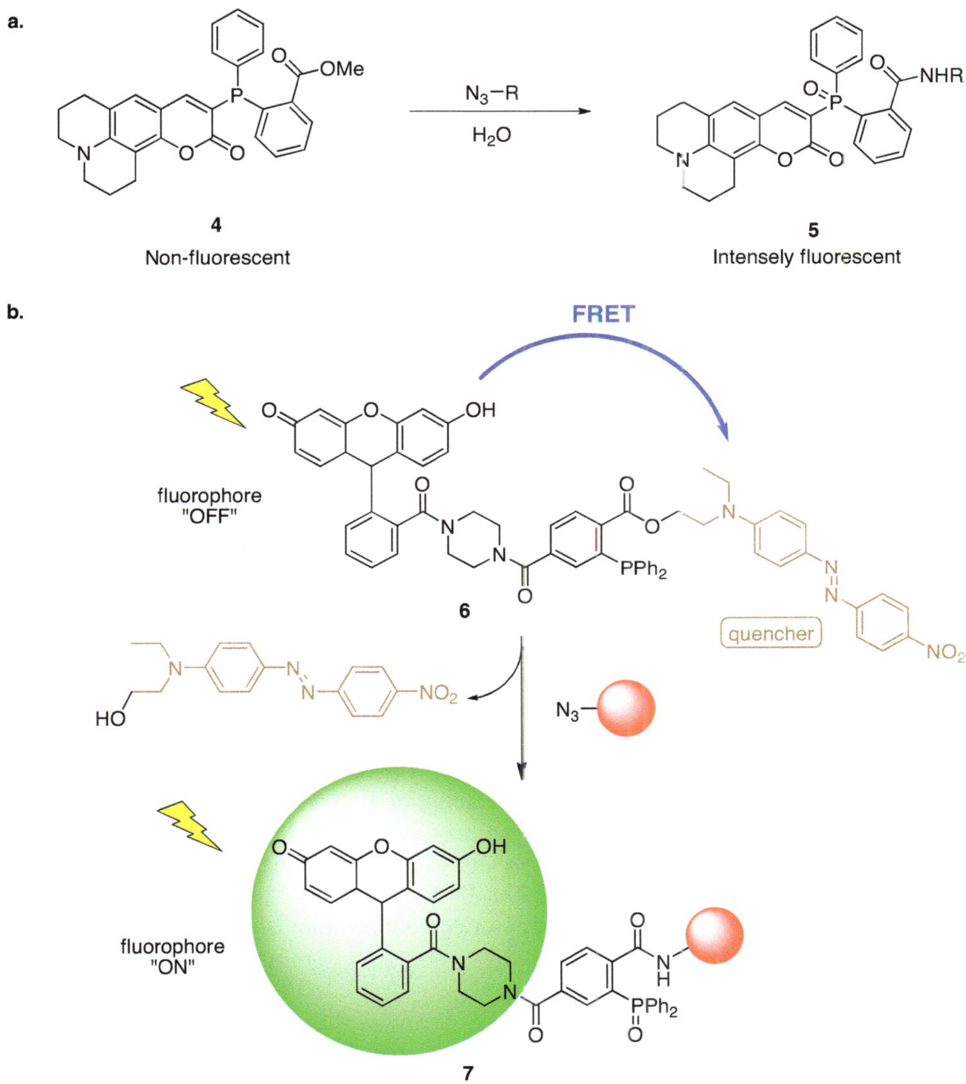

Scheme 14.3 Fluorogenic Staudinger reactions.[43,44] *(b) Reprinted with permission from M. J. Hangauer, C. R. Bertozzi, (2008), A FRET-based fluorogenic phosphine for live-cell imaging with the Staudinger ligation,* Angew. Chem. Int. Ed., ***47****(1). Copyright 2008 Wiley-VCH.*

enhance their fluorescence intensities while addition of electron donating groups at the 4-, 6- or 7-positions or electron-withdrawing groups at the 3-position shift the fluorescence band to longer wavelengths (Figure 14.3).[52–54]

The challenge was the design and synthesis of coumarin derivatives whose fluorescence could be quenched with an azido or alkyne moiety. Zhou and Fahrni synthesized a coumarin-based fluorogenic probe **8** (Scheme 14.4) bearing an alkyne at the 7-position.[51]

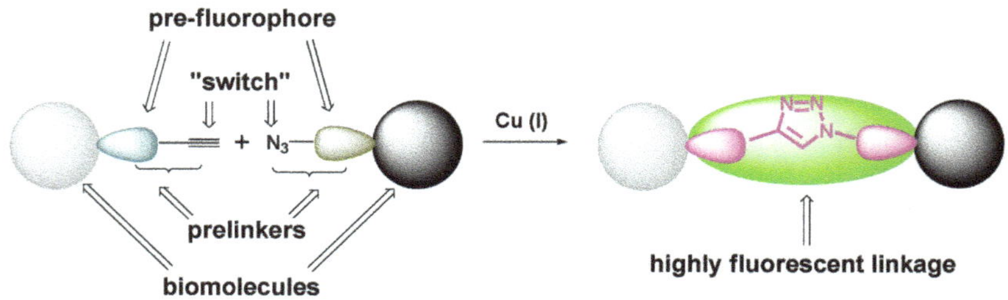

Figure 14.2 Schematic representation of fluorogenic CuAAC reaction. Reprinted with permission from K. Sivakumar et al., (2004), A fluorogenic 1,3-dipolar cycloaddition reaction of 3-azidocoumarins and acetylenes, Org. Lett., **6**, 4603–4606. Copyright 2004 ACS.

The formation of the triazole ring after CuAAC with an azide increases the electron-donating strength at the 7-position and consequently strongly enhances the fluorescence signal of the cycloaddition product. This fluorescence triggering was also confirmed by a semiempirical quantum calculation study on the electronic frontier orbital of starting material **8** and the final triazolo-compound **9**.

Wang and coworkers designed a series of fluorogenic 3-azidocoumarins **10** as shown in Scheme 14.5.[50] The fluorescence of these compounds is quenched due to the electron-rich α-nitrogen of the azido group. After the formation of the triazolo compound via the CuAAC reaction, the electronic density at the 3-position is reduced because the lone pair electrons contribute to the aromatic ring, which strongly enhances fluorescence. Moreover, they have applied these fluorogenic CuAAC to a wide range of alkynes, allowing the synthesis of triazolocoumarin dyes combinatorially. The formation of the triazolocoumarins **11** was directly detected by fluorescence screening of a 96-well plate upon irradiation at 365 nm (Figure 14.4). These triazolocoumarins can be prepared in large quantity by a simple filtration. Because of the high reactivity of aromatic azides used in the synthesis, the cycloaddition can be completed even at 0 °C, which benefits a real application of ligation between biomolecules, for which elevated temperature is usually destructive and physiological conditions are requested.

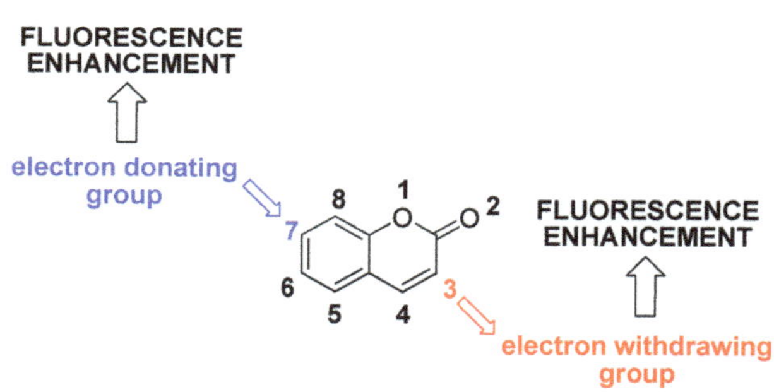

Figure 14.3 Electron donating group at the 7-position and electron-withdrawing groups at the 3-position of the coumarin scaffold enhance its fluorescence.

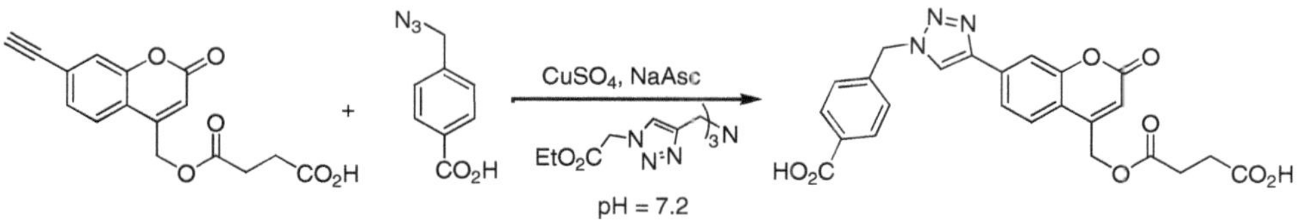

8 ($\Phi_f = 0.014$) **9** ($\Phi_f = 0.25$)

Scheme 14.4 *Fluorogenic CuAAC reaction based on the nonfluorescent 7-alkynylcoumarin **8**.*

Two CuAAC reaction-activated fluorescent probes based on 1,8-naphthalimide were reported by Wong and coworkers (Scheme 14.6).[55] The substitution of 1,8-naphthalimide at the 4-position by an electron donating group is known to strongly affect the fluorescence.[56] The 1,8-naphthalimide derivatives **12a** and **12b**, bearing at the 4-position either an alkyne or an azide, respectively, showed no fluorescence. Upon conjugation with complementary azido- and acetylene-modified L-fucose, **13a** and **13b**, respectively, they afforded strongly fluorescent triazolo-compounds.

Recently, Wang and coworkers proposed a new type of fluorogenic reaction based on the PET (photoinduced electron transfer) process of anthracenes.[57] In their work, an azido group

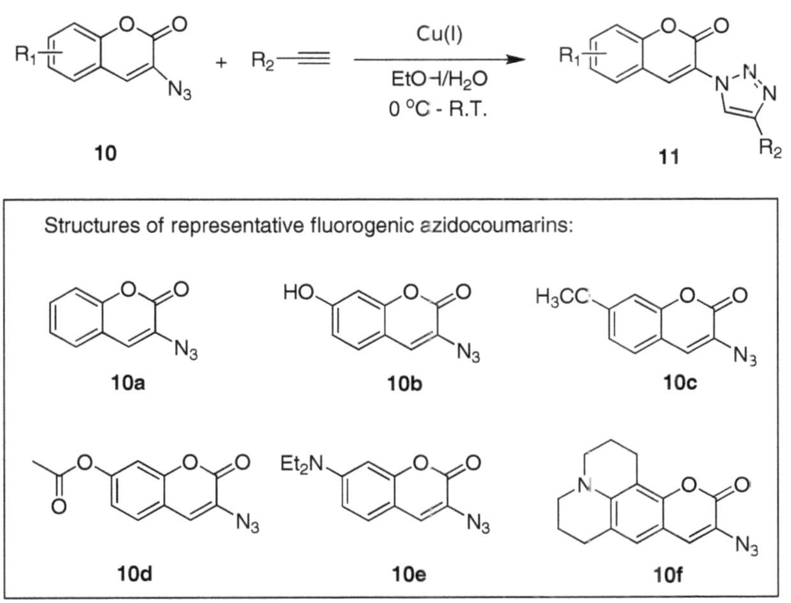

Scheme 14.5 *Fluorogenic 3-azidocoumarins and their CuAAC reaction. Reprinted with permission from K. Sivakumar* et al.*, (2004), A fluorogenic 1,3-dipolar cycloaddition reaction of 3-azidocoumarins and acetylenes,* Org. Lett.*, **6**, 4603–4606. Copyright 2004 ACS.*

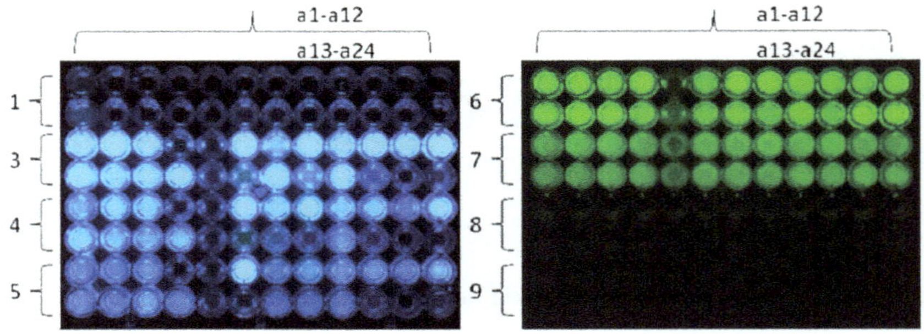

Figure 14.4 *Combinatorial synthesis and screening of triazolocoumarins (11) library in microtiter plates. The colors shown here do not represent the true fluorescent wavelengths due to the use of UV filters. Reprinted with permission from K. Sivakumar* et al., *(2004), A fluorogenic 1,3-dipolar cycloaddition reaction of 3-azidocoumarins and acetylenes,* Org. Lett., ***6****, 4603–4606. Copyright 2004 ACS.*

was introduced close to the anthryl core via a nonconjugated linker to allow a favorable electron transfer from the azido donor to the excited anthryl core inducing quenching of fluorescence. After the CuAAC reaction, the lone pair of electrons of the nitrogen is a part of the aromatic system, thus the nitrogen is a weaker electron donor, which does not permit the PET process and induces fluorescence activation (Scheme 14.7). In pure DMSO, the fluorescent emission intensity of product **15a** (R_2 is a phenyl group) was 75-fold stronger than that of **14a**, while **15a** shows almost the same absorption intensity as **14a** (Figure 14.5). The quantum yield of **15a** was 0.96, much higher than that of **14a** (~0.02). Moreover, there was no shift in emission and excitation wavelength accompanying the change of fluorescence intensity. All these results were consistent with a PET process between the azido group and the anthryl core. This fluorogenic CuAAC was tested between a series of azido-anthracene derivatives and a wide range of alkynes. The mild reaction

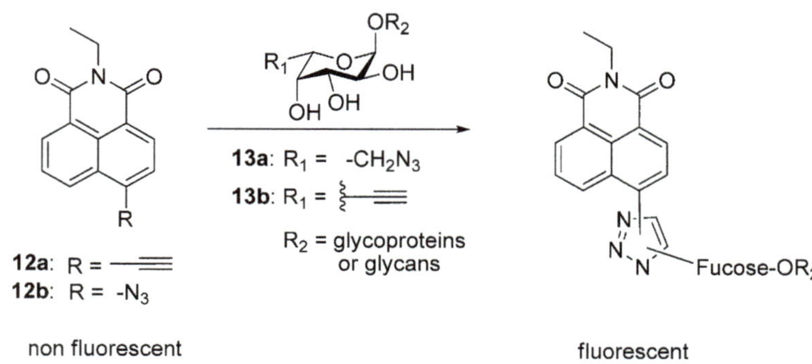

Scheme 14.6 *Fluorogenic ligation between 6-modified fucose analogs and 1,8-naphthalimide. Reprinted with permission from M. Sawa* et al., *(2006), Glycoproteomic probes for fluorescent imaging of fucosylated glycans* in vivo, Proc. Natl Acad. Sci. USA, ***103****, 12371–12376. Copyright 2006 National Academy of Sciences, USA.*

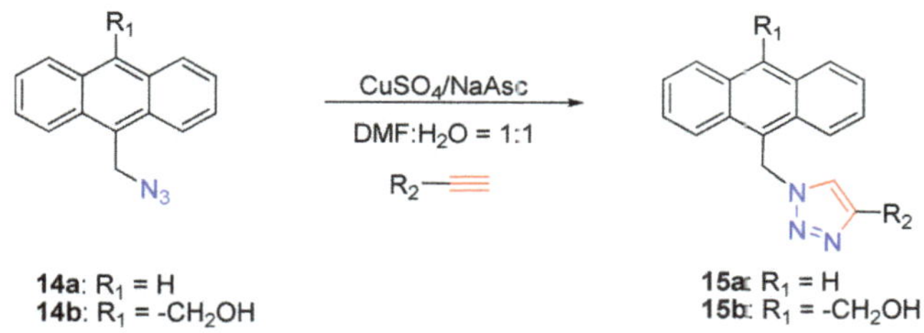

Scheme 14.7 *Fluorogenic reaction of azido-anthracene 14. Reprinted with permission from F. Xie* et al., *(2008), A fluorogenic "click" reaction of azidoanthracene derivatives,* Tetrahedron, ***64*** *(13), 2906–2914. Copyright 2008 Elsevier.*

conditions and high fidelity of the Cu(I)-catalyzed process allowed them to screen the fluorogenic properties of the cycloaddition reactions combinatorially.

A fluorogenic click reaction using borondipyrromethene (BODIPY) as pre-fluorophore has also been developed.[58] Since the fluorescence properties of BODIPY dyes can be changed by modifying the 3- (or 5-) position of the pyrrole ring, an azido group was introduced to the 3-position of the pyrrole ring to quench the fluorescence of BODIPY **16**. The formation of triazole rings via CuAAC with alkynes reduces the electron-donating effect and increases strongly the fluorescence of derivatives **17** (Scheme 14.8). It was also noticed that the fluorescence enhancement is higher when the alkynes bear strong electron-withdrawing than when they bear strong electron-donating groups.[58]

14.4 Applications of CuAAC in Bioconjugation

The very mild conditions of the fluorogenic CuAAC reaction as well as the biocompatibility of the functional groups make it an ideal reaction for a wide range of *in vitro* and *in vivo* bioconjugation applications (Figure 14.6).

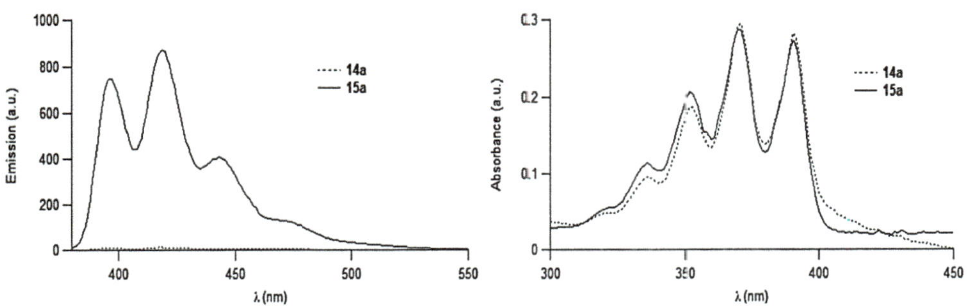

Figure 14.5 *Comparison of fluorescent emission (left) and absorption spectra (right) of* ***14a*** *and* ***15a*** *(R_2 = phenyl) in DMSO (10 μM for emission spectra and 50 μM for absorption spectra). Reprinted with permission from F. Xie* et al., *(2008), A fluorogenic "click" reaction of azidoanthracene derivatives,* Tetrahedron, ***64*** *(13), 2906–2914. Copyright 2008 Elsevier.*

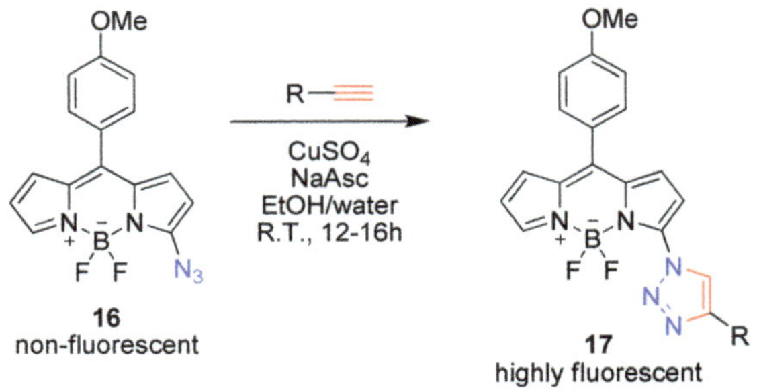

Scheme 14.8 *Fluorogenic CuAAC reaction of a 3-azido-BODIPY compound.*

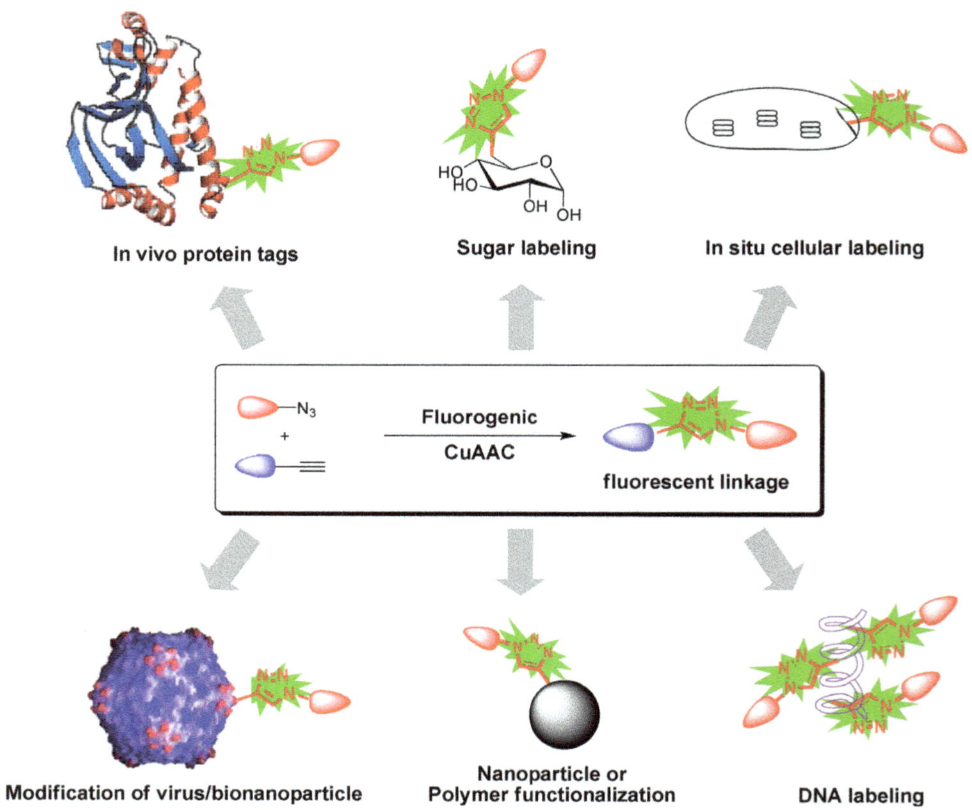

Figure 14.6 *Schematic illustration of the bioconjugate applications of the fluorogenic CuAAC reaction.*

14.4.1 Fluorogenic Probing of Cellular Components

In situ labeling proteins is of particular interest in biology because it allows the localization of the cell–cell interaction and the newly synthesized proteins. Fluorogenic CuAAC can be a useful tool in the imaging of proteins because it allows the use of profluorophore combining advantages, including small size, membrane permeability, intense fluorescence after activation and bioorthogonal reactivities.

Tirrell and coworkers have incorporated noncanonical amino-acids homopropargylglycine (Hpg) or ethylnyl-phenylamine (Eth) into recombinant barstarproteins by cotranslation [Figure 14.7(a)].[59] These two alkynyl amino-acids substituted methionine (Met) and phenylalanine (Phe) residues in the protein, respectively, and provided a triple bond for a possible ligation with an azido-profluorophore. An overnight treatment of cell cultures containing the recombinant barstar protein with the membrane-permeant 3-azido-7-hydroxycoumarin **10b**, Cu(I) and **1** at 4 °C was performed for *in situ* imaging. Excitation at 395 nm of cells gave a very strong fluorescent signal at 470 nm with a fluorescent enhancement up to 14-fold higher. The obtained fluorescence suggests that the protein is localized in inclusion bodies and a study by gel electrophoresis confirms that dye-labeling occurs mainly on the barstar [Figure 14.7(b)].

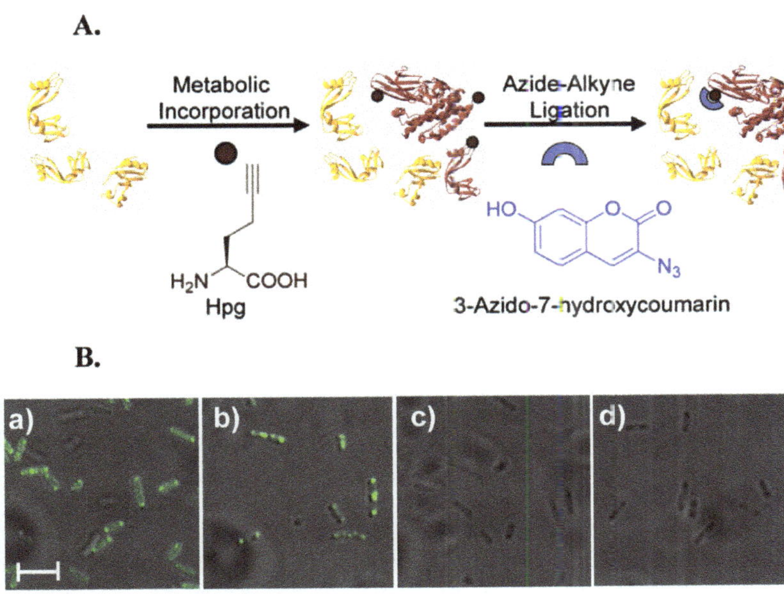

Figure 14.7 (a) Bioorthogonal labeling of newly synthesized proteins. Reprinted with permission from K. E. Beatty et al., (2006), Fluorescence visualization of newly synthesized proteins in mammalian cells, Angew. Chem., Int. Edn, ***45***, 7364–7367. Copyright 2006 Wiley-VCH. (b) Fluorogenic labeling of barstar in E. coli cells after CuAAC reaction with **10b**. Cells were induced in media supplemented with 19 amino acids and one of the following amino acids: Hpg (a); Eth (b); Met (c); Phe (d). Scale bar is 5 μm. Reprinted with permission from K. E. Beatty et al., (2005), Selective dye-labeling of newly synthesized proteins in bacterial cells, J. Am. Chem. Soc., ***127***, 14150–14151. Copyright 2005 American Chemical Society.

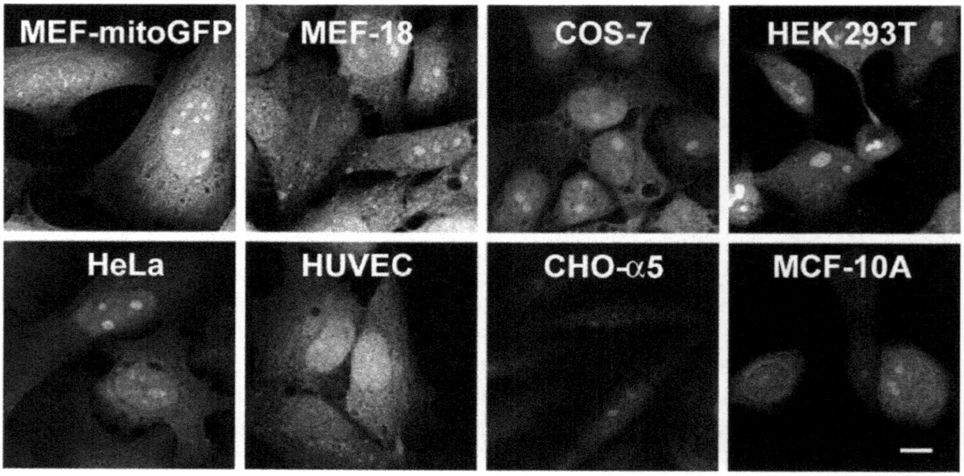

Figure 14.8 *Fluorogenic labeling of proteins in different type of cells. Scale bar is 10 μm. Reprinted with permission from K. E. Beatty* et al.*, (2006), Fluorescence visualization of newly synthesized proteins in mammalian cells,* Angew. Chem., Int. Edn*,* ***45****, 7364–7367. Copyright 2006 Wiley-VCH.*

Using the same method, Tirrell and coworkers also attempted to label newly synthesized proteins in a wide variety of mammalian cells.[60] First, mouse embryonic fibroblasts that express a mitochondrially targeted GFP (MEF-mitoGFP) were pulse-labeled with Hpg for 4 h, then reacted overnight in the dark at room temperature with coumarin **10b**, $CuSO_4$, TCEP and ligand **1** and then washed before visualization. The cell viability does not seem to be affected by incorporation of Hpg. A variety of different parameters have been optimized to find the conditions to visualize coumarin fluorescence by confocal microscopy. This imaging strategy was extended to different kind of mammalian cells (fibroblasts, endothelial and epithelial cells) and to different species (human, mouse, monkey and hamster). Most of the labeled cells show intense fluorescence in nuclear structures where the ribosome biogenesis takes place (Figure 14.8). Evidently, the fluorogenic CuAAC reaction enables a very efficient labeling and *in vivo* imaging of newly synthesized proteins in a wide range of mammalian cells.[60]

Wong *et al.* also reported the application of a fluorogenic CuAAC reaction to label fucosylated glycans *in vivo*.[55,61] Glycosylation is a co- or posttranslational phenomenon which takes place in more than half of eukaryotic proteins. Because L-fucose is the final sugar on glycans and participates in cell–cell interactions and cell migration processes in connection with biological processes such as embryogenesis, lymphocyte trafficking and cancer metastasis, fucosylation is a very important glycosylation process.[62–65] However, due to its high structural complexity of carbohydrates and the diversity of glycans, many functions of fucosylated glycoconjugates remain to be elucidated and a simple strategy for tagging glycans is of particular interest. In this study, acylated 6-azidofucose was incorporated in human hepatoma cell line (Hep3B) through the salvage biosynthetic pathway by incubation for 3 days.[55] Then cells were fixed, washed with PBS buffer and then reacted

a.

b.

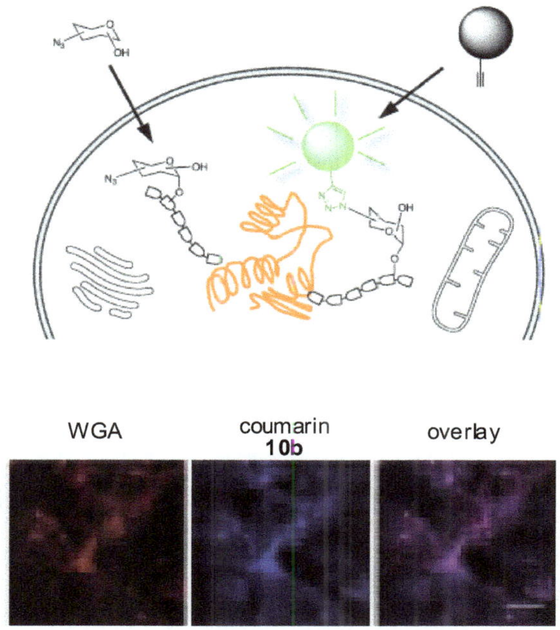

Figure 14.9 *(a) Schematic illustration of specific fluorescent labeling of fucosylated glycans in cells. Reprinted with permission from M. Sawa* et al., *(2006), Glycoproteomic probes for fluorescent imaging of fucosylated glycans* in vivo, Proc. Natl Acad. Sci. USA, ***103****, 12371–12376. Copyright 2006 National Academy of Sciences, USA. (b) Fluorescence imaging of sialyl glycoconjugates in Hep3B cells using CuAAC activated probe* ***10b****. Cells were treated with 25 mm alkynyl ManNAc for 3 days, clicked with* ***10b*** *and stained with WGA lectin (Alexa Fluor 594). Scale bar is 20 μm. Reprinted with permission from T.-L. Hsu* et al., *(2007), Alkynyl sugar analogs for the labeling and visualization of glycoconjugates in cells,* Proc. Natl Acad. Sci. USA, ***104****, 2614–2619. Copyright 2007 National Academy of Sciences, USA.*

with naphthalimide derivative **12a** in the presence of CuBr [Figure 14.9(a)]. After CuAAC, the engendered fluorescence in the cell by the formation of the triazolo-compound was visualized by fluorescence microscopy.[55]

However, acylated 6-azidofucose was found to be quite toxic for the cells, then alkynyl ManNAc was used instead because of its low toxicity.[61] Several human cancer cell lines were treated with it and then CuAAC reaction with **12b** was achieved: fluorescent-labeling of cell surface glycoconjugates was measured by flow cytometry and intracellular glycan labeling was controlled by fluorescence microscopy [Figure 14.9(b)], showing localization in the Golgi.

14.4.2 Fluorogenic Conjugation of DNA

Incorporation of labeled nucleosides into DNA is of particular interest for DNA diagnosis of genetic disorders and for exploration of structure, dynamics and interactions of nucleic acids.[66] Unfortunately incorporation of such tags is very difficult: enzymatic replacement of natural oligonucleotides by labeled ones relies on highly modified protocols and chemical modification of bases by solid-phase synthesis gives poor yields. Postsynthetic introduction of labels has also been tried but coupling yields are very low. Thus click reaction has been recently used as an easy and successful alternative to incorporate tags via introduction of small functional groups on DNA.

Carell *et al.* developed a postsynthetic method for high density labeling of DNA.[67] Alkyne-modified uridine nucleosides **18** and **19** have been prepared, transformed into their corresponding phosphoramidites and incorporated into a series of oligodeoxyribonucleotides (ODNs) via solid-phase synthesis (Figure 14.10). Once triple bonds were introduced, click reactions were performed with azides **10b**, **20** and **21** using standard conditions and Cu(I)-complexing ligand **1** in order to avoid strand breaks in DNA. The high-density reliable modification of all alkyne sites was achieved by using flexible alkynes **19**, whereas rigid alkynes **18** led to partially labeled DNA, showing that linker length plays an important role in the efficiency of the reaction. Bioconjugation by means of CuAAC reaction is very important because it allows efficient introduction of labeled nucleosides into DNA and permits DNA imaging either by fluorescence (fluorescence is triggered off via reaction with **10b** or by introduction of fluorescein via **21**) or by Ag staining (**20** is a sugar enabling Ag staining).

They also managed to functionalize DNA with up to three different labels by successive click reactions.[68] Using the same method as in their previous work they incorporated the cytidine building block **22** and the thymidine building blocks **23a** and **23b** into ODNs (Figure 14.11). The first click reaction was performed directly on the resin by shaking it with a solution of CuBr, ligand **1**, sodium ascorbate and azido-profluorophore. Once modified, the oligonucleotide was cleaved from the support by using aqueous solution of ammonia which also removed the trimethylsilyl (TMS) protecting group [but did not deprotect the triisopropylsilyl (TIPS) group] and then purified by HPLC. The second click reaction was performed in solution followed by its precipitation from ethanol and deprotection of the TIPS group by tetrabutylammonium fluoride. The last click reaction and precipitation afforded the triple-modified oligonucleotide with an overall good yield (50%). Therefore, click chemistry can be used to incorporate very sensitive labels into DNA with good efficiency and simple work-up.

Seela *et al.* thoroughly investigated the conjugation of functionalized nucleoside with nonfluorescent 3-azido-7-hydroxycoumarin **10b**.[69–71] Alkynyl chains were introduced into oligonucleotides and incorporated into ODNs for further tagging. The DNA duplexes obtained with these modified oligonucleotides shows an enhancement of the stability compared with natural oligonucleotides. Functionalization of modified nucleosides or ODNs was easily achieved by reaction with **10b** via CuAAC using *t*-BuOH–H_2O–DMSO–THF mixture in the presence of 1:1 complex of $CuSO_4$-ligand **1** and TCEP to afford strongly fluorescent 1,2,3-triazolyl oligonucleotide conjugates. This bioorthogonal fluorogenic CuAAC reaction which allows incorporation of tags into DNA without destabilizing DNA duplexes is useful for DNA detection in solution or in DNA–protein complexes and can be used for the *in vivo* labeling of DNA. Moreover they demonstrated that enzymatic hydrolysis of 1,2,3-triazolyl oligonucleotide conjugates shows a strong fluorescence quenching for 7-deazapurines compared with pyrimidines. This nucleobase specific quenching, which is probably due to an electron transfer between the nucleobase and the coumarin, can be used to monitor conformational dynamics of nucleotides in solution.

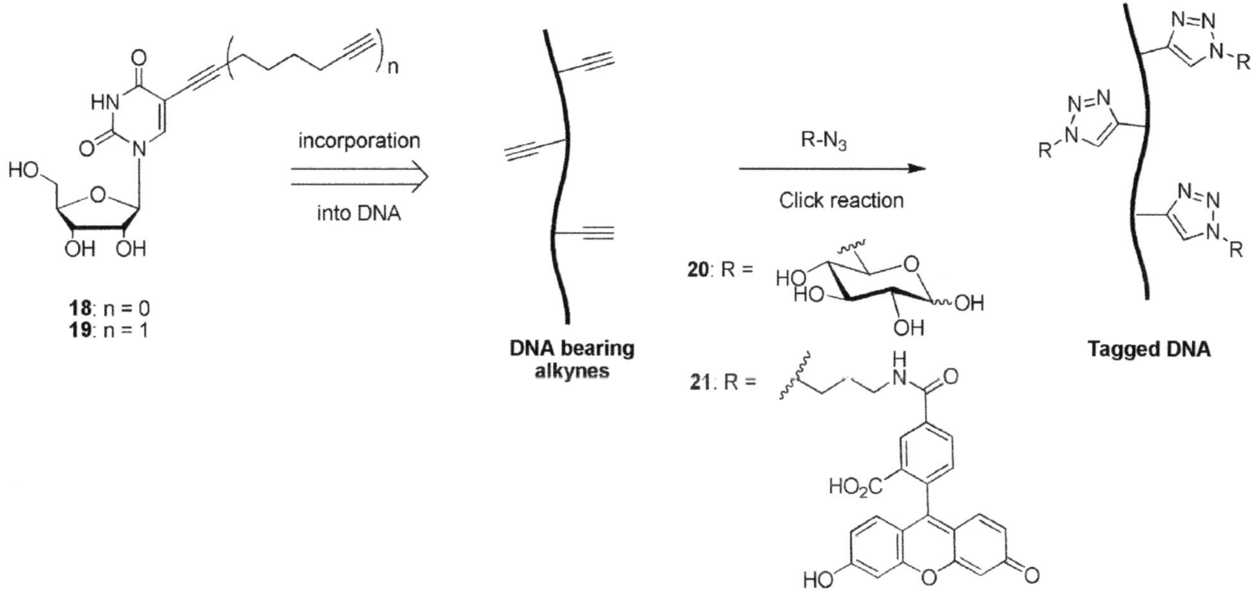

Figure 14.10 *Modified DNA by click reaction and structures of alkyne modified uridine nucleosides (**18** and **19**) and azide labels (**20** and **21**).*

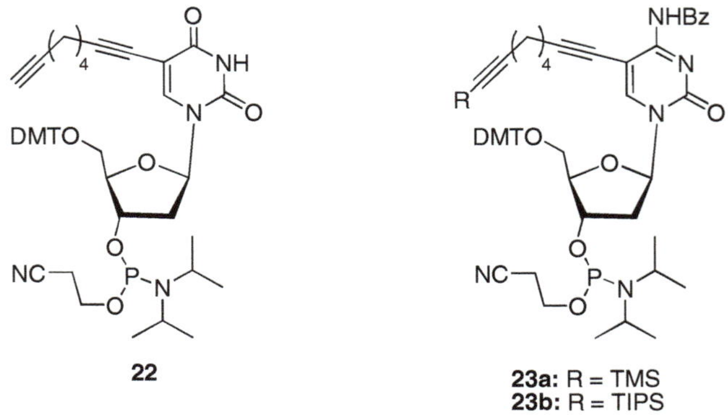

Figure 14.11 *Phosphoramidites **22** and **23**, DMT = 4,4′-dimethoxytriphenylmethyl. TMS = trimethylsilyl; TIPS = triisopropylsilyl; Bz = benzoyl.*

14.4.3 Fluorogenic Conjugation of Viruses

Finn *et al.* succeeded in labeling the cowpea mosaic virus (CPMV) with fluorescein.[5] CPMV is a stable and structurally well-characterized particle available in large quantities. Its capsid is composed by 60 identical copies of a two-protein asymmetric unit which enveloped the single-stranded RNA genome in the core. The outside of the capsid was decorated with azide or alkyne using either amide coupling or thio-ether formation at lysine or cysteine residues. The fluorescein derivatives were then conjugated to these three different virus-azides or virus-alkynes by click reaction under different conditions (Scheme 14.9). Addition of tris(triazolyl)amine **1** in CuAAC reaction showed acceleration of the reaction rate and a quasi-quantitative yield of the modified virus **25** was then obtained.[13]

They also replaced tris(triazolyl)amine **1** by the water-soluble sulfonated bathophenanthroline **2** as a ligand to modify the CPMV surface.[72] Tris(triazolyl)amine **1** is not very water-soluble, which can lead to some damage to the protein if the amount of available ligand is not enough in heterogenous solution. Use of sulfonated bathophenanthroline **2** under the same conditions permitted a decrease in concentration of labeled substrate and

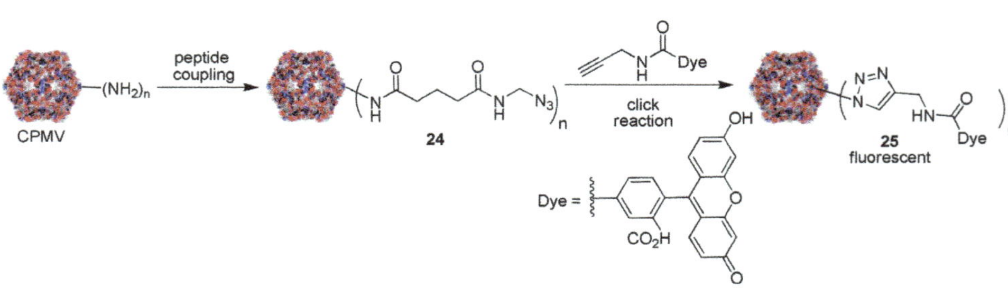

Scheme 14.9 *Bioconjugation of CPMV particle via CuAAC reaction.*

also permitted, with modified procedures, modification of a wide range of molecules such as complex carbohydrates, peptides, proteins or polymers in high yield.[73,74]

Wang *et al.* revisited the surface modification of tobacco mosaic virus (TMV) using CuAAC reaction.[7] TMV is a rod-shaped virus of 300 nm length and 18 nm diameter that can be obtained in large quantity. It is made from 2130 identical protein subunits arranged helically around genomic single-strand RNA that also stabilizes the coat protein assembly. TMV is a very attractive to be used in different fields such as nanoelectronics and energy harvesting devices[75–77] or as a template to grow metal or metal oxide nanowires.[78–80] Tyrosine residues were transformed into alkynes by means of an electrophilic substitution reaction at the *ortho* position of the phenol ring by using a diazonium salt generated *in situ* from the 3-ethynylaniline. CuAAC reactions between alkyne-derivatized TMV and a series of azides using $CuSO_4$ and sodium ascorbate were performed to modify the TMV surface with molecules such as peptides or polymers.[7] A double surface modification was also achieved using a mixture of azides under the same reaction conditions.

In particular, Wang and coworkers demonstrated that fluorogenic CuAAC reaction can be employed to titrate the reactivity of alkyne groups in a polyvalent system. For example, trispropynyloxybenzene **26**, alkyne-derivatized CPMV **28** and alkyne-derivatized TMV **30** containing 3, 60 and 2130 terminal alkyne moiety, respectively, were reacted with nonfluorescent azido-anthracene derivatives **14a** and **14b** [Figure 14.12(a)]. Conversion of **28** into intensely fluorescent triazolo-anthracene **29** was confirmed by gel electrophoresis [Figure 14.12(b)] and its integrity by TEM and size-exclusion chromatography [Figure 14.12(c, d)]. Based on the intensity of the fluorescent emission of the final conjugate product, they can quantitatively determine the reaction efficiency of the polyvalently displayed alkyne groups towards CuAAC reactions.

14.4.4 Fluorogenic Conjugation of Nanoparticles/Polymers

Self-assembly of amphiphilic block copolymers into polymeric micelles with a core–shell type structure is interesting as a drug delivery vehicle, as the hydrophobic core of the micelles can encapsulate a lipophilic molecule, whereas the hydrophilic shell stabilizes the micelles in aqueous media and protect the contents of hydrophobic core from possible degradation. The cross-linking of micelles in the core or in the shell plays an important role for the robustness of nanoparticles as well as relative ratio of block length, composition and molecular weight of amphiphilic block copolymers. Thus with the introduction of appropriate functionalities into specific parts of the copolymer it is possible to enhance their performance as drug delivery systems. The difficulty lies in the control of the presence of the functional group into the polymer.

As a classical example, Wooley, Hawker and coauthors synthesized a new class of block copolymers with acetylene group in the hydrophobic block using reversible addition fragmentation chain transfer techniques.[81] These acetylene-functionalized block copolymers were then self-assembled and cross-linked to give shell cross-linked knedel-like (SCK) nanoparticles **32** with acetylene groups in the core domain. As the presence and, more importantly, the reactivity, of the alkyne within the nanoparticle core cannot be detected by standard analysis such as NMR and MS, fluorogenic CuAAC reaction becomes a unique tool. Therefore, CuAAC with nonfluorescent coumarin derivative **10b** was performed at room temperature for 2 days using an organic copper(I) catalyst [$CuBr(PPh)_3$] and triamine

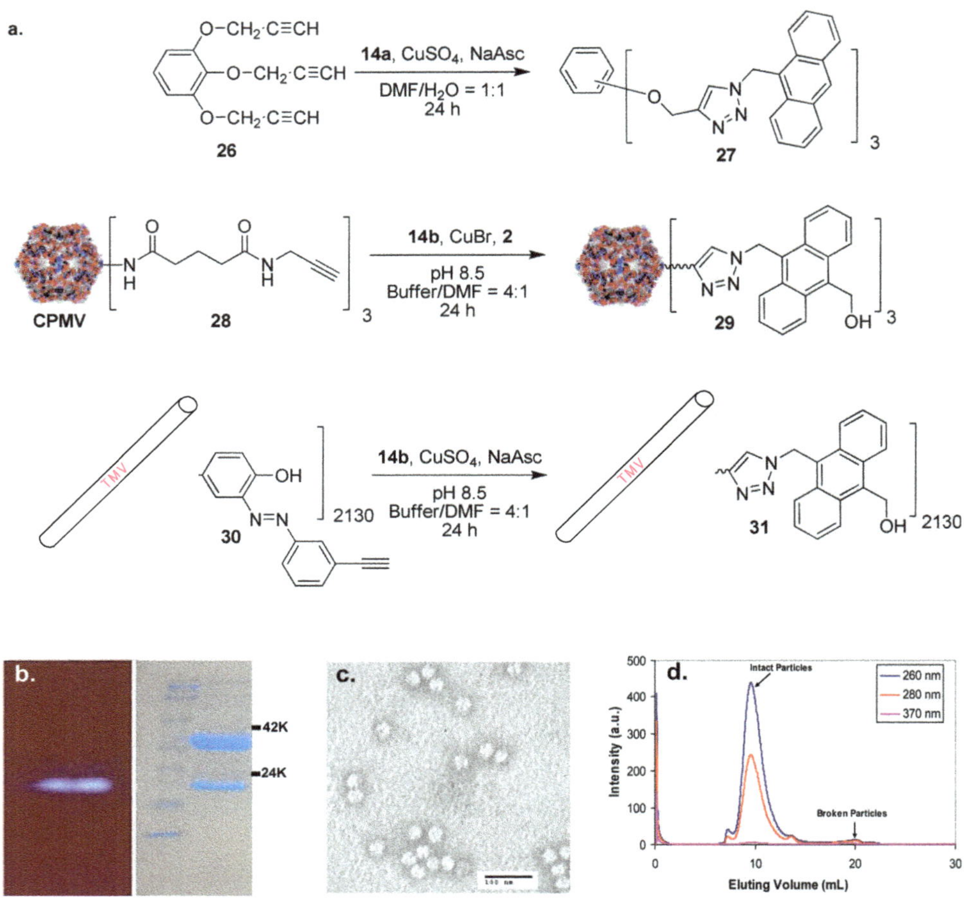

Figure 14.12 *(a) Virus modifications by CuAAC reactions. (b) SDS-PAGE of **29** visualized under UV irradiation (left) and upon staining with Coomassie blue (right). (c) TEM image of **27**. The scale bar is 100 nm. (D) Size exclusion FPLC analysis of **29**. Reprinted with permission from F. Xie* et al., *(2008), A fluorogenic "click" reaction of azidoanthracene derivatives,* Tetrahedron, ***64**, 2906–2914. Copyright 2008 Elsevier.*

to afford the fluorescent particle **33** (Scheme 14.10). The formation of triazolo-compound was easily confirmed by fluorescence measurements and analytical ultracentrifugation proved that acetylene moieties were available within the core domain. In this case, click reaction allows the validation of the presence of triple bonds into the core and to image the nanoparticle.

In another example, O'Reilly *et al.* reported the copolymerization of a terpyridine functionalized styrene monomer with styrene using nitroxide-mediated polymerization.[82] The terpyridine moiety of the nanostructure was selectively located within their hydrophobic core domain and was functionalized by metal complexation to afford novel

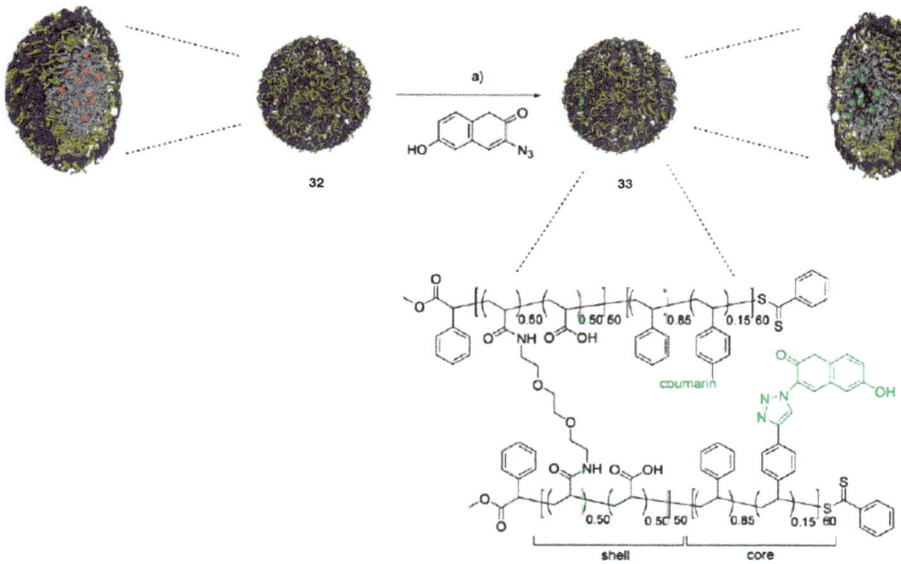

Scheme 14.10 *Fluorogenic click reaction on shell cross-linked knedel-like (SCK) nanoparticles. (a) Dialysis of* ***32*** *into THF/H_2O 4:1 for 3 days, then addition of [CuBr(PPh)$_3$] (0.1 equiv), and DIPEA (1.0 equiv),* ***10b*** *(1.11 equiv to acetylene functionality), RT, 2 days, followed by dialysis against THF–buffered H_2O 1:4 for 10 days, and then dialysis against pH 7.3 phosphate buffered saline, 4 days. Reprinted with permission from R. K. O'Reilly* et al., *(2006), Fluorogenic 1,3-dipolar cycloaddition within the hydrophobic core of a shell cross-linked nanoparticle,* Chem. Eur. J., ***12****, 6776–6786. Copyright 2006 Wiley-VCH.*

metal-functionalized polymer nanostructures. The terpyridine was utilized to complex metal centers (Fe, Cu and Ru) in the core domain. To test if the Cu-tethered metal complex within the nanostructures was an active catalyst, they used the fluorogenic CuAAC reaction by adding phenylethynyl and nonfluorescent 3-azidocoumarin to a solution of the nanoparticles. After 4 h reaction time, a high-fluorescence enhancement was noticed at 550 nm, confirming that Cu–terpyridin complex is active in click catalysis.

Wang and coworkers reported the chemoselective modification of horse spleen apoferritin (apo-HSF) by means of a CuAAC reaction.[83] Apo-HSF is derived from ferritin and its cage contains 24 identical subunits arranged into a hollow and spherical shell with an inner diameter of 8 nm and an outer diameter of 12.5 nm. The modification of lysine residues was achieved by acylation with NHS ester reagent to afford an alkyne on the biomolecule. In order to know that this chemoselective ligation has been achieved, a fluorogenic click reaction with the nonfluorescent coumarin **10b** in the presence of CuBr and ligand **2** (combination of $CuSO_4$–NaAsc or $CuSO_4$–phosphine did not give the triazolo-compound since aggregation and denaturation of apoferritin were observed) was realized to afford a triazolo-derivative, which shows very strong fluorescence at 474 nm upon excitation at 340 nm (Scheme 14.11). The triggering of fluorescence via the CuAAC reaction confirmed the modification of apo-HSF with an alkyne group. Twenty triazolo-coumarins by

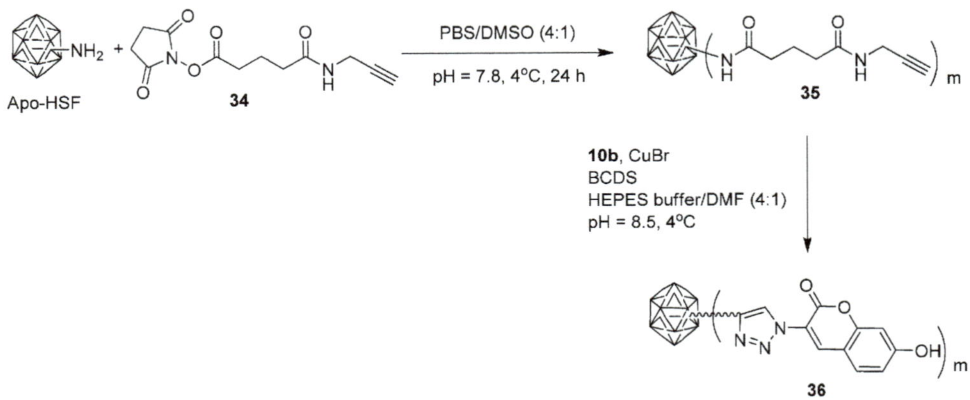

Scheme 14.11 *Bioconjugation of apoferritin by CuAAC reaction.*

apoferritin particle (i.e. about one coumarin by subunit) were found in the protein by measuring the fluorescence intensity of final conjugates.

14.5 Conclusions

The CuAAC reaction has become a major ligation tool in bioconjugation in the past few years. The reaction between alkyne and azide presents several advantages – easy preparation of alkynes and azides, bioorthogonality of starting materials, high yielded reaction and very mild reaction conditions – which make this reaction very suitable to label cellular elements. The fluorogenic CuAAC, particularly, have shown very interesting results for tagging diverse biomolecules (viruses, proteins, sugars, etc.), both *in vitro* and *in vivo*, which enable the localization of biological processes in an intracellular environment. Moreover, obtention of fluorescent triazolo derivatives from nonfluorescent polymers and nanoparticles allows detection of the presence of functionalities such as acetylene or azido groups other than by classical techniques. Finally, fluorogenic CuAAC is an easy way to introduce labels into DNA in order to detect genetic diseases or investigate nucleic acid interactions.

New perspectives have recently emerged for fluorogenic reaction including monitoring the progress of specific reactions by increase in fluorescence. For example, Rozhlov *et al.*[84] have reported a new fluorogenic transformation based on formation of C–C bonds catalyzed by palladium, allowing the screening of reaction variables such as base, ligand, temperature, etc. However the fluorescence of the final product can be quenched by catalysts if used in high quantities, which can lead to detection problems. Tanaka *et al.* have synthesized new fluorogenic imines to detect Mannich-type reactions of phenol in water, which could also be very interesting for the screening of catalysts and conditions to tag reactions of phenol-bearing molecules.[85] Additionally, Marchand *et al.* have implemented this reaction in ionic liquid micro-reactors to evaluate the efficiency of alternative mixing methods on the reaction kinetics, opening up a wide subject.[86] Finally, fluorogenic copper-free Huisgen cycloaddition reactions have recently been developed for intracellular labeling to avoid the potential cytotoxicity of copper,[43,44] which is introduced in Chapter 3.

References

(1) G. T. Hermanson (Ed.), (1996), *Bioconjugate Techniques*, Academic Press: New York.
(2) V. V. Rostovtsev, L. G. Green, V. V. Fokin and K. B. Sharpless, (2002), A stepwise Huisgen cycloaddition process: copper(I)-catalyzed regioselective 'ligation' of azides and terminal alkynes, *Angew. Chem., Int. Edn*, **41**, 2596–2599.
(3) C. W. Tornoe, C. Christensen and M. Meldal, (2002), Peptidotriazoles on solid phase: [1,2,3]-triazoles by regiospecific copper(i)-catalyzed 1,3-dipolar cycloadditions of terminal alkynes to azides, *J. Org. Chem.*, **67**, 3057–3064.
(4) J. M. Baskin, J. A. Prescher, S. T. Laughlin, N. J. Agard, P. V. Chang, I. A. Miller, A. Lo, J. A. Codelli and C. R. Bertozzi, (2007), Copper-free click chemistry for dynamic *in vivo* imaging, *Proc. Natl Acad. Sci. USA*, **104**, 16793–16797.
(5) Q. Wang, T. R. Chan, R. Hilgraf, V. V. Fokin, K. B. Sharpless and M. G. Finn, (2003), Bioconjugation by copper(I)-catalyzed azide-alkyne [3 + 2] cycloaddition, *J. Am. Chem. Soc.*, **125**, 3192–3193.
(6) T. R. Chan, R. Hilgraf, K. B. Sharpless and V. V. Fokin, (2004), Polytriazoles as copper(I)-stabilizing ligands in catalysis, *Org. Lett.*, **6**, 2853–2855.
(7) M. A. Bruckman, G. Kaur, L. A. Lee, F. Xie, J. Sepulveda, R. Breitenkamp, X. Zhang, M. Joralemon, T. P. Russell, T. Emrick and Q. Wang, (2008), Surface modification of tobacco mosaic virus with 'click' chemistry, *ChemBioChem*, **9**, 519–523.
(8) E. Strable, D. E. Prasuhn, A. K. Udit, S. Brown, A. J. Link, J. T. Ngo, G. Lander, J. Quispe, C. S. Potter, B. Carragher, D. A. Tirrell and M. G. Finn, (2008), Unnatural amino acid incorporation into virus-like particles, *Bioconjugate Chem.*, **19**, 866–875.
(9) D. E. Prasuhn, Jr., P. Singh, E. Strable, S. Brown, M. Manchester and M. G. Finn, (2008), Plasma clearance of bacteriophage qb particles as a function of surface charge, *J. Am. Chem. Soc.*, **130**, 1328–1334.
(10) D. E. Prasuhn, J. Kuzelka, E. Strable, A. K. Udit, S.-H. Cho, G. C. Lander, J. D. Quispe, J. R. Diers, D. F. Bocian, C. Potter, B. Carragher and M. G. Finn, (2008), Polyvalent display of heme on hepatitis B virus capsid protein through coordination to hexahistidine tags, *Chem. Biol.*, **15**, 513–519.
(11) E. Kaltgrad, M. K. O'Reilly, L. Liao, S. Han, J. C. Paulson and M. G. Finn, (2008), On-virus construction of polyvalent glycan ligands for cell-surface receptors, *J. Am. Chem. Soc.*, **130**, 4578–4579.
(12) W.-H. Zhan, H. N. Barnhill, K. Sivakumar, H. Tian and Q. Wang, (2005), Synthesis of hemicyanine dyes for 'click' bioconjugation, *Tetrahedron Lett.*, **46**, 1691–1695.
(13) S. Meunier, E. Strable and M. G. Finn, (2004), Crosslinking of and coupling to viral capsid proteins by tyrosine oxidation, *Chem. Biol.*, **11**, 319–326.
(14) W. G. Lewis, F. G. Magallon, V. V. Fokin and M. G. Finn, (2004), Discovery and characterization of catalysts for azide–alkyne cycloaddition by fluorescence quenching, *J. Am. Chem. Soc.*, **126**, 9152–9153.
(15) V. O. Rodionov, S. I. Presolski, S. Gardinier, Y.-H. Lim and M. G. Finn, (2007), Benzimidazole and related ligands for Cu-catalyzed azide–alkyne cycloaddition, *J. Am. Chem. Soc.*, **129**, 12696–12704.
(16) A. E. Speers, G. C. Adam and B. F. Cravatt, (2003), Activity-based protein profiling *in vivo* using a copper(I)-catalyzed azide-alkyne [3 + 2] cycloaddition, *J. Am. Chem. Soc.*, **125**, 4686–4687.
(17) A. E. Speers and B. F. Cravatt, (2004), Profiling enzyme activities *in vivo* using click chemistry methods, *Chem. Biol.*, **11**, 535–546.
(18) A. J. Link and D. A. Tirrell, (2003), Cell surface labeling of Escherichia coli via copper(I)-catalyzed [3 +2] cycloaddition, *J. Am. Chem. Soc.*, **125**, 11164–11165.
(19) A. J. Link, M. K. S. Vink and D. A. Tirrell, (2004), Presentation and detection of azide functionality in bacterial cell surface proteins, *J. Am. Chem. Soc.*, **126**, 10598–10602.
(20) A. Deiters, T. A. Cropp, M. Mukherji, J. W. Chin, J. C. Anderson and P. G. Schultz, (2003), Adding amino acids with novel reactivity to the genetic code of *Saccharomyces cerevisiae*, J. Am. Chem. Soc., **125**, 11782–11783.

(21) A. Deiters, T. A. Cropp, D. Summerer, M. Mukherji and P. G. Schultz, (2004), Site-specific PEGylation of proteins containing unnatural amino acids, *Bioorg. Med. Chem. Lett.*, **14**, 5743–5745.
(22) F. Tian, M. L. Tsao and P. G. Schultz, (2004), A phage display system with unnatural amino acids, *J. Am. Chem. Soc.*, **126**, 15962–15963.
(23) S. Seo Tae, Z. Li, H. Ruparel and J. Ju, (2003), Click chemistry to construct fluorescent oligonucleotides for DNA sequencing, *J. Org. Chem.*, **68**, 609–612.
(24) T. S. Seo, X. P. Bai, H. Ruparel, Z. M. Li, N. J. Turro and J. Y. Ju, (2004), Photocleavable fluorescent nucleotides for DNA sequencing on a chip constructed by site-specific coupling chemistry, Proc. Natl Acad. Sci. USA, **101**, 5488–5493.
(25) D. I. Rozkiewicz, D. Janczewski, W. Verboom, B. J. Ravoo and D. N. Reinhoudt, (2006), 'Click' chemistry by microcontact printing, *Angew. Chem., Int. Edn*, **45**, 5292–5296.
(26) D. I. Rozkiewicz, J. Gierlich, G. A. Burley, K. Gutsmiedl, T. Carell, B. J. Ravoo and D. N. Reinhoudt, (2007), Transfer printing of DNA by 'Click' chemistry, *ChemBioChem*, **8**, 1997–2002.
(27) G. Choy, P. Choyke and S. K. Libutti, (2003), Current advances in molecular imaging: noninvasive *in vivo* bioluminescent and fluorescent optical imaging in cancer research, *Mol. Imag.*, **2**, 303–312.
(28) N. Johnsson and K. Johnsson, (2003), A fusion of disciplines: chemical approaches to exploit fusion proteins for functional genomics, *ChemBioChem*, **4**, 803–810.
(29) B. F. Cravatt and E. J. Sorensen, (2000), Chemical strategies for the global analysis of protein function, *Curr. Opin. Chem. Biol.*, **4**, 663–668.
(30) G. J. Mohr, (2006), New chromogenic and fluorogenic reagents and sensors for neutral and ionic analytes based on covalent bond formation – a review of recent developments, *Anal. Bioanal. Chem.*, **386**, 1201–1214.
(31) G. J. Mohr, (2004), Chromo- and fluororeactands: indicators for detection of neutral analytes by using reversible covalent-bond chemistry, *Chem. Eur. J.*, **10**, 1082–1090.
(32) A. P. de Silva, H. Q. N. Gunaratne, T. Gunnlaugsson, A. J. M. Huxley, C. P. McCoy, J. T. Rademacher and T. E. Rice, (1997), Signaling recognition events with fluorescent sensors and switches, *Chem. Rev.*, **97**, 1515–1566.
(33) G. J. Mohr, D. Citterio, C. Demuth, M. Fehlmann, L. Jenny, C. Lohse, A. Moradian, T. Nezel, M. Rothmaier and U. E. Spichiger, (1999), Reversible chemical reactions as the basis for optical sensors used to detect amines, alcohols and humidity, *J. Mater. Chem.*, **9**, 2259–2264.
(34) J. Vandekerckhove and M. Van Montagu, (1974), Sequence analysis of fluorescamine-stained peptides and proteins purified on a nanomole scale. Application to proteins of bacteriophage MS2, *Eur. J. Biochem.*, **44**, 279–288.
(35) M. A. Bridges, K. M. McErlane, E. Kwong, S. Katz and D. A. Applegarth, (1986), Fluorometric determination of nanogram quantities of protein in small samples: application to calcium-transport adenosine triphosphatase, *Clin. Chim. Acta*, **157**, 73–79.
(36) G. M. Funk, C. E. Hunt, D. E. Epps and P. K. Brown, (1986), Use of a rapid and highly sensitive fluorescamine-based procedure for the assay of plasma lipoproteins, *J. Lipid Res.*, **27**, 792–795.
(37) A. Lorenzen and S. W. Kennedy, (1993), A fluorescence-based protein assay for use with a microplate reader, *Anal. Biochem.*, **214**, 346–348.
(38) V. K. Boppana, C. Miller-Stein, J. F. Politowski and G. R. Rhodes, (1991), High-performance liquid chromatographic determination of peptides in biological fluids by automated pre-column fluorescence derivatization with fluorescamine, *J. Chromatogr.*, **548**, 319–327.
(39) R. Y. Tsien and A. Miyawak, (1998), Seeing the machinery of live cells, *Science*, **280**, 1954–1955.
(40) S. R. Adams, R. E. Campbell, L. A. Gross, B. R. Martin, G. K. Walkup, Y. Yao, J. Llopis and R. Y. Tsien, (2002), New biarsenical ligands and tetracysteine motifs for protein labeling *in vitro* and *in vivo*: synthesis and biological applications, *J. Am. Chem. Soc.*, **124**, 6063–6076.

(41) G. Gaietta, T. J. Deerinck, S. R. Adams, J. Bouwer, O. Tour, D. W. Laird, G. E. Sosinskyl, R. Y. Tsien and M. H. Ellisman, (2002), Multicolor and electron microscopic imaging of connexin trafficking, *Science*, **296**, 503–507.
(42) E. Saxon and C. R. Bertozzi, (2000), Cell surface engineering by a modified Staudinger reaction, *Science*, **287**, 2007–2010.
(43) G. A. Lemieux, C. L. de Graffenried and C. R. Bertozzi, (2003), A fluorogenic dye activated by the Staudinger ligation, *J. Am. Chem. Soc.*, **125**, 4708–4709.
(44) M. J. Hangauer and C. R. Bertozzi, (2008), A FRET-based fluorogenic phosphine for live-cell imaging with the Staudinger ligation, *Angew. Chem., Int. Edn*, **47**, 2394–2397.
(45) H. C. Kolb, M. G. Finn and K. B. Sharpless, (2001), Click chemistry: diverse chemical function from a few good reactions, *Angew. Chem., Int. Edn*, **40**, 2004–2021.
(46) H. C. Kolb and K. B. Sharpless, (2003), The growing impact of click chemistry on drug discovery, *Drug Discov. Today*, **8**, 1128–1137.
(47) Q. Wang, S. Chittaboina and H. N. Barnhill, (2005), Advances in 1,3-dipolar cycloaddition reaction of azides and alkynes – a prototype of 'click' chemistry, *Lett. Org. Chem.*, **2**, 293–301.
(48) P. Wu and V. V. Fokin, (2007), Catalytic azide–alkyne cycloaddition: reactivity and applications, *Aldrichim. Acta*, **40**, 7–17.
(49) C. W. Tornoe, C. Christensen and M. Meldal, (2002), *J. Org. Chem.*, **67**, 3057–3062.
(50) K. Sivakumar, F. Xie, B. M. Cash, S. Long, H. N. Barnhill and Q. Wang, (2004), A fluorogenic 1,3-dipolar cycloaddition reaction of 3-azidoccumarins and acetylenes, *Org. Lett.*, **6**, 4603–4606.
(51) Z. Zhou and C. J. Fahrni, (2004), A fluorogenic probe for the copper(I)-catalyzed azide-alkyne ligation reaction: modulation of the fluorescence emission via 3(n,p)-1(p,p) inversion, *J. Am. Chem. Soc.*, **126**, 8862–8863.
(52) C. E. Wheelock, (1959), Fluorescence of some coumarins, *J. Am. Chem. Soc.*, **81**, 1348–1352.
(53) D. J. Yee, V. Balsanek and D. Sames, (2004), New tools for molecular imaging of redox metabolism: development of a fluorogenic probe for 3a-hydroxysteroid dehydrogenases, *J. Am. Chem. Soc.*, **126**, 2282–2283.
(54) M.-S. Schiedel, C. A. Briehn and P. Bauerle, (2001), Single-compound libraries of organic materials: parallel synthesis and screening of fluorescent dyes, *Angew. Chem., Int. Edn*, **40**, 4677–4680.
(55) M. Sawa, T.-L. Hsu, T. Itoh, M. Sugiyama, S. R. Hanson, P. K. Vogt and C.-H. Wong, (2006), Glycoproteomic probes for fluorescent imaging of fucosylated glycans *in vivo*, *Proc. Natl Acad. Sci. USA*, **103**, 12371–12376.
(56) C. J. McAdam, J. L. Morgan, R. E. Murray, B. H. Robinson and J. Simpson, (2004), Synthesis and fluorescence properties of new enaminenaphthalimides, *Aust. J. Chem.*, **57**, 525–530.
(57) F. Xie, K. Sivakumar, Q. Zeng, M. A. Bruckman, B. Hodges and Q. Wang, (2008), A fluorogenic 'click' reaction of azidoanthracene derivatives, *Tetrahedron*, **64**, 2906–2914.
(58) F. Xie and Q. Wang, Unpublished results.
(59) K. E. Beatty, F. Xie, Q. Wang and D. A. Tirrell, (2005), Selective dye-labeling of newly synthesized proteins in bacterial cells, *J. Am. Chem. Soc.*, **127**, 14150–14151.
(60) K. E. Beatty, J. C. Liu, F. Xie, D. C. Dieterich, E. M. Schuman, Q. Wang and D. A. Tirrell, (2006), Fluorescence visualization of newly synthesized proteins in mammalian cells, *Angew. Chem., Int. Edn*, **45**, 7364–7367.
(61) T.-L. Hsu, S. R. Hanson, K. Kishikawa, S.-K. Wang, M. Sawa and C.-H. Wong, (2007), Alkynyl sugar analogs for the labeling and visualization of glycoconjugates in cells, *Proc. Natl Acad. Sci. USA*, **104**, 2614–2619.
(62) J. Becker Daniel and B. Lowe John, (2003), Fucose: biosynthesis and biological function in mammals, *Glycobiology*, **13**, 41R–53R.
(63) E. Staudacher, (1996), a-1,3-Fucosyltransferases, *Trends Glycosci. Glyc.*, **8**, 391–408.
(64) S. Haltiwanger Robert and B. Lowe John, (2004), Role of glycosylation in development, *Annu. Rev. Biochem.*, **73**, 491–537.
(65) D. H. Dube and C. R. Bertozzi, (2005), Glycans in cancer and inflammation – potential for therapeutics and diagnostics, *Nat. Rev. Drug Discov.*, **4**, 477–488.

(66) D. M. Hammond, A. Manetto, J. Gierlich, V. A. Azov, P. M. E. Gramlich, G. A. Burley, M. Maul and T. Carell, (2007), DNA Photography: an ultrasensitive DNA-detection method based on photographic techniques, *Angew. Chem., Int. Edn*, **46**, 4184–4187.
(67) J. Gierlich, G. A. Burley, P. M. E. Gramlich, D. M. Hammond and T. Carell, (2006), Click chemistry as a reliable method for the high-density postsynthetic functionalization of alkyne-modified DNA, *Org. Lett.*, **8**, 3639–3642.
(68) P. M. E. Gramlich, S. Warncke, J. Gierlich and T. Carell, (2008), Click-click-click: single to triple modification of DNA, *Angew. Chem., Int. Edn*, **47**, 3442–3444.
(69) F. Seela and X. Ming, (2008), Oligonucleotides containing 7-deaza-2'-deoxyinosine as universal nucleoside: synthesis of 7-halogenated and 7-alkynylated derivatives, ambiguous base pairing, and dye functionalization by the alkyne–azide 'click' reaction, *Helv. Chim. Acta*, **91**, 1181–1200.
(70) F. Seela and V. R. Sirivolu, (2008), Pyrrolo-dC oligonucleotides bearing alkynyl side chains with terminal triple bonds: synthesis, base pairing and fluorescent dye conjugates prepared by the azide–alkyne 'click' reaction, *Org. Biomol. Chem.*, **6**, 1674–1687.
(71) F. Seela, V. R. Sirivolu and P. Chittepu, (2008), Modification of DNA with octadiynyl side chains: Synthesis, base pairing, and formation of fluorescent coumarin dye conjugates of four nucleobases by the alkyne–azide 'click' reaction, *Bioconjugate Chem.*, **19**, 211–224.
(72) S. Sen Gupta, J. Kuzelka, P. Singh, G. Lewis Warren, M. Manchester and M. G. Finn, (2005), Accelerated bioorthogonal conjugation: a practical method for the ligation of diverse functional molecules to a polyvalent virus scaffold, *Bioconjugate Chem.*, **16**, 1572–1579.
(73) H.-J. Musiol, S. Dong, M. Kaiser, R. Bausinger, A. Zumbusch, U. Bertsch and L. Moroder, (2005), Toward semisynthetic lipoproteins by convergent strategies based on click and ligation chemistry, *ChemBioChem*, **6**, 625–628.
(74) S. F. M. van Dongen, M. Nallani, S. Schoffelen, J. J. L. M. Cornelissen, R. J. M. Nolte and J. C. M. van Hest, (2008), A block copolymer for functionalisation of polymersome surfaces, *Macromol. Rapid Commun.*, **29**, 321–325.
(75) J. Tseng Ricky, C. Tsai, L. Ma, J. Ouyang, S. Ozkan Cengiz and Y. Yang, (2006), Digital memory device based on tobacco mosaic virus conjugated with nanoparticles, *Nat. Nanotech.*, **1**, 72–77.
(76) S. V. Kalinin, S. Jesse, W. Liu and A. A. Balandin, (2006), Evidence for possible flexo-electricity in tobacco mosaic viruses used as nanotemplates, *Appl. Phys. Lett.*, **88**, 153902/153901–153902/153903.
(77) R. A. Miller, A. D. Presley and M. B. Francis, (2007), Self-assembling light-harvesting systems from synthetically modified tobacco mosaic virus coat proteins, *J. Am. Chem. Soc.*, **129**, 3104–3109.
(78) T. L. Schlick, Z. Ding, E. W. Kovacs and M. B. Francis, (2005), Dual-surface modification of the tobacco mosaic virus, *J. Am. Chem. Soc.*, **127**, 3718–3723.
(79) V. A. Fonoberov and A. A. Balandin, (2005), Phonon confinement effects in hybrid virus-inorganic nanotubes for nanoelectronic applications, *Nano Lett.*, **5**, 1920–1923.
(80) H. Yi, S. Nisar, S.-Y. Lee, A. Powers Michael, E. Bentley William, F. Payne Gregory, R. Ghodssi, W. Rubloff Gary, T. Harris Michael and N. Culver James, (2005), Patterned assembly of genetically modified viral nanotemplates via nucleic acid hybridization, *Nano Lett.*, **5**, 1931–1936.
(81) R. K. O'Reilly, M. J. Joralemon, C. J. Hawker and K. L. Wooley, (2006), Fluorogenic 1,3-dipolar cycloaddition within the hydrophobic core of a shell cross-linked nanoparticle, *Chem. Eur. J.*, **12**, 6776–6786.
(82) A. D. Ievins, X. F. Wang, A. O. Moughton, J. Skey and R. K. O'Reilly, (2008), Synthesis of core functionalized polymer micelles and shell cross-linked nanoparticles, *Macromolecules*, **41**, 2998–3006.
(83) Q. Zeng, T. Li, B. M. Cash, S. Li, F. Xie and Q. Wang, (2007), Chemoselective derivatization of a bionanoparticle by click reaction and ATRP reaction, *Chem. Commun.*, 1453–1455.
(84) R. V. Rozhkov, V. J. Davisson and D. E. Bergstrom, (2008), Fluorogenic transformations based on formation of C–C bonds catalyzed by palladium: an efficient approach for high throughput optimizations and kinetic studies, *Adv. Synth. Catal.*, **350**, 71–75.

(85) H. M. Guo, M. Minakawa and F. Tanaka, (2008), Fluorogenic imines for fluorescent detection of Mannich-type reactions of phenols in water, *J. Org. Chem.*, **73**, 3964–3966.
(86) G. Marchand, P. Dubois, C. Delattre, F. Vinet, M. Blanchard-Desce and M. Vaultier, (2008), Organic synthesis in soft wall-free microreactors: real-time monitoring of fluorogenic reactions, *Anal. Chem.*, **80**, 6051–6055.

15

Synthesis and Functionalization of Biomolecules via Click Chemistry

Christine Schilling, Nicole Jung and Stefan Bräse

15.1 Introduction

The products of the click reaction – the 1,2,3-triazoles – are a desired target class as they provide additional functionality, such as hydrogen bonding and coordination prospects. Furthermore, they possess a broad spectrum of biological properties, not only anti-HIV, anti-allergenic and antibacterial features but also fungicidal and herbicidal activity.[1] As such, the click reaction is a very attractive method in material science as well as in the development of novel biologically active compounds.

There are two possible applications of the click reaction, the first of which is to *construct* novel compounds/materials, in which new modified building blocks are added to the material/compound. The second method is to *modify* existing materials/compounds through chemical alteration.[2] One example of the latter involves the labeling of molecules *in vivo* and *in vitro* for detection and purification purposes. The click reaction also receives substantial attention in the field of polymer sciences.[4] The reaction can be applied to the immobilization/modification of, for example, carbohydrates and proteins on solid surfaces,[3,5] azido-sugars on gold to generate carbohydrate self-assembled monolayers (SAMs),[6] and to the addition of electroactive as well as bioactive recognition elements to electrode surfaces.[7]

Click Chemistry for Biotechnology and Materials Science Edited by Joerg Lahann

15.2 Labeling of Macromolecular Biomolecules

15.2.1 Fluorescent Labeling

A broad range of alternatives are known for the introduction of markers into organic molecules. Of these, the bonding of fluorescent molecules is one of the most widely used techniques. While coumarin, fluorescein and rhodamine are the preferred sources for these fluorescent markers, other less frequently used molecules can be introduced via click reaction, thus enabling the analysis of the coupled compounds via fluorescent measurements. The principle of these fluorescent labeling strategies is demonstrated below, with fluorescein as the example (Figure 15.1).

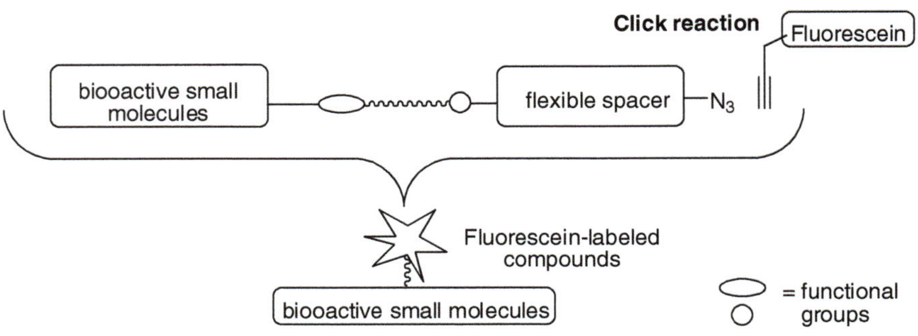

Figure 15.1 *Schematic overview of the fluorescent labeling through click chemistry.*[8]

In principle, two possible fluorescein-derivatives exist, which are useful for click chemistry. The first derivative – compound **3** (Scheme 15.1) – employed by Yao *et al.*[8] differs from the second one (not shown) used by Koberstein *et al.*,[9] Pierters *et al.*[10] and Wooley *et al.*,[11] in terms of the position of the clickable functionalization. Yao *et al.* used *O*-propargylated fluorescein derivatives, whereas the other groups performed click reactions

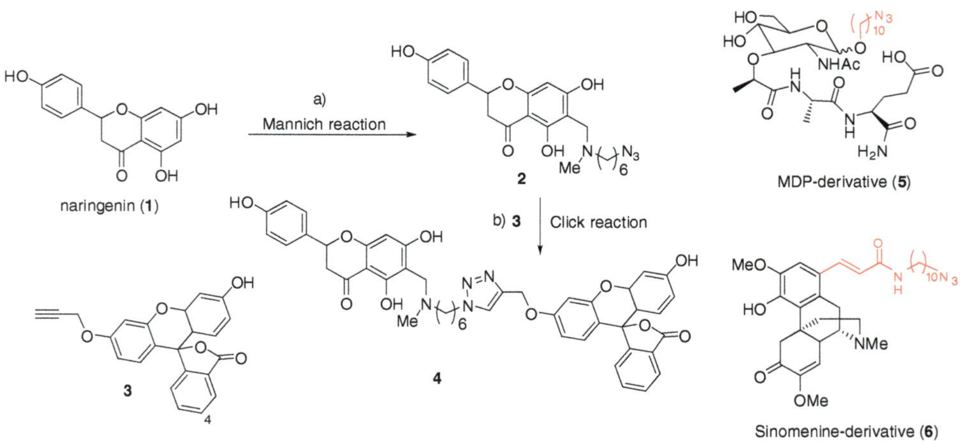

Scheme 15.1 *Fluorescent labeling of naringenin through click reaction: (a) formaldehyde, $MeNH(CH_2)_6N_3$, $ZnCl_2$, EtOH, 65 °C, 3 h, 87%; (b) $CuSO_4{\cdot}5H_2O$ (cat.), sodium ascorbate, fluorescein derivative 3, $^tBuOH/H_2O$, 67%.*[8]

with fluorescein-azides, bearing the functionalized side chain in position 4. As natural products in their original structure contain no azides, the desired functionality has to be introduced through selective azidation of the target bioactive compound or through reaction of the latter with azide-containing molecules. The second method was performed by Yao *et al.*[8] The group chose four bioactive compounds, which were derived in order to react in click reactions and therefore bind the fluorescent label fluorescein. Fluorescein was selected from the available organic dyes because fluorescein and its derivatives are still cheaper than most of the other fluorescent molecules. Scheme 15.1 illustrates the selective attachment of an azide-bearing side chain on naringenin (**1**) via Mannich reaction of the most nucleophilic ring with formaldehyde and the azide-bearing amine.

After the introduction of the side chain, click reaction yielded the desired fluorescein-derived flavone **4**, which could then be analyzed regarding its influence on plant root nodulation. Similar reactions to those presented in Scheme 15.1 generated natural product derivatives **5** and **6**, which were both connected to azide-containing side chains and were used for click reaction with fluorescein-derivative **3** to give labeled analogs of the natural products MDP (muramyl dipeptide) and sinomenine.

Fluorescent labeling with fluorescein is also used for the labeling of proteins in the detection of cancer-linked galectin-3, as shown by Pieters *et al.*[10] Beyond this investigation, there are few publications dealing with the introduction of fluorescein into nanoparticles via click reaction[11] and the labeling of polymer materials. In the latter case, the immobilization of fluorescein-derivatives is undertaken, in order to demonstrate the reactivity of surface-bound alkynes.[9]

Rhodamine is a second compound used for the fluorescent labeling of enzymes and bioactive compounds. Taunton *et al.*[12] demonstrated that the combination of rhodamine incorporation with photo-affinity labeling could be used to identify protein targets of cyclodepsipeptides. For this reason, cyclodepsipeptides containing a propargyl functionality were reacted via photo-affinity labeling with the target proteins and subsequently ‘clicked’ with azide-derived rhodamine. Similar approaches have also been recognized for the tagging of probes with biotin.[13]

As shown in labeling strategies with biotin in Scheme 15.6 and Scheme 15.7 (later on in this chapter), rhodamine derivatives are used for activity-based protein profiling (ABPP). The bioorthogonal coupling reaction can be performed with an alkyne-modified rhodamine, which is covalently attached to an enzyme bearing an azide functionality.[14] The inverse case, the reaction of an azide containing rhodamine with alkynylated enzymes (Scheme 15.2), has also been established (compare to MacKinnon *et al.*,[12]). In the latter case, cytochrome P450 (**7**) was labeled with a 2-ethinylnaphthalene-derived activity-based probe **8**. After metabolization of the probe and anchoring to P450, rhodamine was introduced by click reaction to give **9**.[15]

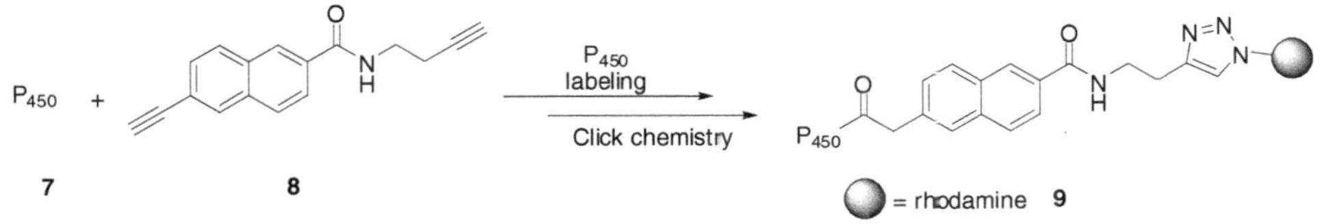

Scheme 15.2 *Acetylene-modified 2-ethinylnaphthalene as a derivative for conjugation via click chemistry to azide modified rhodamine.*[15]

There are several derivatives of fluorescein-type molecules that are applied in fluorescent labeling and in *in vivo* imaging. Figure 15.2 shows four compounds (**10–13**) linkable via alkyne-containing side chains, which were used for click reactions with azide-derived glycans. These fluorescent molecules consist of either a linear alkyne (**12, 13**) or a difluorinated cyclooctyne (**10, 11**). Given the ring strain of cyclooctynes in combination with the presence of the strong electron-withdrawing group, the latter enable copper-free click chemistry.[16]

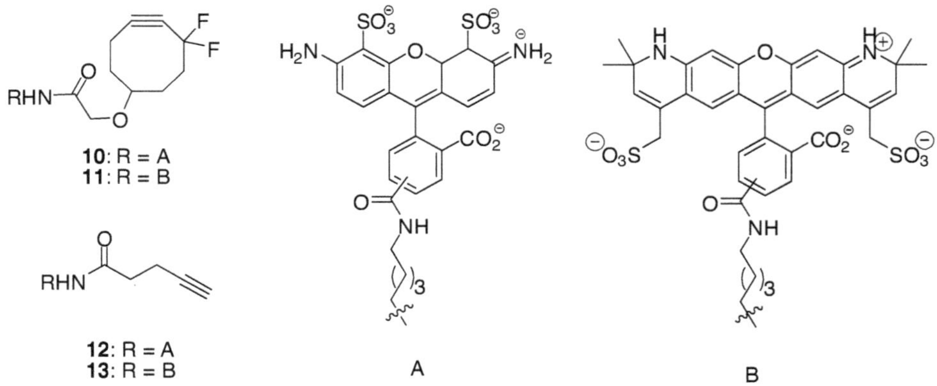

Figure 15.2 *Derivatives of Alexa Fluor 488 and Alexa Fluor 568 for copper-free click chemistry.*[16]

Rhodamine-derived molecules have also been used for the labeling of receptor ligands. Bonnet *et al.* illustrated the utility of Lissamine (**15**) (fluorescent dye) and Patent blue derivatives (**16**) (nonfluorescent dye) for the investigation of ligand interactions with the human muscarin M1 receptor.[28]

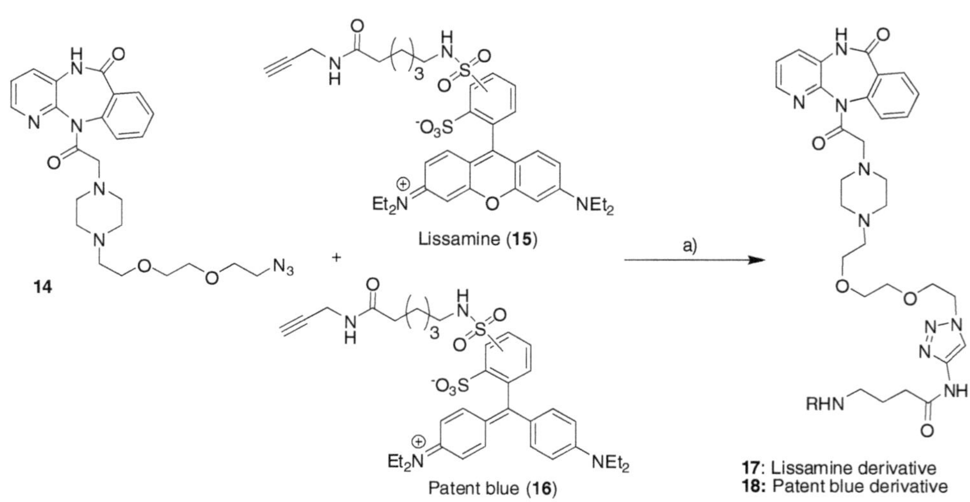

Scheme 15.3 *Labeled pirenzepine derivatives **17** and **18** obtained via click chemistry. (a) $CuSO_4$ (10% Cu) wire, MeCN/H_2O (9:1), 4 h.*[28]

Sawa *et al.* showed that *in vivo* imaging of fucosylated glycans bearing an azide functionality is possible via glycan labeling with 1,8-naphthalimide-derivative **19**.[17] This non-fluorescent imide can be bound via click chemistry to the target 6-modified fucose analogs, thus turning into its fluorescent form (Scheme 15.4). After formation of the triazole ring, the naphthalimide-derivative **21** is strongly fluorescent and the intracellular localization of fucosylated glycoconjugates can be detected by fluorescent microscopy.

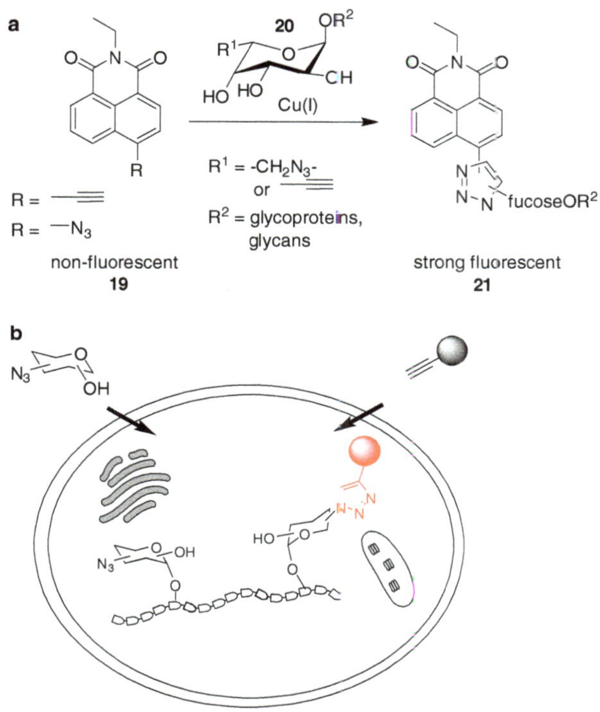

Scheme 15.4 *Glycan labeling: (a) fluorescent adduct **21** is generated through the click reaction of fucoside **20** with the probe **19**; (b) Strategy for specific fluorescent labeling of fucosylated glycans in cells.*[17]

The labeling of sugar derivatives through the attachment of biotin as well as the fluorogenic derivation by coumarin-derivatives has been presented by Wong *et al.*[18] For this reason, 3-azido-7-hydroxycoumarin (**22**) has been synthesized and connected via click reaction with alkynyl-substituted sugar analogs (Figure 15.3).

Other derivatives for fluorescent labeling are those of 5-(dimethylamino)naphthalene-1-sulfonyl-like compound **23** or **24**[19,20] which can be attached, for example, to azide-containing side chains of sugar-derivatives. In order to gain information about the K30 antigen and its mode of operation, Du *et al.*[19] demonstrated the labeling of the K30 antigen repeating unit via click chemistry of **23** with an azide containing a tetrameric sugar-derivative.

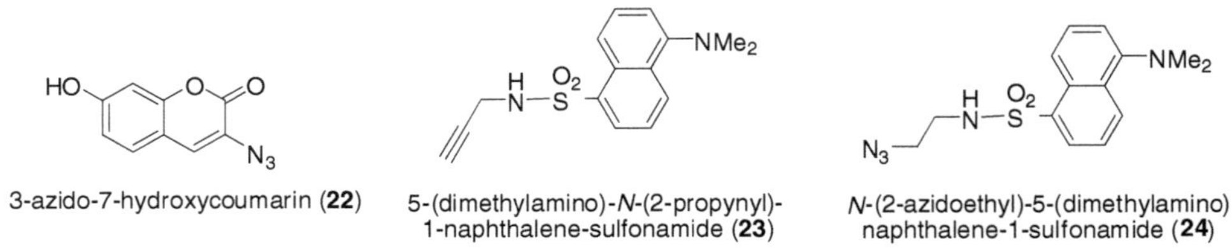

Figure 15.3 Derivatives for fluorogenic labeling via click chemistry.[18–20]

15.2.2 Labeling of Bovine Serum Albumin

Nolte *et al.* developed a method of synthesizing enzyme dimers via click reaction.[21] These dimers consist of one specific enzyme in combination with bovine serum albumin (BSA), which forms a so-called 'protein foot' for the immobilization of the whole surface construct (Scheme 15.5). The BSA-labeled enzyme can be further anchored onto surfaces used for single enzyme studies. The synthesis of the acetylene moiety **26** was built up through amide formation on the enzyme – *Thermomyces lanuginosa* lipase (TLL) – via the addition of pentynecarboxylic acid under peptide coupling conditions. Compound **26** was connected with an azide-functionalized BSA component **28** to give the triazole target compound **29**.

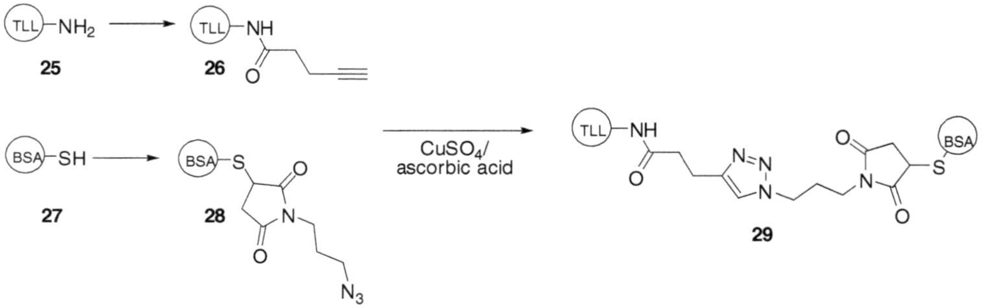

*Scheme 15.5 BSA-labeling of enzymes: synthesis of TLL–BSA dimer **29**.*[21]

Thermomyces lanuginosa lipase has been the target of additional enzyme studies, with the click reaction as the key step. In another approach by Nolte *et al.*, TLL was successfully immobilized on gold nanoparticles through the formation of triazoles as the connecting group.[22] These functional hybrids of lipase and gold nanoparticles were created through the use of azide-functionalized gold nanoparticles that can be prepared via the treatment of citrate-stabilized gold hydrosols with aqueous solutions of an azide-containing thiol-linker. After functionalization of the gold surface, the addition of acetylene-modified enzymes enabled their immobilization via triazole linkage. The connection of BSA to macromolecules is an established method, even in the synthesis of biohybrid amphiphiles.[23] Via click chemistry, Rutjes *et al.* connected terminal azide-functionalized polystyrene and alkyne-functionalized BSA derivatives to produce giant amphiphiles.

15.2.3 Biotin-labeling of Biomolecules: ABPP

Click chemistry is an often-used tool in order to label molecules with biotin, both *in vivo* and *in vitro*. Biotin is a popular label because of its affinity to avidin and streptavidin. Therefore, bioconjugations with biotin (and other markers) are of great interest in reference to ABPP. This technique enables the binding of enzymes to a mechanism-based inhibitor that is bound to a fluorophore, an affinity tag or a bioorthogonal chemical reporter. The bioorthogonal chemical reporter can be reacted with different probes, such as biotin-derivatives, through the use of click reactions. The use of this reaction in order to convert bioorthogonal reporters *in vivo* into easily detectable derivatives facilitates the covalent binding of azide- or alkyne-containing molecules to enzymes. This method is very useful because the often bulky and noncell permeable tags can be formed afterwards *in vitro*. The biotin-labeled proteins are then purified and analyzed (via bonding to avidin or strepatavidin) and can be applied for the investigation of the proteins expressed in cancer proteomes. A schematic presentation of the ABPP-mechanism is illustrated in Scheme 15.6.

In the first step, active enzymes of the proteome can be labeled *in vivo* with alkyne or azide-containing probes (a, Scheme 15.6) that are converted into triazoles through the *in vitro* addition of the counterpart (b). Subsequently, the labeled proteins have to be enriched by binding them to streptavidin (c). Cravatt *et al.*[24] expanded the ABPP concept through combination with tandem orthogonal proteolysis (d), in order to ascertain the parallel characterization of the probe-labeled proteins and the sites of probe modification. Afterwards, proteolysis was performed by on-bead trypsin-digestion. The supernatant is removed after the digestion and the probe-labeled peptides can be cleaved from the beads through incubation with *tobacco etch virus protease* (TEV) to give conjugates **30**. The peptides – eluted after the trypsin and TEV digestion – can be analyzed separately (Scheme 15.6).

One recent example of biotinylation experiments can be illustrated through the proteome analysis of *Pseudomonas aeruginosa* NagZ.[25] Vocadlo *et al.* incubated the investigated proteome with 2AA5FGF (**32**), which was then anchored on the activated enzymes **31** via the formation of an ester bond with the nucleophile of the β-glucosaminidase. Thereafter, ligation with the biotin-containing reporter group was achieved through Staudinger reaction (not shown) or click reaction. The target enzymes **35** were purified and identified via immobilized streptavidin stationary phase (Scheme 15.7).

The essential azido-compound **32** has been synthesized via a five-step protocol including two fluorination steps (using DAST and $AgBF_4$) and the transformation of an phthalimido-protected amine into the corresponding azide through deprotection with N_2H_4, acylation and substitution with NaN_3.

A similar approach for the biotin-labeling of glucosamines has been recently published by Gurcel *et al.*[26] Given its importance for the activity of many nuclear and cytoplasmic proteins, the group investigated dynamic glycosylation. The bioorthogonal click reaction has been used to connect biotinylated acetylenes or azides with *N*-acetylglucosamine (GlcNAc) analogs that contain azides or alkynes. Because of biotinylation and subsequent affinity purification on streptavidin beads, thirty-two *O*-GlcNAc-azido-tagged proteins were identified.

Furthermore, biotin-labeling was used by Tate *et al.*[27] in a specific manner: the group used *N*-myristoyl transferase for the introduction of 'clickable' components into target enzymes

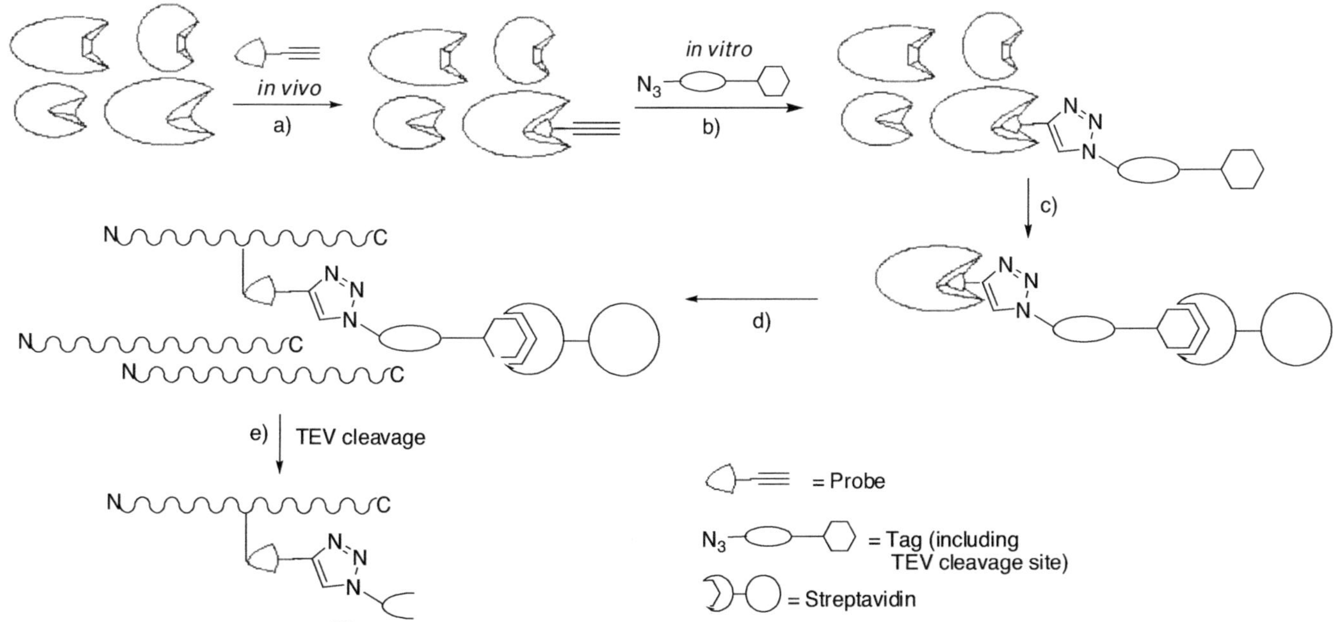

30

Scheme 15.6 *ABPP in combination with tandem orthogonal proteolysis strategy: (a) in vivo labeling; (b) in vitro click reaction; (c) streptavidin addition; (d) trypsin digestion; (e) TEV cleavage.*[24]

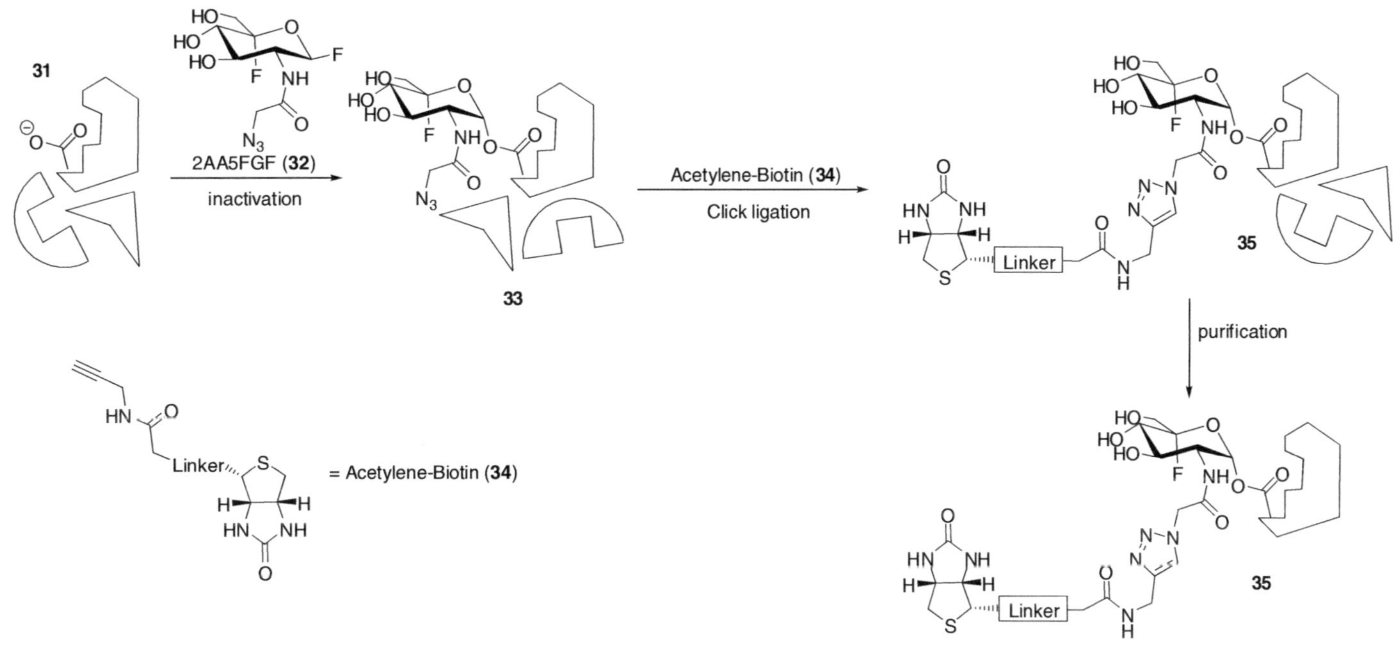

Scheme 15.7 *Activity-based detection and purification strategy using 2-azidoacetamido-2-deoxy-5-fluoro-β-D-glucopyranosyl fluoride (2AA5FGF).*[25]

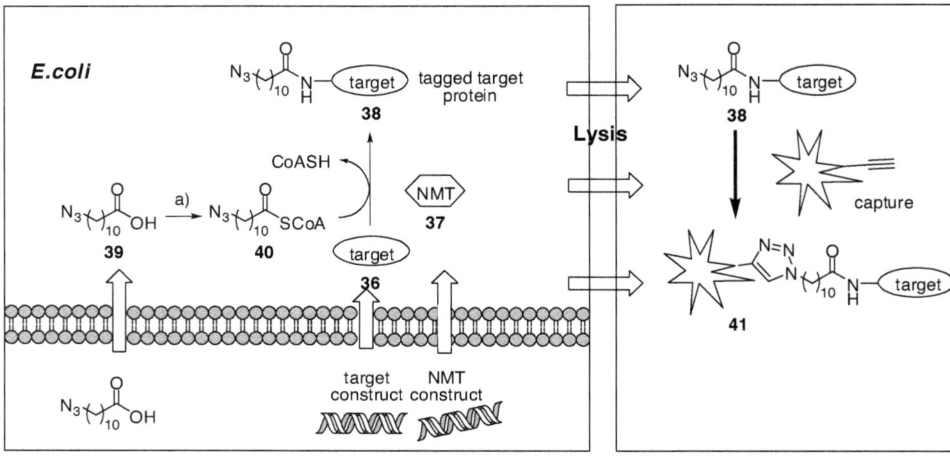

Scheme 15.8 *E. coli co-expression system applied to azide and alkyne tagging of a protein* **36** *in vivo. (a) CoASH, acetyl-CoA synthetase.*[27]

(Scheme 15.8). In their search for methodologies for posttranscriptional labeling concepts, Tate *et al.* found that the use of *N*-myristoyl transferase facilitates the *in vitro* and *in vivo* site-specific generation of *N*-terminal azide-tagged recombinant proteins. These tagged proteins can be 'clicked' after cell lysis with the corresponding acetylenes to produce biotin-labeled triazole linker-containing proteins. The entire procedure is briefly described in Scheme 15.8. The required myristic acid analogs **39** were synthesized starting from bromine precursors that could be transferred into the azide-containing compound via substitution with sodium azide. Incubation of the target enzyme **36** with the functionalized myristic acid derivatives **40** in combination with *N*-myristoyl transferase **37** yielded tagged target proteins **38** in *E. coli*.

Beyond these enzyme labeling methods with biotin, several biologically active molecules have also been labeled. One example is the tagging of enzyme ligands to characterize ligand–receptor interactions. This concept has been extended to ligands that were mentioned in the first chapter of this article, in reference to labeling with fluorescent markers.[28]

15.2.4 Fluorine Labeling

^{18}F-labeling has been used extensively in the observation of peptidic structures. The peptides under investigation can be labeled with ^{18}F-containing substances via click reaction when either ^{18}F-containing azides or alkynes are prepared. Scheme 15.9 illustrates the synthesis of an ^{18}F-labeled alkynyl-chain that has been incorporated into peptides through triazole formation of compounds **44** and **46**.[29] The synthesis has been attained via the integration of ^{18}F into the alkyne-bearing component, as demonstrated by Wuest *et al.* The aforementioned group synthesized ^{18}F-containing alkynylated aryls that have been used for the labeling of neurotensin by click reaction.[30]

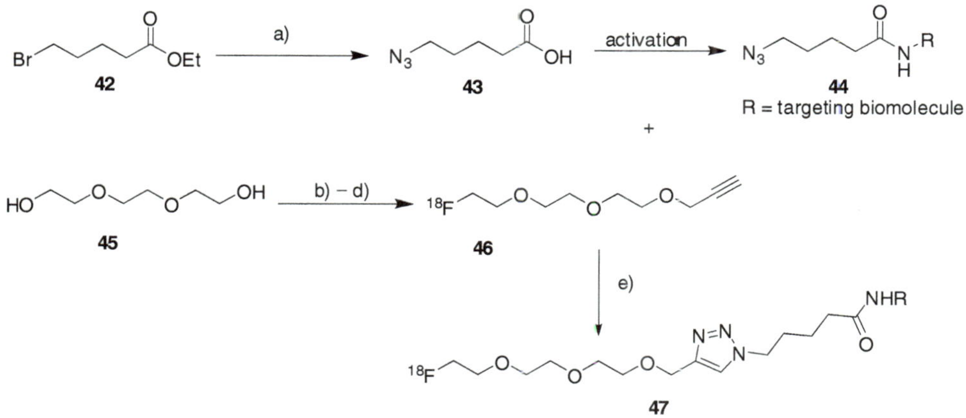

Scheme 15.9 Radiosynthesis of ^{18}F-PEG-alkyne intermediate **46** and click reaction with azide **44**. (a) NaN_3 then H^+ or OH^-; (b) propargyl bromide, NaH, THF, r.t. 18 h; (c) TsCl, NEt_3, MeCN, r.t., 16 h; (d) K_2CO_3, KF, $K^+[^{18}F]F^-$, MeCN, 90 °C, 40 min; (e) Cu^{2+}, ascorbate, THF, r.t., 24 h.[29]

While the formation of the ^{18}F-labeled building block **46** for click reaction can be completed in a three-step synthetic protocol starting with triethyleneglycol (**45**), there are other methods to prepare ^{18}F-modified alkynes via the direct addition of ^{18}F-sources ([^{18}F]KF) to tosylated alkynyl alcohols.[31]

^{18}F-bearing azide-derivatives for click reaction with alkynes can be synthesized following a procedure established by Glaser and Arstad.[32] This approach is favorable because of the substantial number of readily available alkynes that can be used for the click reaction. The synthesis of the fluorinated compound **48** was achieved via nucleophilic fluorination of 2-azidoethyl-4-toluenesulfonate using Kryptofix 222 $K^+[^{18}F]F^-$ (Figure 15.4).

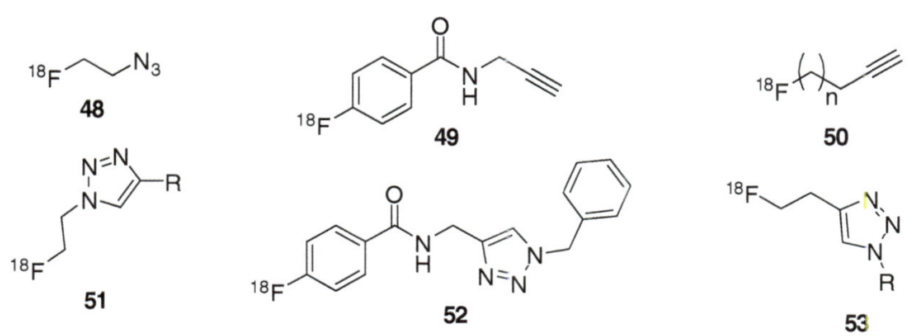

Figure 15.4 ^{18}F-containing azides and alkynes for click chemistry.[30–32]

15.3 Syntheses of Natural Products and Derivatives

To this day, sophisticated knowledge of molecular biology and the advances in biomedical research have increased the number of known targets for therapeutic intervention. Thus, the interest in natural compounds with high bioactive potential for novel pharmacological screenings has risen exponentially. The rapid and well-known regioselective 1,2,3-triazole synthesis can be used to generate an enormous chemical database with potential bioactivity for screening purposes.[33,34]

Glycopeptides, for example, constitute an important class of natural products, most of which are involved in biochemical transformations. Today, the synthesis of different glycopeptidomimetics[35,36] and analogs as well as the synthesis of new macrolide antibiotics,[37] enzyme inhibitors,[38,39] steroids[40] and natural nucleosides[41] remains a great challenge.

Specific problems, such as the instability in *in vivo* applications, the synthesis of glycosidic-linked saccharides and peptides as well as antibiotic resistance, have led to an alternative synthesis in the case of novel peptidomimetics and macrolides, e.g. C-linked analogs or the indirect linkage via a 1,2,3-triazole ring.

In one approach, the peptide chain was replaced by either a carbamate derivative **54** (*in vitro* activity against HL-60 human leukemia, HT-29 human colon carcinoma and antineoplatic activities) or by 2(1*H*)pyrazinones **55** via 1,2,3-triazole in a microwave-enhanced regioselective cycloaddition reaction, thus unveiling a new class of interesting analogs of glycopeptidomimetics (Figure 15.5).[35]

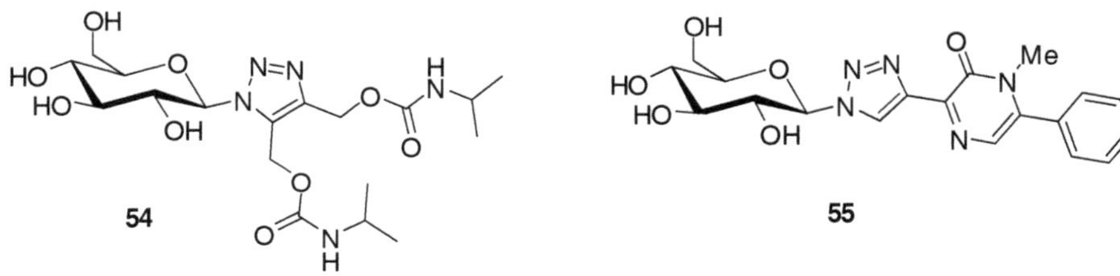

Figure 15.5 *Glycopeptidomimetics: carbamate derivative **54** and 2(1H)pyrazinone **55**.*[35]

Through its mechanism of action, the antibacterial drug linezolid (**60**), which is an oxazolidinone derivative, has prompted further investigation in terms of chemical modification and structure–activity relationship (SAR) studies. Chang and his coworkers[42] designed a variety of new molecules containing an oxazolidinone, a triazole ring as mimic of the benzene ring, and a pyranose component as a surrogate of the morpholine-moiety of linezolid (Scheme 15.10).

SAR studies have unveiled an essential *N*-aryl group directly linked to the oxazolidinone, which was mimicked through triazole-formation (*N*-aryl-type linkage). The compounds were tested for inhibition of various strains of bacteria and fungi but surprisingly no inhibitory activity was observed. In the field of macrolide antibiotics, Omura *et al.* designed several 8,9-anhydroerythromycin A 6,9-hemiketal analogs with anti-MRSA (multiresistent *Staphylococcus aureus*) and -VRE (vancomycin-resistant enterococci) activity using click chemistry.[37] Besides many carbohydrate and nucleotide conjugates with biological activity, natural products based on a simple organic molecule such as resveratrol **61** are also well known. Highly concentrated in wine, resveratrol plays a crucial role in the cardiovascular system.

A remaining challenge is to develop the targets responsible for each individual effect, such as the impact on the lipid metabolism and platelet function of resveratrol, because

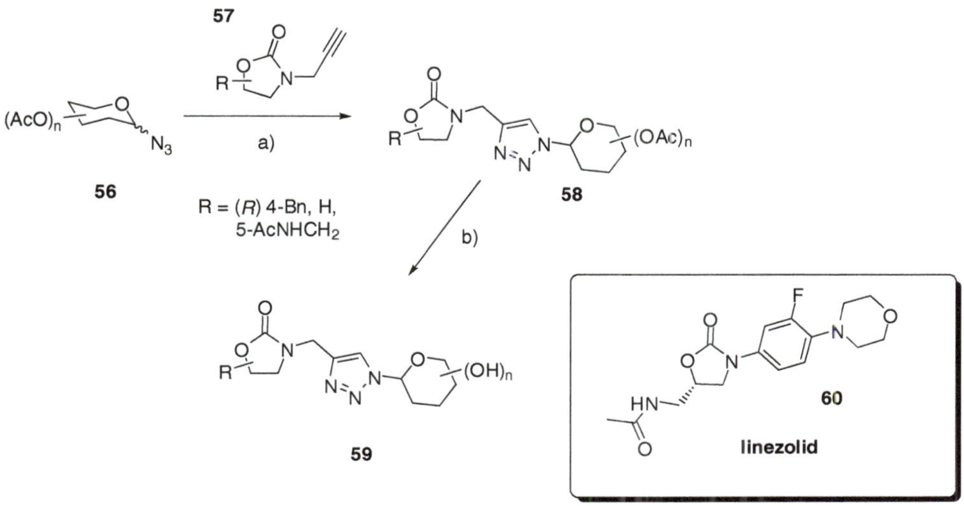

Scheme 15.10 *Synthesis of different linezolid mimics* ***59*** *using click chemistry. (a) Cu(OAc)$_2$, sodium ascorbate, MeOH–THF–H$_2$O, sonication; (b) NaOMe, MeOH.*[42]

a high micromolar concentration of the natural product is necessary to exert its effects.[33] Genazzani and his coworkers observed some triazole analogs whose bioactivities are in fact comparable to a resveratrol-like action (Scheme 15.11).

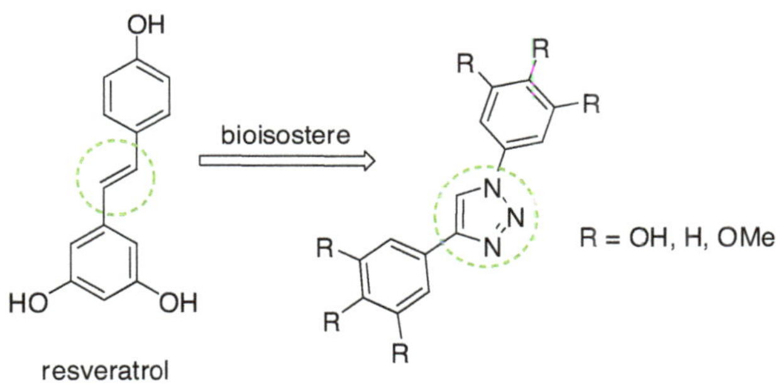

Scheme 15.11 *Resveratrol* ***61*** *and its triazole-modified analogs* ***62***.[33]

Other such examples are the natural products steganacin and podophyllotoxin, both of which inhibit the assembly of tubulin into microtubules and possess cytotoxic activity against several cancer cell lines.[43] Therefore, a lactone group was replaced with 1,5-disubstituted 1,2,3-triazoles, synthesized via well-known ruthenium-catalyzed click chemistry.

In medical chemistry, one of the great challenges is to create novel, effective chemotherapeutic agents with specificity for cancer cells, combined with low systemic toxicity.

Moreover, the need for antitumor drugs continually increases. Novel targets of these drugs are enzymes of NAD (nicotinamide adenine dinucleotide) biosynthesis and recycling pathways, given that tumor cells have a higher turnover rate than healthy cells. Thus, Tron and his coworkers[38] developed bioactive compounds useful for medical chemistry, such as isosteric triazole analogs of FK866, which blocks nicotinamide phosphoribosyltranferase (NMPRTase).

The therapeutic potential of analogs has led to novel syntheses of compounds via click chemistry, in which the amide bond is replaced by a triazole ring, acting as a bioisostere (Scheme 15.12). The required azide was synthesized starting from commercial available 4-piperidine butyric acid hydrochloride, which was reduced to the corresponding alcohol using lithium aluminum hydride, *N*-benzoylated and then converted into an azide with DPPA (diphenylphosphorylazide and sodium azide).

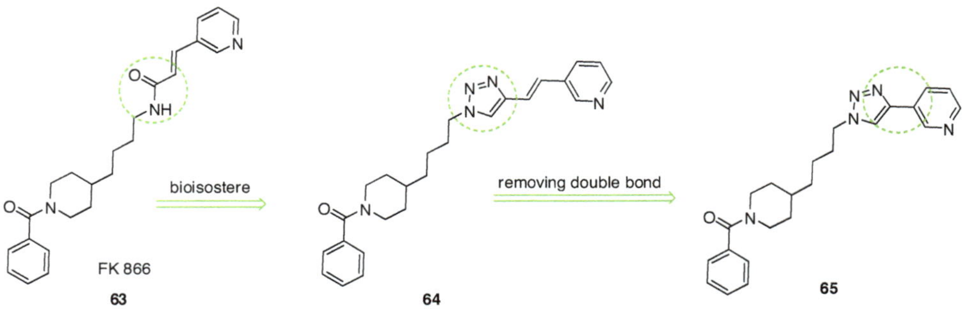

Scheme 15.12 *1,4-Disubstituted triazole analogs of FK866.*[38]

The novel triazole analogs were then tested in a cell viability assay (neuroblastoma cell line, SH-SY5Y) and the intracellular NAD levels were measured with a cycling assay after incubation with compounds **64** and **65** for 24 h.[38] Surprisingly, analog **64** – the closest analog to FKK866 – displayed an IC_{50} value of 3.0 ± 0.2 μM, higher than the analog **65**, which lacks the olefinic moiety (shortened distance between the triazole and the pyridine ring). Jiang *et al.* reported another example of an enzyme inhibitor against H5N1, an avian influenza virus (AIV), synthesized via click chemistry.[44]

15.4 Enzymes and Click Chemistry

The formation of an enzyme inhibitor through chemical reaction of two components bound to the enzyme can be effectively performed if the coupling reaction commonly takes place with high yields. Given this and the aforementioned attributes of the click reaction, the click reaction is the reaction of choice for the development of enzyme inhibitors. Enzyme ligands with known affinity for the active side of enzymes were chosen to bind to two distinct, neighboring binding sides and were connected via a linker that is able to perform click reactions. Given the enzyme-enforced proximity of the two reaction partners, the formation of these triazole-connected twofold ligands should be catalyzed by the enzyme. Through this strategy, the enzyme is used to catalyze the formation of its own inhibitor.

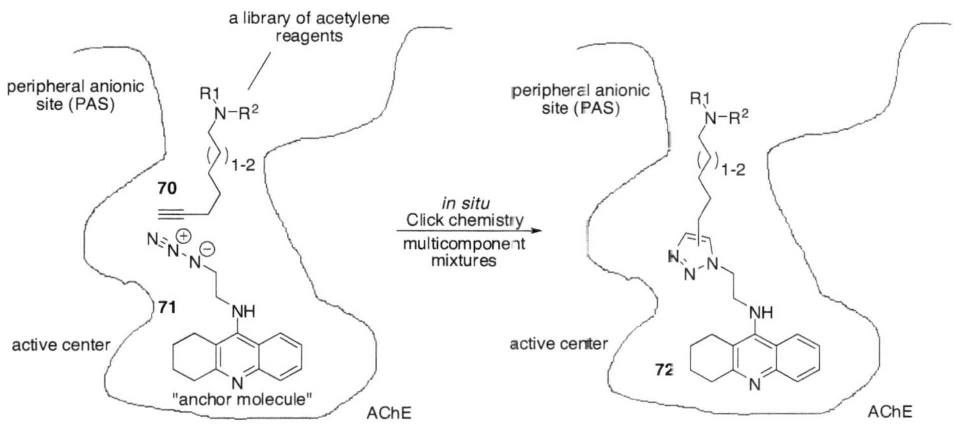

Scheme 15.13 *In situ click chemistry screening for AChE inhibitors with novel peripheral site ligands.*[47]

The most famous example for the exploration of these enzyme inhibitory ligands is the search for acetylcholinesterase inhibitors (AChE). Sharpless and Kolb's respective groups investigated tacrine and phenylphenanthridinium derivatives,[45,46] in terms of their function as inhibitors of *Electrophorus electricus* and mouse AChE. The target acetylcholinesterase was chosen because of its role in the central and peripheral nervous system and its importance in the investigation of acetylcholine in neurotransmission. Two sites of AChE were chosen for the binding of the acetylene and the azide component: the catalytic site of the enzyme at the bottom of a 20 Å deep narrow gorge and another peripheral binding site near the protein surface (Scheme 15.13). The potentially potent inhibitors of AChE were identified through the incubation of several in preliminary experiments selected ligands **66-69** (Figure 15.6, all containing azide as well as alkyne functionalities) and the target enzyme for six days at room temperature (series of 49 binary mixtures incubated). From the theoretically feasible combinations of products, only one was observed. It was shown that, while the enzyme-catalyzed reaction with the triazoles was merely selective, in particular with respect to the formation of the *syn*-isomer, the antitriazole was not detected.

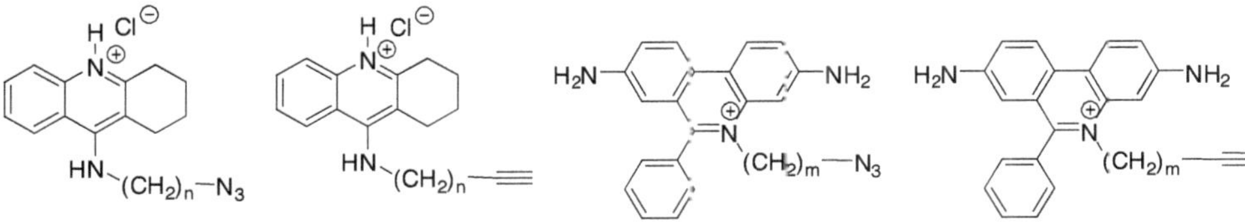

Figure 15.6 *Azide and acetylene building blocks for the formation of 49 binary mixtures and their incubation with AChE.*[45,46]

X-ray studies have proven the interaction of the bivalent ligand with protein binding sites. It has been shown that the tacrine moiety was incorporated into the active center of the enzyme and that the phenylphenanthridinium group is located on the peripheral site, both connected via a triazole linker generated by click reaction. In subsequent experiments, the phenylphenanthridinium group was replaced by phenyltetrahydroisoquinoline building blocks (**70**), which yielded inhibitors with an up-to-three-times enhanced activity in combination with the previously used tacrine moiety (**71**) (Scheme 15.13). As a well-known ligand for the active center of AChE, the tacrine group was used combined with a two-carbon azide linker, given that preceding experiments had proven this to be the optimal distance (see formation of **72**). It is possible to perform the experiments with up to 10 acetylene-containing compounds at one time, allowing the enzyme to choose between different ligands and to select the most suitable one for the click reaction. In these competing experiments, out of 10 structurally different acetylenes, only two were able to react with the enzyme-ligated azide. Interestingly, the enzyme only showed selectivity for the formation of the *syn*-triazole products, but there was no preference exhibited for the formation of one specific enantiomer. Both the *R*- and the *S*-enantiomers are used for enzyme-located click reactions.[47]

Another example of target-guided synthesis (TGS) has been illustrated by Kolb *et al.*, with the synthesis of enzyme-generated inhibitors of carbonic anhydrase (CA) II.[48] Yet again, the click reaction was selected because of its bioorthogonal character in the ligation of the target-bound functionalized ligands. Compound **73** was chosen as acetylene moiety since most inhibitors of CA are aromatic or heteroaromatic sulfonamides that are able to coordinate to the Zn^{2+}-ion at the active site of the enzyme. Acetylenic benzenesulfonamide (**73**) was incubated for 40 h with bovine carbonic anhydrase II and the corresponding azide-containing counterpart **74**. Subsequently, triazole formation was observed (Scheme 15.14). Altogether, the group was able to demonstrate the formation of twelve triazoles **75**, among them triazoles from piperidine azides, bicyclic azides, stilbene azides, phenyl and ethyl azides.

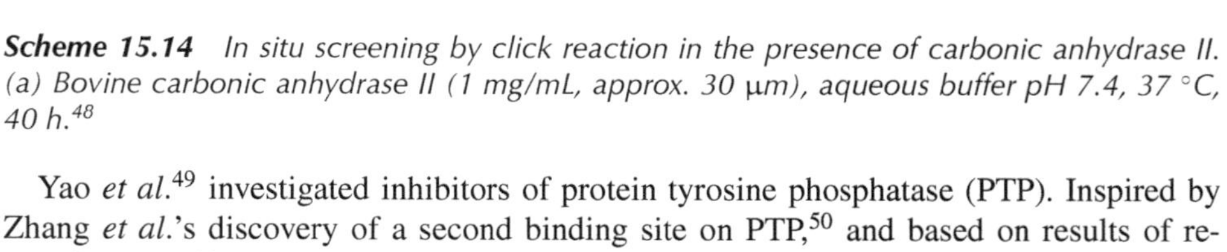

73 74 75

Scheme 15.14 *In situ screening by click reaction in the presence of carbonic anhydrase II. (a) Bovine carbonic anhydrase II (1 mg/mL, approx. 30 μm), aqueous buffer pH 7.4, 37 °C, 40 h.*[48]

Yao *et al.*[49] investigated inhibitors of protein tyrosine phosphatase (PTP). Inspired by Zhang *et al.*'s discovery of a second binding site on PTP,[50] and based on results of researchers at Abbott, who found a cell-permeable bidentate PTP inhibitor, the group synthesized a library of potential PTP ligands with azide and alkyne functionality and thereafter combined the building blocks to 66 bidentate ligands. Subsequent *in situ* enzymatic screenings revealed a potential PTP1B inhibitor with an IC_{50}-value of 4.7 μM.

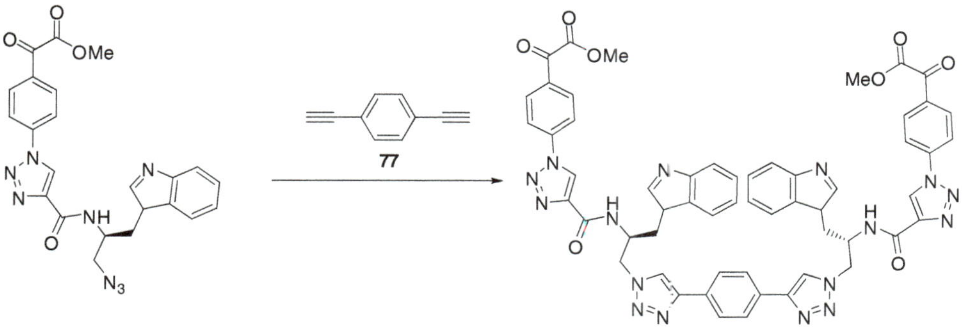

Scheme 15.15 *Structure of the best bis(α-ketocarboxylic acid) inhibitor of protein tyrosine phosphatase.*[52]

Seto *et al.* extended the search for PTP inhibitors and generated two sequential libraries of PTP inhibitors. In the first step, 4-azidobenzoylformate was reacted with fifty-six mono- and diynes. The resulting esters were hydrolyzed and tested against Yersinia PTP and PTP1B. Four selected examples were further investigated and one of them (the precursor of compound **76**) was derived to give the second-generation PTP inhibitors (Scheme 15.15). An alcohol functionality was converted into an azide group via the Mitsunobu reaction with ZnN_3 and subsequently into a triazole group through the addition of the same fifty-six mono- and diynes from the first generation inhibitors. Click reaction of the first-generation inhibitor-derivative **76** with 1,4-diethynylbenzene (**77**) produced the precursor **78,** which was transferred into the target compound via hydrolysis of the ester functionalities. The bis(α-ketocarboxylic acid) inhibitor **78** was found to have an IC_{50}-value of 550 nM against Yersinia PTP (IC_{50} = 710 nM against TCPTP).[51,52]

Beyond the search for phosphatase inhibitors, other enzymes have been under investigation concerning their inhibition abilities. Examples include the exploration of phosphonate inhibitors for the regulation of serine hydrolases,[53] the search for Grb2 SH2 domain-binding macrocycles[54] and the investigation of glycoconjugate benzene sulfonamides as carbonic anhydrase inhibitors.[55]

15.5 Synthesis of Glycosylated Molecular Architectures

As the separation of highly polar compounds by HPLC remains a great challenge, an alternative method with a polar stationary phase and an aqueous mobile phase (hydrophilic interaction liquid chromatography, HILIC) has been developed. A variety of separation materials such as underivatized silica, amido silica, poly(succinimide)-bonded silica, polyhydroxy silica with different retention characteristics and separation selectivity have been described in the literature. Moreover, glucose, maltose and β-CD (Scheme 15.16) with several polar groups and unique structures have been immobilized via click reaction, leading to novel complex separation materials.[56]

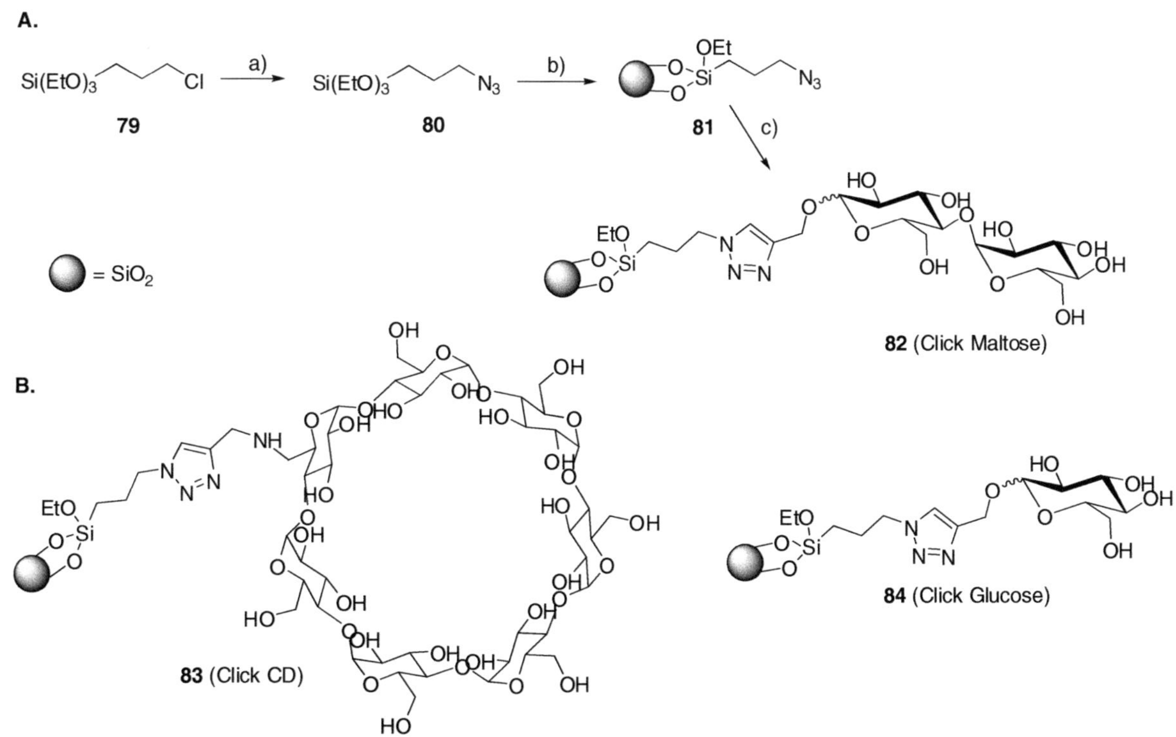

Scheme 15.16 *(A) Synthesis of the model molecule click Maltose (**82**). (a) NaN_3, DMF, KI, 90–100 °C; (b) silica beads, DMF, 100–110 °C; (c) 1-O-propargylmaltose, MeOH–H_2O (1:1), 5 mol% $CuSO_4$, 15 mol% sodium ascorbate, r.t.; (B) click-CD (**83**) and click-glucose (**84**) as separation materials.*[56]

15.6 Synthesis of Nitrogen-rich Compounds: Polyazides and Triazoles

In recent years, the study of polyazides and their highly applicable energetic materials has experienced a renaissance. Numerous homoleptic azides $[M(N_3)_n]$, derivable salts[57] as well as organic azides have been prepared.[58,59] A major challenge, the isolation of tetraazidomethane, which was expected to be a highly explosive compound, was overcome by Banert and coworkers in 2007 (Scheme 15.17). The lower-substituted derivative, the triazidomethane, was obtained via simple nucleophilic substitution from bromoform (CBr_4). However, all attempts to generate the homoleptic azide from halides, as well as a variety of precursors bearing donor or acceptor substituents, through different azide reagents were unsuccessful.[58]

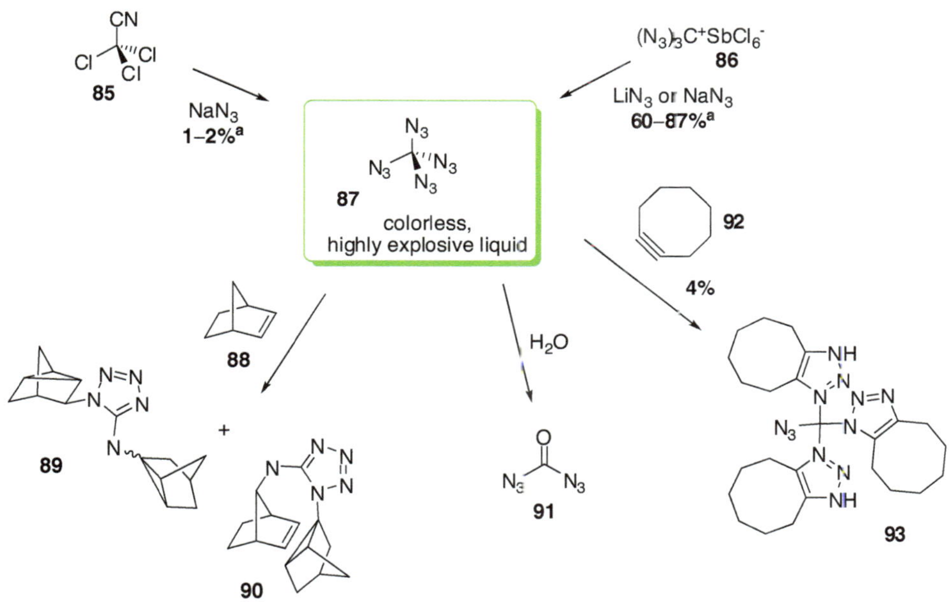

Scheme 15.17 Attempts to prepare the tetraazidomethane (87) and further reactions with this compound. The yield was determined by the integration of the ^{13}C NMR signal.[58]

The only sign of formation was generated through the treatment of $C(CN)Cl_3$ (**85**) with sodium azide in MeCN. Subsequent approaches involved the reaction of the salt **86** with dry lithium or sodium azide, leading to significantly higher yields. Against all expectations, antimonate salts are expensive and explosive starting materials, which are therefore unsuitable for preparative purposes. The workup is less problematic and more reproducible for the first-mentioned procedure.

Special safety precautions have to be followed, because azides are highly energetic and potentially explosive materials, particularly those with low molecular weights. Pure tetraazidomethane (**87**) is extremely dangerous and can explode at any time without a recognizable cause. According to convention, the ratio of nitrogen to carbon (oxygen)

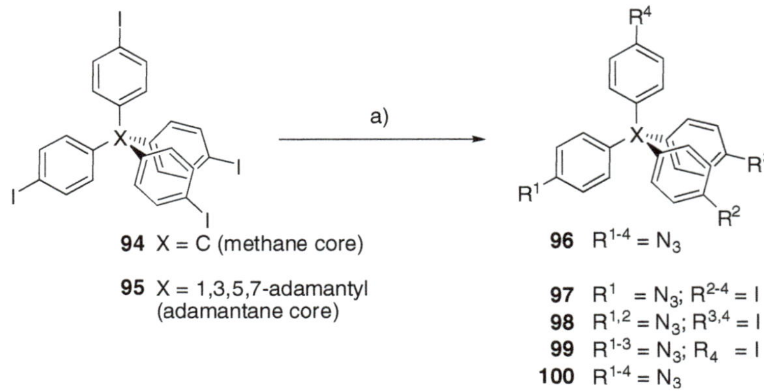

Scheme 15.18 *Ullmann-type coupling reaction leading to polyazides* ***96–100****. (a)* NaN_3*, ligand (diamine), Cu(I), sodium ascorbate, DMSO–*H_2O*, 100 °C, 48 h.*[59]

should be maximum 1:3 for a neat isolation of the yielded azide. Compounds with lower ratios should be handled with extreme caution.

Sodium azide is very toxic, similar to sodium cyanide. The use of additional metals as well as halogenated solvents such as dichloromethane in the presence of the sodium azide must be avoided.[60]

Nowadays, aryl azides are increasingly used in organic synthesis, due to the versatile transformations of the azide functional group. Thus, additional polyazides were synthesized, as previously reported by Bräse *et al.* (Scheme 15.18).[58] For the investigation of novel nanomaterials and compounds for polymer science and polymer processes as well as material and macromolecular sciences, rigid tetrafunctionalized molecules are promising compounds. The tetrasubstituted polyazides were prepared via Ullmann-type coupling reactions.

Furthermore, 1,4-disubstituted 1,2,3-triazoles, easily accessible from polyazides **96–100**, have found application in drug discovery, bioconjugation, surface modification and material development as multifunctional ligands as well as in polymer science. Additional examples, such as hyperbranched poly(1,2,3-triazoles), reported by Tang *et al.*, were inclined to self-oligomerize and failed to yield soluble polymers using Cu(I)-catalyzed click polymerization.[61]

15.7 Conclusions

The benefits of the click reaction, namely high yields and biocompatibility, enable the very efficient application of this cycloaddition reaction to the synthesis of molecular architectures based on biomolecules, the synthesis of labeled biomolecules and their general application in living systems. In this article, the structure of derived markers, biomolecules, natural products and biologically active compounds containing either an azide or an alkyne functionality for click reactions is described. It has been illustrated that the potential of the click reaction lies not exclusively in its application as a labeling method, but that it can also be used to build up novel biologically active compounds. These compounds can be created according to known active targets, but can also be identified via direct formation of triazole derivatives in the active site of enzymes through incubation with several theoretically clickable precursors.

References

(1) A. Krasinski, V. V. Fokin, K. B. Sharpless, (2004), Direct synthesis of 1,5-disubstituted-4-magnesio-1,2,3-triazoles, revisited, *Org. Lett.*, **6**, 1237–1240.

(2) J. A. Johnson, M. G. Finn, J. T. Koberstein, N. J. Turro, (2008), Construction of linear polymers, dendrimers, networks, and other polymeric architectures by copper-catalyzed azide-alkyne cycloaddition click-chemistry, *Macromol. Rapid Commun.*, **29**, 1052–1072.

(3) X.-L. Sun, C. L. Stabler, C. S. Cazalis, E. L. Chaikof, (2005), Carbohydrate and protein immobilization onto solid surfaces by sequential Diels–Alder and azide–alkyne cycloadditions, *Bioconjugate Chem.*, **17**, 52–57.

(4) L. Polito, D. Monti, E. Caneva, E. Delnevo, G. Russo, D. Prosperi, (2008), One-step bioengineering of magnetic nanoparticles via a surface diazo transfer/azide-alkyne click reaction sequence, *Chem. Commun.*, **5**, 621–623.

(5) G. Chen, L. Tao, G. Mantovani, J. Geng, D. Nystroem, D. M. Haddleton, (2007), A modular click approach to glycosylated polymeric beads: design, synthesis and preliminary lectin recognition studies, *Macromolecules*, **40**, 7513–7520.

(6) Y. Zhang, S. Luo, Y. Tang, L. Yu, K.-Y. Hou, J.-P. Cheng, X. Zeng, P. G. Wang, (2006), Carbohydrate–protein interactions by 'clicked' carbohydrate self-assembled monolayers, *Anal. Chem.*, **78**, 2001–2008.

(7) Y. Li, W. Zhang, J. Chang, J. Chen, G. Li and Y. Ju, (2008), 'Click' on conducting polymer coated electrodes: a versatile platform for the modification of electrode surfaces, *Macromol. Chem. Phys.*, **209**, 322–329.

(8) Y.-H. Zhang, Z.-X. Gao, C.-L. Zhong, H.-B. Zhou, L. Chen, W.-M. Wu, X.-J. Peng, Z.-J. Yao, (2007), An inexpensive fluorescent labeling protocol for bioactive natural products utilizing Cu(I)-catalyzed Huisgen reaction, *Tetrahedron*, **63**, 6813–6821.

(9) H. R. Rengifo, L. Chen, C. Grigoras, J. Ju, J. T. Koberstein, (2008), 'Click-functional' block copolymers provide precise surface functionality via spin coating, *Langmuir*, **24**, 7450–7456.

(10) L. Ballell, M. van Scherpenzeel, K. Buchalova, R. M. J. Liskamp, R. J. Pieters, (2006), A new chemical probe for the detection of the cancer-linked galectin-3, *Org. Biomol. Chem.*, **4**, 4387–4394.

(11) M. J. Joralemon, R. K. O'Reilly, C. J. Hawker, K. L. Wooley, (2005), Shell click-crosslinked (SCC) nanoparticles: a new methodology for the synthesis and orthogonal functionalization, *J. Am. Chem. Soc.*, **127**, 16892–16899.

(12) A. L. MacKinnon, J. L. Garrison, R. S. Hegde, J. Taunton, (2007), Photo-leucine incorporation reveals the target of a cyclodepsipeptide inhibitor of cotranslational translocation, *J. Am. Chem. Soc.*, **129**, 14560–14561.

(13) T. Mayer, M. E. Maier, (2007), Design and synthesis of a tag-free chemical probe for photoaffinity labeling, *Eur. J. Org. Chem.*, **28**, 4711–4720.

(14) A. E. Speers, G. C. Adam, B. F. Cravatt, (2003), Activity-based protein profiling in vivo using a copper(I)-catalyzed azide–alkyne [3 + 2] cycloaddition, *J. Am. Chem. Soc.*, **125**, 4686–4687.

(15) A. T. Wright, B. F. Cravatt, (2008), Chemical proteomic probes for profiling cytochrome P450 activities and drug interactions *in vivo*, *Chem. Biol.*, **14**, 1043–1051.

(16) J. M. Baskin, J. A. Prescher, S. T. Laughlin, N. J. Agard, P. V. Chang, I. A. Miller, A. Lo, J. A. Codelli, C. R. Bertozzi, (2007), Copper-free click chemistry for dynamic *in vivo* imaging, *Proc. Natl Acad. Sci. USA*, **104**, 16793–16797.

(17) M. Sawa, T.-L. Hsu, T. Itoh, M. Sugiyama, S. R. Hanson, P. K. Vogt, C.-H. Wong, (2006), Glycoproteomic probes for fluorescent imaging of fucosylated glycans in vivo, *Proc. Natl Acad. Sci. USA*, **103**, 12371–12376.
(18) T.-L. Hsu, S. R. Hanson, K. Kishikawa, S.-K. Wang, M. Sawa and C.-H. Wong, (2007), Alkynyl sugar analogues for the labeling and visualization of glycoconjugates in cells, *Proc. Natl Acad. Sci. USA*, **104**, 2614–2619.
(19) L. Cheng, Q. Chen, J. Liu, Y. Du, (2007), Synthesis of a fluorescence-labeled K30 antigen repeating unit using click chemistry, *Carbohydr. Res.*, **342**, 975–981.
(20) O. S. Wolfbeis, (2007), The click reaction in the luminescent probing of metal ions, and its implications on biolabeling techniques, *Angew. Chem. Int. Edn*, **46**, 2980–2982.
(21) N. S. Hatzakis, H. Engelkamp, K. Velonia, J. Hofkens, P. C. M. Christianen, A. Svendsen, S. A. Patkar, J. Vind, J. C. Maan, A. E. Rowan and R. J. M. Nolte, (2006), Synthesis and single enzyme activity of a clicked lipase-BSA hetero-dimer, *Chem. Commun.*, **19**, 2012–2014.
(22) J. L. Brennan, N. S. Hatzakis, T. R. Tshikhudo, N. Dirvianskyte, V. Razumas, S. Patkar, J. Vind, A. Svendsen, R. J. M. Nolte, A. E. Rowan, M. Brust, (2006), Bionanoconjugation via click chemistry: the creation of functional hybrids of lipases and gold nanoparticles, *Bioconjugate Chem.*, **17**, 1373–1375.
(23) A. J. Dirks, S. S. Van Berkel, N. S. Hatzakis, J. A. Opsteen, F. L. Van Delft, J. J. L. M. Cornelissen, A. E. Rowan, J. C. M. Van Hest, F. P. J. T. Rutjes, R. J. M. Nolte, (2005), Preparation of biohybrid amphiphiles via the copper catalyzed Huisgen [3 + 2] dipolar cycloaddition reaction, *Chem. Commun.*, **33**, 4172–4174.
(24) A. E. Speers, B. F. Cravatt, (2005), A tandem orthogonal proteolysis strategy for high-content chemical proteomics, *J. Am. Chem. Soc.*, **127**, 10018–10019.
(25) K. A. Stubbs, A. Scaffidi, A. W. Debowski, B. L. Mark, R. V. Stick, D. J. Vocadlo, (2008), Synthesis and use of mechanism-based protein-profiling probes for retaining β-D-glucosaminidases facilitate identification of *Pseudomonas aeruginosa* NagZ, *J. Am. Chem. Soc.*, **130**, 327–335
(26) C. Gurcel, A.-S. Vercoutter-Edouart, C. Fonbonne, M. Mortuaire, A. Salvador, J.-C. Michalski, J. Lemoine, (2008), Identification of new *O*-GlcNAc modified proteins using a click-chemistry-based tagging, *Anal. Bioanal. Chem.*, **390**, 2089–2097.
(27) W. P. Heal, S. R. Wickramasinghe, R. J. Leatherbarrow, E. W. Tate, (2008), *N*-Myristoyl transferase-mediated protein labelling *in vivo*, *Org. Biomol. Chem.*, **6**, 2308–2315.
(28) D. Bonnet, B. Ilien, J.-L. Galzi, S. Riche, C. Antheaune, M. Hibert, (2006), A rapid and versatile method to label receptor ligands using click chemistry: validation with the muscarnic M1 antagonist pirenzepine, *Bioconjugate Chem.*, **17**, 1618–1623.
(29) Z.-B. Li, Z. Wu, K. Chen, F. T. Chin and X. Chen, (2007), Click chemistry for ^{18}F-labeling of RGD peptides and microPET imaging of tumor integrin α vβ 3 expression, *Bioconjugate Chem.*, **18**, 1987–1994
(30) T. Ramenda, R. Bergmann, F. Wuest, (2007), Synthesis of ^{18}F-labeled neurotensin(8–13) via copper-mediated 1,3-dipolar [3 + 2]cycloaddition reaction, *Lett. Drug Design Discov.*, **4**, 279–285.
(31) J. Marik, J. L. Sutcliffe, (2006), Click for PET: rapid preparation of [^{18}F]fluoropeptides using CuI catalyzed 1,3-dipolar cycloaddition, *Tetrahedron Lett.*, **47**, 6681–6684.
(32) M. Glaser, E. Arstad, (2007), 'Click labeling' with 2-[^{18}F]fluoroethylazide for positron emission tomography, *Bioconjugate Chem.*, **18**, 989–993.
(33) (a) F. Pagliai, T. Pirali, E. Del Grosso, R. Di Brisco, G. C. Tron, G. Sorba, A. A. Genazzani, (2006), Rapid synthesis of triazole-modified resveratrol analogues via click chemistry, *J. Med. Chem.*, **49**, 467–470; (b) I. A. Inverarity, R. F. H. Viguier, P. Cohen, A. N. Hulme, (2007), Biotinylated anisomycin: a comparison of classical and 'click' chemistry approaches, *Bioconjugate Chem.*, **18**, 1593–1603.
(34) H. Chen, J. L. Taylor, S. R. Abrams, (2007), Design and synthesis of β-methoxyacrylate analogues via click chemistry and biological evaluations, *Bioorg. Med. Chem. Lett.* **17**, 1979–1983.
(35) D. Ermolat'ev, W. Dehaen, E. Van Der Eycken, (2004), Indirect coupling of the 2(^{1}H)-pyrazinone scaffold with various (oligo)-saccharides via 'click chemistry': en route towards glycopeptidomimetics, *QSAR Comb. Sci.*, **23**, 915–918.

(36) (a) S. Quader, S. E. Boyd, I. D. Jenkins, T. A. Houston, (2007), Multisite modification of neomycin B: combined Mitsunobu and click chemistry approach, *J. Org. Chem.*, **72** (6), 1962–1979; (b) C. Paloma, J. M. Aizpurua, E. Balentová, I. Azcune, J. I. Santos, J. Jiménez-Barbero, J. Cañada, J. I. Miranda, (2008), 'Click' saccharide/β-lactam hybrids for lectin inhibition, *Org. Lett.*, **10**, 2227–2230.
(37) A. Sugawara, T. Sunazuka, T. Hirose, K. Nagai, Y. Yamaguchi, H. Hanaki, K. B. Sharpless, S. Ōmura, (2007), Design and synthesis via click chemistry of 8,9-anhydroerythromycin A 6,9-hemiketal analogues with anti-MRSA and -VRE activity, *Bioorg. Med. Chem. Lett.*, **17**, 6304–6344.
(38) U. Galli, E. Ercolano, L. Carraro, C. R. B. Roman, G. Sorba, P. L. Canonico, A. A. Genazzani, G. C. Tron, R. A. Billington, (2008), Synthesis and biological evaluation of isosteric analogues of FK**866**, an inhibitor of NAD salvage, *ChemMedChem*, **3**, 771–779.
(39) P. Van Der Peet, C. T. Gannon, I. Walker, Z. Dinev, M. Angelin, S. Tam, J. E. Ralton, M. J. McConville, S. J. Williams, (2006), Use of click chemistry to define the substrate specificity of Leishmania β-1,2-mannosyltransferases, *ChemBioChem*, **7**, 1384–1391.
(40) (a) B.-C. Suh, H. B. Jeon, G. H. Posner, S. M. Silverman, (2004), Vitamin D side chain triazole analogs via cycloaddition 'click' chemistry, *Tetrahedron Lett.* **45**, 4623–4625; (b) L. Suarez, Z. Gándara, G. Gómez, Y. Fall, (2004), Vitamin D and click chemistry. Part 1: a stereoselective route to vitamin D analogues with triazole rings in their side chains, *Tetrahedron Lett.*, **45**, 4619–4621; (c) T. Pirali, S. Gatti, R. Di Brisco, S. Tacchi, R. Zaninetti, E. Brunelli, A. Massarotti, G. Sorba, P. L. Canonico, L. Moro, A. A. Genazzani, G. C. Tron, C. Gian, R. A. Billington, (2007), Estrogenic analogues synthesized by click chemistry, *ChemMedChem*, **2**, 437–440.
(41) I. Pérez-Castro, O. Caamaño, F. Fernández, M. D. García, C. López, E. De Clercq, (2007), Synthesis of 4-substituted-1,2,3-triazole carbanucleoside analogues of ribavirin via click chemistry, *Org. Biomol. Chem.*, **5**, 3805–3813.
(42) J. Zhang, H.-N. Chen, F.-I Chiang, J. Y. Takemoto, M. Besacci, C.-W. T. Chang, (2007), Sonication-assisted library synthesis of oxazolidinone-carbohydrate conjugates, *Comb. Chem.*, **9**, 17–19.
(43) D. Imperio, T. Pirali, U. Galli, F. Pagliai, L. Cafici, P. L. Canonico, G. Sorba, A. A. Genazzani, G. C. Tron, (2007), Replacement of the lactone moiety on podophyllotoxin and steganacin analogues with a 1,5-disubstituted 1,2,3-triazole via ruthenium-catalyzed click chemistry, *Bioorg. Med. Chem.*, **15**, 6748–6757.
(44) J. Li, M. Zheng, W. Tang, P.-L. He, W. Zhu, T. Li, J.-P. Zuo, H. Liu, H. Jiang, (2006), Syntheses of triazole-modified zanamivir analogues via click chemistry and anti-AIV activities, *Bioorg. Med. Chem. Lett.*, **16**, 5009–5013.
(45) W. G. Lewis, L. G. Green, F. Grynszpan, Z. Radic, P. R. Carlier, P. Taylor, M. G. Finn, K. B. Sharpless, (2002), Click chemistry *in situ*: acetylcholinesterase as a reaction vessel fort he selective assembly of a femtomolar inhibitor from an array of building blocks, *Angew. Chem. Int. Edn*, **41**, 1053–1057.
(46) R. Manetsch, A. Krasinski, Z. Radi, J. Raushel, P. Taylor, K. B. Sharpless, H. C. Kolb, (2004), *In situ* click chemistry: Enzyme inhibitors made to their own specifications, *J. Am. Chem. Soc.*, **126**, 12809–12818.
(47) A. Krasinski, Z. Radic, R. Manetsch, J. Raushel, P. Taylor, K. B. Sharpless, H. C. Kolb, (2005), *In situ* selection of lead compounds by click chemistry: target-guided optimization of acetylcholinesterase inhibitors, *J. Am. Chem. Soc.*, **127**, 6686–6692.
(48) V. P. Mocharla, B. Colasson, L. V. Lee, S. Roeper, K. B. Sharpless, C.-H. Wong, H. C. Kolb, (2005), *In situ* click chemistry: enzyme-generated inhibitors of carbonic anhydrase II, *Angew. Chem., Int. Edn*, **44**, 116–120.
(49) R. Srinivasan, M. Uttamchandani, S. Q. Yao, (2006), Rapid assembly and *in situ* screening of bidentate inhibitors of protein tyrosine phosphatases, *Org. Lett.*, **8**, 713–716.
(50) Z.-Y. Zhang, (2001), Protein tyrosine phosphatases: prospects for therapeutics, *Curr. Opin. Chem. Biol.*, **5**, 416–423.
(51) K. A. Kalesh, P.-Y. Yang, R. Srinivasan, S. Q. Yao, (2007), Click chemistry as a high-throughput amenable platform in catalomics, *QSAR Comb. Sci.*, **26**, 1135–1144.

(52) J. Xie, C. T. Seto, (2007), A two stage click-based library of protein tyrosine phosphatase inhibitors, *Bioorg. Med. Chem.*, **15**, 458–473.
(53) H. P. Dijkstra, H. Spong, B. N. H. Aerts, C. A. Kruithof, M. R. Egmond, R. J. M. Klein Gebbink, (2008), Selective and diagnostic labeling of serine hydrolases with reactive phosphonate inhibitors, *Org. Biomol. Chem.*, **6**, 523–531.
(54) W. J. Choi, Z.-D. Shi, K. M. Woerthy, L. Bindu, R. G. Karki, M. C. Nicklaus, R. J. Fisher, T. R. Burke, Jr, (2006), Application of azide–alkyne cycloaddition 'click chemistry' for the synthesis of Gbr2 SH2 domain-binding macrocycles, *Bioorg. Med. Chem. Lett.*, **16**, 5265–5269.
(55) B. L. Wilkinson, L. F. Bornaghi, T. A. Houston, A. Innocenti, C. T. Supuran, S.-A. Poulsen, (2006), A novel class of carbonic anhydrase inhibitors: glycoconjugate benzene sulfonamides proared by 'click-tailing', *J. Med. Chem.*, **49**, 6539–6548.
(56) Z. Guo, A. Lei, Y. Zhang, Q. Xu, X. Xue, F. Zhang and X. Liang, (2007), 'Click saccharides': novel separation materials for hydrophilic interaction liquid chromatography, *Chem. Commun.*, **24**, 2491–2493.
(57) (a) R. Haiges, J. A. Boatz, A. Vij, V. Vij, M. Gerken, S. Schneider, T. Schroer, M. Yousufuddin, K. O. Christe, (2004), Polyazide chemistry: preparation and characterization of $As(N_3)_5$, $Sb(N_3)_5$ and $[P(C_6H_5)_4][SbN_3)_6]$, *Angew. Chem. Int. Edn*, **43**, 6676–6680; (b) T. M. Klapötke, B. Krumm, R. Ilg, D. Trögel, R. Tackel, (2007), The sila-explosives $Si(CH_2N_3)_4$ and $Si(CH_2ONO_2)_4$: silicon analogues of the common explosives pentaerythrityl tetraazide, $C(CH_2N_3)_4$, and pentaerythritol tetranitrate, $C(CH_2ONO_2)_4$, *J. Am. Chem. Soc.*, **129**, 6908–6915; (c) T. Klapötke, B. Krumm and I. Schwab, (2005), Developments on the chemistry of tellurium and selenium azides, phosphorus, *Sulfur Silicon Relat. Elem.*, **180**, 957–960.
(58) K. Banert, Y.-H. Joo, T. Rüffer, B. Walfort, H. Lang, (2007), The exciting chemistry of tetraazidomethane, *Angew. Chem. Int. Edn*, **46**, 1168–1171.
(59) C. I. Schilling, S. Bräse, (2007), Stable organic azides based on rigid tetrahedral methane and adamantane structures as high energetic materials, *Org. Biomol. Chem.*, **5**, 3586–3588.
(60) S. Bräse, C. Gil, K. Knepper, V. Zimmermann, (2005), Organic azides: an exploding diversity of a unique class of compounds, *Angew. Chem. Int. Edn*, **44**, 5188–5240.
(61) A. Qin, J. W. Y. Lam, C. K. W. Jim, L. Zhang, J. Yan, M. Häussler, J. Liu, Y. Dong, D. Liang, E. Chen, G. Jia, B. Z. Tang, (2008), Hyperbranched polytriazoles: click polymerization, regioisomeric structure, light emission, and fluorescent patterning, *Macromolecules*, **41**, 3808–3822.

16

Unprecedented Electro-optic Properties in Polymers and Dendrimers Enabled by Click Chemistry Based on the Diels–Alder Reactions

Jingdong Luo, Tae-Dong Kim and Alex K.-Y. Jen

16.1 Introduction

Organic second-order nonlinear optical (NLO) materials have been intensively studied for several years due to their promising applications in high-speed and broadband information technologies, THz generation/detection systems, optical circuits integrated in silicon chips, and multifunctional nano-devices.[1–4] Recent advances in the use of nano-scale architectural control and Rational molecular design have led to exceptionally large electro-optic (E-O) activities in organic NLO materials and devices. For instance, dendronized NLO chromophores and polymers have shown significant enhancement in poling efficiency by encapsulation of the chromophore with dendritic substituents that can electronically shield the core, π-electrons and form spherical molecular shapes.[5] Another significant improvement of E-O activity was also shown in the supramolecular self-assembly of NLO chromophores surrounded by dendritic moieties that can be poled and crosslinked into a robust extended network.[6]

The ultimate goal for the NLO materials is to simultaneously achieve very large E-O activity, good thermal stability, high optical transparency, and excellent mechanical

Click Chemistry for Biotechnology and Materials Science Edited by Joerg Lahann

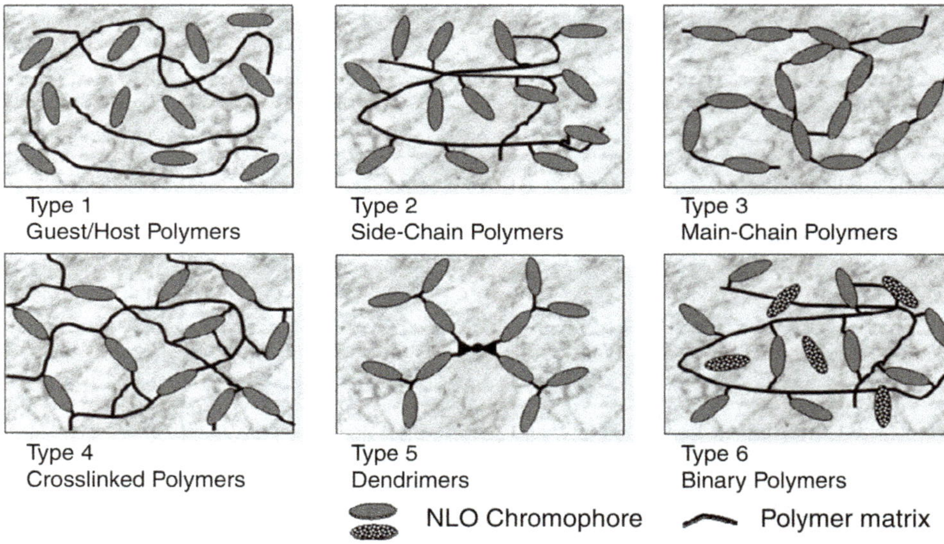

Figure 16.1 *Various types of polymeric NLO materials.*

properties within the same materials. Figure 16.1 represents the various types of polymeric NLO materials developed over two decades. It is highly desirable to covalently incorporate chromophores into polymer networks through postfunctionalization (type 2) and crosslinking reactions (type 4) to improve both their thermal and mechanical properties.[1a,c] The globular shape of dendrimers (type 5) is suitable for obtaining spherical macromolecular structures of encapsulated chromophores and engineering nanoscale macromolecular architectures with pre-organized chromophores.[5] However, low poling efficiency from limited chromophore alignment was found in postfunctionalized and crosslinked NLO polymers and dendrimers due to reduced rotational flexibility of chromophores. Therefore, there is a strong need to establish creative processing/poling protocols using carefully controlled chemistry to overcome the 'nonlinearity–stability trade-off'.

Recently, we have exploited high performance NLO polymers and dendrimers showing unprecedented E-O properties by applying click chemistry based on the Diels–Alder reactions for postfunctionalization and lattice hardening. The reactions meet the stringent requirements of processing/poling protocols with comprehensive material properties. Here we highlight these revolutionary materials achieved by click chemistry, which resulted in ultrahigh E-O coefficients (up to ~380 pm/V), excellent thermal stability and optical transparency, and are suitable for the implementation into novel opto-electronic and photonic devices.

16.2 Diels–Alder Click Chemistry for Highly Efficient Side-chain E-O Polymers[7–10]

Two different synthetic strategies can be used for the preparation of side-chain E-O polymers. One is the polymerization of NLO chromophores attached with reactive groups

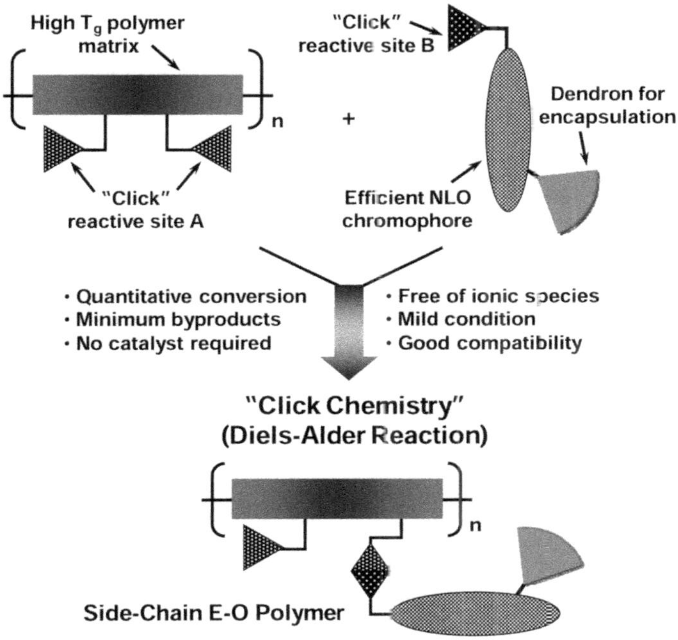

Figure 16.2 *The postfunctionalization method for side-chain E-O polymers by click chemistry based on Diels–Alder reaction.*

by step-growth or chain-growth reactions. The other is the postfunctionalization of the NLO chromophores into conventional linear polymers. The latter approach is a preferred method for systematic molecular engineering of polymers with optimized properties since we can take advantages of well-defined linear polymers as starting materials. Most of the postfunctionalization methods used for making side-chain E-O polymers such as the azo-coupling,[11] tricyanovinylation,[12] Mitsunobu etherification,[13] Knoevenagel condensation[14] and catalyzed esterification,[5c,d] often generate byproducts or trace amount of residual ionic impurities that can significantly attenuate the effective electrical field for poling and cause the DC bias to drift during device operation. To alleviate these problems, the selection criteria for ideal postfunctionalization methods, as illustrated in Figure 16.2, are: (i) quantitative conversion with minimum byproducts, (ii) no catalyst required, (iii) free of ionic or polar species, and (iv) mild reaction conditions and good compatibility of a host polymer with various highly efficient, but chemically sensitive chromophores. In searching of possible solutions, the click chemistry pioneered by Sharpless *et al.* seems to fit most of these criteria.[15]

The click-type reactions, mainly exemplified by Huisgen 1,3-dipolar azide-alkyne or Diels–Alder cycloadditions, have led to reliable and self-directed modular organic reactions to make molecular connections with absolute fidelity. They also inspired a multitude of applications in biology, chemistry and materials science.[16] As a powerful enabling tool, it possesses great potential for meeting the aforementioned stringent criteria for post-functionalization of E-O polymers to generate comprehensive properties. Nevertheless, one drawback for these 1,3-dipolar azide–alkyne reactions is the high reactivity between azide and the cyano-containing acceptors that are commonly used for NLO chromophores and can hinder their general applicability for E-O polymers. To solve this problem, the Diels–Alder cycloaddition was selected as an alternative for generating new series of high-performance E-O polymers. The Diels–Alder reaction involves a ring-forming coupling between a dienophile and a conjugated diene, which can be described by a symmetry-allowed concerted mechanism without forming biradical or zwitterionic intermediates. One of the important features in a Diels–Alder reaction is that the resultant adducts can be thermo-reversibly cleaved to render the starting materials. Wudl and coworkers have reported a thermally amendable polymeric network by the reversible Diels–Alder cycloaddition adapted for fracture self-healing in polymer thermosets.[17] These encouraging results have shown the potential and versatility to create sophisticated properties for polymers.

We first introduce the Diels–Alder cycloaddition reaction as a general and efficient synthetic method for reversibly crosslinkable E-O polymers.[7] Three different functional moieties, including the derivatives of a CLD-type chromophore **1**, a capped maleimide **2** (dienophile) and furanic ring **3** (diene), were sequentially attached onto the polymer backbone as side chains to afford a crosslinkable E-O polymer PSDACLD (Figure 16.3). The maleimide was protected with furan initially to prevent any crosslinking reaction from occurring prior to the poling and lattice hardening step. The poling temperature is very close to the onset temperature of the retro-Diels–Alder reaction (110 °C) and slightly higher than the typical temperature range used for the Diels–Alder crosslinking reaction (60–80 °C). At this temperature, the material possesses the characteristics of a typical thermoplastic polymer. Concurrently, the chromophores can be effectively reoriented under the poling field. After the poling process, a sequential cooling/curing process (85 °C for 1 h, 75 °C for 1 h, and 65 °C for 1 h) was performed to anneal and crosslink the polymer through the Diels–Alder reaction.

By smartly controlling the poling and crosslinking processes through the reversible Diels–Alder reactions, it allows highly polarizable chromophores to be efficiently poled at the stage of a low viscosity linear thermoplastic polymer. The resulting material exhibits a combination of a very large E-O coefficient (r_{33}) value (76 pm/V at 1.3 μm) and good temporal stability at 70 °C. From this novel crosslinking system, E-O materials with optimal physical properties can be achieved by fine-tuning a processing temperature window of the Diels–Alder and retro-Diels–Alder reactions. Furthermore, modifying the electronic properties of the crosslinking reagents of deactivated furan diester or anthryl groups[8] led to the ability to fine-tune the processing temperature and reversibility of these Diels–Alder reactions to optimize thermal stability and processability. Because of the efficient lattice hardening of the Diels–Alder reaction, these materials also exhibited very good temporal alignment stability, retaining ~80% of their original r_{33} values even after baking at 85 °C for 500 h.

The Diels–Alder reactions can be utilized for click postfunctionalization in order to make side-chain E-O polymers, as shown in Figure 16.4. Maleimide-containing NLO chromophores can trigger click reactions to the polymer backbone (**PMMA-AMA**) with pendant anthracenyl diene moieties to afford **PM-1b**, **PM-2** and **PM**-3, respectively.[9] This synthetic approach is very mild, versatile, quantitative and free of ionic species and catalysts. In the side-chain E-O polymers made by using post-esterification, the typical connection point for side-chain or dendron attachment is limited only through the donor- or the acceptor-end of the chromophores (**A**-type, **PS-FTC**). Although the chromophore

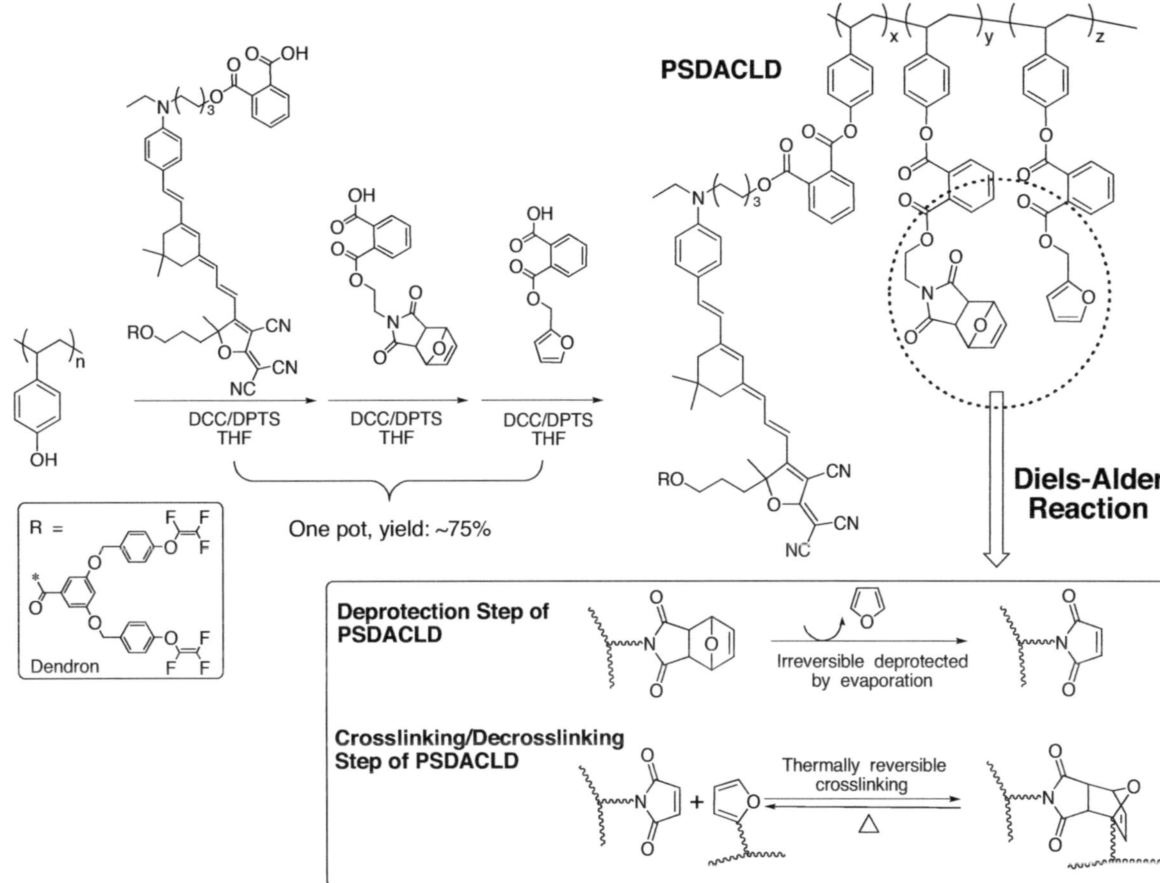

Figure 16.3 *Synthesis of a thermoreversibly crosslinkable side-chain E-O polymer by one-pot postesterification. Reprinted with permission from J. Luo,* et al., *(2003), Highly efficient and thermally stable electro-optic polymer from a smartly controlled crosslinking process,* Advanced Materials, ***15*** *(19), 1635–1638. Copyright 2003 Wiley-VCH.*

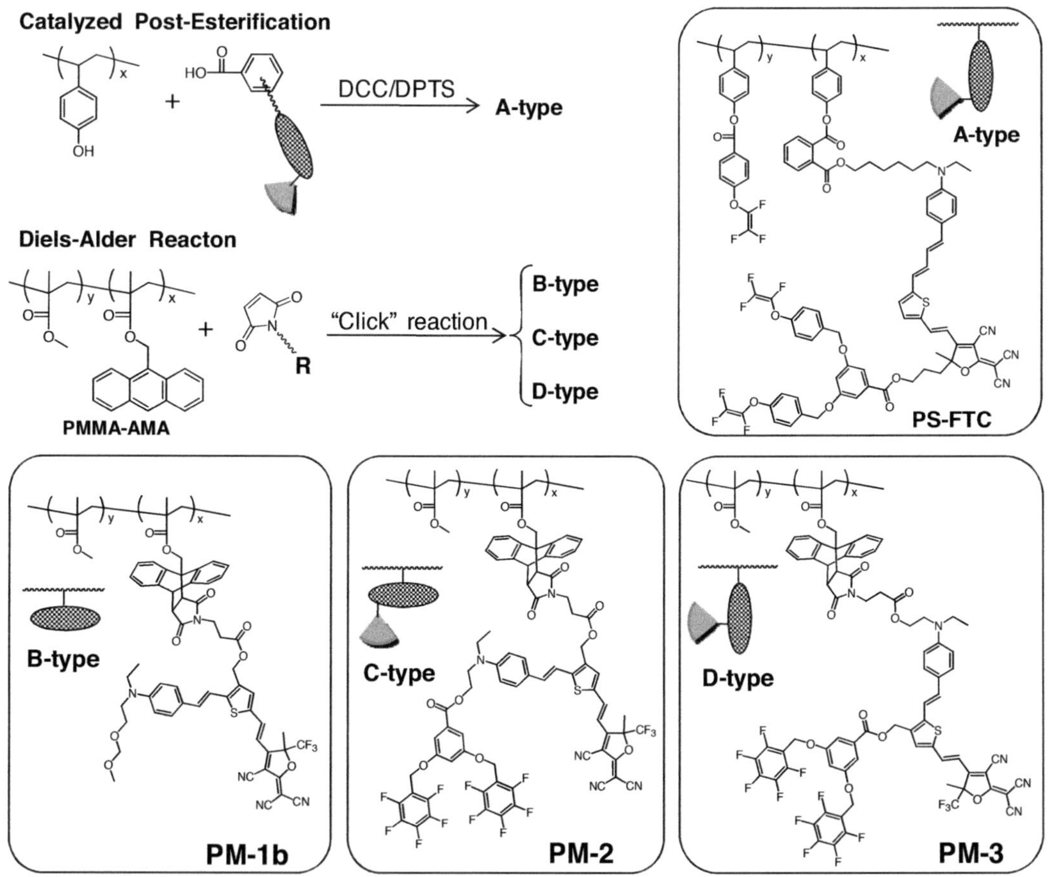

Figure 16.4 *Synthesis of side-chain E-O polymers by using catalyzed post-esterification for* ***PS-FTC*** *and the Diels–Alder click reaction for* ***PM-1b***, ***PM-2*** *and* ***PM-3***. *Inserted illustrations show different anchoring styles of the NLO chromophore into the polymer backbone.*

can be structurally shielded from its neighbors to minimize the electrostatic interactions, this head-to-tail linking style creates a quite bulky shape, which causes substantial steric hindrance during poling. As a result, it often needs very high electric field (~150 V/μm) to align the chromophore dipole. By using the efficient Diels–Alder click reaction, different architectures of macromolecules can be easily created by changing the attaching modes of the chromophore onto the polymers. Since both of the donor and acceptor ends were used in the **A**-type side-chain polymers, the alternative site for linking the rod-shaped chromophore is through the center of its bridge. By activating the center site of chromophoric bridge, it also creates multiple new placements of chromophores in side-chain dendronized E-O polymers. This provides the possibility to molecularly engineer side-chain dendronized E-O polymers with desirable shape for ease of rotation during poling (**B**-, **C**- and **D**-type). **PM-1b** is the one that bears the center-anchored chromophore as a side-chain (**B**-type or side-on type). **PM-2** also adopts the similar anchoring style, while the donor-end of the chromophore is functionalized with a fluorinated dendron (**C**-type). **PM-3** belongs to the **D**-type construction, in which the chromophore is attached to polymer backbone via its donor-end and also incorporated with a fluorinated dendron on its bridge center.

Thermal analysis using differential scanning calorimetry (DSC) showed relatively high glass transition temperatures (T_gs) in these PMMA-type polymers: 152 °C for **PM-1a**, 154 °C for **PM-1b**, 140 °C for **PM-2** and 135 °C for **PM-3**. On the other hand, the T_g of **PS-FTC** only showed ~90 °C. This can be attributed to the rigidity introduced by the bulky anthryl-maleimido Diels–Alder adducts. From thermogravimetric analysis (TGA), the onset decomposition temperatures for the dendronized polymers **PM-2** and **PM-3** are roughly 50 °C higher than those of the nondendronized **PM-1b**, indicating improved thermal properties by encapsulating the NLO chromophore in fluorinated dendrons.

PM-1b, **PM-2** and **PM-3**, with a 20 wt% chromophore loading level, exhibited excellent poling characteristics. The measured E-O activities were 37 pm/V for **PM-1b**, 60 pm/V for **PM-2** and 57 pm/V for **PM-3** at 1.3 μm. The poled **PM-2** and **PM-3** films showed almost twice r_{33} values compared with the nondendronized counterpart, **PM-1b**. These results demonstrate the effect of using dendritic moieties to improve poling efficiency of polymers. Compared with the **A**-type side-chain dendronized polymer, **PS-FTC**, these polymers exhibited much improved electric-field-dependent poling behavior (Figure 16.5).

Within the poling field range between 100 and 160 V/μm, the E-O activities of **PM-2** and **PM-3** increase linearly with the increase in the applied field. For example, under a poling field of 100–125 V/μm, the poled films of **PM-2** and **PM-3** show large r_{33} values of 48 and 40 pm/V, respectively, which are ~70–80% of their optimal E-O activity calculated by using a two-level model. For comparison, we can only obtain 36% of the optimal E-O activity for the **A**-type **PS-FTC**. Moreover, these new side-chain dendronized polymers, **PM-2** and **PM-3**, can achieve high poling efficiency over a much broader range of poling fields (100–160 V/μm). This is because the center site of chromophoric bridge has been activated for polymer or dendron attachment, through which either the chromophore shape is modified (more 3D extended as in the **D**-type **PM-3** to prevent preferential 2D packing of pristine chromophore prolate ellipsoids) or the steric hindrance is reduced for chromophore reorientation (in **C**-type **PM-2**). This clearly demonstrates the advantage of this new center-anchored approach over the previously used head-to-tail connecting side-chain E-O polymers. All of these side-chain dendronized E-O polymers also exhibited

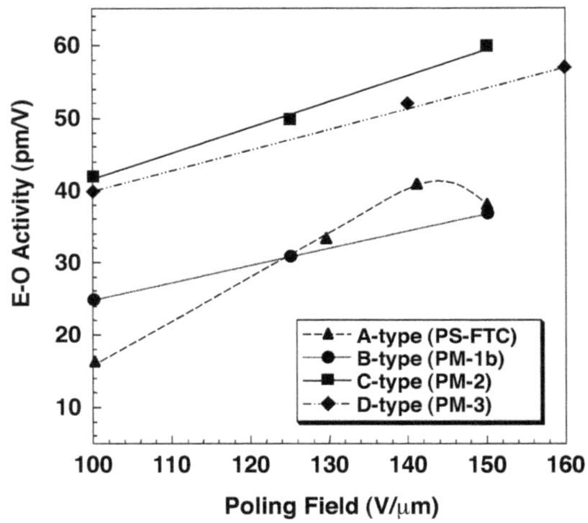

Figure 16.5 *Electric field-dependent poling behavior of different types of side-chain E-O polymers.*

promising alignment stability: more than 80% of their original r_{33} values were retained at 85 °C for more than several hundred hours (Figure 16.6).

Diels–Alder click chemistry can be also utilized in a solid state. As shown in Figure 16.7, the highly reactive anthracenyl **PMMA-AMA** was reacted with maleimide-containing NLO chromophores during the poling process to afford **PA08, PA16**, **PA21**, **PA25**, **PA28**, **PA32**, **PA34** and **PA39** respectively.[10] After adding more than 24 wt% of chromophore contents,

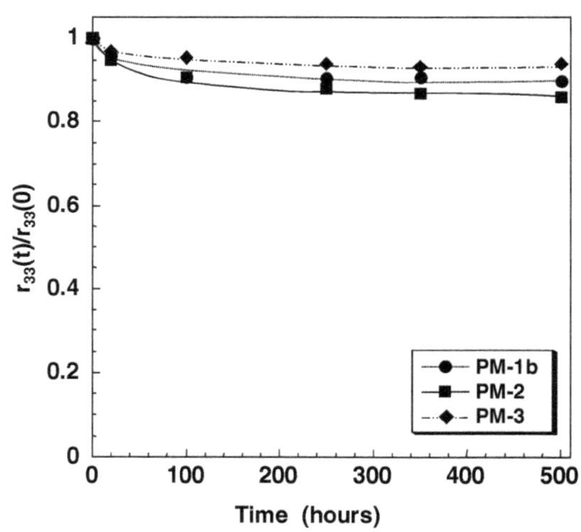

Figure 16.6 *Temporal stability of the side-chain E-O polymers at 85 °C.* $r_{33}(0)$: *initial* r_{33} *values;* $r_{33}(t)$: r_{33} *values at different baking time. Reprinted with permission from T.-D. Kim,* et al., *(2006), Diels–Alder "Click-Chemistry" for highly efficient electrooptic polymers,* Macromolecules, ***39*** *(5), 1676. Copyright 2006 American Chemical Society.*

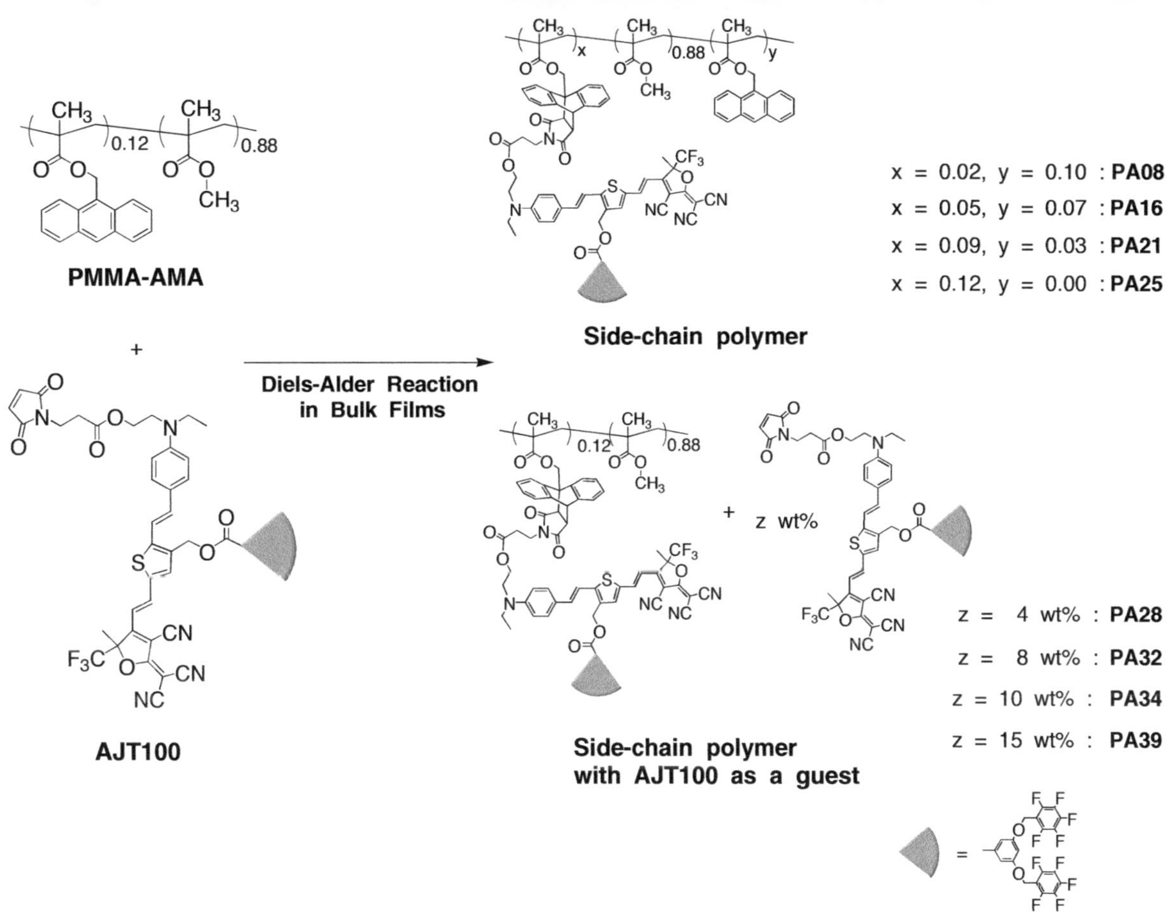

Figure 16.7 In-situ *generated side-chain E-O polymers by solid phase Diels–Alder click reaction. Reprinted with permission from J. W. Kang* et al., *(2005), Very large electro-optic coefficients from* in situ *generated side-chain nonlinear optic polymers,* Applied Physics Letters, ***87*** *(7), 071109/1. Copyright 2005 American Institute of Physics.*

Table 16.1 *Physical and optical properties of the polymers*

Sample	Composition		Chromophore contents[a] (wt%)	Poling temperature (°C)	Applied voltage (V/μm)	r_{33}[b] (pm/V)
	PMMA-AMA	AJT100				
PA08	0.806	0.194	8	135	100	18
PA16	0.625	0.375	16	120	100	32
PA21	0.510	0.490	21	124	100	43
PA25	0.417	0.583	24	110	100	48
PA28	0.357	0.643	28	107	100	74
PA32	0.278	0.722	30	105	110	99
PA34	0.222	0.778	34	90	120	110
PA39	0.100	0.900	39	80	100	87

[a] The donor–bridge–acceptor core part of chromophoric contents (formula $C_{28}H_{21}F_3N_4OS$, molecular weight 518.6);
[b] E-O coefficient measured by simple reflection at the wavelength of 1.3 μm.

the unreacted chromophore remains as a guest in these systems. Very attractive features of this click reaction, which makes side-chain E-O polymers in the bulk state, are the use of no solvent, quantitative yields and regio-specifically formed adducts without any generation of by-products. As a consequence, it is possible to control the chromophore loading concentration in the polymers by adjusting the chromophore ratio. This system provides a means to afford side-chain E-O polymers with different chromophore concentration to determine its critical loading density at which electrostatic interactions occur in the polymer matrix. Furthermore, the clicked side-chain E-O polymers can act as hosts with unreacted guest chromophores above their certain loading density to form binary NLO polymeric systems.

Measured r_{33} values and poling conditions of these poled polymers are summarized in Table 16.1. Note that **PA21** polymer by the solid-phase reaction corresponds with **PM-3** prepared by the solution reaction. The data of r_{33} values and optimal poling temperatures for **PA21** are very reproducible and are in well agreement with **PM-3**. For an example, the r_{33} values of **PM-3** and **PA21** are 41 and 43 pm/V, respectively, under 100 V/μm of the poling field around 125 °C. The poled films of **PA34** have shown the highest r_{33} value (110 pm/V at 120 V/μm of a poling field). This value is extraordinary in such a high chromophore concentration when compared with other guest–host NLO systems. This behavior can be explained as follows: (1) dendron effects preventing the chromophore-chromophore electrostatic interactions; (2) the difficulty of the chromophore aggregation due to the structural shape; and (3) cooperative effects of binary polymeric systems converted to the side-chain E-O polymers with guest chromophores.

16.3 Diels–Alder Click Chemistry for Crosslinkable E-O Polymers Containing Binary NLO Chromophores[18,19]

Many studies of polymeric E-O materials have shown that lattice hardening approaches can significantly improve long-term alignment stability. However, a reduction of 20–40% in E-O activity is usually accompanied by such approaches, since typical poling of conventional thermoset E-O polymers is achieved through sequential lattice hardening and poling process, resulting in severely limited chromophore reorientation. The Diels–Alder cycloaddition reaction for the lattice hardening method can tune temperature windows for crosslinking and poling. This Diels–Alder reaction has been applied successfully in lattice hardening processes of E-O polymers with both high nonlinearity and thermal stability. We have extended the concept of binary systems into crosslinkable E-O polymers through Diels–Alder click reactions to further incorporate highly polarizable NLO chromophores. This combined effort demonstrates that binary mixtures of chromophores can be loaded into side-chain E-O polymers and efficiently poled to give E-O activities higher than the summed value of two added chromophores. These systems can also be mildly cured to ensure a thermally stable E-O response.

The crosslinkable E-O polymers containing binary chromophores exist as a three-component guest–host system (Figure 16.8): **PMMA-AMA** as a host polymer, guest chromophore **AJC146**, and secondary chromophore **1a**, **1b**, or **1c**.[18] The chromophores are functionalized with maleimido moieties, which can act as an active crosslinker to react with the anthracenyl side-chains on **PMMA-AMA**. After solid-state Diels–Alder click reactions, they form chromophore-embedded networks. **BMI**, 1,6-bismaleimidohexane, was also included as a passive crosslinker for a parallel comparison. After curing **PMMA-AMA/BMI/AJC146**, the main absorption bands (two major absorption sub-peaks, centered at the wavelengths of 799 nm and 951 nm) of **AJC146** remained unchanged when the same amounts of anthracenyl and maleimide groups were equivalent. However, it underwent a significant decrease in intensity (25–40%) if an excess amount of maleimide was used. This suggests that the anthracenyl group has a higher reactivity toward maleimide and can serve as a scavenger to prevent the polyenic chromophore from reacting with maleimide. In all cases, the chromophore absorption bands remained almost unchanged throughout the thermal curing and poling, indicating that good chromophore stability was achieved under a mild curing condition and carefully adjusted diene–dienophile ratios.

All poling processes were performed at temperatures of around 110 °C with a poling field ranging from 75 to 125 V/μm. Both poling fields and currents were monitored *in-situ* to optimize the entire process. All of the binary systems exhibited very large r_{33} values (up to 237–263 pm/V). However, it is hard to achieve more than 200 pm/V in a singular chromophore system even at higher chromophore loading levels. These results suggest the structural features for the chromophores and processing control desirable for binary systems. The shape of a guest chromophore, **AJC146**, is a roughly prolate ellipsoid while chromophoric crosslinkers (**1a**, **1b**, or **1c**) are Λ-shaped. During poling and annealing processing, such a combination of chromophores can minimize the formation of antiparallel or close head-to-tail centrosymmetric stacking between chromophores. Furthermore, chromophoric crosslinkers can provide further modification of the polymer hosts, through the *in-situ* Diels–Alder crosslinking, leading to better homogeneity and stability to additional polyenic chromophore dopants. In this process-induced morphological confinement, both guest chromophores and *in-situ* generated active polymer networks could respond cooperatively to the poling field. Side-chain E-O polymers can also crosslink with Diels–Alder reactions in the form of binary chromophoric systems.[19] Figure 16.9 shows a schematic illustration for chromophore aligning/lattice hardening processes from a side-chain E-O polymer **PM-AJL**. Side-chain chromophore contents were adjusted to 8 wt% with an anthracenyl group in PMMA-AMA to further crosslink the binary systems. A guest chromophore **AJL28** and passive crosslinker **TMI** were added with different concentrations to maximize poling efficiency and alignment stability. Direct spin-coating of the dissolved mixture in 1,1,2-trichloroethane (~8 wt%) gave high optical quality thin films,

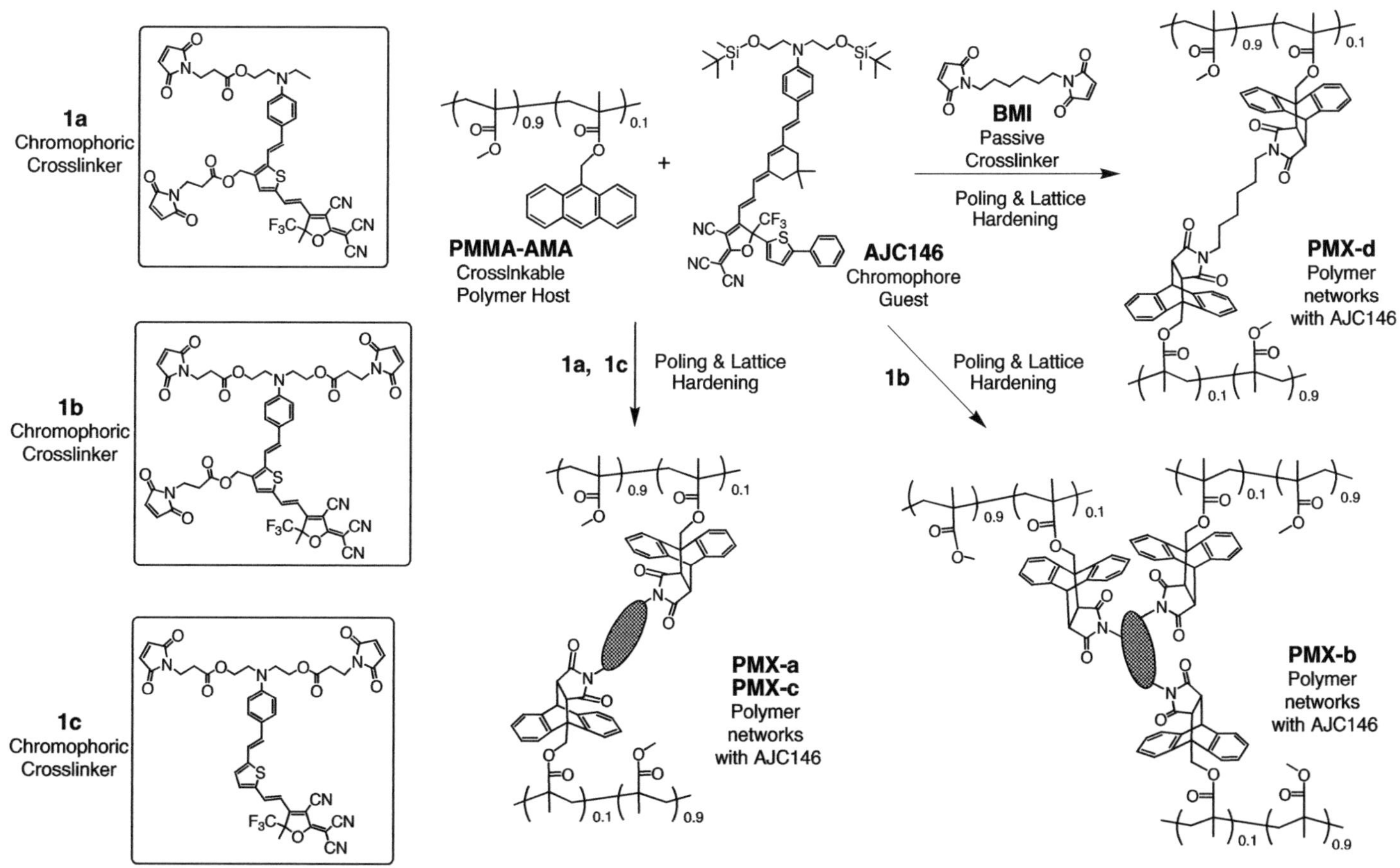

Figure 16.8 *In-situ poling and Diels–Alder click crosslinking of guest–host crosslinkable E-O polymers containing binary chromophores. Reprinted with permission from A. K.-Y. Jen, T.-D. Kim and J. Luo, (2006), Ultralarge and Thermally Stable Electro-optic Activities from Diels–Alder Crosslinkable Polymers Containing Binary Chromophore Systems,* Advanced Materials, ***18*** *(22), 3038–3042. Copyright 2006 Wiley-VCH.*

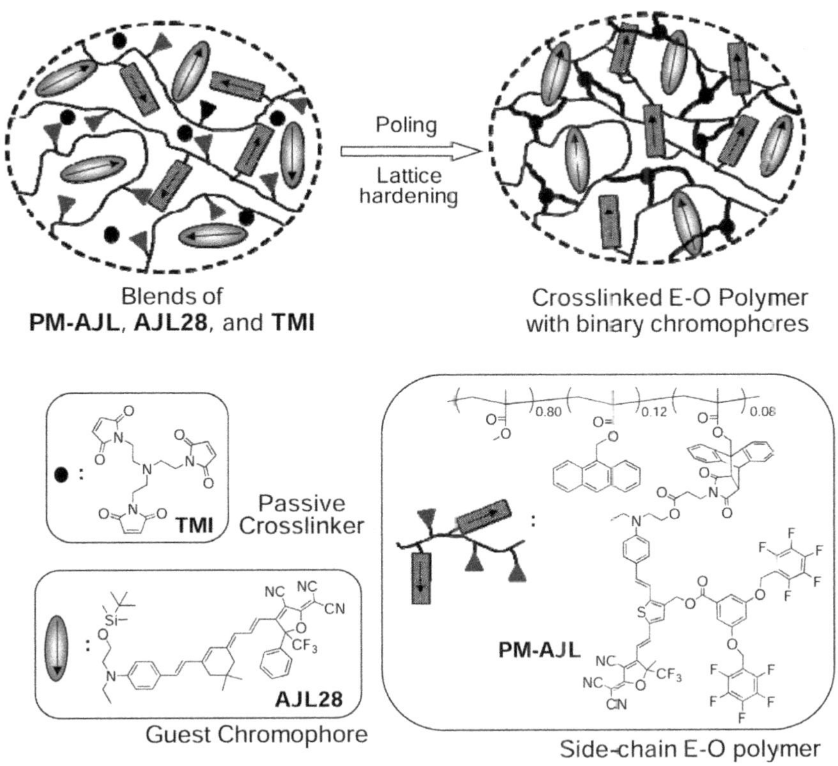

Figure 16.9 A crosslinked E-O polymer in-situ generated by Diels–Alder click reaction in a side-chain E-O polymer with binary chromophores. Reprinted with permission from T.-D. Kim, et al., (2008), Binary chromophore systems in nonlinear optical dendrimers and polymers for large electrooptic activities, Journal of Physical Chemistry, 112 (21), 8091. Copyright 2008 American Chemical Society.

which were then subjected to the electric field to form a poled crosslinked polymer. We observed lowered poling efficiency at an excess amount of **TMI** due to polyenic chromophore decomposition by maleimide groups. The concentration of the side-chain chromophore can also affect poling efficiency in the binary crosslinked polymers. Careful control of **TMI** content (1–2 wt%) and side-chain chromophore concentration (8%) was taken into account to maximize EO activity and lattice hardening. A continuous increase in r_{33} values can be seen with the increase of guest chromophore **AJL28** in **PM-AJL**.

The highest r_{33} value of 387 pm/V was obtained for the film containing 30% of **AJL28**, which is significantly larger than that (198 pm/V) of singular guest chromophore **AJL28** in **PMMA-AMA**. This result suggests that binary chromophores in different morphological confinement can cooperatively respond to the poling field. The binary systems may also provide a unique nanoenvironment for enhancing local field factors. However, further increase in the chromophore contents led to saturation or decrease in E-O coefficients due to severe aggregation of chromophores. In addition, a poling voltage higher than 75 V/μm across the films resulted in catastrophic electrical breakdown. All crosslinked E-O polymers showed good thermal alignment stability. After an initial fast decay, *ca* 75% of these E-O activities could be maintained at 85 °C for over 500 h. This demonstrates that the binary systems can be efficiently poled and cured to form thermally stable E-O lattices. These results are again a great demonstration of the advantages offered by binary chromophoric systems combining with well-controlled lattice hardening and poling methodology via Diels–Alder click reactions.

16.4 Diels–Alder Click Chemistry for NLO Dendrimers[19,20]

E-O activities of NLO materials can be significantly improved by encapsulating the chromophore with substituents that can electronically shield the core and form spherical molecular shapes. In order to create a structurally more well-defined and stable NLO material, we have explored dendrimers with multiple dendritic chromophores branched out from a passive core unit. The most important advantage of NLO dendrimers is that the active volume fraction of chromophore can be maximized without phase separation and aggregation. It occurs easily in the guest–host polymeric systems when the chromophore is highly loaded. The dendrimers are also expected to have reproducible physical and optical properties with well-defined molecular structures.

Three-arm NLO dendrimers (**D3-PS** and **D3-DA**) were synthesized by post-esterification and Diels–Alder click reactions between the three branches of desirable core molecules and the chromophore precursors that are surrounded by perfluorinated-phenyl dendrons as the exterior moieties (Figure 16.10).[19] In case of **D3-PS**, the resulting *N*-acrylurea and anhydride by-products should be removed by repetitive precipitation after the condensation reaction. As described in Chapter 2, a trace amount of residual ionic impurities or by-products can significantly attenuate the effective poling electrical field and possibly cause the DC bias drift during device operation. To alleviate this problem, the Diels–Alder click chemistry was selected as an alternative for generating a new high-performance NLO dendrimers. In order to conduct Diels–Alder click reactions for the NLO dendrimer, a maleimide-containing NLO chromophore and an anthryl-containing diene core were prepared. The poled NLO dendrimer **PS-DA** was constructed by the Diels–Alder reaction during the poling in solid films.

The weight-percent of active chromophore content is 40% for **D3-PS** and 38% for **D3-DA**. The NLO dendrimers can be directly spin-coated to form a monolithic molecular glass without any prepolymerization process. Through an *in-situ* post-functionalization process during the electric field poling, a very large E-O coefficient ($r_{33} = 109$ pm/V) was achieved for **D3-DA**. This is significantly higher than the r_{33} value from **D3-PS** ($r_{33} = 79$ pm/V), which was prepared by catalyzed post-esterification condensation. This is due to easier alignment of dendronized chromophores before forming Diels–Alder adducts covalently bonded with anthracenyl core units. Electric fields were as high as 110 V/μm applied to the NLO dendrimer film. Furthermore the screening effect, provided by the peripheral groups of dendrimer allows the chromophores to be spatially isolated, and the large void-containing structure of dendrimers provides the needed space for efficient reorientation of the chromophores. The globular geometry of dendrimers is ideally suited for the spherical shape modification of chromophores. In terms of chromophore alignment stability, the

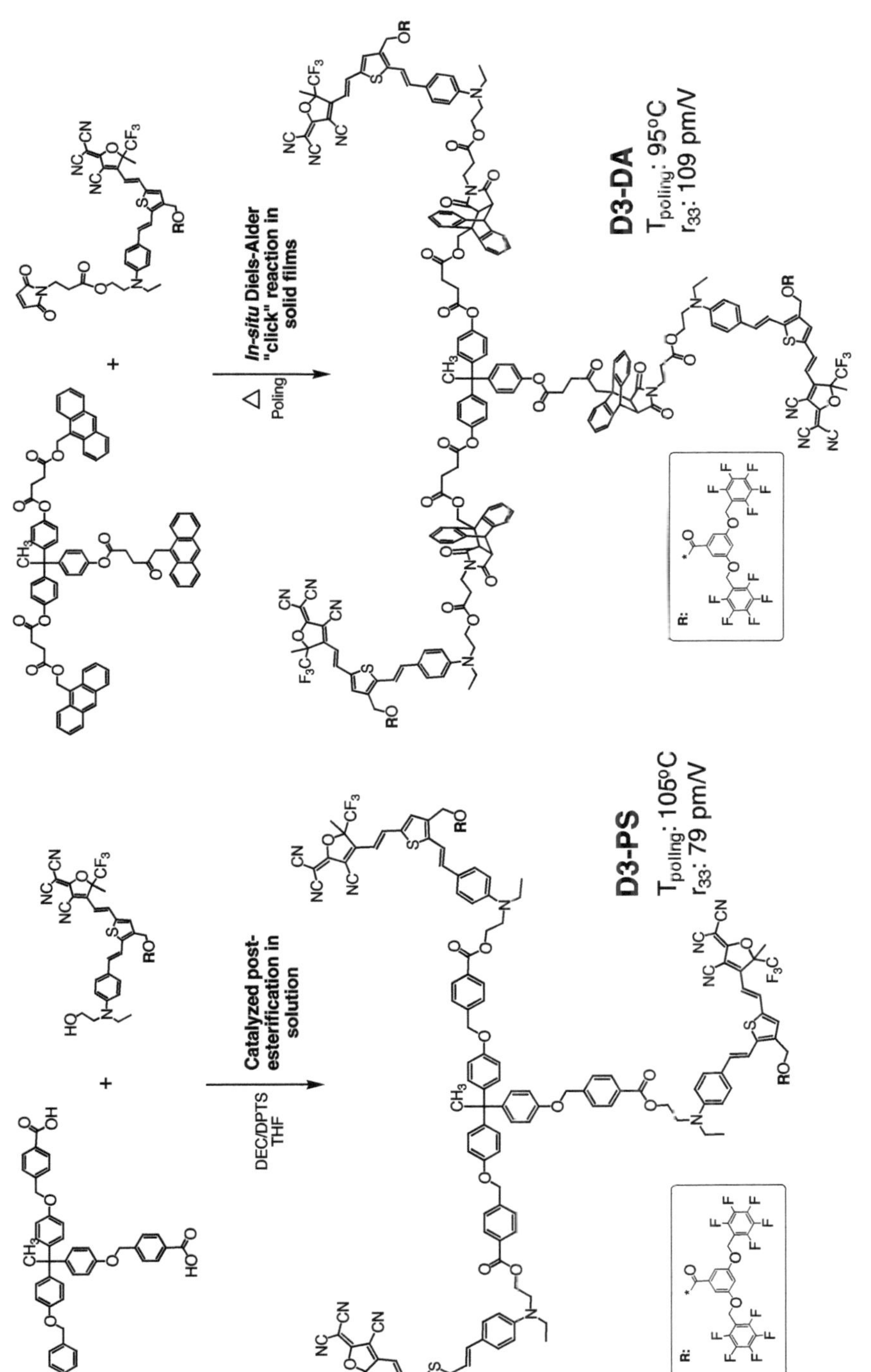

Figure 16.10 *Synthesis and E-O activity of **D3-PS** and **D3-DA** for multiarm NLO dendrimers.*

dendrimers showed very promising results, retaining more than 90% of their original r_{33} values after several hundred hours at 85 °C. Unfortunately we could not achieve any appreciable r_{33} values from **D3-DA** with additional guest polyenic-type chromophores due to chemical sensitivity of the polyenic chromophore towards the Diels–Alder reactions. However, since a polar substitution such as a methoxy group in the middle of the polyenic chromophore has provided significantly reduced dienic reactivity, we expect that improved poling efficiency and chemical stability are highly achievable in the NLO dendrimer systems afforded by Diels–Alder click reaction in the future.

Multidiene functionalized chromophores (**EOD1**, **EOD2**, **EOD3**, and **EOD4**), shown in Figure 16.11, were prepared to utilize Diels–Alder click reactions as a means of *in-situ* crosslinking for NLO dendrimers.[20] **EOD1–4** were obtained using the carbodiimide-mediated esterification between each dendronic acid and the dihydroxylated chromophore. These materials possess clear glass transition temperatures measured by DSC, which indicate amorphous solids. The processing temperature window for crosslinked NLO dendrimers can be controlled by tuning the electronic properties of the diene branches. In addition, the lattice-hardening of **EOD1–4** also converts the electron-rich dendrons into rigidified cyclic structures with far fewer electrons. This leads to a less polarizable environment. As a result, a considerable blue shift could be observed in their absorption spectra. Compared with the absorption maxima (λ_{max}) for **EOD4** (776 nm), the λ_{max} of the blend system (**EOD4** and **TMI**) blue-shifted 40 nm (to 736 nm) even before the thermal curing. This is attributed to the partial reaction between the reactive anthryl and maleimido groups.

The films were baked in a vacuum at 50 °C overnight to ensure the removal of any residual solvent. Although the baking temperature is well below the T_g of the dendrimers, all of the films prereacted to a certain extent during this step because of the highly reactive nature of the dienes and **TMI**. This reaction increases molecular weight and dielectric strength of the materials and leads to films with moderate solubility in organic solvents such as acetone. Moreover, the interchain Diels–Alder adducts can be thermally dissociated through the retro-Diels–Alder reaction. This can be tapped to further enhance the alignment of chromophores during poling. With all these dynamics encountered, the optimal poling temperature is strongly dependent on proper balance between their T_gs (after pre-crosslinking), T_{dis}s (dissociation) and T_xs (crosslinking) to accomplish high poling efficiency (Table 16.2). High-temperature and high-voltage poling could also be applied concurrently to increase lattice hardening. Upon removal of the electric field and cooling to room temperature, the poled films of **EOD1–3** showed large E-O coefficients (r_{33} = 63–99 pm/V), which are much higher than those from the guest–host EO polymer (**AJL8/APC**) or the uncrosslinked **EOD5**, which contains a similar chromophore. Over 90% of the original E-O activities could be retained at 100 °C for more than 500 h.

16.5 Conclusions

Our recent results from organic second-order NLO materials development have shown dramatically enhanced E-O activity compared with that of the past two decades. High-performance E-O polymers and dendrimers were demonstrated by a facile and reliable Diels–Alder click reaction for postfunctionalization and lattice hardening to improve EO

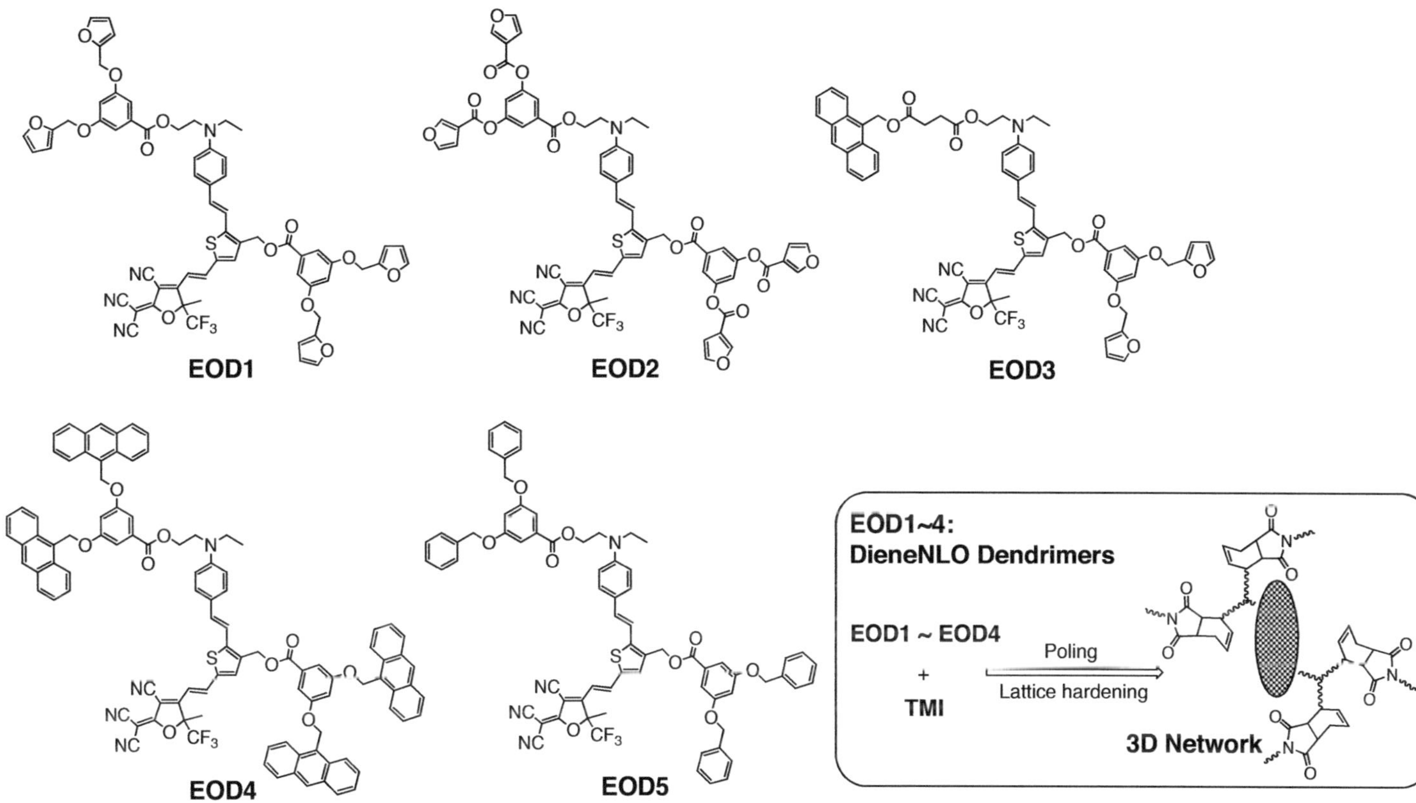

Figure 16.11 *Diene NLO dendrimers (**EOD1**, **EOD2**, **EOD3**, and **EOD4**) and a Fréchet–dendron chromophore (**EOD5**) for Diels–Alder reactions to form a 3D network. Reprinted with permission from A. K.-Y. Jen,* et al., *(2007), Highly efficient Diels–Alder cross-linkable electro-optic dendrimers for electric-field sensors,* Advanced Functional Materials, ***17** (14), 2557–2563. Copyright 2007 Wiley-VCH.*

Table 16.2 *Summary of the physical properties of NLO dendrimers EOD1–5*

	Dye content[a] (wt%)	λ_{max}[b] (nm)	λ_{max}[c] (nm)	T_g[d] (°C)	T_x[d] (°C)	T_{dis}[d] (°C)	T_{dec}[d] (°C)	Poling voltage (V/μm)	Poling temperature (°C)	r_{33} (pm/V)	Temporal stability[e] (%)
EOD1	31.9	721	714	62	93	124	181	110	130	99	99
EOD2	30.9	708	706	102	136	167	169	75	115	63	90
EOD3	34.7	742	716	73	93,134	120	155	120	120	87	94
EOD4	25.3	776	724	154	RT to 65	>200	245	—	—	—	—
EOD5	40.1	719	—	57	—	—	200	60	60	52	—
AJL8/APC	15.8	711	—	140	—	—	—	100	140	40	80

[a] Net weight percentage of chromophore within dendrimers.
[b] The wavelengths of the absorption maxima for the sample without **TMI**.
[c] The wavelengths of the absorption maxima for the sample with **TMI**.
[d] Analytic results of DSC at the heating rate of 10°C/min on thermo-equilibrate samples: T_g, glass transition temperatures; T_x, onset crosslinking temperatures; T_{dis}, onset dissociation temperatures of retro-Diels–Alder reactions; T_{dec}, onset decomposition temperatures.
[e] Temporal stability after isothermal heating at 100°C for 500 h.

activity and thermal stability. The systematic exploitation of a nanoscale architecture control approach and supramolecular self-assembly has created a series of amorphous molecular NLO materials, combining well-defined properties of small molecules with facile processability. With these novel approaches, the E-O coefficients of a series of nanoengineered materials have reached ~380 pm/V, which is more than one order of magnitude higher than that of the state-of-the-art inorganic material, lithium niobate. These materials also possess excellent thermal stability and optical transparency, which are suitable for the implementation into novel opto-electronic and photonic devices. This significant breakthrough will lead to a new paradigm of developing high-bandwidth, low-energy-consumption and light-weight devices for telecommunications, computation and sensing applications.

Acknowledgment

Financial support from the National Science Foundation (NSF-NIRT and the NSF-STC Program under Agreement number DMR-0120967), the Defense Advanced Research Projects Agency (DARPA) MORPH program, and the Air Force office of Scientific Research (AFOSR) under the MURI Center on Polymeric Smart Skins is acknowledged. Alex K.-Y. Jen thanks the Boeing–Johnson Foundation for its support. Tae-Dong Kim acknowledges the Korea Research Foundation Grant supports funded by the Korean Government (MOEHRD, Basic Research Promotion Fund; KRF-2008-331-D00157).

References

(1) (a) Kajzar, F., Lee, K.-S., Jen, A. K.-Y., (2003), *Adv. Polym. Sci.*, **161**, 1; (b) Lee, M., Katz, H. E., Erben, C., Gill, D. M., Gopalan, P., Heber, J. D., McGee, D. J., (2002), *Science*, **298**, 1401; (c) Ostroverkhova, O., Moerner, W. E., (2004), *Chem. Rev.*, **104**, 3267.
(2) Zheng, X., Sinyukov, A., Hayden, L. M., (2005), *Appl. Phys. Lett.*, **87**, 081115.
(3) (a) Xu, Q., Schmidt, B., Pradhan, S., Lipson, M., (2005), *Nature*, **435**, 325; (b) Baehr-Jones, T., Hochberg, M., Wang, G., Lawson, R., Liao, Y., Sullivan, P. A., Dalton, L., Jen, A. K.-Y., Scherer, A., (2005), *Optics Express*, **13**, 5216.
(4) (a) Enami, Y., Meredith, G., Peyghambarian, N., Kawazu, M., Jen, A. K.-Y., (2003), *Appl. Phys. Lett.*, **82**, 490; (b) Kim, S.-K., Hung, Y.-C., Seo, B.-J., Geary, K., Yuan, W., Bortnik, B., Fetterman, H. R., Wang, C., Steier, W. H., Zhang, C., (2005), *Appl. Phys. Lett.*, **87**, 061112; (c) Kuo, Y.-H., Luo, J., Steier, W. H., Jen, A. K.-Y., (2006), *IEEE Photonics Technology Letters*, **18**, 175.
(5) (a) Luo, J., Ma, H., Haller, M., Jen, A. K.-Y., Barto, R. R., (2002), *Chem. Commun.*, 888; (b) Ma, H., Chen, B. Q., Sassa, T., Dalton, L. R., Jen, A. K.-Y., (2001), *J. Am. Chem. Soc.*, **123**, 986; (c) Luo, J., Liu, S., Haller, M., Liu, L., Ma, H., Jen, A. K.-Y., (2002), *Adv. Mater.*, **14**, 1763; (d) Luo, J., Haller, M., Ma, H., Liu, S., Kim, T.-D., Tian, Y., Chen, B., Jang, S.-H., Dalton, L. R., Jen, A. K-Y., (2004), *J. Phys. Chem. B*, **108**, 8523.
(6) (a) Kim, T.-D., Kang, J.-W., Luo, J., Jang, S.-H., Ka, J.-W., Tucker, N., Benedict, J. B., Dalton, L. R., Gray, T., Overney, R. M., Park, D. H., Herman, W. N., Jen, A. K.-Y., (2007), *J. Am. Chem. Soc.*, **129**, 488; (b) Gray, T., Kim, T.-D., Knorr, D. B. Jr, Luo, J., Jen, A. K.-Y., Overney, R. M., (2008), *Nano Lett.*, **8**, 754.
(7) Luo, J., Haller, M., Li, H., Kim, T.-D., Jen, A. K.-Y., (2003), *Adv. Mater.*, **15**, 1635.
(8) Haller, M., Luo, J., Li, H., Kim, T.-D., Liao, Y., Robinson, B. H., Dalton, L. R., Jen, A. K.-Y., (2004), *Macromolecules*, **37**, 688.

(9) Kim, T.-D., Luo, J., Tian, Y., Ka, J.-W., Tucker, N. M., Haller, M., Kang, J.-W., Jen, A. K. Y., (2006), *Macromolecules*, **39**, 1676.
(10) Kang, J.-W., Kim, T.-D., Luo, J., Haller, M., Jen, A. K.-Y., (2005), *Appl. Phys. Lett.*, **87**, 071109/1.
(11) (a) Wang, X., Chen, J.-I., Marturunkakul, S., Li, L., Kumar, J., Tripathy, S. K., (1997), *Chem. Mater.*, **9**, 45; (b) Schilling, M. L., Katz, H. F., Cox, D. I., (1988), *J. Org. Chem*, **53**, 5538.
(12) Jen, A. K.-Y., Liu, Y.-J., Cai, Y., Rao, V. P., Dalton, L. R., (1994), *J. Chem. Soc., Chem. Commun.*, 2711.
(13) Chen, T.-A., Jen, A. K.-Y., Cai, Y., (1995), *J. Am. Chem. Soc.*, **117**, 7295.
(14) Luo, J., Qin, J., Kang, H., Ye, C., (2001), *Chem. Mater.*, **13**, 927.
(15) Kolb, H. C., Finn, M. G., Sharpless, K. B., (2001), *Angew. Chem. Int. Edn*, **40**, 2004.
(16) (a) Fazio, F., Bryan, M. C., Blixt, O., Paulson, J. C., Wong, C. H., (2002), *J. Am. Chem. Soc.*, **124**, 14397; (b) Agard, N. J., Prescher, J. A., Bertozzi, C. R., (2004), *J. Am. Chem. Soc.*, **126**, 15046; (c) Parrish, B., Breitenkamp, R. B., Emrick, T., (2005), *J. Am. Chem. Soc.*, **127**, 7404; (d) Stocking, E. M., Williams, R. M., (2003), *Angew. Chem. Int. Ed.*, **42**, 3078; (e) Wu, P., Feldman, A. K., Nugent, A. K., Hawker, C. J., Scheel, A., Voit, B., Pyun, J., Fréchet, J. M. J., Sharpless, K. B., Fokin, V. V., (2004), *Angew. Chem. Int. Ed.*, **43**, 3928; (f) Binder, W. H., Kluger, C., (2004), *Macromolecules*, **37**, 9321; (g) Tsarevsky, N. V., Bernaerts, K. V., Dufour, B., Prez, F. E. D., Matyjaszewski, K., (2004), *Macromolecules*, **37**, 9308; (h) U. Scherf, K. Mullen, (1995), *Adv. Polym. Sci*, **123**, 1; (i) M. Grigoras, G. Colotin, (2001), *Polym. Int*, **50**, 1375.
(17) Chen, X., Dam, M. A., Ono, K., Mal, A., Shen, H., Nutt, S. R., Sheran, K., Wudl, F., (2002), *Science*, **295**, 1698.
(18) Kim, T.-D., Luo, J., Ka, J.-W., Hau, S., Tian, Y., Shi, Z., Tucker, N. M., Jang, S.-H., Kang, J.-W., Jen, A. K.-Y., (2006), *Adv. Mater.*, **18**, 3038.
(19) Kim, T.-D., Luo, J., Cheng, Y.-J., Shi, Z., Hau, S., Jang, S.-H., Zhou, X.-H., Tian, Y., Polishak, B., Huang, S., Ma, H., Dalton, L. R., Jen, A. K.-Y., (2008), *J. Phys. Chem. C*, **112**, 8091.
(20) Shi, Z., Hau, S., Luo, J., Kim, T.-D., Tucker, N. M., Ka, J.-W., Sun, H., Pyajt, A., Dalton, L. R., Chen, A., Jen, A. K.-Y., (2007), *Adv. Funct. Mater.*, **17**, 2557.

Index

Note: Page numbers in **bold** refer to figures, schemes and tables.

Click Chemistry for Biotechnology and Materials Science Edited by Joerg Lahann

www.ingramcontent.com/pod-product-compliance
Ingram Content Group UK Ltd.
Pitfield, Milton Keynes, MK11 3LW, UK
UKHW050842101026
15624UKWH00018B/432

9 780470 699706